TRANSMISSIBLE SUBACUTE
SPONGIFORM ENCEPHALOPATHIES:
PRION DISEASES

TRANSMISSIBLE SUBACUTE SPONGIFORM ENCEPHALOPATHIES: PRION DISEASES

IIIrd INTERNATIONAL SYMPOSIUM ON TRANSMISSIBLE SUBACUTE SPONGIFORM ENCEPHALOPATHIES: PRION DISEASES

18-20 MARCH 1996
VAL-DE-GRÂCE, PARIS, FRANCE

Edited by
Louis Court, Betty Dodet

ELSEVIER

Amsterdam, Oxford, Paris, New York, Tokyo

A member of Elsevier Science

Cover: Creutzfeldt-Jakob disease: spongiform changes in the molecular layer of the cerebellum (hematoxylin and eosin staining, original magnification x 192).
Cover figure provided by Dr Larisa Cervenáková, National Institutes of Health, Bethesda, MD, USA.

© 1996 Elsevier, Paris
Éditions scientifiques et médicales Elsevier, 141, rue de Javel, 75747 Paris cedex 15

Imprimé en France par SNI Jacques & Demontrond, 25220 Thise   *Printed in France*
Dépôt légal : 12687 – ISBN 2-906077-91-7

CONTENTS

PATHOGENESIS

IN VITRO EXPRESSION

NON PRION MODELS

EPIDEMIOLOGY AND HUMAN CLINICAL ASPECTS

IATROGENIC CREUTZFELDT-JAKOB DISEASE: RISKS AND CONTROLS

Transmissible Subacute Spongiform Encephalopathies:
Prion Diseases
L Court, B Dodet, eds
© 1996, Elsevier, Paris

Foreword

No frontier between clinical and veterinary medicine

A relationship between scrapie and Creutzfeldt-Jakob disease should be considered seriously. Another disease that does not occur spontaneously, but is observed in mink breeding farms, results from accidental infection via contaminated sheep meat. A similar origin for Creutzfeldt-Jakob disease is not unlikely. This hypothesis should, however, be considered with the utmost caution due to the economic consequences that, if proven, it could have on cattle breeders. Moreover, if a relationship between the two diseases does exist, it is certainly not a straightforward link because of the rarity of clinical cases and the absence of a clear-cut predominance of cases among high-risk occupations.

This statement was made by Françoise Cathala on 17 September 1978 during a meeting on slow viruses organized by the Marcel Mérieux Foundation with Raymond Latarjet. This meeting also included a talk given by D Carleton Gajdusek on kuru disease. The Foundation published the proceedings of this meeting, which are now out of print, but the pertinence of these words must be noted.

We would like to congratulate Louis Court and Dominique Dormont on the complete success of this year's already historic meeting held at the prestigious Val de Grâce Hospital. The Marcel Mérieux Foundation organizes annual meetings on comparative pathology in this impressive setting under the presidency of Henri-Michel Antoine because we have long been concerned with the role of doctors of veterinary medicine in cooperation with medical doctors, leaving no frontier between clinical and veterinary medicine.

Charles Mérieux

Dominique Dormont, François Lacroute, Mick Tuite and Christophe Cullin are sincerely thanked for their help in establishing the respective nomenclatures that were used in this volume.

The editorial staff – Marie-Christine Clugnet, Janet Jacobson, Sharon Lahaye and Valerie Libert – are to be commended for their professionalism.

Transmissible Subacute Spongiform Encephalopathies:
Prion Diseases
L Court, B Dodet, eds
© 1996, Elsevier, Paris

Introduction

The International Symposium on Transmissible Spongiform Encephalopathies held 18–20 March 1996 to which this book is devoted, was the third meeting organized by the *Centre de recherche du service de santé des armées (CRSSA)* with the *Direction de la recherche et de la technologie*, and the *Commissariat à l'énergie atomique (CEA)*, bringing together most of the scientists working on this subject.

On the occasion of the first meeting in December 1981, notable scepticism and indifference prevailed as Stanley Prusiner first introduced the concept of the 'prion'. The second meeting, in November 1986, focused on epidemiological and experimental pathological studies and showed the common archaic protein in its native form, becoming 'infectious' in its pathological form, although none of us learned much about the origin and the mechanism of this transformation. The third meeting started off with the same indifference or doubt in spite of renewed, refined interest; however, the only advance revealed by the epidemiologists was the dramatic increase in the number of patients presenting with iatrogenic, and especially human growth hormone-related Creutzfeldt-Jakob disease.

Although everyone was aware of the story of bovine spongiform encephalopathy (BSE) in Great Britain, on the third day of the Val de Grâce meeting, the unexpected departure of our British colleagues drew our attention*. We learned with the official announcement by the British Government of the new problems posed by the emergence of Creutzfeldt-Jakob disease in young subjects, with no immediate scientific support and with no clear definition of the type of problems involved in this new expression of both the human and animal diseases, or of its economic consequences.

The international meeting organized on 17 September 1978 by the Marcel Mérieux Foundation and the Society for Comparative Pathology at the European Center of the Tufts University in Talloires (France), bringing together specialists under the chairmanship of Raymond Latarjet and D Carleton Gajdusek, had already evoked numerous questions. The physical properties of these 'agents', and consequently their nature, such as their nosographical classification, had been the subject of much discussion. Nothing had been done in the meantime, except perhaps admire the work of the San Francisco school and the Lasker Prize received by its head, Stanley Prusiner, and to rediscover, with the emergence of iatrogenic Creutzfeldt-Jakob disease, the true, striking properties of this 'agent' or of this mysterious protein.

Here and there, and even in neurosurgery where surgeons pretended to discover the danger of using surgical instruments that had been disinfected by procedures invalid for prion decontamination despite multiple publications and warnings, the potential risk of disease transmission via contaminated blood or via placenta-derived products was overemphasized. Known risks such as the highly probable alimentary contamination at the origin of kuru disease, the experimental oral contamination at the origin of Creutzfeldt-Jakob disease in primates and the new risk of contamination via animal protein-enriched feeds that had been distributed ad libitum, were not thought to be relevant.

The opening lecture by D Carleton Gajdusek, the history of kuru and the parallel between its endemicity and that of BSE in regard to the diagnosis of unusual cases of Creutzfeldt-Jakob disease centers the question, readily. The first section of this book is devoted to the current epidemiology of animal transmissible spongiform encephalopathies such as scrapie, BSE, rare and old sporadic forms and current endemicity, after passage in cattle and domestic animals. The basic neuropathological features that are often forgotten but essential for the differential diagnosis of new forms of the disease are presented in the second section. The pathogenesis section brings to light the contributions of molecular biology and the use of transgenic animals. The molecular biology section aims at describing the conformation of the normal and pathological proteins and various models. The fifth section covers in vitro expression, cell culture and non protein models, and the viral hypotheses. Current research, environmental risks, human disease, inactivation procedures, and the management of risks are addressed in the final sections.

Fundamental but also menacing problems remain for public health and common medical practice. Are we confronted with a new human disease with increasing incidence? What are its etiopathogenesis, risk factors, genetic features and predispositions? What is its place in neuro-degenerative diseases of the central nervous system? Despite statements and excited argument in both the media and scientific circles, answers to this complex biological and pathological question, development of new effective screening, diagnostic and therapeutic procedures, and of simple measures for eradication can be found only by comparative epidemiology of human and animal transmissible spongiform encephalopathies (although difficult to implement), the continuation or the resumption of experimental pathology studies, calm reflection and the means to conduct research. Let us hope that this book will be a modest contribution.

We would like to acknowledge Charles Mérieux and all his coworkers, and our friend, Jacques de Saint-Julien who in the prestigious Val de Grâce Hospital made this meeting possible, enabling us to enforce and enrich our reflection.

Betty Dodet   Louis Court

Robert Will and Richard Kimberlin left before they had the opportunity to make their planned presentation. They participated however in these proceedings.

Transmissible Subacute Spongiform Encephalopathies:
Prion Diseases
L Court, B Dodet, eds
© 1996, Elsevier, Paris

Kuru in childhood: implications for the problem of whether bovine spongiform encephalopathy affects humans

D Carleton Gajdusek

Laboratory of Central Nervous System Studies, National Institutes of Health, Building 36, Room 5 B21, Bethesda, MD 20892, USA

Summary – The kuru epidemic in New Guinea started at the beginning of this century, peaked only in mid-century, and has subsided to extinction at the end of the century. It killed over 3,000 people in the affected population of about 30,000. In the village of highest prevalence, over half of all deaths were from kuru. During the late 1950s when kuru research started, some 15% of kuru patients were prepubertal children and another 15% adolescents under 20 years of age. Thus, about one-third of the patients were children and adolescents. The youngest patients were 4.5 years old. With the cessation of contamination of the skin and mucosae of infants and children with brain and visceral tissues during the cannibalistic rituals, no one born after cannibalism ceased has ever died of kuru. Age at disease onset must have been related to dose and route of infection through skin, mucosal membrane, or gut. Children and adolescents often had a shorter course than many adults, but their disease was so similar to kuru in adult patients that, had there been cases in 1 to 4.5 year olds, they could not have been missed. Most children developed spastic strabismus early in the course. Neuropathology of child patients was the same as in adults. Some 80% of children had profuse kuru amyloid plaques in their brains. The occurrence of Creutzfeldt-Jakob disease (CJD) in a single patient under 15 years of age, particularly in a prepubertal child, in the United Kingdom at this time, would indicate a relationship between the bovine spongiform encephalopathy outbreak and such CJD, never seen before in child patients. However, we will not know whether bovine meat or milk, porcine tissues, products using tallow, or chicken manure has been the source of the human contamination.

Introduction

About 15% of the over 3000 kuru deaths were in prepubertal children. Adolescents comprised over 15% more of the patients, thus about one-third of the patients were children and adolescents under 20 years of age. The disease in the smallest children who were 4.5 to 5 years old mimicked surprisingly closely that in adults. It was thus amply clear that if the disease occurred also in 1, 2, or 3 years old, we would have no trouble recognizing it [1–4].

Thus the child cases clearly indicated a minimum incubation period of 4 to 5 years, even with a huge dose of the infectious agent inoculated peripherally. Since opportunities for contact with infectious tissue in children were minimal, except during the cannibalistic rituals, we presumed transcutaneous or mucosal inoculation was most likely, yet the oral route remained

Key words: kuru / Creutzfeldt-Jakob disease / infectious amyloidoses / nucleation / bovine spongiform encephalopathy / scrapie / fatal familial insomnia / Gerstmann-Sträussler-Scheinker disease

a possibility. Children rarely participated in the consumption of the tissues of the deceased, and infants and toddlers never. However, the unwashed hands of the women who did the cannibalistic dissections, including removal of the brain, could easily have contaminated infant foods. We never implied that the oral route was the source of infection from the cannibalistic rituals and believed the peripheral route through scratching and mucosal injury was more likely. Subsequent study of late outbreaks after long incubation periods clearly implicated infection in infancy. In infancy, the oral route would be unlikely, whereas rubbing of the eyes and other mucosae and scratching of scabies and insect bites could clearly account for the transcutaneous or transmembranous inoculation [1–5].

Kuru, whether in adults or children, presents quite a different picture from sporadic Creutzfeldt-Jakob disease (CJD). Paul Brown and I have noted that the peripherally inoculated patients who received CJD-contaminated human growth hormone (hGH) have all presented ataxia and tremors, with an unusually slow appearance and evolution of the dementia and impaired cortical function. This is clinically more like kuru than the usual sporadic CJD in older patients [6].

Since there have been only two patients under 20 years of age in the world of the 6000 diagnosed CJD patients, and none under 15 years old, we have pointed out that the appearance at this time in Great Britain of CJD in adolescents and prepubertal children would be a clear-cut link with the bovine spongiform encephalopathy (BSE) epidemic. This would not mean that beef or sausage made from viscera of slaughtered cattle was the cause, nor could it clearly implicate milk and milk products. All would remain possibilities [7].

Pigs held for eight years after intracerebral inoculation with BSE develop a transmissible spongiform encephalopathy. We do not yet know whether the BSE agent takes orally in pigs. Contaminated bone meal was fed to pigs. Thus, meat and other tissues of the pig might harbor low-level infectivity at the time of slaughter at an early age.

We also do not know whether the agent replicates in chicken or other poultry. BSE-contaminated bone meal was fed to chickens. Poultry would be expected to shed massive quantities of the infectious amyloid in their feces. Chicken manure is widely used as fertilizer on vegetable crops. This means that vegetarians might be at risk. It is best to admit our ignorance rather than to imply that we have information about this.

Any protein chemist would assume that the extremely hydrophobic scrapie agent would enter the hydrophobic lard or tallow in a rendering plant and would appear in very low titer in the hydrophilic bone meal. Experiments to date ruling out contamination of the lard are certainly not adequate. I do not know what effect the presence in fat would have on absorption of the infectious protein or its processing in a cell, and thus the virus titer. I do not know how to titrate a virus in butter.

Clinical picture of kuru in children

The onset of kuru in children between 4.5 and 5 years old was with insidious locomotor ataxia. The child patients were aware of their probable kuru before they told their parents thereof as they noticed their stumbling and ineptness at play. The ataxia progressed rapidly to interfere with play producing a broad-based haltering gait, and they soon developed the kuru tremor. This unusual tremor is really a titubation akin to that of shivering from cold and trembling from fear. The tremor was accentuated by cold or arousal, and if the child patient was fully supported and cuddled, the tremor subsided and disappeared, as it did in sleep [2–4, 8].

An emotionalism was reported by the parents [in pidgin English: *fasun blong kuru*. It was a levity and propensity to giggling and laughter or hilarity, at times inappropriate, similar to that seen in most adult patients and was often accompanied by a true euphoria and a slow relaxation of facial expressions of emotion. Thus, a smile or laugh would linger longer than normally. About one-tenth of the patients developed an antagonism or angry withdrawal. This was common in male adolescent patients and rare in prepubertal children and adolescent girls. Adolescent boys and young men rather frequently resented their condition and became morose, antagonistic, and then withdrawn from their family and friends. The majority of child patients expressed the levity and euphoria of most adult kuru patients.

Children almost always developed a spastic strabismus which disappeared or changed with arousal and tone differences shifted from one set of extraocular muscles to another.

Children developed early pyramidal signs including brisk deep tendon reflexes and ankle jerk, ankle and patellar clonus. A positive Babinski sign did not appear. There were no Kernig or Brudzinksi signs. Radioperiosteal and biceps and triceps jerks were brisk, but usually not as hyperexcitable as those in the lower extremities. Abdominals and cremasteric usually disappeared. Small children often demonstrated a dramatically wide-based gait early in the disease. None of the child patients demonstrated early dementia and signs of dementia began to appear with severe dysphagia and total aphonia late in the disease. Most of the patients remained good informants throughout more than half of their clinical course of 3 months to 1 year. Many aspects of kuru in children amenable to photography are shown in figures 1 through 11.

Two dramatic aspects of this terrible disorder for a European physician were self-diagnosis of kuru, even in children 5 years of age before their parents recognized that their child was ill and juvenile, and well-informed requests for euthanasia. Kuru had become the first cause of death in their communities and children had seen full evolution of the disorder in relatives who progressed to terminal stages while still participating out of doors in the social scene. The other aspect was the fact that prepubertal children and adolescents asked to be sent home when they became incapacitated and completely aphonic, requiring gavage feeding and intravenous feeding, no longer able to swallow liquids poured into their mouth or premasticated food without aspiration. They asked us to send them home with their parents before they were so incapacitated, knowing that they would be smothered promptly. A few older children explicitly stated this well-informed request for euthanasia at a time well before their verbal dysarthria crippled their communication. Others watching us tube feed and deliver intravenous nutrients to patients who otherwise aspirated food and drink and watching us dress deep ducubitous ulcers of extremely cachectic patients, simply made a more direct request: "Send me home. Don't let me get like that." After the first year of research work, we regularly sent the patients home in the late stages, where death ensued quickly.

Pathologic picture of kuru in children

The pathology of kuru and that of CJD do not differ greatly, except for the presence of the large amyloid plaques of kuru in 80% of kuru brains. The intracytoplasmic vacuolation, particularly of the large neurons of the striatum, is extensive, whereas status spongiosus is usually absent [9, 10]. Kuru amyloid plaques are present, even in the youngest child patients (fig 12). However, 20% of the children have no amyloid plaques at all. Scrapie associated fibrils seen by electron microscopy in density gradients of brain suspensions are present in all kuru brains, even in those lacking amyloid plaques by light microscopy [11].

Fig 1. Four women and one girl, victims of kuru, in a South Fore village. A girl is held in erect sitting posture by her mother; she would otherwise fall over. The woman in the lower left has the sticks which she requires to walk lying on the ground before her. The three women standing in upper right are still ambulatory with early kuru. The woman on the far right stands with her elbow flexed and hands raised in a position characteristic of kuru patients trying to control involuntary movements.

Fig 2. Kuru victim Yakurimba, a boy of 7 years, from Agakamatasa village, South Fore, arriving at the Kuru Hospital after two days of being carried by his agemates on a stretcher across the ranges to Okapa Patrol Post.

Fig 3. Six young kuru victims: two girls and four boys at the Okapa Kuru Hospital. The adolescent boy preventing the girl on the right from falling over, claimed at the time of the photo that he was starting with kuru. We could not confirm this, but he was dead of kuru within a year. The 5-year-old boy standing is in early kuru, self-diagnosed at 4.5 years of age. He required a broad base to stand or walk and had kuru tremors in addition to locomotor ataxia. Another kuru victim stands beside the window in the background.

Fig 4. A girl with advanced kuru who requires support to stand. She has a lingering smile of the early and mid-stages of kuru with true euphoria, a propensity to giggling and laughter, and slow relaxation of facial expression.

TRANSMISSIBLE SUBACUTE SPONGIFORM ENCEPHALOPATHIES:
PRION DISEASES

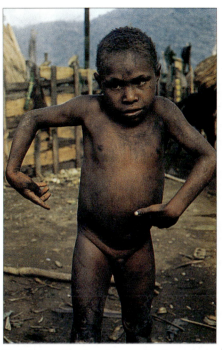

Fig 5. A small boy, kuru victim, from Miarasa village, South Fore, who can no longer stand or walk without support. He stands on a wide base and holds his forearms rigidly to control involuntary atheoid movements.

Fig 6. Yakurimba, a boy of seven with advanced kuru, caught by the camera during involuntary choreoathetoid movements precipitated by his attempts to remain standing. He has a left internal spastic strabismus. He is shown in figure 2 carried on a litter by his peers to the Kuru Hospital at the Okapa Patrol Post from his South Fore village of Agakamatasa.

Fig 7. Haida'abo, a Gimi boy with advanced kuru, from Amuraisa village held in standing posture by his father. He is trying to execute finger-to-finger tests.

Fig 8. A boy with advanced kuru lying on a bark cape before his mother and infant sibling, Okapa Kuru Hospital. He spoke with marked dysarthria, still replied intelligently and was well-oriented. He had marked spastic strabismus.

Fig 10. A small boy in mid-stage of kuru showing a severe left intermittent strabismus.

Fig 9. A girl with kuru who can no longer sit or stand without support is held by her brother.

Fig 11. A preadolescent 10-year-old North Fore girl, Atana, totally incapacitated by kuru in 1957 at the Pintogori Kuru Hospital. The child had such severe dysarthria that she could no longer communicate by word, but was still intelligent and alert. She had spastic strabismus. She could not stand, sit without support, or even roll over. She had been ill for less than 6 months and died within a few months of the photography. Such wasting and cachexia with deep decubitus ulcers did not develop in the village, where patients were suffocated at an earlier stage. Usually, parents fed children premasticated food and spit water into their mouth. At the hospital, gavage and intravenous feedings were instituted, until we later abandoned such practices at the insistence of the parents.

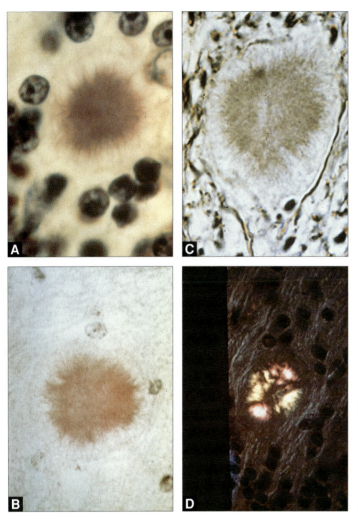

Fig 12. Kuru plaques in the cerebellum of a 5-year-old kuru victim (Yani). **A.** Metachromatic plaque in granule layers (toluidine blue stain; × 1000); **B.** Plaque in molecular layer (periodic acid-Schiff (PAS) stain; × 400); **C.** Argentophilic plaque in granule layer (hematoxylin-eosin, × 1000); **D.** Birefringent plaque in granular layer (cresyl violet, polarized light; × 400). Modified from Klatzo et al [10].

Origin and spread of kuru

Unanswered crucial questions posed by all the infectious amyloids are related to their biologic origin and mode of survival in nature. The diseases they evoke are not artificial diseases produced by researchers tampering with cellular macromolecular structures, although this is now one of our goals. They are naturally occurring diseases for which we do not understand either the mode of dissemination or maintenance adequately to explain their long-term persistence. For kuru, we have a full explanation of the unique epidemiological findings and their change over the past several decades: the contamination of close kinsmen within a mourning family group by the opening of the skull of dead victims in a rite of cannibalism, during which all girls, women, babes in arms, and toddlers of the kuru victim's family were thoroughly smeared with infectious brain. The disease has almost disappeared with the cessation of cannibalism; it first disappeared in children, then adolescents, and later in young adults, with progressively increasing age of the youngest victims (fig 13). However, this does not provide us with a satisfactory explanation for the origin of kuru. Was it the unlikely event of a sporadic case of worldwide CJD which, in the unusual cultural setting of New Guinea, produced an unique epidemic? A spontaneous case of CJD was seen in a native 26-year old Chimbu New Guinean from the Central Highlands, whose clinical diagnosis was proved by light and electron microscopy of brain biopsy specimen [12].

Passage in man in successive cannibalistic rituals might have resulted in a change in the clinical picture of the disease, with modification of the virulence of the original agent. On the other hand, the peripheral route of inoculation – percutaneous, mucosal membrane, or oral – as opposed to intracerebral route may determine the ataxic, slowly dementing illness. We have evidence for this in the experience with CJD cases resulting from percutaneous injection of contaminated hGH which were ataxic and slowly dementing, resembling kuru rather than classical sporadic CJD [6].

Small infants were often infected with kuru [2–4, 8, 13, 14] and were not fed kuru-infected brain. Through scabies and impetigo and cleaning their orifices, infants were surely inoculated with kuru by the unwashed brain-handling hands of their caretakers. Furthermore, a chimpanzee was given by nasogastric tube 20% suspensions of kuru brain containing 10^8 to 10^9 chimpanzee intracerebral LD_{50} infectious doses of kuru without producing disease over 18 years of observation, and this animal showed no brain pathology when it died of other causes. Only squirrel monkeys have been successfully infected orally by self-feeding of human kuru, CJD, and mouse scrapie brain tissue, respectively. The incubation period of kuru was 36 months, those of CJD were 23 and 27 months, and those of scrapie were 25 and

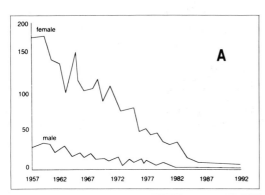

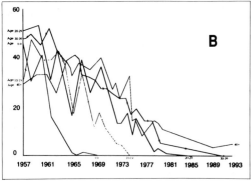

Fig 13. A. The overall incidence of kuru deaths in male and female patients by year since its discovery (in 1957) through 1992. More than 2,500 patients died of kuru during this 26-year period of surveillance, and there has been a slow, irregular decline in the number of patients to one-tenth the number seen in the early years of the kuru investigation. The incidence in males has declined significantly only during the last few years, whereas in females, it started to decline over a decade earlier. This decline in incidence has occurred during the period of acculturation from a Stone Age culture in which endocannibalistic consumption of dead kinsmen was practiced as a rite of mourning to a modern coffee-planting society practicing cash economy. Because the brain tissue with which the officiating women contaminated both themselves and all their infants and toddlers contained more than 1 million infectious doses per gram, self-inoculation through the eyes, nose and skin as well as by mouth, was a certainty whenever a victim was eaten. The decline in incidence of the diseases has followed the cessation of cannibalism which occurred between 1957 and 1962 in various villages. **B.** Kuru deaths by age group from 1957 through 1993. By 1969, it had disappeared in prepubertal children under 10 years of age. By 1974, it disappeared in the 20- to 29-year-old age group. Currently, it has almost fully disappeared, and no one under 40 years of age has had kuru for the past 5 years. The number of adult patients has declined to less than one-tenth that seen in the early years of investigation. These changes in the pattern of kuru incidence can be explained by the cessation of cannibalism during the late 1950s. [No child born since cannibalism ceased in his village has developed the disease.] Data for the last decade has been less accurate than before since dates of deaths are often imprecise. No patient under 30 years of age has had kuru since 1985 and no one under 40 since 1991.

32 months. Kuru and CJD agents were given as whole chimpanzee-passaged brain tissue and scrapie agent as whole hamster brain tissue in their hamster brain passage [15]. Transmission by nasogastric tube administration of infected suspensions of human and scrapie mouse brains was attempted in squirrel, capuchin, woolly, and Cynomolgus monkeys, but no animals became ill after up to 18 years of observation. These observations make it clear that enormous doses of scrapie-, kuru-, or CJD-infected tissues rarely, if ever, produce disease in monkeys or chimpanzees. We are not sure by what route the squirrel monkeys eating the infected tissues naturally inoculated themselves. It could just as well have been through breaks in the skin or mucous membranes as via the gastrointestinal tract.

For sporadic CJD, we believe de novo spontaneous generation of the infectious agent as a rare stochastic event is sufficient explanation. There is no infectious chain. For familial transmissible cerebral amyloidoses (TCA) (CJD, Gerstmann-Sträussler-Scheinker disease, fatal familial insomnia), this spontaneous event is much more likely because of the mutational change in the structure of the precursor. For the BSE epidemic, the introduction of bone meal, at first contaminated by sheep scrapie and later by the infectious amyloid adapted to cattle with a cow amino acid sequence, and thus a truly high-tech neocannibalism, provides adequate explanation of its origin and its dissemination.

Is scrapie pathogenic for man?

We do not know and have no ethical way of determining whether scrapie is pathogenic for man. The epidemiology of CJD in the United Kingdom will tell us in due course whether BSE has spread to man. If it has, we should, as in the kuru epidemic, expect to find cases in children.

Scrapie has now been found to cause a disease clinically and neuropathologically indistinguishable from experimental CJD in three species of New World monkeys and two species of Old World monkeys [16, 17]. This disease occurs via either intracerebral or peripheral routes of inoculation. Natural sheep scrapie strains, as well as experimental goat and mouse scrapie strains have caused disease in the monkeys. The Compton strain of scrapie virus, as a result of such passage through primates, develops an altered host range, for it no longer produces disease in inoculated mice, sheep, or goats [18]. There is a similar situation when scrapie is produced in ferrets or mink: the mink or ferret brain agent of natural transmissible mink encephalopathy which presumably had its origin in the feeding of scrapie-infected sheep or cattle carcasses to minks on commercial mink farms [18, 19] will no longer take in mice.

CJD or kuru agents may produce, after asymptomatic incubation for more than 2 years, an acute central nervous system disease in the squirrel monkey, with death ensuing in a few days; even sudden death without previously noted clinical disease has been seen. The same strains of kuru or CJD agents produce chronic clinical disease in the spider monkey, closely mimicking the human disease, after incubation periods of 2 years or more. The time sequence of disease progression also mimics that in man, ranging from several months to more than a year until death. A single strain of kuru or CJD virus may cause severe status spongiosus lesions in many brain areas, particularly the cerebral cortex in chimpanzees and spider monkeys with minimal or no involvement of the brain stem or spinal cord, whereas, in the squirrel monkey, this same strain of nucleating protein may cause extensive brain stem and spinal cord lesions. Both kuru and CJD infectious proteins inoculated intracerebrally cause scrapie-like disease in goats but not in sheep [18].

For kuru and CJD, it appears that there is no transovarian infection or transmission in utero or during parturition, or from a milk factor or some other neonatal source [8]. Kuru epidemiological studies, which found no kuru in children born to kuru-affected mothers since the cessation of cannibalism, provide no evidence for such transmission [5, 8, 13, 19]. The concept of de novo spontaneous generation of the infectious amyloid in sporadic CJD, as well as in familial cases of TCA, needs no infectious chain [7, 8, 13, 20–24]. An infectious chain is seen only in the highly technologic neocannibalism of the more recent outbreaks of iatrogenic cases, such as the debacles of the contaminated hGH and dura mater, and also in the economic disaster of contaminated bone meal feed causing BSE in the United Kingdom and the human disaster of kuru producing contamination during primitive cannibalism.

We talk much about strains of CJD, scrapie and even BSE which propagate their own properties of incubation period, presence of amyloid plaques or white matter lesions, lesion profile, route of infection and host range. In microbiology, we see these phenomena as replication of genetically independent strains. In crystallography, they are looked upon as nucleation events that set or seed the lattice pattern. Crystallographers or polymer chemists refer to crystal growth or acceleration or augmentation of fibril growth and to elongation and condensation following the grid pattern set by the seed crystal or nucleant. It becomes autocatalytic when the baby crystals continue the pattern, determining a nucleation process often initiated by a polymer, microcrystal or microfibril, but sometimes by a foreign nidus such as glass particles, as when crystallization is initiated by scratching the side of a beaker with a glass rod.

All β-pleated proteins can also form vitreous-like three-dimensional crystals, as well as two-dimensional sheets and one-dimensional crystals (fibrils). They are, rather, semisolid, amorphous, vitreous, mucilaginous pseudocrystals, thick emulsions, or gels which are insoluble precipitates in the hydrophilic media of the *milieu intérieur*. On heating, drying, annealing, or aging, they may increase their hydrogen bonding and decrease their hydration to assume increasingly a β-pleated structure. Such non-crystalline or semisolid states occur in glasses, clays, ceramics of all kinds, cements; glues and pastes, and in plastics, gels, and emulsions. This is the heart of amyloidology and the basis of much of the biochemical alteration of proteins in cuisine and the aging of leather, parchment, rubber, natural fabrics, mucilage, and even the aging of cartilage and the lens of the eye.

In World War II, an industrial infection with a molecular nucleant which caused precipitation of ethylene diamine tartrate in a new crystalline state was started as though it were a transmissible replicating microbe. This infection crippled the industry and could not be cured. Kurt Vonnegut used this event as the basis for his imaginative "ice-nine" in his novel *Cat's Cradle* in which he hypothesized the nucleation of configural change which would crystallize water at body temperature and thereby generate an infectious nucleating agent. His older brother was a physical chemist who worked on problems of nucleation causing industrial infections. In 1952, a film, *The Man in a White Suit*, starring Alexander Guiness, showed a novel configurational fibril of polyglutamate producing a textile that could not wear out and repelled dirt. Ephraim Katchalski-Kazir in his role as a biophysicist of 'protein fibrilization' (not yet President Kazir of Israel) was a scientific advisor to the production of this film. These dreamers have priority on many ideas of today's amyloidology.

The process of conformational change may well be an induced nucleation and epitaxic pattern-setting for crystalline or fibril growth. The quantum dynamics of the transition to the lower energy state of the cross-β-pleated conformation in the semicrystalline, amorphous or vitreous state of sheets or fibrils is now being studied for transthyretin and βA4 amyloid formation. How this process may be hetereonucleated by minerals or replication initiated with ghost replicas or casts is also under consideration [20, 25–28]. This could have implications for the low level of infectivity often found to survive after long exposure to temperatures about 300 °C.

It appears likely that a coordinate-covalent or covalent alteration in the precursor may be induced or a strongly hydrogen-bonded one-dimensional crystal-packing may be formed, which endows the infectious nucleant with great stability. We thus have reason to anticipate that the infectious form of the scrapie amyloid precursor may eventually be induced in vitro, even from the synthetic polypeptides.

References

1 Farquhar J, Gajdusek DC, eds. *Kuru: Early Letters and Field Notes From the Collection of D Carleton Gajdusek.* New York: Raven Press; 1980
2 Gajdusek DC, Zigas V. Degenerative disease of the central nervous system in New Guinea: the endemic occurrence of "kuru" in the native population. *N Engl J Med* 1957;257:974-81
3 Gajdusek DC, Zigas V. Kuru: clinical, pathological and epidemiological study of an acute progressive degenerative disease of the central nervous system among natives of the Eastern Highlands of New Guinea. *Am J Med* 1959; 26 442-69
4 Zigas V, Gajdusek DC. Kuru: clinical study of a new syndrome resembling paralysis agitans in natives of the Eastern Highlands of Australian New Guinea. *Med J Aust* 1957;2:745-54
5 Gajdusek DC. Unconventional viruses and the origin and disappearance of kuru. *Science* 1977;197:943-60
6 Brown P. Transmissible human spongiform encephalopathy (infectious cerebral amyloidosis): Creutzfeldt-Jakob disease, Gerstmann-Sträussler-Scheinker syndrome, and kuru. In Calne DB, ed. *Neurodegenerative Diseases.* Philadelphia: WB Saunders; 1993;839-76
7 Gajdusek DC. Evaluation of the risk posed by BSE to human health. Joint OIE/WHO Consultation on Bovine Spongiform Encephalopathy. 2 September 1994, Paris: Office International des Épizooties; 1994
8 Gajdusek DC. Infectious amyloids: spongiform subacute encephalopathies as transmissible cerebral amyloidoses. Chapter 91. In: Fields BN, Knipe DM, Howley PM et al, eds. *Fields Virology*, 3rd edition, Philadelphia: Lippincott-Raven Publishers, 1996;2851-900
9 Beck E, Daniel PM, Davey A, Gajudsek DC. A comparison between the neuropathological changes in kuru and scrapie, a system degeneration. In: Luthy F, Bischoff A, eds. *Proceedings of the Fifth International Congress of Neuropathology.* International Congress Series No. 100. Amsterdam: Excerpta Medica; 1966;213-8
10 Klatzo I, Gajdusek DC, Zigas V. Pathology of kuru. *Lab Invest* 1959;8:799-847
11 Merz PA, Rohwer RG, Somerville RA, Wisniewski HM, Gibbs CJ Jr, Gajdusek DC. Scrapie associated fibrils in human Creutzfeldt-Jakob disease. *J Neuropathol Exp Neurol* 1983;42:327
12 Hornabrook RW, Wagner F. Creutzfeldt-Jakob disease. Papua New Guinea. *Med J* 1975;18:226-8
13 Gajdusek, DC. Infectious and noninfectious amyloidoses of brain: Systemic amyloidoses as predictive models in transmissible and nontransmissible amyloidotic neurodegeneration of Creutzfeldt-Jakob disease and aging brain and Alzheimer's disease. In: Calne DB, ed. *Neurodegenerative Diseases.* Philadelphia: WB Saunders. 1993;301-17
14 Klitzman RL, Alpers MP, Gajdusek DC. The natural incubation period of kuru and the episodes of transmission in three clusters of patients. *Neuroepidemiology* 1985;3:3-20
15 Gibbs CJ Jr, Amyx HL, Bacote A, Masters C, Gajdusek DC. Oral transmission of kuru, Creutzfeldt-Jakob disease and scrapie to nonhuman primates. *J Infect Dis* 1980:142:205-8
16 Gibbs CJ Jr, Gajdusek DC. Transmission of scrapie to the cynomolgus monkey (Macaca fascicularis). *Nature* 1972;236:73-4
17 Gibbs CJ Jr, Gajdusek DC, Amyx H. Strain variation in the viruses of Creutzfeldt-Jakob disease and kuru. In: Prusiner SB, Hadlow WJ, eds. *Slow Transmissible Diseases of the Nervous System,* vol 2. New York: Academic Press, 1979;87-110
18 Marsh RF, Hartsough GR. Is there a scrapie-like disease in cattle? Proceedings of the Seventh Annual Western Conference for Food Animal Veterinary Medicine, University of Arizona, March 17-19, 1986
19 Marsh RF, Hanson RP. On the origin of transmissible mink encephalopathy. In: Prusiner SB, Hadlow WJ, eds. *Slow Transmissible Diseases of the Nervous System,* vol 1. New York: Academic Press, 1979;451-60
20 Gajdusek DC. Fantasy of a "virus" from the inorganic world: pathogenesis of cerebral amyloidoses by polymer nucleating agents and/or "viruses." In: Neth R, Gallo RC, Greaves M et al, eds. *Modern Trends in Human Leukemia VIII.* New York: Springer-Verlag; 1989:481-99

21 Gajdusek DC. Genetic control of de novo conversion to infectious amyloids of host precursor proteins: kuru-CJD scrapie. In: Kurth R, Scwerdtfeger W, eds. *Current Topics in Biomedical Research.* Proceedings of Paul Erlich Institute Scientific Conference: Concepts in Biomedical Research, Frankfurt, May 3-5, 1990. Berlin: Springer-Verlag; 1991;95-123

22 Gajdusek DC. Nucleation of amyloidogenesis in infectious and noninfectious amyloidoses of brain. Slow infections of the central nervous system. *Ann NY Acad Sci* 1994;724:173-90

23 Gajdusek DC. Spontaneous generation of infectious nucleating amyloids in the transmissible and nontransmissible cerebral amyloidoses. *Mol Neurobiol* 1994;8:1-13

24 Gajdusek DC. Genetic control of nucleation of precursors to infectious amyloids in the transmissible amyloidoses of brain. *Br Med Bull* 1994;49:913-31

25 McPherson A, Shlichta P. Heterogeneous and epitaxial nucleation of protein crystals on mineral surfaces. *Science* 1988;239:385-7

26 Shlichta P. Heterogeneous epitaxial (HE) nucleation of protein crystals in relation to the formation of amyloid fibrils. In: Gajdusek DC, Beyreuther K, Brown P, eds. *Regulation and Genetic Control of Brain Amyloid. Brain Res Rev* 1991;16:105-6

27 Vonnegut K Jr. *Cat's Cradle.* New York: Bantam Doubleday Dell Publishing Group Inc, 1963

28 Weiss A. Replication and evolution in inorganic systems. *Angew Chem* 1981;20:850-60

29 Fukatsu R, Gibbs CJ Jr, Gajdusek DC. Cerebral amyloid plaques in experimental murine scrapie. In: Tateishi J, ed. *Proceedings of the Workshop on Slow Transmissible Diseases.* Research Committee on Slow Virus Infections, Japanese Ministry of Health, August 31, Tokyo, 1984;71-84

NATURAL SCRAPIE AND BOVINE SPONGIFORM ENCEPHALOPATHY

Transmissible Subacute Spongiform Encephalopathies:
Prion Diseases
L Court, B Dodet, eds
© 1996, Elsevier, Paris

Clinical aspects of scrapie in sheep

Jeanne Brugère-Picoux[1], Hélène Combrisson[2], Gilberte Robain[2], Jacqueline Chatelain[3], Jean-Louis Laplanche[3], Henri Brugère[1]

[1]*École Nationale Vétérinaire d'Alfort, 94704 Maisons-Alfort cedex;* [2]*Hôpital Jean Rostand, 94200 Ivry-sur-Seine;* [3]*Laboratoire de Biologie Cellulaire, Faculté de Pharmacie, 4, avenue de l'Observatoire, 75006 Paris, France*

Summary – Clinical characteristics of natural scrapie in sheep were investigated so as to determine which biological tests would allow an antemortem indication of the disease, particularly with regard to genetic susceptibility. In addition to classically described symptoms, urodynamic investigation confirmed urethra and bladder dysfunction, and a reduced functional bladder capacity. To obtain a differential diagnosis, a specific antemortem test can detect protease-resistant prion protein (PrP-res) in contaminated tissue, but this requires biopsy of specific tissues. Using body fluids, an electrochemical analysis of urine (voltametry) appears able to diagnose scrapie antemortem by identifying a specific chemical.

Scrapie is a non febrile, insidious disease in sheep caused by progressive damage to the nerve cells which gives a spongiform appearance to the brain tissue. In the national veterinary school of Alfort, we received 135 sheep infected by natural scrapie (mostly from two flocks) during the past years. The objective was to perform clinical studies of scrapie in order to determine which biological tests would allow an antemortem indication of this disease, especially with regard to genetic susceptibility [1]. This study confirms the information from previous reports about the duration of the evolution of the disease, which always results in death within 2 to 12 months, as well as the age of the infected animals (18 months to 5 years) [2].

Clinical signs

The onset of clinical signs was often marked by a slight change in behavior (nervosity, hypersensitivity, apprehension) noted in 90% of the cases. Other sensory signs soon followed. Intense pruritus led to sheep rubbing against fixed objets (fig 1), resulting in wool loss (fig 2), excoriation and inflammation of the skin (fig 3). The affected areas were bilaterally symmetrical. During the rubbing movements, the sheep displayed exaggerated signs of ecstasy, making nibbling movements of the lips or excessive licking or trembling. Other signs included head tremor and muscular fasciculations, grinding of teeth and abnormal posture. Motor abnormality appeared later (2 to 3 months before death) and often included a high-stepping (trotting) gait of the forelimbs and a 'bunny hop' movement of the back legs (especially when

Key words: scrapie / clinical signs / diagnosis / urodynamic study / urine test / voltametry

Fig 1. Scrapie-infected Ile-de-France sheep exhibiting vigorous pruritus of the hind quarters.

Fig 2. Scrapie-infected Romanov sheep showing wool loss due to intense pruritus.

the animal was made to run). As the disease progressed, severe ataxia (falling and recumbency) was observed.

Autonomic nervous system disturbances (urinary and digestive disorders) were also noted. An increase of micturition frequency was observed in scrapie-infected ewes. Urodynamic studies were performed to evaluate this dysfunction [3]. As urodynamic parameters of healthy ewes have not been described in vivo, it was necessary to establish reference parameters.

In healthy ewes (fig 4), during bladder filling, there was no increase of bladder pressure and urethral pressure was constant. When the bladder volume was close to functional bladder capacity, the ewes began to move and large fluctuations of the urethral pressure appeared. When

30

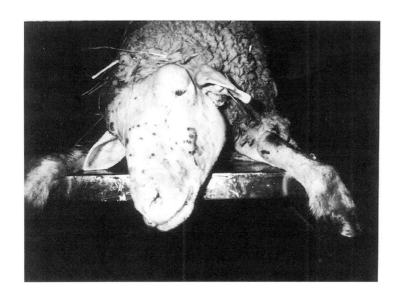

Fig 3. Cutaneous abrasions on the head due to intense pruritus in a scrapie-infected Ile-de-France sheep.

the functional bladder capacity was obtained, a detrusor contraction and a dramatic decrease in urethral pressure appeared and micturition was achieved.

In scrapie-infected ewes (fig 5), during bladder filling, the detrusor developed spontaneous contractions, allowing recording of increases in bladder pressure; the urethral tone varied and large fluctuations with high frequency were noted throughout bladder filling. The micturition

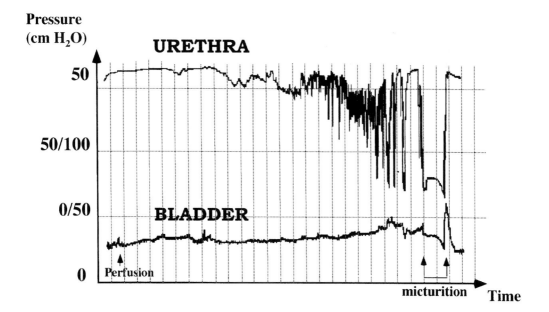

Fig 4. Urodynamic investigation in a healthy ewe.

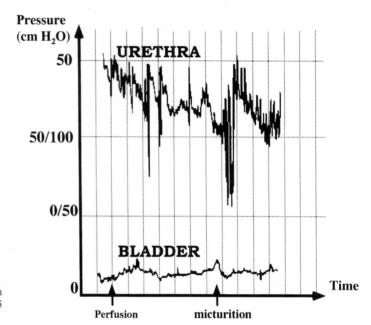

Fig 5. Urodynamic investigation in a scrapie-infected ewe 105 days before death.

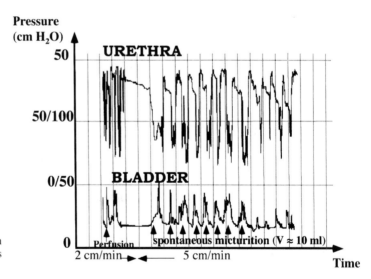

Fig 6. Urodynamic investigation in a scrapie-infected ewe 70 days before death.

was similar in healthy and scrapie-infected ewes; however, the functional bladder capacity was reduced in the latter. Functional bladder capacity decreased markedly as the neurological status deteriorated (fig 6). For some animals, it was not possible to analyze the curves due to unrest or abnormal movements.

Evolution

The weight loss (observed in 50% of the animals) became perceptible 2 to 3 months before death and scrapie seemed to evolve without hyperthermia reaction, while appetite was not severely affected. Not all sheep showed all the signs of the disease, but the signs were characteristically progressive, ending in recumbency and death.

Differential diagnosis

Scrapie must be differentiated from: cutaneous and/or pruriginous diseases (ectoparasites such as lice and mites, photosensitization); nervous diseases (listeriosis, pregnancy toxemia, hypomagnesemia, Aujeszky's disease, rabies, organophosphorous poisoning); and chronic wasting syndromes such as paratuberculosis or ovine progressive pneumonia (maedi–visna).

Antemortem diagnosis

Apart from the clinical suspicion, it is difficult to diagnose scrapie before the onset of clinical signs because no serological method is available. An antemortem diagnosis is therefore difficult. A specific diagnosis (preclinical and clinical) can be made via the detection of protease-resistant prion protein (PrP-res) by biopsy of explorable lymph nodes or in placenta [4].

A new perspective for the diagnosis of scrapie resulted from the study of urine with voltametry, an electrochemical method [5]. This method makes it possible to characterize chemicals exhibiting oxidoreduction properties. During the late 1980s, it was shown that voltamograms obtained in 110 cases of Alzheimer's disease showed a peculiar peak that was not seen in those cases with dementias of vascular origin or in the healthy elderly [6]. This peak resulted from the presence of an unidentified substance in the urine. An investigation in sheep at the clinical stage of spontaneous scrapie revealed a similar observation [7]. The same electroanalytical characteristics were seen in urine from sick animals and led to the conclusion that the same substance was found in the urine of sheep with scrapie. The electrochemical criteria taken into account were the voltage for the substance to be oxidized (1,200 mV) and the voltage for the derived product to be reduced (850 mV) (fig 7). The unknown substance could perhaps be considered as a marker of some neurodegenerative brain diseases. A subsequent study was conducted in cows with bovine spongiform encephalopathy (BSE). Urine samples from 110 cows imported from the United Kingdom into France were submitted to electroanalysis. The study was a fully double-blind procedure, as the operators of the electroanalysis

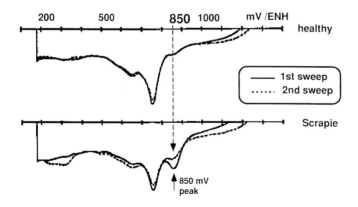

Fig 7. Voltametric analysis of urine in sheep.

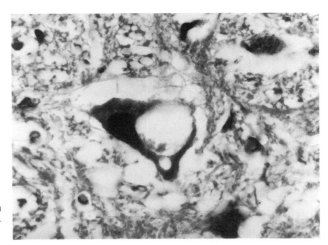

Fig 8. Vacuolated nerve cell in the brain stem of a scrapie-infected sheep (hematoxylin–eosin, × 350).

were not aware of the status of the cows (healthy or BSE). The results showed that a majority of cows with BSE were found to have the 850 mV peak. In view of a diagnostic application, detection of diseased cows was positive with a 95% success rate [8]. At the same time, the opportunity was given to obtain urine samples from two patients with Creutzfeldt-Jakob disease. Both had positive urine tests. Studies to identify and characterize the 'urine marker' are in progress.

Postmortem diagnosis

Histopathological examination of the brain is the only practical method of supporting a clinical diagnosis of scrapie. The characteristic changes are vacuolization, degeneration and necrosis of nerve cells (fig 8), spongy vacuolar transformation of the neuropil and astrocytic reaction in the brain.

The location, extent and severity of these changes vary. Vacuolization is sometimes minimal in clinical cases of scrapie or, conversely, spongiosis in the brain can be seen in other diseases such as pregnancy toxemia [9].

More specific methods of postmortem diagnosis of scrapie are immunohistochemical detection of PrP-res not only observed in brain regions with vacuolization but also in regions with minimal or no vacuolization [10], electron microscopical demonstration of 'scrapie-associated fibrils' and detection of PrP-res using immunoblotting techniques in the brain tissue.

References

1 Laplanche JL, Chatelain J, Westaway D et al. PrP polymorphisms associated with natural scrapie discovered by denaturing gradient gel electrophoresis. *Genomics* 1993;15:30-7
2 Rocher O. Suivi clinique de moutons atteints de tremblante naturelle. Thesis, Alfort, France, 1993
3 Combrisson H, Robain G, Brugère-Picoux J, Chatelain J, Brugère H. Modifications des paramètres urodynamiques chez la brebis atteinte de tremblante. *Bull Acad Vet Fr* 1991;64:257-66
4 Ikegami Y, Ito M, Isomura H et al. Pre-clinical and clinical diagnosis by detection of PrP protein in tissues of sheep. *Vet Rec* 1991;128:271-5
5 Banissi-Sabourdy C, Planques B, David JP et al. Electroanalytical characterization of Alzheimer's disease and ovine spongiform encephalopathy by repeated cyclic voltametry at a capillary graphite paste electrode. *Bioelectrochem Bioenerg* 1992;28:127-47

6 Planques B, Banissi-Sabourdy C, Jeannin C, Bizien A, Buvet R. Caractérisation électroanalytique des démences de type Alzheimer. In: *IV Congrès International de Gérontologie*, Montreal and Paris: Edisem and Masson, 1990;262-6

7 Brugère H, Banissi C, Brugère-Picoux J, Chatelain J, Buvet R. Recherche d'un témoin biochimique urinaire de l'infection du mouton par la tremblante. *Bull Acad Vet Fr* 1991;61:139-45

8 Brugère H, Banissi-Sabourdy C, Brugère-Picoux J. Electrochemical analysis of urine from sheep with scrapie and cows with BSE. In: Bradley R, Marchant B, eds. *Transmissible Spongiform Encephalopathies.* Proceedings of a consultation on BSE with the Scientific Veterinary Committee of the Commission of the European Communities Brussels, Belgium, 14–15 September 1993;359-68

9 Jeffrey M, Higgins RJ. Brain lesions of naturally occurring pregnancy toxemia of sheep. *Vet Pathol* 1992;29: 301-7

10 van Keulen LMJ, Schreuder BEC, Meloen RH et al. Immunohistochemical detection and localization of prion protein in brain tissue of sheep with natural scrapie. *Vet Pathol* 1995;32:299-308

Transmissible Subacute Spongiform Encephalopathies:
Prion Diseases
L Court, B Dodet, eds
© 1996, Elsevier, Paris

Scrapie in France: data and recent hypotheses

Jacqueline Chatelain

Cell Biology Laboratory, Faculty of Pharmaceutical and Biological Sciences, 4, avenue de l'Observatoire, 75006 Paris, France

Summary – Data obtained during a survey of scrapie in France in 1981 are reviewed. Similar data were sought in 1995 for breeds reared industrially, such as the Lacaune and Romanov. Two recent hypotheses for the possible causes of the disease are presented.

Introduction

Infectious subacute spongiform encephalopathies belong to a group of diseases known as 'prion' diseases which affect both humans (Creutzfeldt-Jakob disease [CJD] in its familial, spontaneous and iatrogenic forms; Gerstmann-Sträussler-Scheinker disease and fatal familial insomnia) and domestic and wild animals (eg, chronic wasting disease in wild deer) [1]. Pets, such as cats, are susceptible to a similar disease, called feline spongiform encephalopathy [2]. Cattle are affected by bovine spongiform encephalopathy and domestic and wild sheep (mouflons) and goats are the main targets of scrapie [3]. Scrapie in sheep has been known in Europe for more than 250 years.

In order to ascertain whether there is a geographical relationship between cases of CJD and scrapie in sheep, a survey was conducted in France in 1982 to localize cases of scrapie [4]. In 1995, a similar survey was conducted of economically important breeds of sheep: Lacaune, Manech, Ile-de-France, Romanov and Texel.

This study indicated two important risk factors in these breeds: stress, in agreement with the chaperone theory [5]; and the composition of the gene pool of the flock, ie, the percentages of animals carrying sensitivity and resistance factors. Certain breeders and veterinary surgeons have found that stress associated with a particular genotype in a certain environment can lead to the appearance of a clinical form of the disease.

Results and discussion

Table I gives a classification of the 92 *départements* of France (except for those in the Paris region) in which scrapie was diagnosed between 1968 and 1979. The diagnosis was classified as 'definite' in nine *départements*, with anatamopathological confirmation and transmission to mice, 'probable' in 35 other *départements* and 'possible' in a further 16. A diagnosis of scrapie was made at least once during this period in 58 (63%) of the 92 *départements* in which sheep and goats are reared.

Key words: scrapie / sheep / stress / genetics

Table I. Occurrence of scrapie in the rural *départements* of France between 1968 and 1979.

Diagnostic category	Départements	No of contaminated flocks
Definite (histological and/or clinical diagnosis in at least one flock)	Aveyron	10
	Côte-d'Or	6
	Nièvre, Hautes-Alpes	5 each
	Aisne	2
	Cantal, Charente-Maritime, Corrèze, Marne	1 each
Probable (clinical diagnosis in two or more flocks)	Pyrénées-Atlantiques	8
	Gironde	6
	Alpes-de-Haute-Provence, Calvados, Sarthe	5 each
	Bas-Rhin, Gard, Haute-Garonne, Lot, Lozère, Tarn	4 each
	Drome, Hérault, Ille-et-Vilaine, Lot-et-Garonne, Maine-et-Loire, Vaucluse	3 each
	Allier, Ariège, Bouches-du-Rhône, Dordogne, Doubs, Haute-Loire, Haute-Savoie, Indre, Isère, Moselle, Nord, Seine-Maritime, Tarn-et-Garonne, Vienne	2 each
Possible (clinical diagnosis in a single flock)	Ain, Alpes-Maritimes, Ardèche, Aude, Charente[1], Cher, Côtes-du-Nord, Deux-Sèvres, Eure, Gers[1], Haute Corse, Hautes-Pyrénées, Haute-Vienne, Loire-Atlantique, Oise, Pyrénées-Orientales, Saône-et-Loire[1], Yonne[1]	1 each

[1]In goats.

Table II shows the occurrence of scrapie in 1982 and in 1995 in various breeds of sheep. Préalpes, Lacaune, Ile-de-France, Texel and Suffolk sheep have been endemically affected for at least 25 years. Romanov sheep, which were scarce in 1982, were not then reported to be diseased; between 1990 and 1995, however, numerous cases were diagnosed and verified in various flocks. Lacaune and Romanov sheep are reared industrially, the former for milk production and the latter for cross-breeding with meat-producing breeds. The economically viable life span of a Lacaune ewe is 4 years, whereas in other breeds it is up to 10 years.

The incidence of scrapie in these flocks is still very difficult to determine. In a survey carried out by our laboratory on Lacaune and Romanov flocks, the disease occurred in very young animals. In 31 diseased sheep, the mean age at which the disease was seen was 26 months (range: 19–42 months). In other breeds, such as the Ile-de-France, the mean age at which scrapie appears was about 3.5 years, and disease developed in sheep up to about 7 years of age [6].

If the incidence of the disease in these industrially reared flocks proves to be higher than in other types, stress could be considered to be partially responsible for its appearance. The observation of a shortened incubation period in these flocks, however, indicates a genetic predisposition. The existence of this type of phenomenon would tend to confirm that the disease has been present in an endemic state for a long time at a variable incidence, which is difficult to evaluate precisely [7].

The role of stress in initiating the disease is difficult to quantify; however, an evaluation of the percentage of sensitive animals and of those with a resistant genotype can predict the future of the flock. A study of three flocks of Romanov sheep located at considerable distances from one another showed the extreme diversity that exists within one breed. In the Romanov

Table II. Occurrence of scrapie in 1982 and 1995 in various breeds of sheep.

Region	Breed of sheep	Type of breeding	Economic use	Diagnosis of scrapie 1982	Diagnosis of scrapie 1995	Genetic search
Mediterranean						
Languedoc-Roussillon,	Préalpes	Extensive	Milk	+	+	+
Provence, Côte d'Azur	Merinos d'Arles	Traditional	Meat	+		
	Caussenarde		Milk, meat	+		
	Corse		Milk	+	?	
Eastern central						
Rhônes, Alpes, Auvergne	Noire du Velay	Extensive	Milk	+	?	
	Rava	Traditional	Meat	+	?	
	Blanche du MC		Meat	+	?	
Southwest						
Aquitaine, Midi-Pyrénées,	Lacaune	Industrial	Milk, cheese	+	+	+
Limousin	Pyrénées	Traditional	Milk		+	Draft
	Tarasconaise		Meat			
Parisian basin						
Champagne, Ardennes,	Ile-de-France	Intensive	Meat	+	+	+
Picardie, Haute Normandie,	South Down	Industrial	Meat	+	?	
Centre, Basse Normandie,	Texel		Meat	+	+	+
Bourgogne	Suffolk		Meat	+	+	+
	Hampshire		Meat	+	?	
East						
Alsace, Lorraine,	Mérinos de l'Est	Intensive	Wool, meat	?	?	
Franche-Comté	Romanov		Cross-breeding	?	+	+

+: With anatamopathological confirmation; ?: without anatamopathological confirmation.

breed, codon 136 on the *Prnp* gene (encoding the prion protein) carries either two valines, two alanines or an alanine and a valine. A sheep that carries a codon with two alanines is resistant to scrapie, while one with a two-valine codon is susceptible to the disease [8].

Table III shows the frequencies of the alleles of codon 136 in the three flocks of Romanov sheep. None of the animals in these samples was diseased. When the three flocks are compared two-by-two, those in the Indre-et-Loire and the Cher have significantly different percentages of the alanine and valine alleles at codon 136. The flocks in the Cher and the Vosges are also statistically significantly different, whereas there is no significant difference in

Table III. Frequencies of alleles of codon 136 in three different flocks of Romanov sheep in France.

Département	No	Codon 136 Alanine (%)	Codon 136 Valine (%)	χ^2 (with Yates' correction)	P
Indre-et-Loire	27	76	24		
				19.48	< 0.001
Cher	26	98	2		
				29.92	< 0.001
Vosges	10	70	30		

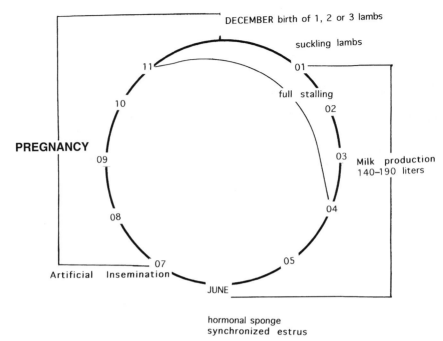

Fig 1. Annual cycle of a milk-producing ewe.

the percentages of the different alleles between those in the Indre-et-Loire and the Vosges. The only flock in which no case of scrapie has been seen is that in the Cher, in which 98% of the codons 136 carry two alanines.

These results indicate that the genetic content of resistance and susceptibility alleles in a flock are the most important factors in determining the occurrence of scrapie, which may be 'triggered' by stress. The term 'stress' is used in its broadest sense. Figure 1 shows the hormonal cycle of a Lacaune ewe [9], in which there is no period of hormonal rest and no anestrus; thus, the role of hormonal stress in the appearance of the disease cannot be confirmed.

References

1 Kirkwood JK, Cunningham AA, Austin AR, Wells GAH, Sainsbury AW. Spongiform encephalopathy in a greater kudu (*Tragelaphus strepsiceros*) introduced into an affected group. *Vet Rec* 1994;134:167-8

2 Fraser H, Pearson GR, McConnell I, Bruce ME, Wyatt JM, Gruffydd-Jones TJ. Transmission of feline spongiform encephalopathy to mice. *Vet Rec* 1994;134:449

3 Wood JLN, Lund LJ, Done SH. The natural occurrence of scrapie in moufflon. *Vet Rec* 1992;130:25-7

4 Chatelain J, Cathala F, Brown P, Raharison S, Court L, Gajdusek DC. Epidemiologic comparisons between Creutzfeldt-Jakob disease and scrapie in France during the 12-year period 1968–1979. *J Neurol Sci* 1981;51: 329-37

5 Doh-ura K, Perryman S, Race R, Chesebro B. Identification of differentially expressed genes in scrapie-infected mouse neuroblastoma cells. *Microb Pathog* 1995;18:1-9

6 Chatelain J, Baron H, Baille V, Bourdonnais A, Delasnerie-Laupêtre N, Cathala F. Study of endemic scrapie in a flock of 'Ile de France' sheep. *Eur J Epidemiol* 1986;2:31-5

7 Foster JD, Dickinson AG. Age at death from natural scrapie in a flock of Suffolk sheep. *Vet Rec* 1989;125:415-7

8 Laplanche JL, Chatelain J, Westaway D et al. PrP polymorphisms associated with natural scrapie discovered by denaturing gradient gel electrophoresis. *Genomics* 1993;15:30-7

9 Perret G. *Races ovines*. Institut technique de l'élevage ovin et caprin, Paris. 1986

Transmissible Subacute Spongiform Encephalopathies:
Prion Diseases
L Court, B Dodet, eds
© 1996, Elsevier, Paris

Scrapie outbreak in sheep after oral exposure to infective gastrointestinal nematode larvae

Jean-Louis Laplanche[1], Jean-Michel Elsen[2], Francis Eychenne[2], François Schelcher[3], Sophie Richard[1], Yves Amigues[4], Jean-Marie Launay[1]

[1]*Laboratoire de Biologie Cellulaire, Faculté de Pharmacie, 4, avenue de l'Observatoire, 75006 Paris;* [2]*Centre INRA, BP 27, 31326 Castanet-Tolosan;* [3]*Ecole Nationale Vétérinaire, 23, chemin des Capelles, 31076 Toulouse;* [4]*Labogena, INRA, Domaine de Vilvert, 78352, Jouy-en-Josas, France*

Summary – Scrapie suddenly appeared in Romanov sheep after oral exposure to infective gastrointestinal nematode larvae and spread to non experimentally parasited sheep which became severely affected. The characteristics of this outbreak are described. *Prnp* genotypes encoding the prion protein were determined in both groups for the three polymorphic codons 136, 154 and 171. Distribution of genotypes in scrapie-affected sheep was indicative of exposure to a very high level of scrapie infectivity. This observation also confirms the reduced susceptibility to scrapie of sheep encoding arginine at codon 171 living in a heavily infected environment and points out a rather similar effect of histidine at codon 154.

Scrapie is a fatal neurodegenerative disorder of unknown etiology endemically affecting sheep. It is one of the transmissible subacute spongiform encephalopathies, also named 'prion diseases', which are diseases affecting both humans and animals. Scrapie was the first spongiform encephalopathy whose transmissibility was demonstrated [1]. The biochemical hallmark of these diseases is the accumulation of an abnormal protease-resistant form of a host-encoded protein, the prion protein (PrP), in the central nervous system of affected humans and animals. This pathological accumulation could result from a protein conformational change under the influence of unknown factors (for review see [2]). Experimentally inoculated or ingested sheep brain fractions containing this protein transmit the disease, but the nature of the pathogen naturally present in the field is still unknown. The mode of transmission of the agent is not exactly known. For many years, several lines of evidence have indicated that sheep susceptibility to scrapie is under genetic control. *Sip* has been defined as the major gene controlling the susceptibility of Cheviot sheep, either experimentally challenged by the scrapie source SSBP/1 [3] or naturally affected [4]. A close association between experimental incubation times in Cheviot and flanking polymorphisms of the sheep *Prnp* gene suggested that *Sip* and the *Prnp* gene were tightly linked, if not identical [5]. To date, seven polymorphisms in the sheep *Prnp* gene coding sequence have been found at codons 112, 136, 137, 141, 154, 171 and 211 in different breeds [6–16]. It has been shown that variations at codons 136 (alanine-A/valine-V), 154 (arginine-R/histidine-H) and 171 (glutamine-Q/histidine-H/arginine-

---

Key words: transmissible subacute spongiform encephalopathies / scrapie / prion protein

R) influence scrapie susceptibility. Five alleles have been described: ARQ ($A^{136}R^{154}Q^{171}$), VRQ, ARR, ARH and AHQ. In V^{136}-coding sheep breeds, the VRQ allele is associated with high scrapie susceptibility. In all sheep breeds studied to date, the ARR allele appears to be associated with reduced susceptibility to scrapie. As long as the exact nature and mode of transmission of the scrapie agent remains unknown, thus precluding effective prevention, attempts to define genetic factors reducing scrapie susceptibility will help to limit the incidence of natural scrapie.

We previously reported a sudden outbreak of scrapie in a subflock of Romanov sheep (flock 'P') selected to study their resistance or susceptibility to common gastrointestinal nematode parasites following oral exposure to infective larvae [12]. It was hypothesized that the artificial parasitic infestation by large amounts of hematophagous larvae had suddenly increased the sheep susceptibility to a scrapie agent present at a low level in the fields. Interestingly, scrapie also appeared in the main flock, non experimentally parasited (the 'NP' flock), from which the P flock issued. As an extension of our previous study, we now present preliminary observations of this epidemic in the two flocks and relationships between *Prnp* genotypes and the disease incidence.

History of the scrapie outbreak

The Romanov sheep included in this study came from a single experimental closed flock bred for various scientific purposes. Natural scrapie seemed to be absent from the initial flock since its creation in 1973–1974. The origin of the P flock has been described elsewhere [12]. Briefly, 204 sheep born in October 1991 were orally challenged with high amounts of infective third-stage larvae (three times 20,000 larvae), obtained in vitro from eggs of gastrointestinal nematodes *(Teladorsagia circumcincta)*, between May and August 1992. These sheep were kept on the same farm as the main flock from which it was issued, but were maintained in a specific enclosure. Seventy-one sheep were kept for reproduction after 1 year of age and constituted the P flock. scrapie unexpectedly appeared in the P flock in April 1993 and, until present, has caused the death of 35 sheep (histopathologically verified cases) with a peak incidence in 1993 ($n = 22$) and 1994 ($n = 10$). Since the first observation of scrapie in the P flock, the disease also appeared in the NP flock and, up to present, 236 histopathologically confirmed scrapie cases were recorded among this flock, out of 1,000 sheep. The scrapie onset was not precisely recorded, but it seems that the disease occurred a few months later than in the P flock (August 1993). Thirty-six scrapie cases were recorded in the NP flock in 1993, after the first case was diagnosed. This number increased nearly three times in 1994 to 102 cases and remained constant in 1995, with 85 cases. It seemed to decrease in 1996, as only 13 cases were observed between January and July, and could be a consequence of the death of the most susceptible sheep during the previous 2 years. Except for sheep born in 1983 ($n = 1$), 1984 ($n = 2$) and 1985 ($n = 5$), which did not die of scrapie but were not examined histologically, all birth cohorts from 1986 to 1994 were affected. The average age at mortality of scrapie-affected animals was 1,078 days (range: 352–3,425 days). Sheep born between 1986 and 1988 seemed to have a prolonged incubation time when compared to the others, as none of them developed scrapie in 1993, nor even in 1994 for sheep born in 1986, presumably reflecting age-related susceptibility. Conversely, the animals born in 1993 and 1994 had a shorter incubation period and over 70% of the affected animals ($n = 55$ and 51, respectively) developed scrapie by the age of 18 months.

Scrapie-affected sheep were removed from the fields and penned separately, when the signs became clear-cut, until they were culled. Scrapie was confirmed by pathological examination of the brain sections and, in some cases, detection of the abnormal proteinase K-resistant form of PrP in brain extracts. *Prnp* genotypes at codons 136, 154 and 171 were determined as previously described [8, 12]. Information on *Prnp* genotypes at the three codons was available for 43 sheep from the P flock (31 scrapie cases), 164 scrapie-affected and 492 asymptomatic sheep from the NP flock. Comparisons of genotype distributions and allelic frequencies between scrapie-affected and asymptomatic sheep were performed with the χ^2 test of association.

Prnp genotypes in the P flock

Data concerning the P flock were updated and are presented in table I. Nine of 71 sheep (13%) were still alive on August 1996 and their survival time was up to 1,760 days. Four alleles corresponding to variations at codons 136, 154 and 171 were found with the following frequencies: ARQ: 0.48, VRQ: 0.38, ARR: 0.08 and AHQ: 0.06 (n = 86 chromosomes). As previously described, all but sheep encoding at least one arginine at codon 171 were equally and strongly susceptible to scrapie, regardless of their genotype at codon 136. However, it should be noted that two animals, one ARQ/ARQ and one ARQ/VRQ sheep, were still alive. One possible interpretation could be the existence of other rare undetected polymorphisms in their *Prnp* gene or other loci responsible for the lengthening of the incubation time. Only one of the three VRQ/AHQ sheep developed scrapie. Its survival time was two times longer than that of the ARQ/VRQ affected animals, suggesting a delaying effect of H[154] on scrapie onset.

Prnp genotypes in the NP flock

The allelic frequencies observed in the whole NP flock were: ARQ: 0.46, VRQ: 0.32, ARR: 0.13 and AHQ: 0.09 (n = 1,312 chromosomes). No departure from the Hardy-Weinberg equilibrium was found. These frequencies did not differ significantly from those observed in the P flock (P = 0.32), indicating that the higher incidence of scrapie in the parasited sheep could not be due only to an excess of the VRQ susceptibility allele.

As in the P flock, scrapie developed predominantly in sheep homozygous for R at codon 154 and Q at codon 171. These sheep represented 97% of the affected animals in the NP flock

Table I. *Prnp* genotypes in scrapie-affected and healthy surviving Romanov sheep after oral challenge with nematode larvae (P flock).

Prnp *genotypes*	*Scrapie* (n = 31)	*Age at death (days)* Mean (range)		*Healthy*[1] (n = 9)
ARQ/ARQ[2]	11	856	(697–968)	1
ARQ/VRQ	11	707	(687–1,232)	1
VRQ/VRQ	8	716	(704–848)	0
VRQ/AHQ	1	1,422		2
ARQ/ARR[3]	0	NA		2
AHQ/ARR	0	NA		2
VRQ/ARR[4]	0	NA		1

[1]Time from birth to 1,760 days; [2]one ARQ/ARQ dead at 1,444 days without diagnosis; [3]one ARQ/ARR dead at 1,178 days with no histological lesions of scrapie; [4]one VRQ/ARR dead at 1,264 days with no histological lesions of scrapie; NA: not applicable.

(table II). In order to obtain an approximation of the disease penetrance for each *Prnp* geno-type, proportions of affected sheep were calculated by using as the denominator the number of NP sheep that were exposed to the scrapie agent during at least 1 year after the outbreak. This limit was chosen because it represented the shortest scrapie incubation time observed in this flock. The proportions of scrapie-affected sheep were the highest in the VRQ/VRQ (78%) and ARQ/VRQ (54%) groups (table II). These proportions were probably still underestimated because some sheep were culled or died too young for scrapie to have developed. However, it could be noted that in some birth cohorts (1987 and 1992), in which no intercurrent deaths had occurred in the VRQ/VRQ group since the onset of the outbreak, all sheep with this genotype died of scrapie, suggesting complete penetrance of the disease in this group. A high incidence of scrapie in the homozygous VRQ and ARQ/VRQ groups was in good agreement with that previously reported in different V136-encoding breeds [8, 11, 16] and constituted a further example of a strong association between the VRQ allele and scrapie susceptibility ($P < 0.0001$ for VRQ allele carriers and allelic frequency between affected and healthy sheep). Affected sheep born between 1986 and 1988 carried susceptibility genotypes, suggesting that the delayed disease onset seen in these animals was due to other factors, developmental or immunological, rather than their PrP structure.

Interestingly, 31% of the ARQ/ARQ sheep also died of scrapie. These sheep were represen-ted in all birth cohorts with the exception of the 1990 birth cohort. This last phenomenon is unclear and needs further investigation. ARQ/ARQ sheep remained apparently unaffected in other single-flock studies of different V136-encoding sheep breeds [15, 17, 18]. These obser-vations suggest that animals carrying this genotype could only tolerate a low level of scrapie infection without developing the disease in natural conditions. Therefore, the higher frequency of ARQ/ARQ scrapie-affected sheep, as observed in both affected P and NP flocks, could be indicative of exposure to a very high level of infectivity or to a different scrapie agent strain with an enhanced tropism.

Clearly, not only ARR but also AHQ alleles, as previously suggested [8], appeared to be associated with reduced susceptibility to scrapie in both flocks, even in such an epizootologi-

Table II. *Prnp* genotypes in the NP flock.

Prnp *genotypes*	NP flock[1] (n = 534)	Scrapie cases (n = 164)	
	n	n	%
VRQ/VRQ	54	42	78
ARQ/VRQ	147	79	54
ARQ/ARQ	122	38	31
ARQ/ARR	53	2	4
VRQ/AHQ	38	1	3
ARQ/AHQ	55	1	2
VRQ/ARR	46	1	2
AHQ/ARR	9	0	0
ARR/ARR	5	0	0
AHQ/AHQ	5	0	0

[1]Including scrapie-affected sheep (*n* = 164) and only asymptomatic sheep which lived at least 1 year in the flock after the outbreak (*n* = 370).

cal context ($P < 0.0001$ for ARR or AHQ allele carriers and allelic frequencies between healthy and affected NP sheep). Although a survival advantage was conferred by all the possible combinations with other alleles, this advantage seemed to be most prominent for homozygous ARR/ARR or AHQ/AHQ and heterozygous AHQ/ARR sheep (no scrapie case observed until now, table II). This observation indicated that both H^{154}- and R^{171}-encoding sheep kept a low susceptibility to scrapie even in the case of an increased exposure to infection. However, the recent report of an ARR/ARR scrapie-affected sheep in Japan [13] showed that these sheep are not fully resistant and can become susceptible to scrapie under certain circumstances.

Taken together with other data on V^{136}-non coding breeds [10, 12, 13] these observations show that a policy for lowering scrapie incidence in sheep would require discarding sheep harboring VRQ and/or ARQ alleles.

How and when could sheep be infected by scrapie?

One of the plausible hypotheses to explain the occurrence of scrapie in the P flock was contamination by the scrapie agent through the lesions of the gastrointestinal tract created by the infective nematode larvae. This hypothesis supposes that the scrapie agent was present in the field for several years but at a very low level of infectivity [12]. The manner in which the NP flock was contaminated by the scrapie agent is unclear. The most obvious source of infection was the P flock. It is known that placenta from scrapie-affected or incubating ewes contains the scrapie agent and is a very likely source of infection after parturition [19, 20]. Twenty ewes from the P flock were lambing in July 1992, before exposure to the nematode larvae, and in March 1993, a few weeks and 5 months before the outbreak of scrapie in the P and NP flocks, respectively. Lambs were not kept. Scrapie-incubating ewes could have disseminated the agent in the environment which, in turn, could have contaminated the NP sheep. However, even in the hypothesis that the last lambing has played a role in spread of the disease, it could not account for contamination of the NP sheep which developed scrapie only a few months later.

Another hypothesis could link the outbreak of scrapie in the NP flock to the experimental parasitism. As a consequence of the artificial infestation of sheep by the parasite, between May and August 1992, a considerable build-up of nematode eggs, then infective larvae, could have occurred in the environment in the following months. A large number of NP sheep could have been naturally contaminated by the parasite during this period. In a scenario similar to that supposed to have occurred in the P flock, this rise in the number of naturally acquired infective larvae could have enhanced the NP sheep susceptibility to the scrapie agent, creating the conditions for the outbreak of the epidemic in the NP flock.

References

1 Cuille J, Chelle PL. La tremblante du mouton est bien inoculable. *CR Acad Sci Paris* 1938;206:78-9
2 Prusiner SB. Genetic and infectious prion diseases. *Arch Neurol* 1993;50:1129-53
3 Dickinson AG, Stamp JT, Renwick CC, Rennie JC. Some factors controlling the incidence of scrapie in Cheviot sheep infected with a Cheviot-passaged scrapie agent. *J Comp Pathol* 1968;78:313-21
4 Foster JD, Dickinson AG. The unusual properties of CH1641, a sheep-passaged isolate of scrapie. *Vet Rec* 1988;123:5-8
5 Hunter N, Foster JD, Dickinson AG, Hope J. Linkage of the gene for the scrapie-associated fibril protein (PrP) to the *Sip* gene in Cheviot sheep. *Vet Rec* 1989;124:364-6
6 Goldmann W, Hunter N, Foster JD, Salbaum JM, Beyreuther K, Hope J. Two alleles of a neural protein gene linked to scrapie in sheep. *Proc Natl Acad Sci USA* 1990;87:2476-80

7 Goldmann W, Hunter N, Benson G, Foster JD, Hope J. Different scrapie-associated fibril proteins (PrP) are encoded by lines of sheep selected for different alleles of the *Sip* gene. *J Gen Virol* 1991;72:2411-7

8 Laplanche JL, Chatelain J, Westaway D et al. PrP polymorphisms associated with natural scrapie discovered by denaturing gradient gel electrophoresis. *Genomics* 1993;15:30-7

9 Hunter N, Goldmann W, Benson G, Foster J, Hope J. Swaledale sheep affected by natural scrapie differ significantly in *PrP* genotype frequencies from healthy sheep and those selected for reduced incidence of scrapie. *J Gen Virol* 1993;74:1025-31

10 Westaway D, Zuliani V, Cooper C et al. Homozygosity for prion protein alleles encoding glutamine-171 renders sheep susceptible to natural scrapie. *Genes Dev* 1994;8:959-69

11 Belt PBGM, Muileman IH, Schreuder BEC, Bos-de Ruijter J, Gielkens ALJ, Smits MA. Identification of five allelic variants of the sheep *Prp* gene and their association with natural scrapie. *J Gen Virol* 1995;76:509-17

12 Clouscard C, Beaudry P, Elsen JM et al. Different allelic effects of the codons 136 and 171 of the prion protein gene in sheep with natural scrapie. *J Gen Virol* 1995;76:2097-101

13 Ikeda T, Horiuchi M, Ishiguro N, Muramatsu Y, Kai-Uwe GD, Shinagawa M. Amino acid polymorphisms of PrP with reference to onset of scrapie in Suffolk and Corriedale sheep in Japan. *J Gen Virol* 1995;76:2577-81

14 Belt PBGM, Schreuder BEC, Muileman IH, Bossers A, Smits MA. PrP genotype and modulation of disease onset in natural scrapie. In: *International Symposium on Prion Diseases*, Göttingen, Germany, 26–27 October, 1995 (abstract 8)

15 Hunter N, Foster JD, Goldmann W, Stear MJ, Hope J, Bostock C. Natural scrapie in closed flock of Cheviot sheep occurs only in specific *PrP* genotypes. *Arch Virol* 1996;141:809-24

16 Hunter N, Goldmann W, Smith G, Hope J. The association of a codon 136 *PrP* gene variant with the occurrence of natural scrapie. *Arch Virol* 1994;137:171-7

17 Hoinville L, Dawson M, Peace S et al. Analysis of the incidence of scrapie in selected flocks and the effect of risk factors on the occurrence of clinical disease. In: *IIIrd International Symposium on Transmissible Subacute Spongiform Encephalopathies: Prion Diseases*, Paris, France, 18–20 March 1996 (abstract)

18 Bossers A, Schreuder BEC, Muileman IH, Belt PBGM, Smits MA. PrP genotype and modulation of disease onset in natural scrapie. In: *IIIrd International Symposium on Transmissible Subacute Spongiform Encephalopathies: Prion Diseases*, Paris, France, 18–20 March 1996 (abstract)

19 Pattison IH, Hoare MN, Jebbett JN, Watson WA. Spread of scrapie to sheep and goats by oral dosing with foetal membranes from scrapie-affected sheep. *Vet Rec* 1972;90:465-8

20 Onodera T, Ikeda T, Muramatsu Y, Shinagawa M. Isolation of scrapie agent from the placenta of sheep with natural scrapie in Japan. *Microbiol Immunol* 1993;37:311-6

Transmissible Subacute Spongiform Encephalopathies:
Prion Diseases
L Court, B Dodet, eds
© 1996, Elsevier, Paris

Prion protein (*Prnp*) genotypes and natural scrapie in closed flocks of Cheviot and Suffolk sheep in Britain

Nora Hunter

Institute for Animal Health, BBSRC/MRC Neuropathogenesis Unit, West Mains Road, Edinburgh EH9 3JF, Scotland

Summary – NPU Cheviot sheep form a unique closed flock founded in 1960 and selected in two lines, positive and negative, on the basis of response to experimental challenge with a source of scrapie known as SSBP/1. Positive line animals develop disease following subcutaneous challenge with SSBP/1, whereas negative line animals are resistant. Natural scrapie has been occurring in positive line sheep at an apparently low incidence for a number of years and can be differentiated from the experimental disease in terms of clinical signs and histopathology. *Prnp* gene variants are associated with differing incidences of both experimental and natural scrapie in sheep and genotyping has revealed that 80% of NPU Cheviot sheep affected by natural scrapie encode valine at codon 136 on both *Prnp* alleles (VV^{136}). Indeed all animals of this genotype develop natural scrapie at 7–900 days of age if they live that long. The remaining 20% of affected sheep form a subgroup of heterozygotes at codon 136, the subgroup being defined by the precise amino-acid codon at each of two other positions in the *Prnp* sequence. Detailed records are available for the flock and no major effect (other than *Prnp* genetics) of scrapie in parents, twins or offspring has been found to influence the appearance of scrapie in individual sheep. In Suffolk sheep, the *Prnp* allele encoding V^{136} is vanishingly rare and the genetics of scrapie susceptibility are different. A closed flock of Suffolk sheep founded in the early 1980s has had 87 scrapie cases between November 1990 and February 1996. Although most of the scrapie-affected sheep have *Prnp* genotype QQ^{171}, some animals of genotype RQ^{171} have also developed scrapie. All these sheep are AA^{136}. In both flocks, the most susceptible animals are of homozygous genotype (VV^{136} in Cheviots and QQ^{171} in Suffolks) with apparently 100% penetrance. This could mean that scrapie is simply the result of a *Prnp* gene mutation or that the *Prnp* genotype controls susceptibility to an infecting agent.

Introduction

In sheep, scrapie has been known for many decades to be familial but there is disagreement about whether scrapie is simply caused by a genetic mutation (a genetic disease) or whether it is an infection which only occurs in genetically susceptible animals. Because scrapie is transmissible experimentally to sheep, goats and laboratory rodents, an infectious agent is implicated, however none has as yet been characterised.

In order to study sheep scrapie genetics, the NPU Cheviot flock was founded in 1960 and bred into two lines (positive and negative) which differed in their response to injection with a source of scrapie known as SSBP/1 [1]. All foundation animals were injected, usually subcutaneously, with SSBP/1 scrapie after mating (males) and after lamb weaning (females). Sheep succumbing to challenge by around 500 days post-injection formed the basis of the positive

*Key words: natural scrapie / sheep / **Prnp** genotype / UK sheep flocks*

line and animals were assigned to the negative line if experimental scrapie did not develop within 2 years of inoculation.

Natural scrapie has been occurring in positive line sheep at an apparently low incidence for a number of years and can be differentiated from the experimental disease in terms of clinical signs and histopathology [2].

Genetic control of incubation period of experimental scrapie in NPU Cheviot sheep

The work of Dickinson and his colleagues established that the incubation period of SSBP/1 scrapie in NPU Cheviots was controlled by a single gene, the *Sip* gene, with two alleles, *sA* and *pA*. Polymorphisms of the sheep *Prnp* gene are so closely associated with incubation time differences (and the alleles of *Sip*) in NPU Cheviots [3, 4] that the *Prnp* and *Sip* genes are believed to be synonymous. In NPU Cheviot sheep, valine at codon 136 (V^{136}) is linked to Sip^{sA}, whereas alanine at codon 136 (A^{136}) is linked to Sip^{pA}. Positive line animals with short incubation periods (167 ± 5 days) following subcutaneous inoculation with SSBP/1 are VV^{136} (Sip^{sAsA}), whereas those with longer incubation periods (322 ± 16 days) are VA^{136} (Sip^{sApA}). Negative line animals are all AA^{136} (Sip^{pApA}) and these resist subcutaneous challenge with SSBP/1. Challenge with CH1641 or the bovine spongiform encephalopathy (BSE) agent causes disease in only a proportion of both positive and negative lines and is not associated primarily with codon 136 variation. Instead, animals encoding glutamine at codon 171 (Q^{171}) on both *Prnp* alleles (QQ^{171}) succumb to intracerebral inoculation, whereas those having one allele with arginine (R^{171}) have a much longer incubation period (RQ^{171}) and those with two (RR^{171}) may be resistant [3, 4].

Although SSBP/1 targets animals carry V^{136}, a minor influence of codon 171 is seen, since VA^{136} animals which are also RQ^{171} have a longer incubation period (364 ± 17 days) than those which are QQ^{171} (260 ± 15 days). Similarly, with CH1641 and the BSE agent, although the major effect on incubation period depends on the codon 171 genotype, animals encoding V^{136} have longer incubation periods than AA^{136} sheep [4].

Genetics of natural scrapie in NPU Cheviot sheep

Natural scrapie was first seen in the NPU Cheviot flock in 1970 [5] in 15 animals (1967 and 1968 birth cohorts) which had not been challenged with experimental scrapie, The natural disease continued to appear until the 1978 birth cohort with 172 cases, all in the positive line. A policy of slaughter of family lines of all scrapie cases was successful in eliminating scrapie in that there were no cases in the 1979, 1980 and 1981 birth cohorts. However, in 1986 scrapie reappeared and since then, there have been (up to May 1995) 45 cases in the positive line and none in the negative line.

The *Prnp* genotypes of 35 natural scrapie cases occurring between 1986 and 1995 (table I) were 77% $VV^{136}RR^{154}QQ^{171}$ and 23% $VA^{136}RR^{154}QQ^{171}$ [2]. As there is only one V^{136} allele in the NPU Cheviot flock, no variation is seen at other codons within the VV^{136} group. However, only a subgroup of VA^{136} animals was targeted and no $VA^{136}RQ^{171}$ animals developed natural scrapie between 1986 and 1995 in observation periods up to 3,534 days of age. This is despite the fact that, like all V^{136} carriers, this genotype is susceptible to experimental challenge with SSBP/1. VV^{136} scrapie cases occurred from 1986 to 1995 in the 1982–1993 birth cohorts at a mean age of 907 days (standard deviation, SD = 265). Scrapie-affected VA^{136} animals in the 1983–1991 birth cohorts had significantly longer survival times ($P < 0.0005$) dying at a mean age of 1,482 days (SD = 426). VA^{136} scrapie cases always had

Table I. *Prnp* genotypes of 35 natural scrapie cases in NPU Cheviot sheep flock.

Prnp *genotype*	n	%	Age at death (days)*		
			Mean	*Range*	*SD*
VV^{136}RR^{154}QQ^{171}	27	77	907	497–1631	265
VA^{136}RR^{154}QQ^{171}	8	23	1,462	1107–2250	426

*Ages at death of VV^{136} and VA^{136} animals differ significantly ($P < 0.0005$). *n*: number of animals; SD: standard deviation.

longer lifespans than VV^{136} cases of the same birth cohort. There is also a clear shortening of lifespan of scrapie-affected animals from the 1982 and 1983 birth cohorts up to cases occurring in the 1986 birth cohort (not shown); thereafter, lifespans in the two groups remain essentially constant, with VV^{136} and VA^{136} cases in the last few years occurring at 700–900 and 1,100–1,200 days of age, respectively. These are now taken to be the standard 'at risk' age range for sheep of these genotypes in this flock.

No natural scrapie cases have ever been recorded in AA^{136} animals from either the positive or negative lines. These animals can have long lifespans, for example 18 positive line AA^{136} animals (born between 1980 and 1990) survived beyond 1,900 days (mean age 2451 days) [2].

Detailed records are available for the flock and no major effect (other than PrP genetics) of scrapie in parents, twins or offspring has been found to influence the appearance of scrapie in individual sheep.

Genetics of natural scrapie incidence in Suffolk sheep in a closed flock in Scotland

In Suffolk sheep, the *Prnp* allele encoding V^{136} is vanishingly rare. Two previous studies of Suffolk sheep in the USA [6] and Japan [7] included scrapie affected sheep from many different flocks, although single-flock case control studies tend to give less complex results. The Japanese Suffolks, in addition, were genetically different from the USA Suffolks. In order to clarify the situation for British Suffolks, a study was carried out in Scotland on a single scrapie-affected Suffolk flock in which, between November 1990 and January 1996, there had been 93 scrapie cases.

Genotypes of sheep were assessed at three different amino-acid codons (136, 154 and 171) which have shown clear association with scrapie incidence in other studies. Genotype at codon 136 (valine or alanine) did not vary in the Suffolk sheep: all 153 healthy animals and 35 scrapie cases tested were homozygous for alanine at codon 136 (AA^{136}). There was some variation at codon 154 (arginine or histidine), but there was no association of either allele with incidence of scrapie. The histidine encoding allele (H^{154}) was present in both scrapie and healthy animals at an approximate frequency of 5% (not shown).

Analysis of the frequencies of the genotype at codon 171 (arginine, glutamine or histidine) revealed a great deal of variation, and all three possible amino-acid codons were found. Data from 493 sheep from the 1985–1994 birth cohorts, both healthy and scrapie-affected are presented in table II.

In total, the genotypes of 64 scrapie-affected sheep were tested and all had at least one *Prnp* allele encoding Q^{171}, with 97% homozygotes (QQ^{171}) plus 3% (two animals) heterozygotes

Table II. Codon 171 genotypes in Suffolk sheep flock.

| Group | Frequencies (%) of codon 171 genotypes | | | | | |
	QQ	RQ	QH	RH	RR	n
Total	27	44	3	2	24	493
Healthy	17	50	4	2	28	429
Scrapie	97	3	0	0	0	64

Total: healthy + scrapie sheep; healthy: not affected by scrapie; scrapie: scrapie-affected; n: number of sheep.

(RQ[171]). These frequencies were significantly different from the healthy samples ($P < 0.0005$) and indicate a strong association of QQ[171] with natural scrapie in this flock [8].

Conclusions

In the NPU Cheviot flock, scrapie was clearly associated with the *Prnp* genotype and occurred in VV[136]RR[154]QQ[171] and VA[136]RR[154]QQ[171] animals only. In the Suffolk flock, scrapie occurred primarily in AA[136]RR[154]QQ[171] sheep, but also rarely in AA[136]RR[154]RQ[171] sheep.

In both flocks, the most susceptible animals were of homozygous genotype (VV[136]RR[154]QQ[171] in Cheviots and AA[136]RR[154]QQ[171] in Suffolks) with apparently 100% incidence. Heterozygote Cheviots were also at high risk with around 90% of VA[136]RR[154]QQ[171] animals developing scrapie. However, heterozygote Suffolks were at much less risk of scrapie with around 0.5% of known AA[136]R[154]RQ[171] sheep developing scrapie.

These results do not differentiate between the two etiologies of genetic disease and genetic control of susceptibility to an infectious agent. However, this study does emphasize the importance of the *Prnp* genotype in assessing the risk to a sheep of developing scrapie.

Acknowledgments

The work reported above was also carried out by JD Foster, W Goldmann (IAH/NPU), L Moore, BD Hosie, WS Dingwall, A Greig (SAC).

References

1 Dickinson AG, Outram GW. Genetic aspects of unconventional virus infections: the basis of the virino hypothesis. In: Bock G, Marsh J, eds. *Novel Infectious Agents and the Central Nervous System.* Chichester: Wiley-Interscience, 1988;63-83
2 Hunter N, Foster J, Goldmann W, Stear MJ, Hope J, Bostock C. Natural scrapie in a closed flock of Cheviot sheep occurs only in specific *PrP* genotypes. *Arch Virol* 1996;141:809-24
3 Goldmann W, Hunter N, Benson G, Foster JD, Hope J. Different scrapie-associated fibril proteins (PrP) are encoded by lines of sheep selected for different alleles of the *Sip* gene. *J Gen Virol* 1991;72:2411-7
4 Goldmann W, Hunter N, Smith G, Foster J, Hope J. *PrP* genotype and agent effects in scrapie – change in allelic interaction with different isolates of agent in sheep, a natural host of scrapie. *J Gen Virol* 1994;75:989-95
5 Dickinson AG. Natural infection, 'spontaneous generation' and scrapie. *Nature* 1974;252:179-80
6 Westaway D, Zuliani V, Cooper CM. Homozygosity for prion protein alleles encoding glutamine-171 renders sheep susceptible to natural scrapie. *Genes Dev* 1994;8:959-69
7 Ikeda T, Horiuchi M, Ishiguro N, Muramatsu Y, Kai-Uwe G, Shinagawa M. Amino acid polymorphisms of PrP with reference to onset of scrapie in Suffolk and Corriedale sheep in Japan. *J Gen Virol* 1995;76:2577-81
8 Hunter N, Moore L, Hosie BD, Dingwall WS, Greig A. Natural scrapie in a flock of Suffolk sheep in Scotland is associated with *PrP* genotype. *Vet Rec* (in press)

Transmissible Subacute Spongiform Encephalopathies:
Prion Diseases
L Court, B Dodet, eds
© 1996, Elsevier, Paris

Experimental transmission of bovine spongiform encephalopathy

Ray Bradley

Central Veterinary Laboratory, Woodham Lane, New Haw, Addlestone, Surrey KT 15 3NB, UK

Summary – The natural host range for bovine spongiform encephalopathy (BSE) includes seven species of captive ruminants, three species of captive wild Felidae and domestic cats. The experimental host range for BSE includes transmission to cattle, sheep, goats, mink and mice by oral and parenteral challenge, to pigs by parenteral challenge only and to marmosets by parenteral challenge. Transmission has not occurred to hamsters or chickens. From affected, natural, clinical cases of BSE, infectivity has been found only in the brain, spinal cord and retina, but not in over 40 other neural and non-neural tissues including muscle (meat), milk, semen and embryos. In an incomplete pathogenesis study in cattle challenged experimentally by the oral route, infectivity has been found to date only in the distal ileum (intestine) of animals between 6 and 18 months post-challenge.

Introduction

Bovine spongiform encephalopathy (BSE) is a new disease of cattle. The first clinical case occurred in April 1985 and the first confirmation by microscopical examination of the brain was in November 1986 [1]. Epidemiological studies [2] show that cattle suddenly became effectively exposed to a scrapie-like agent in ruminant-derived feed in the form of meat and bone meal (MBM) in 1981–82 [3]. The trigger for the disease was commercially motivated changes in the rendering industry prior to this date and particularly to the sudden phasing out of hydrocarbon solvent recovery of fat from a significant proportion of the MBM production in Great Britain at this time [3]. The modal age of occurrence of BSE is 5 years and equates also very closely with the modal incubation period since most cases were exposed as calves. The introduction of a ban prohibiting the feeding of ruminant-derived protein to ruminant animals from 18 July 1988 was thus introduced to break the cycle of infection via feed, but not before a total of 18 other Bovidae of seven (ruminant) species in zoos and wildlife parks had been exposed via the same feed as given to cattle and succumbed to spongiform encephalopathy [4].

Public health was protected by compulsory slaughter and complete destruction of suspect cases of BSE (August 1988) and by the specified bovine offals (SBO) ban (November 1989). This enforced the removal from the human food chain of SBO (brain, spinal cord, tonsil, thymus, spleen and intestine) from all cattle over six months of age. The SBO ban was extended to protect all species of animals and birds from September 1990, but not before a small

Key words: bovine spongiform encephalopathy / prion disease / spongiform encephalopathy transmission

number of domestic cats [5] (70 to February 1996) and captive wild Felidae [4] (a total of seven cases in three species) had been exposed via feed, the latter to raw central nervous tissue from fallen cattle presumably infected with the BSE agent.

Once it was recognized that the first occurrence of BSE was not a scientific curiosity but an epidemic disease of cattle, it was necessary to do two things: firstly, to identify a susceptible laboratory species in which to bioassay infectivity, study BSE strain types and other aspects of the agent and its transmission; secondly, to determine the susceptibility of various farm animal species to challenge via the parenteral and oral routes. The results of the latter studies are reported below.

The choice of central nervous and major lymphoreticular tissues for inclusion in the definition of SBO was based on historical knowledge of the pathogenesis of natural scrapie in goats [6] and Suffolk sheep [7]. The successful transmission of BSE to mice [8] enabled the use of this species for the bioassay of infectivity. The validity of the choice of the same tissues as those that were infected during the incubation period of scrapie as SBO has now been checked by bioassay of a wide range of tissues from cattle naturally clinically-affected and confirmed to have BSE, the worst scenario situation. Furthermore, an experimental pathogenesis study in cattle using susceptible mice for bioassay of infectivity in an equivalent range of tissues is in progress. The results of these studies, final or interim as appropriate, are reported below.

Experimental host range studies

These are summarized in table I. All species listed that were challenged parenterally were inoculated with bovine brain material from confirmed cases of BSE by the intracerebral route and usually simultaneously via one or more other routes such as intravenously or intraperitoneally. Details can be found in the respective papers focusing on mice [8, 9], cattle [10], sheep and goats [11], pigs [12, 13], marmosets [14], mink [15], hamsters [16] and chickens [13]. Hamsters and chickens did not succumb to spongiform encephalopathy; the other species did.

Challenge by the oral route using inocula of a much larger volume prepared from bovine brain material from confirmed cases of BSE, and sometimes administered repetitively, has been attempted in mice [17], cattle [18], goats and Cheviot sheep [11], pigs [13], mink [15] and chickens [13]. Transmission has been successful in all of these species, except pigs (which are now over 5 years post-challenge and healthy) and chickens. The occurrence of disease in sheep is dependent upon the *Sip* and *Prnp* genotype.

The experimental disease produced in cattle resembles the natural disease clinically and pathologically. The experimental disease in mink differs from that of transmissible mink encephalopathy [15], despite having some common features, and that seen in sheep and goats is not readily distinguishable from natural scrapie in these species [11].

Tissue transmission studies

Frozen, unfixed tissues from up to nine cattle from different parts of Great Britain, naturally clinically-affected and later confirmed to have BSE, were used to supply over 40 tissues for bioassay in susceptible mice. Each tissue was collected in a manner which prevented cross contamination with other tissues from the same or different animals or species. In the case of intestinal tissue, microscopic examination of adjacent tissue pieces was undertaken to verify the presence or absence of Peyer's patches. Each tissue was prepared as a 1 in 10 homogenate (except liquid tissues which were administered undiluted) and inoculated by the combined

Table I. Experimental bovine spongiform encephalopathy: host range following oral or parenteral challenge with brain from cattle with confirmed BSE (minimum incubation period - months)

Recipient host	Oral	Parenteral
Mouse	Positive (15)	Positive (9.7)
Cattle	Positive (35)*	Positive (18)
Sheep	Positive (18)	Positive (14)
Goat	Positive (31)	Positive (17)
Pig	Ip	Positive (16)
Marmoset	Not done	Positive (46)
Mink	Positive (14)	Positive (12)
Hamster	Not done	Nt
Chicken	Ip	Ip

Ip: in progress; Nt: no transmission; *: provisional.

intracerebral (0.02 ml) and intraperitoneal (0.10 ml) routes into 24 mice. Embryos [19] collected from a large number of cows clinically affected with BSE and washed in accordance with International Embryo Transfer Society (IETS) protocols were ultrasonicated before inoculation into 51 mice. Each mouse received an inoculum made of about 20 embryos extracts (AE Wrathall, personal communication). Uterine flushings were also inoculated into mice. Mice were observed clinically for up to more than 650 days and then brains were examined microscopically for evidence of spongiform encephalopathy.

Only brain (on 9/9 occasions), retina, cervical and terminal spinal cord transmitted disease. Tissues which showed no detectable infectivity are listed in table II. Uterine flushings also showed no detectable infectivity (AE Wrathall, personal communication).

No detectable infectivity has been found in milk [9, 17, 20] or placenta [9, 17] by either feeding or parenteral inoculation into mice. Neither has infectivity been found in placenta over 6 years following oro-nasal challenge of cattle [13, M Dawson, GAH Wells, personal communications].

Embryo transfer

Embryos derived from cows clinically affected by BSE and washed according to IETS protocols have failed to produce disease in recipient females imported from New Zealand. The offspring themselves remain healthy up to 4 years of age [19, AE Wrathall, personal communication]. The experiment will not be finished until 2001.

Pathogenesis

In a pathogenesis study in cattle challenged orally with 100 g of brain, no infectivity was found in any of 44 tissues following bioassay in mice 2 months after challenge. The next four kills at approximately 4-month intervals thereafter showed infectivity in the distal ileum, but in no other tissue [18, GAH Wells, personal communication].

Other studies

An attack rate study (oral bioassay of BSE infectivity in cow brain) in cattle is in progress and shows so far that BSE has been transmitted to all groups following oral challenge with 1, 10,

Table II. Tissues from clinically affected cattle with no detectable infectivity by parenteral inoculation of mice.

Nervous system	Lymphoreticular system and	Gastrointestinal tract
Cerebrospinal fluid	Spleen	Oesophagus
Cauda equina	Tonsil	Reticulum
Peripheral nerves:	Lymph nodes	Rumen (pillar)
N sciaticus (proximal)	– prefemoral	Rumen (oesophageal groove)
N tibialis	– mesenteric	Omasum
N splanchnic	– retropharyngeal	Abomasum
		Proximal small intestine
		Distal small intestine
		Proximal colon
		Distal colon
		Rectum
Blood	Male reproductive tissues	Female reproductive tissues
Clotted blood	Testis	Ovary
Buffy coat	Prostate	Uterine caruncle (pregnant cow)
Serum	Seminal Vesicle	Placental cotyledon
Foetal calf blood	Epididymis	Placental fluids:
	Semen	– amniotic fluid
		– allantoic fluid
Fat muscle skin and bone marrow	Abdominal and respiratory organs	Udder
		Milk
Midrum fat	Liver	
Musculus (M.) semitendinosus	Kidney	Embryos
M diaphragma	Heart	
M longissimus	Pancreas	
M masseter	Lung	
Skin	Trachea	
Bone marrow		

100 or 3×100 g of pooled brain tissue from clinically affected confirmed cases of BSE [13, GAH Wells, personal communication]. The study is incomplete.

A comparative intracerebral titration of a brain pool from five confirmed cases of BSE and intracerebral bioassay of a pool of several lymph nodes (table II) and of a pool of the spleens from the same cases showed no detectable infectivity in the spleens or lymph nodes bioassayed in mice (experiment completed) or so far in cattle up to 36 months post-inoculation. The titre of infectivity in mice is 3.3 $\log_{10}$ ID$_{50}$/g and is likely to be 100–1000 greater in cattle (experiment incomplete, GAH Wells, personal communication). This suggests that although, as expected, the mouse is less sensitive than cattle for detecting BSE infectivity, the difference in infectivity between that in brain and that in lymphoreticular system tissues is some 100–1000 times more than that revealed by bioassay in mice. Any infectivity in other tissues would be expected to be even lower.

Conclusion

BSE is transmissible experimentally to a range of different species, but some appear more resistant than others. Sometimes, variation in the ease with which transmission occurs (in sheep for example) seems attributable to variation in the *Sip* or *Prnp* gene which probably controls the length of the incubation period. Pigs do not appear to be susceptible to BSE via the oral route up to 5 years post challenge. The $Sinc^{s7s7}$ mouse is uniformly susceptible to infection with the BSE agent and, whatever its source, it seems to be biologically invariable. The mouse is an acceptable animal for meaningful bioassay of infectivity, though cattle are more sensitive. A small amount (1 g) of brain administered orally to a calf is sufficient to cause BSE. Infectivity in cattle clinically affected with BSE is confined to the central nervous system (CNS) (which includes the retina). During the incubation period, infectivity has so far (18 months post-challenge) been detected only in the distal ileum. In these respects, cattle incubating BSE are reminiscent of mink incubating transmissible mink encephalopathy. In both diseases there is little evidence of replication outside the CNS during the incubation period. This contrasts with the situation in natural scrapie in sheep and goats and experimental scrapie in several models in mice.

Acknowledgments

I am indebted to many colleagues at the Central Veterinary Laboratory, but especially to GAH Wells and AE Wrathall, and to H Fraser and DM Taylor of the Institute for Animal Health BBSRC/MRC Neuropathogenesis Unit for access to privileged information on the interim results of important experiments. Any views on the interpretation of these unpublished results are entirely the responsibility of the author.

References

1   Wells GAH, Scott AC, Johnson CT et al. A novel progressive spongiform encephalopathy in cattle. *Vet Rec* 1987;121:419-20
2   Wilesmith JW, Wells GAH, Cranwell MP, Ryan JBM. Bovine spongiform encephalopathy: epidemiological studies. *Vet Rec* 1988;123:638-44
3   Wilesmith JW, Ryan JBM, Atkinson MJ. Bovine spongiform encephalopathy: epidemiological studies on the origin. *Vet Rec* 1991;128:199-203
4   Kirkwood JK, Cunningham AA. Epidemiological observations on spongiform encephalopathies in captive wild animals in the British Isles. *Vet Rec* 1994;135:296-303
5   Wyatt JM, Pearson GR, Gruffydd-Jones TJ. Feline spongiform encephalopathy. *Feline Practice* 1993;21:7
6   Hadlow WJ, Kennedy RC, Race RE, Eklund CM. Virologic and neurohistologic findings in dairy goats affected with natural scrapie. *Vet Pathol* 1980;17:187-99
7   Hadlow WJ, Kennedy RC, Race RE. Natural infection of Suffolk sheep with scrapie virus. *J Infect Dis* 1982;146
8   Fraser H, McConnell I, Wells GAH, Dawson M. Transmission of bovine spongiform encephalopathy to mice. *Vet Rec* 1988;123:472
9   Fraser H, Foster JD. Transmission to mice, sheep and goats and bioassay of bovine tissues. In: Bradley R, Marchant B, eds. *Transmissible Spongiform Encephalopathies.* Proceedings of a Consultation on BSE with the Scientific Veterinary Committee of the Commission of the European Communities held in Brussels from 14–15 September 1993. CEC Brussels, 1993;145-59
10 Dawson M, Wells GAH, Parker BNJ. Preliminary evidence of the experimental transmissibility of bovine spongiform encephalopathy to cattle. *Vet Rec* 1990;126:112-3
11 Foster JD, Hope J, Fraser H. Transmission of bovine spongiform encephalopathy to sheep and goats. *Vet Rec* 1993;133:339-41
12 Dawson M, Wells GAH, Parker BNJ, Scott AC. Primary parenteral transmission of bovine spongiform encephalopathy to the pig. *Vet Rec* 1990;127:338
13 Dawson M, Wells GAH, Parker BNJ et al. Transmission studies of BSE in cattle, pigs and domestic fowl. In: Bradley R, Marchant B, eds. *Transmissible Spongiform Encephalopathies. Proceedings of a Consultation on BSE with the Scientific* Veterinary Committee of the Commission of the European Communities held in Brussels from 14–15 September 1993. CEC Brussels, 1993;145-59
14 Baker HF, Ridley RM, Wells GAH. Experimental transmission of BSE and scrapie to the common marmoset. *Vet Rec* 1993;132:403-6
15 Robinson MM, Hadlow WJ, Huff TP et al. Experimental infection of mink with bovine spongiform encephalopathy. *J Gen Virol* 1994;75:2151-5
16 Dawson M, Wells GAH, Parker BNJ, Scott AC. Transmission studies of BSE in cattle, hamsters, pigs and domestic fowl. In: Bradley R, Savey M, Marchant B, eds. *Subacute Spongiform Encephalopathies.* Proceedings

of a Seminar in the CEC Agriculture Research Programme held in Brussels from 12-14 November 1990. CEC Brussels, 1990;25-32

17 Middleton DJ, Barlow RM. Failure to transmit bovine spongiform encephalopathy to mice by feeding them with extraneural tissues of affected cattle. *Vet Rec* 1993;132:545-7

18 Wells GAH, Dawson M, Hawkins SAC et al. Infectivity in the ileum of cattle challenged orally with bovine spongiform encephalopathy. *Vet Rec* 1994;135:40-1

19 Bradley R. Embryo transfer and its potential role in control of scrapie and bovine spongiform encephalopathy (BSE). *Livestock Prod Sci* 1994;38:51-9

20 Taylor DM, Ferguson CE, Bostock CJ, Dawson M. Absence of disease in mice receiving milk from cows with bovine spongiform encephalopathy. *Vet Rec* 1995;136:592

Transmissible Subacute Spongiform Encephalopathies:
Prion Diseases
L Court, B Dodet, eds
© 1996, Elsevier, Paris

Bovine spongiform encephalopathy control measures in the United Kingdom

Kevin C Taylor

Ministry of Agriculture, Fisheries and Food, Government Buildings, Hook Rise South, Tolworth, Surbiton, Surrey KT6 7NF, UK

Summary – The eradication of bovine spongiform encephalopathy (BSE), and the protection of animal and human health, are dependent on preventing effective exposure of susceptible species. The preventative measures applied in the United Kingdom, and the effect of these measures on the BSE epidemic, are described.

Introduction

This paper is the exception to most of the papers in this book, since the subject is the application of scientific information to construct effective control measures which will protect both animal and public health, rather than the science itself. Although different in type and intent from the other papers, it does serve as a reminder that research, however esoteric, has a practical objective: in this case the control and eradication of bovine spongiform encephalopathy (BSE). It is, perhaps, paradoxical, that an animal disease which is a serious problem in only one country should have given such impetus to research in the transmissible spongiform encephalopathy field worldwide, not because of the intrinsic importance of BSE as a disease of cattle, but because of the possibility that the agent responsible can infect man. Without this human dimension, interest in BSE would, I suspect, be much reduced.

BSE was first recognized by pathologists at the Central Veterinary Laboratory, Weybridge, UK, in November 1986. The lesions observed in the brains of two cows from different parts of England were familiar from the study of the brains of sheep clinically affected with scrapie, but had not been seen before in cattle. This suggested that BSE was a new member of the group of diseases collectively known as transmissible subacute spongiform encephalopathies (TSSE), but research was required to confirm transmissibility.

An epidemiological study was started, initially to collect data from 200 cases, and early in 1988 cattle feed containing ruminant protein was identified as the only factor common to all cases, leading to the hypothesis that BSE was a new disease of cattle caused by infected feed. The most likely source of infection was meat and bonemeal which had been manufactured from material which included the waste from sheep infected with scrapie. The logical method of preventing such infection was to prohibit the feeding of ruminant protein to ruminant animals, and it is satisfying to see how the course of the epidemic has been influenced by such controls introduced in July 1988 after study of so few cases.

Key words: control measures / animal health / public health

Preventing infection

UK control measures for BSE address two quite different problems. The first is animal health, where we are seeking to eradicate BSE from the cattle population and to prevent the infection being transmitted to other animal species. The second is a potential public health problem, where we are aiming to protect humans against a risk which might or might not exist. The dramatic fall in the number of BSE cases has made human exposure less likely by reducing the amount of BSE agent in circulation, and so the animal health controls have also protected human health indirectly. This indirect benefit is, however, much less important than the enforcement of measures which are in place specifically to protect human health, which have been designed to be effective irrespective of what happens to the epidemic in cattle.

Whether an animal becomes infected with BSE depends on whether 'effective exposure' occurs, and this in turn is dependent on a number of factors: 1) the dose: that is, the amount of tissue and infectivity titre of that tissue; 2) the route of exposure: oral or parenteral; and 3) the size of the species barrier, which has two major components: i) the prion protein gene sequences of the infected and exposed species, and ii) the strain of TSSE agent.

Within species there is, by definition, no species barrier; it follows that BSE is most easily transmitted to cattle and that such transmission is most difficult to prevent. If transmission is to a very different species, there is always a species barrier, but because it depends on more than one component the size of the barrier cannot be predicted. If it is too high, transmission will be impossible: this may explain why pigs which were experimentally fed BSE-infected cattle brain have not developed a spongiform encephalopathy more than 6 years after being challenged. If it is lower, transmission may be possible if the dose is high enough, as seems to have occurred with some domestic cats. Of the three factors which govern whether effective exposure occurs, only the dose can be controlled, and control of dose provides the basis for all the measures which are in place to protect animal or human health. In each case the objective is to prevent exposure to enough BSE agent to cause infection and disease. The principle is easily stated, but the practical application and enforcement are more complicated.

Protecting human health

BSE has been a notifiable disease since 21 June 1988. All cases identified by farmers, veterinary surgeons or others, whether on the farm or elsewhere (such as at markets or slaughterhouses) must be notified to the State Veterinary Service. Veterinary staff examine all reported animals and prevent the movement of any which they suspect have BSE. Until 7 August 1988 suspect cattle could be slaughtered for human consumption (although the head was removed to obtain the brain to confirm the clinical diagnosis). An expert working party, chaired by Richard Southwood to advise the government on all aspects of BSE, recommended that the carcases of affected cattle should be condemned and destroyed. The advice was received on 21 June, and a policy of compulsory slaughter and destruction of carcases of suspect cattle, with payment of compensation, was introduced on 8 August 1988.

Since that date any animals which are believed, on the basis of clinical examination, to be affected with BSE are destroyed. Repeat visits may be made to observe the development of clinical symptoms, but approximately 80% of suspects are slaughtered on the day they are first examined. Particular care is taken to ensure that suspect cattle are not slaughtered for human consumption: all cattle are inspected in the slaughterhouse lairage before being slaughtered, and in the case of adult animals (which are the only cattle likely to be clinically affected), the inspection must be carried out by an official veterinary surgeon. The signs of BSE are exacerbated by transport and unfamiliar surroundings, making it extremely unlikely that cattle which show clinical signs will escape detection.

The slaughter policy should ensure that no clinically affected animal enters the human food chain. All carcases are now cremated in purpose-built incinerators. Some brain tissue is removed from all heads prior to cremation in order to permit laboratory confirmation of clinical diagnosis.

Preventing any risk from infected but clinically normal, and so unidentifiable, cattle demanded a different approach. The vast majority of British cattle have not contracted BSE: even at the peak of the epidemic no more than 1% of adult cattle were affected in any year, so any approach had to be selective and, at the same time, effective. The solution was to identify tissues and organs which might contain a significant titre of BSE, and prevent their use. In 1989, however, little was known about the distribution of BSE agent in cattle affected with BSE or incubating the disease. Rather than wait for the results of bioassay of cattle, it was decided that the bovine tissues to be excluded from human food should be selected on the basis of knowledge of the pathogenesis of scrapie in sheep, not least because BSE was considered to have arisen from scrapie.

The highest levels of scrapie infectivity in ovine and caprine tissues from naturally infected animals are found in the central nervous system and in lymphoreticular tissues. It was therefore decided that the brain, spinal cord, tonsil, thymus, spleen and intestine of cattle over 6 months of age at slaughter should no longer be used for human consumption. These tissues were named the specified bovine offal (SBO). Tissues from calves aged 6 months and younger were excluded because scrapie is undetectable in the tissues of infected lambs at this age, indicating that offal from young cattle would not present a risk to humans via the relatively inefficient oral route. The validity of the assumptions made has been vindicated by the results of bioassay of tissues from clinically affected cattle, and from cattle which were experimentally fed BSE-infected brain.

The use of SBO in human food was prohibited in November 1989. The order was amended in November 1994 to include intestines and thymus from calves of any age, and changes were made in August 1995 principally to protect animal – particularly cattle – health, which had the effect of requiring eyes to be treated as SBO, even though not defined as such. The use of bovine vertebrae in the production of mechanically recovered meat was prohibited from December 1995 to ensure that even small pieces of spinal cord could not enter human food by that route. All these controls are designed to prevent human consumption of potentially infected tissues from an apparently healthy animal. In conjunction with the controls on suspect animals, any bovine tissue likely to contain significant infectivity should be removed from the human food chain.

Protecting animal health

BSE in the United Kingdom is a feed-borne epidemic, and the most important animal health measure is the 'prohibition on feeding ruminant-derived protein* to ruminant animals (the so-called ruminant feed ban)'. If scrupulously observed, this prevents the transfer of infectious

---

*Milk, milk products and dicalcium bone phosphate are not included in the definition of 'ruminant-derived protein', and may still be fed to ruminant animals.

agent from sheep to cattle and, indeed, from any ruminant to any other ruminant animal. Most importantly, it prevents recycling of infection from cattle to cattle via feed. Since there is no species barrier when transmission is to the same species, this is of critical importance. The measure was implemented on 18 July 1988 and if there is no other significant route of transmission, this measure, alone, will eradicate BSE. Obviously, because of the long incubation period of BSE, such measures cannot be reflected by a change in the incidence of the disease for several years after they are introduced, since it is inevitable that infection established prior to the introduction of the feed ban will result in clinical disease in those cattle which survive sufficiently long.

Although the epidemiological evidence that infected feed was the source of infection has long been overwhelming, the ability of the BSE agent to survive processing of animal waste (rendering) has now been demonstrated experimentally. The ability of different rendering processes used throughout the European Union (EU) to inactivate BSE and scrapie agents has been assessed in a collaborative study. Animal waste was 'spiked' with infected cattle or sheep brains, and the raw mix before processing, and the meat and bonemeal, and in some cases tallow, which are the products of the rendering process, were bioassayed. The results indicate that a number of plants in the United Kingdom used processes which did not inactivate the BSE agent, and in one process the titre in the meat and bonemeal after rendering was little different from that in the unprocessed raw mix. No infectivity was detected in tallow produced by this process, however. As a result of this study, standards have been set which renderers throughout the EU have to meet if they are processing ruminant material: all UK renderers have complied with these standards since 1 January 1995. In addition, on 2 November 1994, legislation was introduced to ban the use of 'mammalian' protein in ruminant feedstuffs, in order to prevent any confusion that may have arisen in attempting to differentiate between meat and bonemeal of ruminant and porcine origin.

The BSE study was limited by the sensitivity of the bioassay system, and could only identify processes which reduced the infectivity titre less than 80-fold. As a consequence, there can be no certainty that the rendering standards now in force will 'completely' inactivate the BSE agent. Neither can we be sure that scrapie agent is completely inactivated, since results from the scrapie spike study are not yet available. The situation is therefore as it was in 1988, when the option of treating potentially infected material to make it safe was considered, but rejected in favour of the ruminant feed ban: the safer option was chosen and will continue to be implemented, despite the change in rendering standards.

In September 1990, following the experimental transmission of BSE to pigs by concurrent intracerebral, intraperitoneal and intravenous injection of bovine brain tissue, the prohibition which prevented the use of SBO in human food was widened to prevent their use, or the use of any protein derived from them, in food for any animal. The human and animal health protection measures differed subtly, in that the latter did not prohibit the use of tallow derived from SBO in animal feed, whereas the former did so in food. The measures were more closely aligned in August 1995, when major changes to the controls on SBO were introduced. Since that date, tallow derived from SBO must not be used in animal feed, although fatty acids extracted from SBO-derived tallow which has been subject to thermal hydrolysis at hyperbaric pressure in the oleochemical industry are exempted from the general prohibition.

The use of meat and bonemeal derived from rendered SBO as fertilizer was prohibited in November 1991, in order to prevent the remote risk of infection if ruminants should come into contact with such material after it had been applied to pasture. From this time disposal of the finished products, primarily by incineration or burial, has been subject to Ministry of Agriculture, Fisheries and Food (MAFF) licensing.

Until 1 April 1995, SBO, like all other material produced at a slaughterhouse which was unfit for human consumption, had to be stained with a black dye before being moved to be rendered. Since that date it has been a legal requirement that SBO be stained with a specified blue dye to distinguish it from non-SBO material, which is also unfit for human consumption, but can legitimately be included in feed for monogastric species after being processed. A test has been developed which will detect the blue dye, even in processed material.

New measures were introduced in November 1994, after the BSE agent had been detected in the distal ileum of calves 6 months after experimental oral challenge with brain from confirmed clinical BSE cases. SBO was redefined to include the intestines and thymus of calves under the age of 6 months. The ban now applies to intestines and thymus of calves of any age which are slaughtered for human consumption, or from calves older than 2 months which die on the farm or in transit. The decision to extend the definition of SBO was taken in spite of advice that any risk from such material could only be minuscule, as the government concluded that any tissues in which infectivity might potentially occur would have to be removed from the animal and human food chains.

The primary animal health purpose of the SBO ban is to prevent the transmission of BSE agent to any non ruminant animal species by removing from the feed chain any tissue in which a significant amount of the agent might be present. The revised SBO rules which have been in place since 15 August 1995 cover all those bovine tissues in which the presence of the BSE agent has so far been detected, either in natural disease (brain, spinal cord and retina) or experimental infection (distal ileum), as well as three (tonsil, thymus and spleen) in which it has not.

Like any other such measure, the SBO ban can only be effective from the time that it is implemented: it can have no retrospective effect. Naturally occurring spongiform encephalopathies, occurring contemporaneously with the BSE epidemic, have been reported in seven species of captive wild ruminants and in three species of wild Felidae in zoological collections and in 70 domestic cats in the United Kingdom. In all these species the putative source of infection is infected feed, although it is possible that other routes of transmission may also occur in greater kudu.

Although introduced to protect human health, and extended to protect non ruminant animal species, the SBO ban should, in theory, also suffice to protect ruminant species. In reality, however, it is better regarded as a valuable adjunct which reinforces the effect of the ruminant feed ban, and reduces the risk if ruminant protein is fed to cattle or other ruminants in contravention of the statutory ban. There is circumstantial evidence that this has happened, and a test has now been developed to identify the presence of ruminant protein in the prepared feed or feed constituents, and this is being used to check compliance with the controls.

Although there is no evidence that BSE can be transmitted maternally or horizontally, precautions have been taken to prevent infection being spread by a cow which is showing clinical signs of BSE and which calves before being slaughtered. Since 21 June 1988, successive BSE orders have required that a suspect which is due to calve must be isolated while calving and for 72 h afterwards in premises which have been approved for the purpose by a veterinary officer. The placenta, bedding and discharges must be burned or buried, and the isolation premises then cleaned and disinfected with hypochlorite. These precautions are intended to prevent transmission to other cattle in the herd: they have nothing to do with preventing maternal transmission.

Effect of control measures

The ruminant feed ban was designed to prevent infection from cattle, sheep or any other ruminant animal being transmitted in feed. The measure had no effect on cattle which had been infected before it was introduced, and the 5-year mean incubation period of BSE made it inevitable that the visible course of the epidemic would continue unchanged until 1992 or 1993. In fact, the British epidemic reached its peak in the winter of 1992–1993, and is now clearly in decline. Figure 1 shows the number of confirmed cases of BSE in Great Britain by month and year of clinical onset, whilst the number of suspect cases reported and placed under restriction since the disease was made notifiable in 1988 is shown in figure 2. The latter figure is plotted using a 4-week rolling mean to minimize the effect of weekly fluctuations in the report rate. The number of suspected cases reported and placed under restrictions in January 1993, when the epidemic was at its peak, averaged more than 1,000 per week. In January 1996, the equivalent figure was about 300 – a 70% improvement.

Perhaps the clearest demonstration of the effect of the ruminant feed ban, and also its failure to completely prevent further transmission of infection, is provided in figure 3. Confirmed cases of BSE with known dates of birth are plotted by month of birth and although recent figures are inevitably incomplete and subject to amendment, the contrast between the steadily rising incidence in cattle born before the ban, and the falling incidence in those born after, is striking. It is equally clear, however, that the feed ban has not prevented a considerable number of cattle born after it was introduced from being infected. As of 31 January 1996, 157,360 cases of BSE had been confirmed in Great Britain since the epidemic began, and 24,870 of these had been born after the introduction of the feed ban. The source of infection in these cases has been investigated.

A number of studies have failed to find any evidence that either maternal or horizontal transmission was causing BSE to occur in cattle born since 18 July 1988. A handful of cases in cattle born after this date may have been attributable to these routes, but it is clear that feed

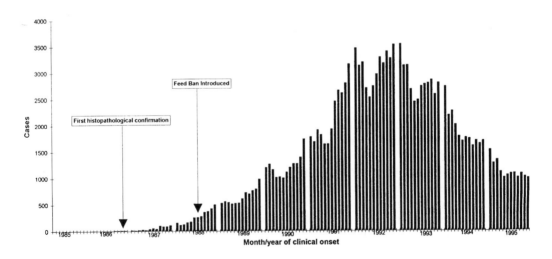

Fig 1. Epidemic curve of confirmed cases of BSE plotted by month and year of clinical onset. Data valid to end of December 1995; produced 4 June 1996.

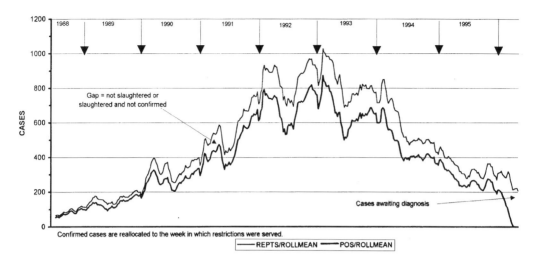

Fig 2. Rolling mean of suspected (REPTS) and confirmed (POS) cases. Data valid to week ending 31 May 1996.

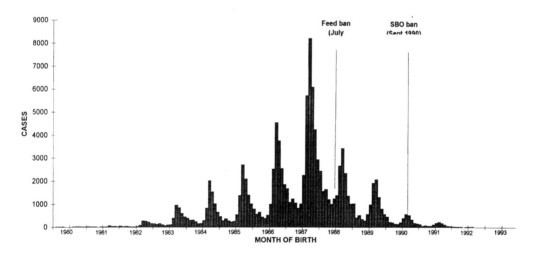

Fig 3. Confirmed cases of BSE plotted by month of birth. Data valid to week ending 10 May 1996.

continued to be the only important route of transmission. The ruminant feed ban had significantly reduced the risk, but had not entirely eliminated it: it has been calculated, for example, that the risk of infection for cattle born in December 1990 was 10% of that for cattle born in December 1987, when the amount of infectivity in feed was approaching its peak.

The ban on feeding SBO, or protein derived from SBO, to animals and birds was intended to prevent exposure of non ruminant species to the BSE agent. If properly observed, it should have incidentally reinforced the effectiveness of the ruminant feed ban by requiring the removal and destruction of all cattle tissues which might contain significant infectivity: if ruminant protein was accidentally fed to ruminants, thereafter the risk of delivering an infectious dose should have been much reduced. Attention was therefore focused on the effective enforcement of the SBO controls. Comparison in July 1994 of the amount of SBO being rendered with the number of cattle being slaughtered suggested that a significant quantity was not being properly separated, but was being rendered with non-SBO waste and so could be present in meat and bone meal which could be fed to poultry and pigs. Research also confirmed that as little as 1 g of unprocessed BSE-infected brain could infect cattle by the oral route, and observation showed that such amounts could easily remain in the cranial cavity when the skull was split and the brain removed for disposal as SBO, as was common practice. Finally, it was clear that the epidemic was declining more slowly in those parts of Britain where the proportion of pigs and poultry to cattle was greatest, suggesting that cross-contamination might be occurring in mills used to produce feed for ruminant and non-ruminant species. Further measures have therefore been introduced to strengthen those already in place requiring the separation and destruction of SBO, and so to tighten the measures which protect animal health.

Conclusion

The measures taken to control the BSE epidemic in the United Kingdom, and to prevent transmission to other animal species, have been effective but – particularly in respect to cattle, where transmission is not across a species barrier and effective exposure is consequently most difficult to prevent – it has been necessary to take further measures to ensure that SBO cannot leak back into animal feed. Even without these changes the epidemic was declining, and this decline is expected to continue. The effect of the changes introduced in 1995 will not be apparent for another 5 years, but on the basis of current knowledge should ensure that BSE is eradicated from British cattle, probably during the first decade of the 21st century.

Note

A number of further control measures have been implemented in the United Kingdom since this paper was presented. Mammalian protein can no longer be used in any feed for farm animals, or as fertilizer on agricultural land, and the whole head (with the exception of the tongue) is now defined as specified bovine material, in addition to the former SBO tissues. Beef from cattle over 30 months old is currently being destroyed. These measures, which further strengthen the measures in place to protect animal and human health, have been i ntroduced following recognition of a variant form of Creutzfeldt-Jakob disease in ten individuals aged 42 years or younger.

Transmissible Subacute Spongiform Encephalopathies:
Prion Diseases
L Court, B Dodet, eds
© 1996, Elsevier, Paris

Surveillance for bovine spongiform encephalopathy in the USA may be complicated by the existence of strains

Doris Olander, Richard F Marsh

University of Wisconsin, Departrnent of Animal Health and Biomedical Sciences, 1655 Linden Drive, Madison, Wl 53706, USA

Summary – The recognition of strains of transmissible spongiform encephalopathies (TSE), as distinguished by different physicochemical and clinical properties, complicates the surveillance of the US cattle herd for the presence of these agents. Potential sources of TSE include bovine spongiform encephalopathy, sheep scrapie, chronic wasting disease of deer and elk, and a native cattle TSE as manifested by transmissible mink encephalopathy. The host species for all of these agents are included in meat and bone meal and cattle rations. Therefore, the route of exposure is in place. These factors combined with rising concern about the zoonotic potential of these agents make surveillance a necessity. Despite the many remaining mysteries about TSE, there is sufficient scientific information available to design an effective surveillance program. The two most important points to integrate into such a program are a broad clinical case definition that can accommodate the demonstrated experimental variability of bovine TSE and the use of diagnostic methods based on the detection of the protease-resistant prion protein (PrP-res), for diagnosis of preclinical or those lacking the typical spongiform lesions.

Introduction

The bovine spongiform encephalopathy (BSE) epizootic in Great Britain and the possibility of zoonotic transmission has increased awareness of transmissible spongiform encephalopathies (TSE) [1]. The TSE encompass a group of progressive neurodegenerative diseases that are characterized by long incubation periods and the presence of a unique etiological agent. Animal TSE include scrapie in sheep and goats, TME and chronic wasting disease (CWD) of deer and elk as well as BSE [2]. In humans Creutzfeldt-Jakob disease is the most commonly recognized TSE.

Major advances as to the nature of the TSE agent have been made in the last 15 years since the prion ('protein only') hypothesis was proposed. This hypothesis states that the modification of the host-encoded, protease-sensitive prion protein (PrP-sen) to its disease-associated protease-resistant isoform (PrP-res) is an important, and perhaps the only, event required for TSE transmission [3, 4]. Multiple TSE 'strains' have been isolated from infected sheep and mink upon passage in mice and hamsters [5, 6]. These strains are distinguished by the host range, the incubation period, the nature and distribution of pathological lesions, and, in some cases, the clinical expression of the disease [6–8]. It is reasonable to assume that other TSE-

---

Key words: transmissible spongiform encephalopathy / bovine spongiform encephalopathy cattle strains surveillance

infected animals harbor one or more TSE strains. Although the molecular basis of strain behavior is not fully understood within the context of the prion hypothesis, recent work suggests that strains reflect different conformations of the PrP-res polypeptide [9].

Although the original source of BSE (sheep scrapie or a bovine TSE) remains unknown, it is generally accepted that the large scale of the epizootic in Great Britain can be attributed to feeding contaminated meat and bone meal (MBM) [10]. If BSE originated from sheep scrapie, then a species barrier was crossed. Experimentally, initial passage of a TSE into a new hos species produces a dramatically lengthened incubation period called the 'species barrier effect' [11, 12]. Therefore, in the case of cross-species transmission of TSE, infected animals can remain clinically normal for extremely long periods of time. Once established in the new host species, subsequent passage of the TSE results in a decreased, but stabilized, incubation period [7].

For economic and public health reasons, it is important to prevent the emergence of TSE infection in American cattle. The British epidemic serves both as a warning and a model for BSE biology and epidemiology. Although animal agriculture practices in the United States (US) are not identical to those present in Great Britain prior to the recognition of BSE, there are similarities. In particular, American cattle are currently fed MBM derived from cattle, and to a lesser extent, sheep and other mammalian species. This provides a mechasnism for the transmission of the agent from one animal to another. In addition, the majority of cattle are slaughtered before the end of their natural lifespan, allowing accumulation of infectious agent prior to the expression of the clinical disease. Repeated feeding of by-products that include tissues derived from the central nervous system (CNS) of clinically normal, but infectious, animals would result in the amplification of the TSE agent. These are crucial factors affecting the emergence and transmission of BSE or a BSE-like disease.

The United States Department of Agriculture (USDA) began active and passive surveillance for BSE in 1990 [13]. Under this program, rabies-negative cattle presenting with BSE-like clinical symptoms or other CNS disorders are examined histopathologically for spongiform lesions in the brain tissue. This strategy is ideally designed to detect BSE as it is manifested in Great Britain. It does not address the possible emergence of one or more non-BSE strains that produce markedly different clinical and histopathological phenotypes of bovine TSE.

The objective of this paper is to describe TSE to which American cattle may be exposed, by dietary exposure, and to discuss how current knowledge of strain behavior can be integrated into surveillance programs. The following four TSE represent potential sources of infection to American cattle.

Bovine spongiform encephalopathy

Between 1981 and 1989, 499 heads of cattle were imported to the US from the United Kingdom. As of May 1996, the status of the imports was 113 alive (location known), 8 exported, 35 location unknown, and 343 dead [14]. It is the latter two groups that are of greatest concern with respect to the introduction of BSE into the US. By the slaughter, rendering, and inclusion in MBM of apparently healthy, but preclinically infected, animals from these groups, exposure of American cattle may have already occurred.

Experimental work in Great Britain has provided strong evidence that BSE represents a single major strain [15]. Therefore, surveillance for BSE in American cattle will be less difficult than for other, unrecognized, strains of bovine TSE for three reasons. First, the clinical presentation of BSE is well described (the USDA has circulated educational materials describing the hallmarks of BSE including the temperament changes, postural abnormalities, incoordination, and terminal recumbency) [16, 17]. Second, the histopathological lesions are readily recognizable by pathologists using conventional tissue preparation and staining methods [18]. Third, as stated earlier, the USDA has an on-going surveillance program examining rabies-negative brains from domestic cattle exhibiting signs of CNS diseases.

Scrapie

In Great Britain, typing of sheep scrapie in panels of inbred mice has identified over 20 distinct strains, none of which resembles the BSE strain [8, 19]. In the US, similar typing experiments of five separate scrapie sources have identified differences indicating the presence of multiple strains [20]. In addition, the transmission of a number of US scrapie sources has been tested in cattle.

Two studies examined the experimental transmission of American scrapie from sheep and goats to cattle. The first study, in Mission, Texas, [21] involved two groups of five calves: one group was inoculated intracerebrally with goat scrapie and the other with sheep scrapie. Two animals in the goat group and one in the sheep group became clinically affected, with neurologic disorders after 27, 36 and 48 months, respectively. In the second study, using a different source of American sheep scrapie than in the Mission trial, 18 calves were inoculated intracerebrally with pooled inocula from the brains of nine scrapie-affected sheep [22]. At one year post-inoculation, 9 calves (none exhibiting clinical signs of disease) were sacrificed. The remainder were retained and all exhibited advanced clinical disease at 14 to 18 months post-inoculation. The clinical manifestations of disease common to these two experiments included difficulty in rising, stiffness, abnormal pelvic gait and recumbency. No hyperaesthesia or changes in temperament were noted. In both experiments, conventional histopathological examination of brain revealed few lesions. Most notably absent were the prominent spongiform changes in the grey matter nuclei that characterize the histopathological changes of BSE. In the first study, Clark et al [21] described the microscopic changes as 'slight and subtle' and in the second, Cutlip et al [22] reported that: 'brain lesions were equivocal and inconsistent'. In both studies, however, PrP-res was detected in the brains of all affected animals and in all asymptomatic animals sacrificed at 1 year post-inoculation in the second experiment [22, 23].

An additional study investigated the second passage of American scrapie in cattle [24]. The inoculum consisted of brain homogenates from the three clinically affected cattle in the Mission study. Four calves were inoculated intracerebrally and all exhibited clinical disease between 18 and 23 months post-inoculation. The clinical description of these animals was similar to that of the affected animals in the previous experiments, except that the authors describe variable excitability and auditory hyperaesthesia. The histological changes described in this study were also characterized as subtle, but with 'slightly greater spongiform changes' when compared to the lesions observed in the brains of the cattle from which the inoculum was derived [24].

In summary, cattle are susceptible to American scrapie infection by intracerebral inoculation, and neither the clinical signs nor the histopathological changes in the CNS correspond to those observed in BSE-affected cattle. Oral transmission of American scrapie to cattle has not been described.

Chronic wasting disease

CWD has been recognized in captive mule deer *(Odocoileus hemionus hemionus)* and elk *(Cervus elaphus nelsoni)* since 1967 and 1979 respectively [25, 26]. More recently, it has also been documented in free-living white-tailed deer *(Odocoileus virginianus)* within a limited geographic area of the western United States (ES Williams, personal communication). In some regions, free-living deer killed by motor vehicles are included in MBM. For example, in Wisconsin 46,443 white-tailed deer were killed on the road in 1995 with 26,488 picked up by contractors for disposal at rendering facilities [27]. Future TSE research should include characterizion of the epidemiology of CWD and its pathogenicity for cattle.

Transmissible mink encephalopathy

TME is a rare disease of ranch-raised mink. In the US there have been five episodes of TME since it was first identified in 1947 [28]. Four of the five were in Wisconsin, reflecting the geographic concentration of the mink-ranching industry. Unlike sheep scrapie which is typically present at variable rates over many years and geographically widespread in the US, TME occurs explosively in animals linked by a common food source or farm of origin, suggesting that mink function as sentinels of TSE infection.

Mink are fed a variety of protein-based diets and, until the emergence of BSE, the predominant hypothesis was that TME outbreaks were the result of feeding scrapie-infected sheep to mink. This hypothesis has not been confirmed epidemiologically or experimentally. Epidemiological investigations revealed that, in half of the North American TME outbreaks, dead and non ambulatory cattle were reported as a major feed constituent, and for two of these, the mink ranchers involved were confident that sheep tissues had not been fed [29]. Experimental transmission studies also suggest that scrapie is not the source of TME in the US. Two different sources of American sheep scrapie, one from American Suffolk sheep and the other from American Cheviot sheep, were inoculated orally and intracerebrally into mink [30, 31]. Only the intracerebral American Suffolk inoculum resulted in neurologic disease. In contrast, mink are readily infected by both BSE and cattle-passaged TME [24, 32]. Most importantly, transmission was achieved following oral inoculation in both studies.

TME has been successfully transmitted to cattle via the intracerebral route in three separate studies using three different sources of TME [24, 29, 33]. The TME-affected cattle did not display several of the clinical signs associated with BSE, particularly the characteristic behavioral changes such as aggression. Instead, many of the clinical signs were non specific: sudden collapse, apprehension, inappetence, varying degrees of hyperexcitability. Later hypermetria and hind limb ataxia were observed in those that remained ambulatory. Following transmission from mink to cattle, one isolate was orally transmitted back into mink and resulted in clinical and neurohistopathological signs indistinguishable from TME [28].

Inoculation of TME into Syrian hamsters resulted in the identification of two strains of the agent [6]. The first strain, termed 'hyper', has a short incubation period (65 days) and is characterized by hyperaesthesia and ataxia. The second, 'drowsy', has a much longer incubation period (165 days) and a clinical picture dominated by lethargy.

Implications for surveillance

Surveillance for a bovine TSE in the US can no longer be focused on the detection of a BSE-like disease. Experimental transmission of TSE from sheep, goats and mink to cattle has demonstrated that a bovine-adapted TSE strain can display a clinical and histopathological disease that is distinct from BSE. Outside the laboratory, a multitude of variables positively and negatively impact on the potential for an epizootic of a bovine TSE in the US. However, the emergence of BSE in Great Britain has provided a stark warning that strain selection and amplification by animal passage is not a theoretical possibility limited to the confines of the laboratory.

Active surveillance is most effective when a disease is well-characterized and a case definition is available. In the US, three criteria are used to designate cattle as BSE suspects: age ≥ 2 years, documented signs of neurological disease and ingestion of protein supplements as a substantial portion of the feed ration [34]. The detection of a new strain of bovine TSE presents greater problems due to the unpredictability of both the clinical and neurohistopathological manifestations.

Experimental observations suggest three strategies to enhance surveillance of American cattle that are not BSE suspects, as defined above. First, because of the long incubation periods, animals older than 4 years of age should be the primary target of surveillance. Second, experimental transmission studies of American scrapie and TME have demonstrated that cattle infected with strains of these agents do not manifest the striking behavioral changes that characterize BSE, but present with clinical signs that are more non specific. Of particular concern is the onset of non ambulatory status with few or no other specific signs of CNS disease [22, 29]. These non ambulatory animals would be handled as one of the thousands of animals resulting from other conditions unrelated to TSE. Therefore, in addition to older animals, non ambulatory cattle both on the farm and at the abattoir should be actively included in testing programs. Finally, the neurohistopathological abnormalities observed in natural and experimental bovine TSE vary widely from detectable spongiform lesions to subtle changes that are barely discernible, even in the late stages of clinical disease [21, 22, 29]. Thus, the detection of PrP-res, but not the presence of spongiform changes in brain tissue, is a more reliable diagnostic criteria that should be adopted in addition to conventional histopathology for the detection of a novel bovine TSE. Currently, based on immunoanalysis, there are several laboratory methods of detecting PrP-res, that have been developed and successfully tested in field cases of scrapie and BSE [35–37].

In conclusion, American cattle are fed animal tissues from a variety of species and sources some of which may harbor unique strains of TSE. Should transmission to cattle occur, neither the clinical nor histopathological expression of the disease can be predicted. Therefore, surveillance must rely on a broad case definition and specific tests for the disease specific protein PrP-res.

Acknowledgements

The authors would like to thank Peter Evans and Debbie McKenzie for their help in preparing this manuscript.

References

1 Will RG, Ironside JW, Zeidler M. A new variant of Creutzfeldt-Jakob disease in the UK. *Lancet* 1996;347: 921-5
2 Bradley R, Mathews D. Sub-acute, transmissible spongiform encephalopathies: current concepts and future needs. In: Bradley R, Mathews D, eds. *Transmissible Spongiform Encephalopathies of Animals.* Paris: Office International des Epizooties, 1992;69-17
3 Prusiner SB. Novel proteinaceous infectious particles cause scrapie. *Science* 1982;216:136-44
4 Prusiner SB. Prions. In: Fields BN, Howley PM, Knipe DM, eds. *Fields Virology.* Philadelphia: Lippincott-Raven, 1996;2901-50
5 Bruce ME, Fraser H, McBride PA, Scott JR, Dickinson AG. The basis of strain variation in scrapie. In: Prusiner SB, Collinge J, Powell J, Anderton B, eds. *Prion Diseases of Humans and Animals.* New York: Ellis Horwood, 1992;497-508

6   Bessen RA, Marsh RF. Identification of two biologically distinct strains of transmissible mink encephalopathy in hamsters. *J Gen Virol* 1992;73:329-34

7   Kimbedin RH, Walker CA. Evidence that the transmission of one source of scrapie agent to hamsters involves separation of agent strains from a mixture. *J Gen Virol* 1978;39:487-96

8   Bruce ME, McConnell I, Fraser H, Dickinson AG. The disease characteristics of different strains of scrapie in *Sinc* congenic mouse lines: implications for the nature of the agent and host control of pathogenesis. *J Gen Virol* 1991;72:595-603

9   Bessen RA, Marsh RF. Biochemical and physical properties of the prion protein from two strains of the transmissible mink encephalopathy agent. *J Virol* 1992;66:2096-101

10 Wilesmith JW, Ryan JB, Atkinson MJ. Bovine spongiform encephalopathy: epidemiological studies on the origin. *Vet Rec* 1991;128:199-203

11 Pattison IH. Experiments with scrapie with special reference to the nature of the agent and the pathology of the disease. In: Gajdusek DC, Gibbs CJ, Alpers MP, eds. *Slow, Latent and Temperate Virus Infections,* NINDB Monograph 2. Washington, DC: US Government Printing Office, 1965;249-57

12 Bartz JC, McKenzie Dl, Bessen RA, Marsh RF, Aiken JM. Transmissible mink encephalopathy species barrier effect between ferret and mink: PrP gene and protein analysis. *J Gen Virol* 1994;75:2947-53

13 USDA:APHIS:VS. Bovine spongiform encephalopathy update. *Foreign Animal Disease Report* 1995;3: 22-4

14 USDA:APHIS Web. *Charts and graphs on BSE Surveillance.* http://.aphis.usda.gov/oa/bsesurv.html, May 1996

15 Bruce ME. Strain typing studies of scrapie and BSE. In: Baker HF, Ridley RM, eds. *Prion Diseases.* Totowa: Humana Press, 1996;223-36

16 Kimberlin RH. Bovine spongiform encephalopathy. In: Bradley R, Mathews D, eds. *Transmissible Spongiform Encephalopathies of Animals.* Paris: Office International des Epizooties, 1992;347-90

17 USDA:APHIS:VS. *Bovine Spongiform Encephalopathy Fact Sheet.* 1995

18 Simmons MM, Harris P, Jeffrey M, Meek SC, Blamire IW, Wells GA. BSE in Great Britain: consistency of the neurohistopathological findings in two random annual samples of clinically suspect cases. *Vet Rec* 1996;138:

19 Bruce M, Chree A, McConnell I, Foster J, Pearson G, Fraser H. Transmission of bovine spongiform encephalopathy and scrapie to mice: strain variation and the species barrier. *Philos Trans R Soc Lond* 1994;343:405-11

20 Carp RI, Callahan SI. Variation in the characteristics of 10 mouse-passaged scrapie lines derived from five scrapie-positive sheep. *J Gen Virol* 1991;72:293-8

21 Clark WW, Hourrigan JL, Hadlow WJ. Encephalopathy in cattle experimentally infected with the scrapie agent. *Am J Vet Res* 1995;56:606-12

22 Cudip RC, Miller JM, Race RE et al. Intracerebral transmission of scrapie to cattle. *J Infect Dis* 1994;169: 814-20

23 Gibbs CJ, Safar J, Ceroni M, Di Martino A, Clark WW, Hourrigan JL. Experimental transmission of scrapie to cattle. *Lancet* 1990;335:1275

24 Robinson MM, Hadlow WJ, Knowles DP et al. Experimental infection of cattle with the agents of transmissible mink encephalopathy and scrapie. *J Comp Pathol* 1995-113:241-51

25 Williams ES, Young S. Chronic wasting disease of captive mule deer: a spongiform encephalopathy. *J Wildl Dis* 1980;16:89-98

26 Williams ES, Young S. Spongiform encephalopathy of Rocky Mountain elk. *J Wildl Dis* 1982;18:465-71

27 Wisconsin Department of Natural Resources Bureau of Law Enforcement. Report of car killed deer, seized deer, and deer picked up on contract: 07-01-1994 through 06-30-1995. 1996

28 Marsh RF, Bessen RA. Epidemiologic and experimental studies on transmissible mink encephalopathy. In: Brown F, ed. *Transmissible Spongiform lncephalopathies – Impact on Animal and Human Health.* Basel: Dev Biol Stand 1993;111-8

29 Marsh RF, Bessen RA, Lehman S, Hartsough GR. Epidemiological and experimental studies on a new incident of transmissible mink encephalopathy. *J Gen Virol* 1991;72:589-94

30 Hanson RP, Eckroade RJ, Marsh RF, ZuRhein GM, Kanitz CL, Gustafson DP. Susceptibility of mink to sheep scrapie. *Science* 1971;172:859-61

31 Marsh RF, Hanson RP. On the origin of transmissible mink encephalopathy. In: Prusiner SB, Hadlow WJ, eds. *Slow Transmissible Diseases of the Nervous System.* New York: Academic Press, 1979;451-60

32 Robinson MM, Hadlow WJ, Huff TP et al. Experimental infection of mink with bovine spongiform encephalopathy. *J Gen Virol* 1994;75:2151-5

33 Marsh RF, Burger D, Eclcroade R, ZuRhein GM, Hanson RP. A preliminary report on the experimental host range of the transmissible mink encephalopathy agent. *J Infect Dis* 1969;120 713-9

34 Miller LD, Davis AJ, Jenny AL. Surveillance for lesions of bovine spongiform encephalopathy in US cattle. *J Vet Diagn Invest* 1992;4:338-9

35 Mohri S, Farquar CF, Somerville RA, Jeffrey M, Foster J, Hope J. Immunodetection of a disease specific fraction in scrapie-affected sheep and BSE-affected cattle. *Vet Rec* 1992;131:537-9
36 Race R, Emst D, Jenny A, Taylor W, Sutton D, Caughey B. Diagnostic implications of detection of proteinase K-resistant protein in spleen, lymph nodes, and brain of sheep. *Am J Vet Res* 1992;53:883-9
37 Miller JM, Jenny AL, Taylor WD, Marsh RF, Rubenstein R, Race RR. Immunohistochemical detection of prion protein in sheep with scrapie. *J Vet Diagn Invest* 1993;5:309-16

NEUROPATHOLOGY OF
CREUTZFELDT-JAKOB DISEASE

Transmissible Subacute Spongiform Encephalopathies:
Prion Diseases
L Court, B Dodet, eds
© 1996, Elsevier, Paris

Immunomorphology of human prion diseases

Johannes A Hainfellner, Herbert Budka*

Institute of Neurology, University of Vienna, 1097 Vienna, Austria

Summary – We systematically describe the immunomorphological spectrum and distribution of cerebral deposition of pathological prion protein (PrP) as a baseline in a series of 78 consecutive Creutzfeldt-Jakob disease (CJD) cases and five Gerstmann-Sträussler-Scheinker (GSS) disease patients without exposure to bovine spongiform encephalopathy. Characteristic patterns of PrP deposition are synaptic, patchy/perivacuolar and plaque types, which may overlap in the individual case. Most frequently, PrP accumulates in a synaptic pattern in both the cerebral and cerebellar cortices. Plaques are more frequent in the cerebellum and patchy/perivacuolar deposits are more frequent and prominent in the cerebral cortex. Plaques are the only type of PrP deposits extending to the subcortical white matter. Synaptic-type deposits and unicentric PrP plaques occur in both CJD and GSS disease, while abundant multicentric plaques are particular to GSS disease. All neuropathological features of the new variant of CJD in the United Kingdom and France (v-CJD) occur infrequently in our CJD series. The combination and prominence of rare features of PrP deposition appear to be unique to v-CJD.

Introduction

The hallmark of transmissible spongiform encephalopathies or prion diseases is accumulation of pathological prion protein (PrP) in the central nervous system which is immunocytochemically detectable [1]. The morphology, amount and distribution of PrP are variable and have been studied extensively in selected and/or small groups of patients for genotype-phenotype correlations [1–4]. Few reports systematically describe the immunomorphology of PrP deposition in small groups of sporadic Creutzfeldt-Jakob disease (CJD) patients [4–6] and in a few cases of Gerstmann-Sträussler-Scheinker (GSS) disease [7].

Recently, a new variant of CJD (v-CJD) with a distinct neuropathological profile has been reported in the United Kingdom [8] and France [9], and a possible causal link to bovine spongiform encephalopathy (BSE) has been suggested [8].

We report the study of a large consecutive CJD series from Austria and some cases of the original Austrian GSS disease family to systematically describe the immunomorphological spectrum and distribution of cerebral PrP deposition as a baseline in a series of human prion diseases without exposure to BSE.

Correspondence and reprints.

Key words: Creutzfeldt-Jakob disease / Gerstmann-Sträussler-Scheinker disease / prion protein / immunocytochemistry

Materials and methods

A total of 78 CJD cases was investigated. This group includes all autopsied Austrian CJD cases between 1969 and 30 September 1995 with definite neuropathological diagnosis. This series was collected by kind cooperation with K Jellinger, Vienna; P Pilz, Salzburg; R Kleinert, Graz; and H Maier, Innsbruck. Clinical features were obtained retrospectively from medical case records and have been recently reported in detail [10]. One recently deceased GSS disease patient was studied in comparison with four cases in which archival brain tissue was retrieved and newly analysed [11]. In each case, one tissue block of the cerebral and cerebellar cortices (in four CJD cases, only the cerebral cortex was available), respectively, was immunostained with a monoclonal antibody against PrP (RJ Kascsak, Staten Island, NY, USA) [12] and a polyclonal anti-PrP rabbit serum against a synthetic peptide of human PrP (H Diringer, Berlin, Germany) [13].

Results

Creutzfeldt-Jakob disease (CJD)

Clinical features

The patient population consisted of 42 female and 36 male CJD patients. Age at death ranged from 27 to 82 years with a mean of 64 years. Two female patients died at the unusually young ages of 27 and 30 years in 1969 and 1982, respectively. The duration of the disease ranged from 1 to 30 months; in more than half of the patients, the duration was 1 to 4 months. No case had an unequivocal family history nor known risk for accidental (iatrogenic) transmission. *PRNP* genotyping, however, has not yet been performed in this series [10].

Immunomorphology of PrP deposition

PrP accumulates both in the cerebral and cerebellar cortices in three major patterns which overlap in a portion of cases: synaptic, patchy/perivacuolar and plaque types (figs 1–3) according to previous descriptions [14]. Synaptic-type deposits are the most frequent pattern in both cortices (table I), patchy/perivacuolar-type deposits are more prominent and frequent in the cerebral cortex, while plaques are more frequent in the cerebellar cortex (table I). The PrP deposition pattern is the same in both cortices in 53% of the cases.

Distribution of PrP accumulation in the cerebral cortex

Synaptic-type PrP is distributed diffusely and accentuated in some cases in the deep and/or middle cortical layers, and perineuronally and/or periaxonally. The extent of deposition ranges from a few dots to dense accumulations.

Patchy/perivacuolar PrP deposits (fig 1) accumulate focally or diffusely and are confined to the cortex.

Plaques are unicentric and distinct (kuru type) or indistinct. Indistinct plaques are also irregularly shaped. Plaques are distributed diffusely and extend as the only deposition type to the subcortical white matter in half of the cases.

In four cases the plaques are frequent and widely distributed. Plaques surrounded by a zone of spongiform change ('florid' plaques of v-CJD) are rare and are not detected in hematoxylin–eosin stained sections, but only with anti-PrP immunocytochemistry.

Distribution of PrP accumulation in the cerebellar cortex

Synaptic-type PrP accumulates focally or diffusely in the molecular and/or granular layers, and is confined to the cortex (fig 2). Most frequently, the extent of deposition is similar in the molecular and granular layers (table II), while deposition is usually coarser in the granular layer.

Table I. Frequency (%) of synaptic, patchy/perivacuolar- and plaque-type PrP deposition patterns in cerebral and cerebellar cortex of the Austrian series of Creutzfeldt-Jakob disease cases.

Type of deposits	Cerebral cortex	Cerebellar cortex
Synaptic	91	94
Patchy/perivacuolar	38	6
Plaque	10	19

Patchy/perivacuolar PrP deposits are a minor feature and occur primarily in the molecular layer in 6% of the cases.

Plaques are unicentric and distinct (kuru-type) (fig 3) or indistinct. Indistinct plaques are also irregularly shaped. Plaques occur in the molecular and/or granular layers. Most commonly, the plaque density is higher in the granular layer (table II). In only two cases, the prominence of deposits in the molecular layer is comparable to that of v-CJD [8].

PrP deposition in young CJD patients
PrP deposition in two young Austrian CJD patients (27 and 30 years of age) does not show any peculiarities. Both cases have synaptic-type PrP deposits with additional infrequent and indistinct plaques in the cerebellum of one patient [10].

Gerstmann-Sträussler-Scheinker disease

Immunomorphology and distribution of PrP deposition
The brains of the recently deceased patient and of four formerly deceased patients of the Austrian GSS disease family show a similar pattern of prominent and widespread deposition of PrP plaques (fig 4). The plaques are variable in size, ranging between small, coarse deposits ('miniplaques') and a diameter up to 150 μm, and variable in morphology, including multicentric (clusters of several to many amyloid cores) and unicentric deposits with and

Table II. Distribution of synaptic (syn) and plaque (pl)-type prion protein (PrP) deposits in the cerebellar cortex of the Austrian series of Creutzfeldt-Jakob disease cases.

PrP deposits	Percentage
syn(m)	6
syn(g)	0
syn(m) = syn(g)	77
syn(m) > syn(g)	4
syn(m) < syn(g)	13
pl(m)	29
pl(g)	14
pl(m) = pl(g)	7
pl(m) > pl(g)	7
pl(m) < pl(g)	43

syn(m): synaptic-type PrP deposits in the molecular layer of the cerebellar cortex; syn(g): synaptic-type PrP deposits in the granular layer of the cerebellar cortex; pl(m): plaques in the molecular layer; pl(g): plaques in the granular layer.

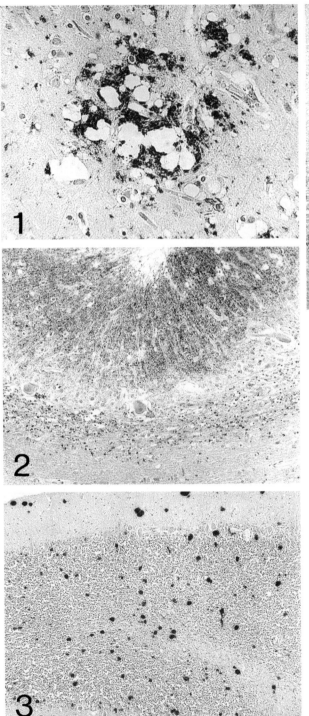

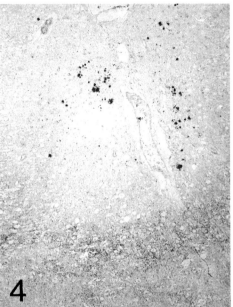

Fig 1. Patchy/perivacuolar-type PrP deposition in the cerebral cortex in Creutzfeldt-Jakob disease (× 300).

Fig 2. Diffuse accumulation of fine granular synaptic-type PrP in molecular and granular layers of the cerebellar cortex in Creutzfeldt-Jakob disease (× 150).

Fig 3. Numerous unicentric PrP plaques distributed in granular and molecular layers of the cerebellar cortex in Creutzfeldt-Jakob disease (× 75).

Fig 4. PrP deposition in the cerebral cortex in Gerstmann-Sträussler-Scheinker disease. Diffuse synaptic-type PrP deposits show laminar accentuation in deep cortical layers. In addition, the cortex harbors plaque-type PrP deposits (× 100).

without (kuru type) small satellite cores; other deposits have a granulofibrillary appearance. Plaques occur in all gray matter structures, except for the substantia nigra, the medulla oblongata and the spinal cord. They are scattered in all layers of the cerebral and cerebellar cortices and are also numerous in the subcortical white matter; they are most prominent in the cerebellar cortex.

In addition to plaques, synaptic-type PrP deposits (fig 4) are detectable only in the recent GSS disease case. Synaptic-type deposits are diffusely distributed in the gray matter structures and are accentuated in a laminar fashion in the middle and deep layers of the cerebral cortex [11].

Summary of the immunomorphological findings and distribution of PrP deposition in the series of Austrian CJD and GSS disease patients

– PrP accumulates both in the cerebral and cerebellar cortices in three characteristic patterns which may overlap in the individual case: synaptic, patchy/perivacuolar and plaque types.

– The PrP deposition pattern is the same in both the cerebral and cerebellar cortices in 53% of CJD cases. The synaptic pattern is the most frequent type of PrP accumulation in both cortices.

– PrP accumulations differ between the two (cerebral and cerebellar) cortices: plaques are more frequent in the cerebellum and patchy/perivacuolar deposits are more frequent and prominent in the cerebral cortex.

– Plaques are the only type of PrP deposits extending to the subcortical white matter.

– Two young Austrian CJD cases do not show the pattern of PrP deposition of v-CJD.

– All neuropathological features of v-CJD occur infrequently in CJD cases without exposure to BSE. The combination and prominence of rare features of PrP deposition appear to be unique to v-CJD.

– Synaptic-type PrP deposits and unicentric PrP plaques are features common to CJD and GSS disease; the abundance of multicentric plaques is particular to GSS disease.

Acknowledgments

This study was carried out within the EU Concerted Action "The human prion diseases" (project leader: H Budka).

References

1 Tateishi J, Kitamoto T. Inherited prion diseases and transmission to rodents. *Brain Pathol* 1995;5:53-9
2 Kitamoto T, Doh-ura K, Muramoto T, Miyazono M, Tateishi J. The primary structure of the prion protein influences the distribution of abnormal prion protein in the central nervous system. *Am J Pathol* 1992;141: 271-7
3 de-Silva R, Ironside JW, McCardle L et al. Neuropathological phenotype and "prion protein" genotype correlation in sporadic Creutzfeldt-Jakob disease. *Neurosci Lett* 1994;179:50-2
4 Parchi P, Castellani R, Capellari S et al. Molecular basis of phenotypic variability in sporadic Creutzfeldt-Jakob disease. *Ann Neurol* 1996;39:767-78
5 Hayward PAR, Bell JE, Ironside JW. Prion protein immunocytochemistry: reliable protocols for the investigation of Creutzfeldt-Jakob disease. *Neuropathol Appl Neurobiol* 1994;20:375-83
6 Doi-Yi R, Kitamoto T, Ogomori K, Mehraein P, Tateishi J. Distribution of prion protein in German patients with Creutzfeldt-Jakob disease is different from that in Japanese patients. *Acta Neuropathol (Berl)* 1994;87:481-3
7 Ghetti B, Piccardo P, Frangione B et al. Prion protein amyloidosis. *Brain Pathol* 1996;6:127-45
8 Will RG, Ironside JW, Zeidler M et al. A new variant of Creutzfeldt-Jakob disease in the UK. *Lancet* 1996;347:921-5

9 Chazot G, Brousolle E, Lapras CI, Blättler T, Aguzzi A, Kopp N. New variant of Creutzfeldt-Jakob disease in a 26-year-old French man. *Lancet* 1996;347:1181

10 Hainfellner JA, Jellinger K, Diringer H et al. Creutzfeldt-Jakob disease in Austria. *J Neurol Neurosurg Psychiatry* 1996;61:139-42

11 Hainfellner JA, Brantner-Inthaler S, Cervenáková L et al. The original Gerstmann-Straüssler-Scheinker family of Austria: divergent clinicopathological phenotypes but constant *PrP* genotype. *Brain Pathol* 1995;5:201-11

12 Goodbrand IA, Ironside JW, Nicolson D, Bell JE. Prion protein accumulation in the spinal cords of patients with sporadic and growth hormone associated Creutzfeldt-Jakob disease. *Neurosci Lett* 1995;183:127-30

13 Oberdieck U, Xi YG, Pocchiari M, Diringer H. Characterisation of antisera raised against species-specific peptide sequences from scrapie-associated fibril protein and their application for post-mortem immunodiagnosis of spongiform encephalopathies. *Arch Virol* 1994;136:99-110

14 Budka H, Aguzzi A, Brown P et al. Neuropathological diagnostic criteria of Creutzfeldt-Jakob disease (CJD) and other human spongiform encephalopathies (prion diseases). *Brain Pathol* 1995;5:459-66

Transmissible Subacute Spongiform Encephalopathies:
Prion Diseases
L Court, B Dodet, eds
© 1996, Elsevier, Paris

Neuropathology of sporadic Creutzfeldt-Jakob disease

Jacqueline Mikol

Department of Pathology, Hôpital Lariboisière, 2, rue Ambroise Paré, 75475 Paris Cedex 10, France

Summary – Neuronal loss, spongiform changes and astrocytosis are the main features of Creutzfeldt-Jakob disease (CJD), often designated as the characteristic triad. The topography and degree of changes vary from case to case. In 10% of the cases, amyloid plaques are predominantly found in the cerebellum. Different variants have been described: the Heidenhain's, striatonigral, thalamic, cerebellar, pancephalopathic variants and the prion disease with normal neuropathology. Recently, a new UK variant characterized by large plaques surrounded by many vacuoles has been reported. Relationships with Alzheimer's disease and progressive subcortical gliosis are still a matter of debate. The nosology of the different forms of CJD might be resolved when precise correlations between the genotype and phenotype are established, unless other relevant factors are found.

Introduction

Since 1922 when W Spielmeyer [1] associated the names of Creutzfeldt and Jakob to refer to Creutzfelt-Jakob disease (CJD) [2, 3], many terms have been used and many neuropathological reviews have been published [4–9]. CJD, a sporadic, familial or iatrogenic disease, is still considered as the major human disorder among prion diseases, also called spongiform encephalopathies or transmissible amyloidoses.

Macroscopic studies

Sporadic CJD neuropathology may vary according to the duration of the disease and global or focal cortical atrophy. Enlarged ventricules are mainly observed in patients surviving several years. Cerebellar atrophy is also found in 10% of the ataxic forms [10].

Histological studies

The three main histological features or triad are neuronal loss, spongiform changes and gliosis.

'Neuronal loss' may be difficult to recognize, especially in biopsy. It is prominent in deep layers of the cortex, more frequent in the frontotemporal and insular cortices, but parietal and occipital cortices may also be involved: the degree of neuronal loss increases with survival duration (mean ratio, 54%) [4]. The hippocampus is often preserved.

Neuronal loss is also noticeable in the thalamus, especially in the dorsomedial nucleus (47% of cases, [4]), neostriatum (26%, [4]) and granular cells of the cerebellar cortex (71%, [11]; 82%, [10]; [12, 13]). Vacuolated [14] or swollen achromasic neurons [15, 16] may be

---

***Key words:** Creutzfeldt-Jakob disease / spongiosis / prion protein*

observed. Vacuolation and axonal swelling of trigeminal ganglia cells have also been recognized, indicating disturbances of neuroaxonal transport [17]. The retina may also be involved [18]. Synaptic disorganization and/or neuronal loss representing a more than 30% reduction in the synaptic index could be the primary event preceding the typical triad [19].

'Spongiosis' is frequent (49%, [4]) or nearly constant (99% of cases, [20]). It is the most useful sign to arrive at a diagnosis, but it may be absent and therefore is one of the potential pitfalls of stereotactic biopsy. Spongiosis consists of coalescent small vacuoles different from the non specific, coarse loosening of brain in status spongiosis; however, in severe forms it becomes massive.

Masters et al [21] have observed cases with prolonged survival in whom less spongiosis and more neuronal loss was found, with diffuse damage of the cerebral cortex. These lesions have to be differentiated from nonspecific spongiosis [22]. Careful examination is necessary to recognize the fine vacuolation surrounding the neurons. It is sparse in the neuropil [7], in the subcortical grey matter [6], or in the head of the caudate nucleus which appears to be a constant location [23]. Ultrastructural studies have shown that vacuoles arise from the astrocytic and presynaptic axodendritic and axosomatic processes [24–26], and from the cisternae of the neuronal smooth endoplasmic reticulum. Frequently, many membrane fragments can be seen within vacuoles. They may also be the result of intracellular digestion in the lysosomal compartment [26].

'Astrocytic proliferation' appears either as clusters of nuclei or enlarged proliferating cells as previously demonstrated by Holzer staining. It has been shown by immunolabeling with glial fibrillary acidic protein (GFAP) to be comparable to the increased glial fibrillary acidic mRNA. The increase in gliosis is concomitant to neuronal loss, with a score of 62% in the cerebral cortex, 50% in the dorsomedial nucleus of the thalamus, 42% in the neostriatum [4]. In CJD with ataxia, the loss of granule cells parallels the significant increase in astrocytes, with strict preservation of the spatial arrangement of all astroglial subtypes [13].

There is no inflammatory process and microglial proliferation is not described as prominent at any stage of the disease [6], but recent immunocytochemical studies have shown activation of the grey matter microglia at an early stage, decreasing as the disease progresses [28]. Microglial cells surround the vacuoles in the neuropil.

Recent data emphasized the presence of amyloid plaques in CJD shown by previous studies [29, 30]. Masters et al [21] pointed out that kuru plaques are seen more frequently in patients with long clinical course. These plaques are difficult to recognize with usual Congo red staining and polarization; the rate of histochemically demonstrable amyloid plaques is 5 to 12% [8, 31].

Using a new antibody directed against amyloid plaque cores from a patient with Gerstmann-Sträussler-Scheinker disease, Kitamoto and Tateishi obtained a percentage of 59% of immuno-histochemically demonstrable small amyloid plaques and confirmed that they are a hallmark of CJD with a long clinical course [31]. Some plaques are surrounded by activated microglial cells which may play the same role as they do in Alzheimer's disease. Kuru-like plaques or ill-defined, unicentric amyloid plaques are mostly found in the cerebellum, but are also distributed in the cerebral cortex, white matter, thalamus, basal ganglia and brain stem with wide variations from case to case [31].

Creutzfeldt-Jakob variants

According to the clinical data and the main topography of the lesions, different variants have been and are still described.

The 'occipital variant' in which visual symptoms are prominent was called the Heidenhain's syndrome [32] by Meyer, Leigh and Bagg [33]. Later on, the Nevin variant [34] in which vascular factors were suspected has been linked to the same group. The importance of spongiosis in these cases contributed to individualize 'subacute spongiform encephalopathies'.

The cortex is mainly atrophic, with severe neuronal loss ('burnt out' cortex); the parenchyma is destroyed by spongiosis, and astrocytosis is massive.

The 'thalamic form' is very rare, but is interesting in regard to fatal familial insomnia. As already described, thalamic lesions are frequently observed in most cases of CJD, but the three initial cases with a 'thalamic form' of the disease [35–37] almost exclusively involved the thalamus.

Clinically, the case reported by Garcin et al [37] (case IV) was a patient with dementia, severe memory disorders and abnormal movements. There was complete neuronal loss in the anterior nucleus, and the other nuclei were also affected.

The 'striatal and striatonigral variants' are frequent. They are more or less associated with cortical and cranial nerve lesions. In 48 cases [38], degeneration of the neostriatum (71% of cases), pallidum (69% of cases), subthalamic nucleus (43% of cases) and substantia nigra (34% of cases) was noted.

The 'cerebellar variant' occurs in varying combinations in all three cerebellar cortical layers and in the nucleus dentatus. The ataxic form of CJD comprises two different patterns: the first one is characterized by predominant granule cells loss and absence of kuru-like plaques [11, 12, 39], the other (mainly reported in Japan) by predominant Purkinje cell loss and widespread kuru-like plaques [40, 41]; both forms may include white matter degeneration. The limit beween these two cerebellar variants is not always clear. Brain stem lesions may be associated, involving the dentate nucleus, inferior olive or cranial nerve nuclei.

Whether an 'amyotrophic form' of the disease exists as an entity is still a matter of debate; in most cases, it seems related to motor neuron disease [42].

'Other rare variants' include forms with oculomotor disorders [43, 44], a unilateral form at the early stage of the disease [45] and the 'panencephalopathic type'. In this latter form, akinetic mutism and tetraplegia are frequent, and the average duration of the disease is 17 months. Atrophy is severe and cortical changes are extensive with ballooned cells; there are white matter changes consisting of diffuse pallor [40, 46, 47] and circumscribed destructive foci. Additional lesions are found in the thalamus, basal ganglia, cerebellum, and brain stem, disclosing combined degeneration of several systems. Non Japanese forms have also been reported [48–52].

Relationships with Alzheimer's disease

The association of CJD and Alzheimer's disease has been suspected in some cases with amyloid plaques and/or spongiosis [53], but only immunostaining can identify the presence of either prion or protein $\beta A4$ [54]. The coexistence of neuritic plaques and cerebral amyloid angiopathy in patients aged 63 or more is age-related [55], however further genetic studies must be carried out. In a sporadic case characterized by both the presence of small plaques in the cerebellum, basal ganglia and cortex, and a 216-base pair insertion in the *PRNP* gene, Duchen et al [56] demonstrated a transition between the pathology of prion disease and that of Alzheimer's disease. Furthermore, CJD may be associated with cerebral $\beta A4$ amyloid angiopathy without Alzheimer's disease [57–59].

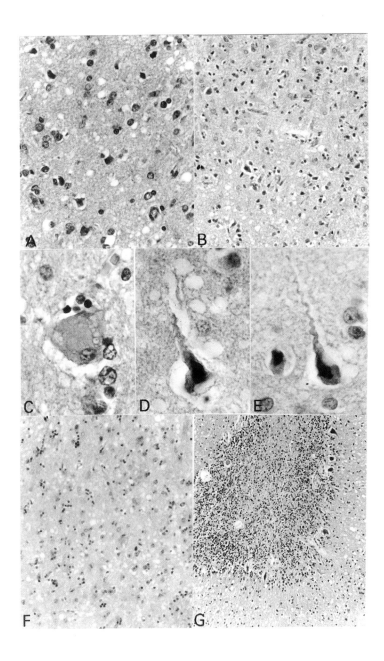

Fig 1. A. Diffuse spongiosis - moderate gliosis. Frontal biopsy, hematin eosin (He × 375). **B**. Massive neuronal loss with diffuse astrocytosis. Deep occipital cortex (Heindehain variant) (× 187.5). **C**. Vacuolated neuron. Insula (He × 875). **D, E**. Perineuronal spongiosis in frontal cortex (He × 875). **F**. Diffuse spongiosis in the caudate nucleus (He × 187.5). **G**. Severe loss of granule cells, rarefaction of Purkinje cells and spongiosis of the molecular layer of a cerebellar folium (cerebellar variant) (He × 94).

Recent problems

'Prion dementia' without characteristic pathology has been reported in patients issued from a single family [60]. The authors concluded "thus CJD cannot always be excluded on neuropathological grounds in an individual dying of a demented condition". This observation was not confirmed in a subsequent study [61].

Prion protein has been demonstrated in 'progressive subcorticalis gliosis' [62, 63]. Extensive studies involving a large series are in progress [21, Budka personal communication].

'The new variant observed in UK' (not presented at the Val de Grâce conference): unusual consistent features include the patients' young age, clinical findings such as a prolonged course, dysesthesia, ataxia, and extensively distributed plaques surrounded by spongiform changes [64].

Nosology

In one study [65], the presence of valine at codon 129 appeared to promote the formation of unicentric amyloid plaques, but this has not yet been conclusively demonstrated.

The nosology of the different forms of CJD may be fully understood when precise correlations between genotype and phenotype are established, unless other relevant factors are found.

Acknowledgments

Supported by the European Union Biomed-1 Concerted Action "The human prion diseases: from neuropathology to pathobiology and molecular genetics".

The author wishes to thank K Keohane for reviewing the English text, J Dellalave and P Castagnet for technical assistance, A Chazalet for the photos and C Martin for preparing of the manuscript.

References

1 Spielmeyer W. Die histopathologische Forschung in der Psychiatrie. *Klin Wochenschr* 1922;2:1817-9

2 Creutzfeldt HG. Uber eine eigenartige herbformige Erkrangkung des Zentral Nervensystems. *Z Ges Neurol Psychiatr* 1920;57:1-18

3 Jakob A. Uber eigenartige Erkrankung des Zentral Nervensystems mit bemerkenswertem anatomischen Befunde. *Z Ges Neurol Psychiatr* 1921;64:47-228

4 Masters CL, Richardson EP. Subacute spongiform encephalopathy (Creutzfeldt-Jakob disease). The nature and progression of spongiform change. *Brain* 1978;101:333-4

5 Brown P, Cathala F, Castaigne P, Gajdusek D. Creutzfeldt-Jakob disease: clinical analysis of a consecutive series of 230 neuropathologically verified cases. *Ann Neurol* 1986;20:597-602

6 Tomlinson BE. Creutzfeldt-Jakob disease. In: Adams JH, Duchen LW, eds. *Greenfield's Neuropathology* 5th edn. London: Edward Arnold, 1992;1366-75

7 Bell JE, Ironside JW. Neuropathology of spongiform encephalopathies in humans. *Br Med Bull* 1993;49: 738-77

8 Prusiner SB. Prion diseases of humans and animals. *J R Coll Phys Lond* 1994;28:1-30

9 Richardson EP Jr, Masters CL. The nosology of Creutzfeldt-Jakob disease and conditions related to the accumulation of PrP^{CJD} in the nervous system. *Brain Pathol* 1995;5:33-41

10 Hauw JJ, Gray F, Baudrimont M, Escourolle R. Cerebellar changes in 50 cases of Creutzfeldt-Jakob disease with emphasis on granule cell atrophy variant. *Acta Neuropathol (Berl)* 1981;Suppl VII :196-8

11 Jellinger K, Heiss WD, Deisenhammer E. The ataxic (cerebellar) form of Creutzfeldt-Jakob disease. *J Neurol* 1974;207:289-305

12 Brownell B, Oppenheimer DR. The ataxic form of subacute presenile polioencephalopathy (Creutzfeldt-Jakob disease). *J Neurol Neurosurg Psychiatry* 1965;28:350-61

13 Lafarga M, Berciano MT, Saurez I, Andres MA, Berciano J. Reactive astroglia-neuron relationships in the human cerebellar cortex: a quantitative, morphological and immunocytochemical study in Creutzfeltd-Jakob disease. *Int J Dev Neurosci* 1993;11:199-213

14 Boellaard JW, Schlote W, Tateishi J. Neuronal autophagy in experimental Creutzfeldt-Jakob disease. *Acta Neuropathol (Berl)* 1989;78:410-8

15 Nakazato Y, Hirato J, Ishida Y, Hoshi S, Hasegawa M, Fukuda T. Swollen cortical neurons in Creutzfeldt-Jakob disease contain a phosphorylated neurofilament epitope. *J Neuropathol Exp Neurol* 1990;49:197-205

16 Kato S, Hiran A, Umahara T, Llena JF, Herz F, Ohama E. Ultrastructural and immunohistological studies of ballooned cortical neurons in Creutzfeld-Jakob disease: expression of alpha B-crystallin, ubiquitin and stress-response protein 27. *Acta Neuropathol (Berl)* 1992;84:443-8

17 Guiroy DC, Shankar SK, Gibbs CJ, Jr, Messenheimer JA, Das S, Gajdusek DC. Neuronal degeneration and neurofilament accumulation in the trigeminal ganglia in Creutzfeldt-Jakob disease. *Ann Neurol* 1989;25: 102-6

18 Lesser RL, Albert DM, Bobowick AR, O'Brien FH. Creutzfeldt-Jakob disease and optic atrophy. *Am J Ophthalmol* 1979;87:317-21

19 Clinton J, Forsyth C, Royston MC, Roberts GW. Synaptic degeneration is the primary neuropathological feature in prion disease: a preliminary study. *Neuroreport* 1993;4:65-8

20 Brown P, Gibbs CJ Jr, Rodgers-Johnson P et al. Human spongiform encephalopathy; the National Institutes of Health series of 300 cases of experimentally transmitted disease. *Ann Neurol* 1994;35:513-29

21 Masters CL, Gajdusek DC, Gibbs CJ Jr. Creutzfeldt-Jakob disease virus isolations from the Gerstmann-Sträussler syndrome. With an analysis of the various forms of amyloid plaque deposition in the virus-induced spongiform encephalopathies. *Brain* 1981;104:559-88

22 Budka H, Aguzzi A, Brown P et al. Neuropathological diagnostic criteria for Creutzfeldt-Jakob disease (CJD) and other human spongiform encephalopathies (prion diseases). *Brain Pathol* 1995;5:459-66

23 Ribadeau-Dumas JL, Escourolle R. The Creutzfeldt-Jakob syndrome: a neuropathological and electron microscopic study. In: Subirana A, Espalader JM, ed. *Neurology*. Amsterdam: Elsevier, 1974;316-29

24 Brion S, Mikol J, Isidor P. Étude anatomo-clinique d'un cas de maladie de Creutzfeldt-Jakob. *Rev Neurol (Paris)* 1969;121:165-79

25 Chou SM, Payne WN, Gibbs CJ, Jr, Gadjusek DC. Transmission and scanning electron microscopy of spongiform change in Creutzfeldt-Jakob disease. *Brain* 1980;103:885-904

26 Bastian FO. *Creutzfeldt-Jakob Disease and Other Transmissible Human Spongiform Encephalopathies*. St Louis: Mosby Year Book, 1991

27 Laslo L, Lowe J, Self et al. Lysosomes as key organelles in the pathogenesis of prion encephalopathies. *J Pathol* 1992;66:333-41

28 Sasaki A, Hirato J, Nakazato Y. Immunohistochemical study of microglia in the Creutzfeldt-Jakob diseased brain. *Acta Neuropathol (Berl)* 1993;86:337-44

29 Chou SM, Martin JD. Kuru-plaques in a case of Creutzfeldt-Jakob disease. *Acta Neuropathol (Berl)* 1971;17: 150-5

30 Boellaard JW, Schlote W. Subakute spongiforme Encephalopathy mit multiformer Plaquebildung. *Acta Neuropathol (Berl)* 1980;49:205-12

31 Kitamoto T, Tateishi J. Immunohistochemical confirmation of Creutzfeldt-Jakob disease with a long clinical course with amyloid plaque core antibodies. *Am J Pathol* 1988;131:435-43

32 Heidenhain A. Klinische und anatomische Untersuchungen über die eigeneratige organische Erkrankung des Zentral Nervensystems in Praesenium. *Z Ges Neurol Psychiatr* 1929;118:49-114

33 Meyer A, Leigh D, Bagg CE. A rare presenile dementia associated with cortical blindness (Heidenhain's syndrome). *J Neurol Neurosurg Psychiatry* 1954;17:128-33

34 Nevin S, McMenemey WH, Behrman S, Jones DP. Subacute spongiform encephalopathy – a subacute form of encephalopathy attributable to vascular dysfunction (spongiform cerebral atrophy). *Brain* 1960;83:519-64

35 Stern K. Severe dementia associated with bilateral symmetrical degeneration of the thalamus. *Brain* 1939;62:157-71

36 Schulman S. Bilateral symmetrical degeneration of the thalamus. *J Neuropathol Exp Neurol* 1957;16:446-70

37 Garcin R, Brion S, Khochneviss A. Le syndrome de Creutzfeldt-Jakob et les syndromes cortico-striés du presenium (à l'occasion de 5 observations anatomo-cliniques). *Rev Neurol (Paris)* 1963;109:419-41

38 Van Rossum A. Spastic pseudosclerosis (Creutzfeldt-Jakob disease). In: Vinken PJ, Bruyn GW, ed. *Handbook of Clinical Neurology, Diseases of the Basal Ganglia*. Amsterdam, North-Holland: 1968;726-60

39 Boudin G, Pepin B, Milhaud M. Maladie de Creutzfeldt-Jakob à symptomatologie cérébelleuse dominante (Étude anatomoclinique d'un cas). *Rev Neurol (Paris)* 1965;113:73-5

40 Mizutani T. Neuropathology of Creutzfeldt-Jakob disease in Japan with special reference to the panencephalopathic type. *Acta Pathol Jpn* 1981;31:903-22

41 Yagishita S, Iwabuchi K, Amano N, Yokoi S. Further observation of Japanese Creutzfeldt-Jakob disease with widespread amyloid plaques. *J Neurol* 1989;236:145-8

42 Masters CL, Gajdusek DC. The spectrum of Creutzfeldt-Jakob disease and the virus-induced subacute spongiform encephalopathies. In: Smith WT, Cavanagh JB, ed. *Recent Advances in Neuropathology*. Edinburgh: Churchill Livingstone, 1982;139-64

43 Russel RWR. Supranuclear palsy of eyelid closure. *Brain* 1980;103:71-82

44 Brunet P, Viader F, Henin D. Encéphalopathie d'évolution rapide avec ophtalmoplégie supranucléaire et neuropathie périphérique. *Rev Neurol (Paris)* 1986;142:159-66

45 Heye N, Henkes H, Hansen ML, Gosztonyi G. Focal-unilateral accentuation of changes observed in the early stage of Creutzfeldt-Jakob disease. *J Neurol Sci* 1990;95:105-10

46 Tateishi J, Ohta M, Koga M, Sata Y, Kuroiwa Y. Transmission of chronic spongiform encephalopathy with kuru plaques from humans to small rodents. *Ann Neurol* 1979;5:581-4

47 Kawata A, Suga M, Oda M, Hayashi H, Tanabe H. Creutzfeldt-Jakob disease with congophilic kuru plaques; CT and pathological findings of the cerebral white matter. *J Neurol Neurosurg Psychiatry* 1992;55:849-51

48 Reggiani R, Solimé F, Nizzoli V, Guazzi GC. La nécrobiose et la lipophanérose, gliale et neuronale, dans la maladie de Creutzfeldt-Jakob-Heidenhain-Nevin : existe-t-il des altérations de la substance blanche ? *Acta Neuropathol (Berl)* 1967;Suppl III :76-9

49 Capon A, Flament J, Guazzi GC. Le rôle de la barrière hémato-encéphalique dans la maladie de Creutzfeldt-Jakob-Heidenhain-Nevin. *Acta Neuropathol (Berl)* 1967;Suppl III :47-53

50 Park TS, Kleinman GM, Richardson EP. Creutzfeldt-Jakob disease with extensive degeneration of white matter. *Acta Neuropathol (Berl)* 1980;52:239-42

51 Macchi G, Abbamondi AL, Di Trapani G, Sbriccoli A. On the white matter lesions of the Creutzfeldt-Jakob disease: Can a new subentity be recognized in man? *J Neurol Sci* 1984;63:197-206

52 Berciano J, Berciano MT, Polo JM, Figols J, Ciudad J, Lafarga M. Creutzfeldt-Jakob disease with severe involvement of cerebral white matter and cerebellum. *Virchows Arch (A)* 1990;417:533-8

53 Powers JM, Liu Y, Hair LS, Kascsacl RJ, Lewis LD, Levy LA. Concomitant Creutzfeldt-Jakob disease and Alzheimer diseases. *Acta Neuropathol (Berl)* 1991;83:95-8

54 Barcikowska M, Kwiecinski H, Liberski PP, Kowalski J, Brown P, Gajdusek DC. Creutzfeldt-Jakob disease with Alzheimer-type A β-reactive amyloid plaques. *Histopathology* 1995;26:445-50

55 Watanabe R, Duchen LW. Cerebral amyloid in human prion disease. *Neuropathol Appl Neurobiol* 1993;19:253-60

56 Duchen LW, Poulter M, Harding AE. Dementia associated with a 216 base pair insertion in the prion protein gene: clinical and neuropathological features. *Brain* 1993;116:555-67

57 Keohane C, Peatfield R, Duchen LW. Subacute spongiform encephalopathy (Creutzfeldt-Jakob disease) with amyloid angiopathy. *J Neurol Neurosurg Psychiatry* 1985;48:1175-8

58 Tateishi J, Kitamoto T, Doh-ura K, Boellaard JW, Peiffer J. Creutzfeldt-Jakob disease with amyloid angiopathy: diagnosis by immunological analysis and transmission experiments. *Acta Neuropathol (Berl)* 1992;83:559-63

59 Gray F, Chrétien F, Cesaro P et al. Creutzfeldt-Jakob disease and cerebral amyloid angiopathy. *Acta Neuropathol (Berl)* 1994;88:106-11

60 Collinge J, Owen F, Poulter M et al. Prion dementia without characteristic pathology. *Lancet* 1990;336:7-9

61 Brown P, Kaur P, Sulima MP, Goldfarb LG, Gibbs CJ Jr, Gajdusek DC. Real and imagined clinicopathological limits of "prion dementia". *Lancet* 1993;341:127-9

62 Petersen RB, Tabaton M, Chen SG, Monari L, Richardson SL, Lynches T, et al. Familial progressive subcortical gliosis: presence of prions and linkage to chromosome 17. *Neurology* 1995;45:1062-7

63 Revesz T, Daniel SE, Lees AJ, Will RG. A case of progressive subcortical gliosis associated with deposition of abnormal prion protein (PrP). *J Neurol Neurosurg Psychiatry* 1995;58:759-60

64 Will RG, Ironside JW, Zeidler M, Cousens SN, Estibeiro K, Alperovitch A et al. A new variant of Creutzfeldt-Jakob disease in UK. *Lancet* 1996;347:921-5

65 De Silva R, Ironside JW, McCardle L, Esmonde T, Bell J, Will R et al. Neuropathological phenotype and "prion protein" genotype correlation in sporadic Creutzfeldt-Jakob disease. *Neurosci Lett* 1994;179:50-2

Transmissible Subacute Spongiform Encephalopathies:
Prion Diseases
L Court, B Dodet, eds
© 1996, Elsevier, Paris

Automatic computerised image analysis of neuropathology in Creutzfeldt-Jakob disease

Kenneth Sutherland, James W Ironside

CJD Surveillance Unit, Western General Hospital, Crewe Road, Edinburgh EH4 2XU, Scotland

Summary – Quantitative image analysis methods have been developed to assist in the interpretation of the neuropathology observed in cases of Creutzfeldt-Jakob disease (CJD). In this study we have quantified the amount of prion protein (PrP) deposition observed in five cases of human pituitary-derived growth hormone (hGH) associated CJD and nine cases of sporadic CJD. The results have revealed a significantly higher level of PrP deposition in the hGH cases. We have also demonstrated that the PrP deposition observed in the hGH cases occurs in a laminar pattern within the cerebral cortex.

Introduction

Creutzfeldt-Jakob disease (CJD) in humans is characterised in pathological terms by spongiform change, neuronal loss, astrocytosis and amyloid plaque formation. The severity and distribution of these lesions are markedly variable in different cases of this disease. Even within the brain of a single individual, the different regions targeted by the disease do not follow a uniform pattern. Hence, unlike some other prion diseases, eg, bovine spongiform encephalopathy [1], it seems that no constant pattern of lesions exists. There are a substantial number of factors which could possibly contribute to the diversity of lesion patterns that have been noted in the study of this disorder. Given the experience of research into other prion diseases in animals, it is possible that host genotype and the strain of the infective agent could influence the pathological profile of the lesions observed [2, 3]. In human cases of CJD, the genotype at codon 129 of the prion protein gene *(PRNP)* (a site of common polymorphism) has been shown to influence the susceptibility of an individual to developing the disease [4]. It may be true that there are different interrelationships between the various pathological features which have implications for the pathogenesis of this disorder and the other prion diseases.

   Within the archives of the United Kingdom (UK) National CJD Surveillance Unit Laboratory, there are a substantial number of cases of sporadic, familial and iatrogenic CJD. The majority of the iatrogenic cases are associated with the administration of human pituitary-derived growth hormone (hGH) [5]. In all cases, the clinical records are available and genetic data have been amassed for a substantial number of these cases. In this way it is possible to

Key words: quantitation / human spongiform encephalopathies / prion protein immunocytochemistry / growth hormone-associated Creutzfeldt-Jakob disease

relate the observed pathology of a particular case to the clinical and genetic information available. In the iatrogenic cases, the route of infection is also known. Hence the intensive analysis of the pathological features of these cases should assist in the construction of an overall picture of the differing influences of these varying factors.

In order to investigate the interrelationships between the different pathological features, it is necessary to have an objective and reproducible method of assessing the severity of each finding. A substantial investment at the CJD unit has been made in the development of a number of image analysis techniques as a means of meeting this primary objective.

Materials and methods

In order to maximise the quantity of material that could be examined using quantitative means, a highly mechanised microscope assembly has been employed. The set-up includes a modified Leitz Ergolux microscope with a Marzhauser motorised scanning slide stage capable of accommodating up to eight slides at any one time. A stabilised light source is also used to eliminate any possible variation in lighting levels. To capture high-definition images a Xillix MicroImager 1400 still video camera has been used. The different software modules developed here employ the Medical Research Council's image processing programming library developed at the MRC Human Genetics Unit in Edinburgh. The suite of programmes developed to quantify CJD neuropathology are run on a Sun IPX computer. This machine is a UNIX workstation and is thus very well suited to the particular demands of a highly automated image processing system.

Using the equipment described above, a novel software system has been developed and tested to recognise spongiform change [6], the most striking feature of the pathology observed in many cases of CJD. The software programme is able to produce a visual output in which areas of vacuolation are highlighted, as shown in fig 1, or a numeric output giving the number of vacuoles observed within a particular area and the percentage tissue area occupied by vacuolation. Using slightly different software techniques we are able to quantify astrocytic proliferation which is seen in many different cases of CJD [7]. Using a glial fibrillary acidic protein-stained section, astrocyte cell bodies are relatively easily distinguished from the surrounding tissue (fig 2). The system also provides a count of the number of cells present and statistical information about the size of each individual cell.

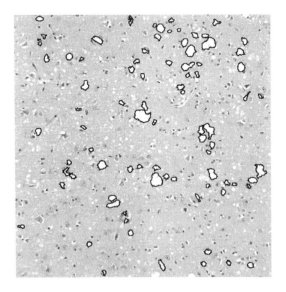

Fig 1. Automated recognition of vacuolation in the cortex of a case of sporadic CJD, areas of spongiform change are outlined in black.

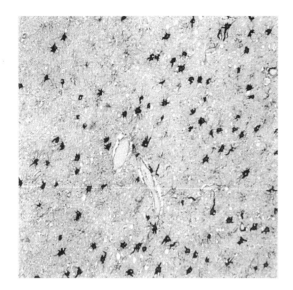

Fig 2. Automated astrocyte counting, candidate astrocytes are coloured black.

With the development of reliable immunocytochemical techniques for the localisation of prion protein (PrP), the deposition of PrP can be investigated and compared with the severity and location of the other observed features of this disease. Quantitative immunocytochemistry of this nature is quite difficult when the deposits lack any clear morphological patterns. In most cases, an interactive stage is required to set an image intensity threshold which represents the transition between foreground staining and background counterstain. In our initial work on amyloid plaque formation, the identification and quantification of plaque staining could be performed quite easily using an operator-defined image intensity threshold [8]. In the spinal cord in CJD [9], we have been able to use fully automatic image thresholding as a means of obtaining completely objective measurements of positively stained tissue areas.

In selecting tissue to use for quantitation, the values obtained from each different brain must be comparable; in order to satisfy this condition, we have laid out a protocol listing the different brain regions to be sampled and the particular anatomical structures to be sampled from each one of those [7]. By using these quantitative techniques, it has been possible to investigate the performance of different antibodies to the PrP. In this research [7], we have shown that there are qualitative morphological differences between the staining patterns observed with particular antibodies, but that overall there are very few quantitative disparities. Further work has concentrated on the effect of host genotype on pathological profile [10], using an example set of nine cases of histologically confirmed CJD for which the codon-129 *PRNP* genotype information was known. This research showed that there was an apparent targeting of the deeper grey matter structures of this brain in the valine homozygote (V/V) group of cases, while in contrast the methionine homozygotes (M/M) showed the opposite effect with more cortical pathology observed.

In an earlier study, we investigated PrP staining patterns in the spinal cord in hGH-related cases of CJD [11]. In order to investigate further the likelihood that this observation could be related to the route of infection of the CJD agent, we have performed a detailed investigation of five cases of hGH-related CJD. Details of these cases and those of a group of nine sporadic cases of CJD are listed in table I. Using the methods described above, we have quantified spongiform change, astrocytosis and PrP deposition. For these five cases, we have studied four different brain regions representing the frontal and occipital lobes, the basal ganglia and the cerebellum. The antibodies and pre-treatment protocols used for this work have been described in earlier work of the CJD unit [12].

Table I. Case details of all 14 tissue samples used in this study.

Case	Age (range)	129 Genotype
1, 2, 3	49–64	V/V
4, 5, 6	65–80	M/V
7, 8, 9	61–78	M/M
hGH 1	26	NA
hGH 2	25	V/V
hGH 3	33	NA
hGH 4	30	NA
hGH 5	31	V/V

V/V: valine homozygotes; M/M: methionine homozygotes; M/V: valine/methionine heterozygotes; NA: not available.

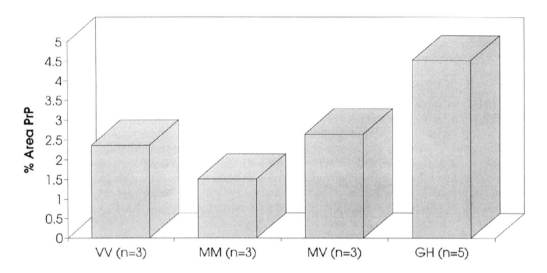

Fig 3. Overall PrP staining averaged across four different brain areas for the different groups used in this study. VV: valine homozygotes; MM: methionine homozygotes; MV: methionine-valine heterozygotes; GH: growth hormone.

Subjective assessment of PrP staining patterns in hGH CJD has suggested that plaque staining is observed and that a substantial amount of pericellular staining is also noted in the deeper cortical lamina. To investigate this phenomenon further, we have quantified PrP staining across the depth of the occipital and frontal cortices in two of the five hGH cases available.

Results

Quantitative assessment of the PrP staining seen in the five hGH cases and the nine sporadic cases reveals a higher level of overall staining for the four different areas examined in this study. This finding is illustrated in figure 3 for the 3F4 antibody. Analysis of the profile of staining areas observed across the four areas studied (fig 4) shows apparent targeting of the

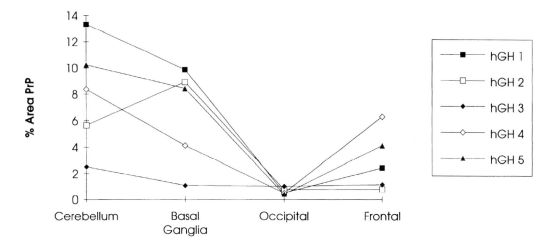

Fig 4. PrP staining profiles for five hGH-associated CJD cases.

cerebellum and the basal ganglia in most cases, with remarkably little staining in the occipital cortex.

A macroscopic representation of PrP deposition in one block of frontal tissue illustrates very clearly the localisation of staining to the deeper cortical areas (fig 5). Spongiform change analysis for this same tissue shows very little co-localisation of PrP and vacuolation, but it does reveal a possible area of focal vacuolation which does not appear to be PrP positive. A macroscopic representation of spongiform change for a serial section partner of figure 5 is shown in figure 6.

Further analysis of a discrete strip of cortical tissue reveals how variable the PrP deposition can be across the depth of the cortex (fig 7a). In the graphical representation of this quantitative data (fig 7b), a small peak can be seen in approximately layer III of the cortex cor-

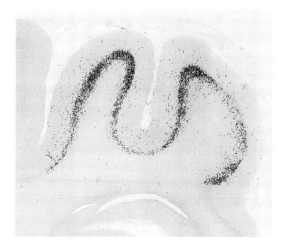

Fig 5. Computer-enhanced PrP staining in a section of tissue from the frontal lobe in a case of hGH-associated CJD.

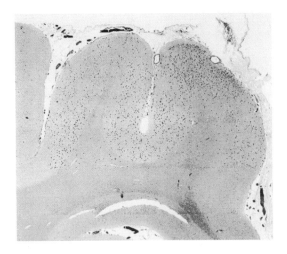

Fig 6. Spongiform change in a serial section to figure 5, each dark dot represents a single spongiform vacuole in the cerebral cortex.

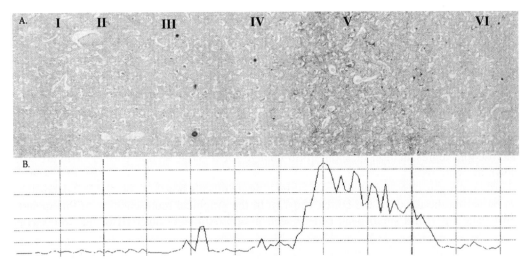

Fig 7. A single strip of cerebral cortex (**A**) and a graph of PrP deposition across that strip (**B**) showing the variation of staining observed across the different layers of the cortex (I–VI).

responding to plaque-like staining in this area and a more substantial peak can be seen in layer V. Analysis of other sporadic cases of CJD does not appear to show such a regular pattern of PrP deposition.

Discussion

Using image analysis techniques, we have illustrated that PrP deposition follows a laminar pattern within the cortex in hGH-associated CJD. Staining for PrP appears to occur in two different morphologies. In approximately layer III of the cortex, some plaque-like structures can be observed, but the staining in layer V appears to be mainly pericellular in nature. While there is some spongiform change within these layers, the strong laminar pattern of PrP staining does not co-localise with vacuolation.

In the five hGH cases studied here, the overall area of tissue stained positively with PrP appears to be higher than that observed for sporadic cases of CJD. The analysis of four different brain areas performed here shows that the pattern of staining observed across these different areas is very similar for all the cases studied. The uniformity of pathological findings is in accordance with earlier observations by other researchers [13].

It is now accepted that peripheral inoculation with infected hGH caused these iatrogenic cases of CJD [5]. The suggestion is that some pituitary glands from individuals infected with CJD were used in the production of the hGH product. It is interesting to note that we have shown such uniformity of pathology between these five different cases. There are two likely reasons for this : the route of infection would determine the pattern of pathology observed in the affected brain; there would be several different strains of CJD, and only one strain would have been transmitted to the five cases in this study.

It seems very unlikely that pathology alone can determine which of these two reasons is the most likely and perhaps transmission studies to mice using material from these cases may be able to determine whether they are all from a single strain of CJD or not. Further detailed studies to identify underlying patterns of CJD pathology will help further our understanding of the pathogenesis of this disease in humans.

Acknowledgments

The authors would like to acknowledge the support of all at the CJD Surveillance Unit, in particular the technical assistance of L McCardle and CA Barrie. K Sutherland's work is supported by the BBSRC (Grant no 15/BS204814).

References

1 Wells GAH, Hawkins SAC, Hadlow WJ et al. The discovery of bovine spongiform encephalopathy and observations on the vacuolar changes. In: Prusiner SB, Collinge J, Powell J, Anderton B, eds. *Prion Diseases of Humans and Animals.* Chichester: Ellis Horwood Limited, 1992;23:256-74

2 Goldmann W. *PrP* gene and its association with spongiform encephalopathies. *Br Med Bull* 1993;49:839-59

3 Bruce ME. Scrapie strain variation and mutation. *Br Med Bull* 1993;49:822-38

4 de Silva R, Ironside JW, McCardle L et al. Neuropathological phenotype and 'prion protein' genotype correlation in sporadic Creutzfeldt-Jakob disease. *Neurosci Lett* 1994;179:50-2

5 Brown P, Preece MA, Will RG. 'Friendly fire' in medicine: hormones, homografts, and Creutzfeldt-Jakob disease. *Lancet* 1992;340:24-7

6 Sutherland K, Ironside JW. Novel application of image analysis to the detection of spongiform change. *Anal Quant Cytol Histol* 1994;16:430-4

7 Sutherland K, Macdonald ST, Ironside JW. Quantification and analysis of the neuropathological features of Creutzfeldt-Jakob disease. *J Neurosci Meth* 1996;64:123-32

8 Sutherland K, Barrie C, Ironside JW. Automatic quantification of amyloid plaque formation in human spongiform encephalopathy. *Neurodegeneration* 1994;3:293-300

9 Sutherland K, Goodbrand IA, Bell JE, Ironside JW. Objective quantification of prion protein in spinal cords of cases of Creutzfeldt-Jakob disease. *Anal Cell Pathol* 1996;10:25-35

10 Macdonald ST, Sutherland K, Ironside JW. A quantitative and qualitative analysis of prion protein immunocytochemical staining in Creutzfeldt-Jakob disease. *Neurodegeneration* 1996;5:87-94

11 Goodbrand IA, Ironside JW, Nicolson D, Bell JE. Prion protein accumulation in the spinal cords of patients with sporadic and growth hormone associated Creutzfeldt-Jakob disease. *Neurosci Lett* 1995;183:127-30

12 Hayward PA, Bell JE, Ironside JW. Prion protein immunocytochemistry: reliable protocols for the investigation of Creutzfeldt-Jakob disease. *Neuropathol Appl Neurobiol* 1994;20:375-83

13 Deslys JP, Lasmezas C, Dormont D. Selection of specific strains in iatrogenic Creutzfeldt-Jakob disease. *Lancet* 1994;343:848-9

Transmissible Subacute Spongiform Encephalopathies:
Prion Diseases
L Court, B Dodet, eds
© 1996, Elsevier, Paris

Cell death in prion disease

Hans A Kretzschmar[1], Armin Giese[1], David R Brown[1], Jochen Herms[1],
Bernhard Schmidt[2], Martin H Groschup[3]

[1]*Institut für Neuropathologie;* [2]*Abteilung Biochemie II, Universität Göttingen Robert Koch Strasse 40,
37075 Göttingen;* [3]*Bundesforschungsanstalt für Viruskrankheiten der Tiere, Tübingen, Germany*

Summary – The in situ end-labeling (ISEL) technique and electron microscopy were used to study cell death in an experimental scrapie system in the mouse. Apoptotic nuclei were found in the brains and retinae of mice infected with the 79A strain of scrapie, whereas no labeling was found in control animals. In the retina, the highest number of labeled nuclei was found in the outer nuclear layer 120 days post-infection followed by massive cell loss in this layer. In the brain labeled nuclei were mainly found in the granular layer of the cerebellum of terminally ill mice. These results support the hypothesis that neuronal loss in spongiform encephalopathies is due to apoptosis, which may explain the almost complete absence of inflammatory response in these diseases. Cell cultures from mice devoid of the protease-sensitive prion protein, PrP^C (*Prnp$^{0/0}$* or *Prnp* knock-out mice) and wild-type mice were used to investigate mechanisms leading to apoptosis. Using a fragment of human prion protein (PrP) consisting of amino acids 106–126 it was shown that the toxic effect of this peptide requires the presence of microglia which respond to PrP 106–126 by increasing their oxygen radical production. The combined direct and microglia-mediated effects of PrP 106–126 are toxic to normal neurons but are insufficient to destroy neurons from mice not expressing PrP^C.

Neuronal cell death in prion diseases is due to apoptosis

It is generally believed that there are two basic mechanisms of cell death: necrosis and apoptosis [1–3]. Necrosis often results from severe and sudden injury and leads to rapid cell lysis and a consecutive inflammatory response. In contrast, apoptosis proceeds in an orderly manner following a cellular suicide program involving active gene expression in response to physiological signals or types of stress. Morphologically, apoptosis is characterized by chromatin condensation and aggregation, cellular and nuclear shrinkage and formation of apoptotic bodies. It is usually not accompanied by an inflammatory response. Judging from previous studies, indirect indications of apoptosis occur in prion disease in vivo. Using laser scanning microscopic analysis of a single-cell gel assay, increased DNA damage was found in three scrapie-infected sheep brains as compared to controls, while no attempt was made to control for necrosis, autolysis or age [4]. Hogan et al [5] gave a detailed morphological description o f retinal degeneration and showed an electron micrograph of a photoreceptor cell which today would be interpreted as apoptosis. Additionally, the almost complete absence of an inflammatory response in the pathology of spongiform encephalopathies [6, 7] suggests that nerve cell loss may be due to apoptosis. On the other hand, in a large study investigating apoptosis in the nervous system using the in situ end-labeling (ISEL) technique, apoptosis was not identified in terminally ill scrapie-infected mice [8]; however, the authors give no details

Key words: apoptosis / microglia / neurotoxicity

concerning the strains of mice and scrapie used and the brain regions investigated. Both disease-specific prion protein (PrPSc) [9] as well as a peptide fragment of prion protein (PrP) [10] have been shown to cause apoptosis of neurons in vitro. However, apoptotic cell death has only recently been shown to occur in prion diseases in vivo [11].

The use of the ISEL technique has facilitated the recognition of apoptotic cells in tissue sections [12, 13]. ISEL is based on the incorporation of labeled nucleotides in fragmented DNA by the enzyme terminal transferase which is considered to be highly characteristic of apoptosis. We used this technique to identify apoptotic cells in mice infected with the 79A strain of scrapie, a well-defined experimental scrapie model [14–17]. To evaluate the pattern and time course of cell death, various brain regions were analyzed at different times in the course of the disease. Since cell loss in scrapie-infected rodents is often quite striking in the photoreceptor layer of the retina [5, 18–21], the eyes were included in this study. Electron microscopy was used to independently assess apoptotic cell death.

Retina

Sections taken up to 90 days after inoculation showed no pathological changes of the retina. When compared to controls, eyes removed from animals 120 days after scrapie infection showed a dissolution of rod inner and outer segment regions. This process was accompanied by the presence of macrophages in this layer. In two out of four animals the outer nuclear layer appeared considerably thinner. No changes were detectable in the inner nuclear and ganglion cell layers. At 150 and 166 days post-infection (dpi), the outer nuclear layer had degenerated to a single cell layer with slight variations of two to three cells in some regions and a total loss of photoreceptor cells in others. The photoreceptor inner and outer segments had almost completely vanished. These changes were similar to those described previously [5].

The ISEL assay showed that labeled nuclei were practically absent from retinae of control mice of all ages (fig 1A). Labeled nuclei were virtually absent from scrapie-infected mice up to day 90 after inoculation. At 120 dpi several labeled nuclei were found scattered in the outer nuclear layer (fig 1B). Some labeled nuclei appeared shrunken, and labeling was often most intense at the periphery of the nucleus. At 150 and 166 dpi isolated labeled nuclei were found in the remainder of the outer nuclear layer. However, considering the extreme loss of cells in this time interval, the percentage of apoptotic cells to the overall number of remaining cells may even have increased.

Cerebellum

The ISEL technique revealed nuclei containing fragmented DNA in the granular layer in scrapie-infected mice from day 120 onwards. No labeling was found in control mice at any times and in scrapie-infected mice up to day 90 after inoculation with the exception of one scrapie-infected animal showing one labeled nucleus in ten high power visual fields 90 days after inoculation. The number of labeled cells increased from day 120 to 150 and was highest in terminally ill mice (day 166). Labeled nuclei often appeared shrunken and showed most intense staining at the periphery of the nucleus. Corresponding to the ISEL assay, small, dark, round, occasionally fragmented nuclei with eosinophilic cytoplasm were found in sections from scrapie-infected mice stained with hematoxylin and eosin from day 120 onwards.

Electron microscopic investigation of specimens from the cerebellum of terminally ill mice identified several cells which showed homogeneously condensed chromatin, dark cytoplasm, membrane blebbing and, occasionally, nuclear fragmentation, all of which are morphological changes characteristic of apoptosis [1, 2, 22].

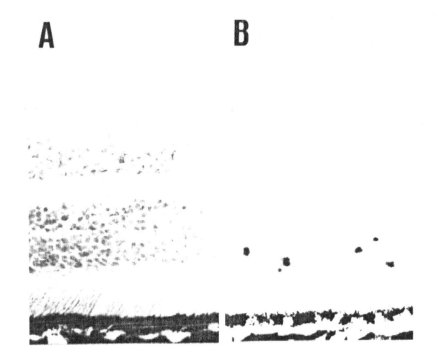

Fig 1. Neuronal cell death in prion disease. In situ end-labeling in the retina of scrapie-infected mice (control [A], 120 days dpi [B]) identifies apoptotic cells almost exclusively in the outer nuclear layer. The highest absolute number of labeled cells is observed at 120 dpi. At 166 days after inoculation only a minor fraction of nuclei is conserved, among which there is a relatively high number of labeled nuclei.

Other regions of the brain

In the other regions of the brain labeling was less obvious. As described previously [8], labeled nuclei were observed occasionally in the ependyma and in a subependymal location along the lateral ventricles both in control and scrapie-infected mice. In scrapie-infected mice at days 150 and 166 a small number of unequivocally labeled nuclei were observed in the basal ganglia, in the granule cell layer of the olfactory bulb, and in the cerebral cortex.

Pattern and time course of apoptotic cell death in scrapie

We used the ISEL technique, which has recently been employed by a great number of investigators, to identify apoptotic cells in histological sections. It appears, however, that the results obtained with this technique are not absolutely specific. DNA fragmentation is also observed in necrosis or autolysis, but in these instances seems to occur at random, in contrast to the specific internucleosomal cleavage which is characteristic of apoptosis [23]. Therefore, DNA laddering in gel electrophoresis, which shows the different size of oligonucleosomal fragments, is often used to distinguish apoptosis from mechanisms leading to random DNA cleavage. In cases where only a small percentage of cells undergoes apoptosis at any given moment, as expected in our experimental system, DNA laddering would not be applicable and hence was not attempted. Therefore classical morphological criteria were employed [1, 2] to validate data obtained by the ISEL technique. These included chromatin condensation, nuclear fragmentation, eosinophilic and electron microscopically dark cytoplasm and membrane blebbing. Since all of these were present, our results convincingly demonstrate the occurrence of neuronal apoptosis in an experimental in vivo scrapie system using two different techniques.

C57Bl6 mice infected with the 79A strain of scrapie are known to show diffuse widespread accumulation of PrP [14] without amyloid plaques [24, 25] as well as moderate vacuolation affecting most gray matter areas, including the molecular layer of the cerebellum, to a fairly similar degree. Other strains of scrapie such as Me7 show comparatively little spongiform change in the cerebellum [17]. We found the most obvious labeling of apoptotic cells in the granular cell layer of the cerebellum and in the outer nuclear layer of the retina. These structures contain a high number and density of morphologically uniform neuronal cells and therefore are ideally suited for detection of a small percentage of apoptotic cells. However, other regions similar in this respect, such as the granule cell layer of the olfactory bulb or hippocampus, showed few or no labeled nuclei, indicating a different degree of cell death in these structures. It is interesting to note, however, that the highest number and density of foci of apoptotic cell death were observed exactly in those areas that did not show spongiform changes, ie, the retina and the granule cell layer of the cerebellum. It will be interesting to see if other experimental scrapie models such as the Me7 strain of scrapie in LM mice, which is associated with different pathological features [26], show a different pattern of cell death, affecting a different population of neurons. This should help establish more detailed lesion profiles and elucidate the correlation between PrP accumulation, spongiform change and neuronal cell death.

The small percentage of labeled cells that were found even in areas of massive on-going cell death is in no way surprising since apoptosis has been shown to be completed within hours [27] and cells can be expected to become apoptotic asynchronously. This indicates that it will be difficult to demonstrate apoptotic cells in conditions where cell loss is less rapid. The demonstration of apoptosis in an animal model of scrapie, therefore, is of considerable importance both for the pathophysiology of spongiform encephalopathies and the validation of cell culture experiments using neurotoxic PrP fragments as well as in the broader context of understanding the mechanisms of neurodegenerative diseases.

Since many common neurodegenerative diseases such as Alzheimer's disease, Parkinson's disease and amyotrophic lateral sclerosis are characterized by a gradual loss of neurons without obvious inflammatory response, it has been hypothesized that cell death in these disorders is due to apoptosis [28]. However, there is only limited sound data available on apoptosis in the adult central nervous system [22, 29]. Our data show that apoptosis of neurons can be demonstrated in the adult nervous system in a well characterized animal model of prion disease, which shows massive and highly predictable cell death.

PrP peptides and neurodegeneration

The mechanisms leading to neuronal cell death by apoptosis, which has now been shown to occur in vivo in scrapie-infected mice [11], are yet to be elucidated. This problem has been tackled using PrP^{Sc} [9] and various synthetic PrP peptides in cell culture [10]. Forloni et al [10] used cultured cells from embryonic rat hippocampus to investigate the neurotoxicity of PrP peptides based on the N-terminus of the human sequence. Two peptides based on amino residues 106–126 and 127–147 had a high β-sheet content, but only PrP 106–126 reduced cell survival in these cultures. The neurotoxicity of this peptide was confirmed by us [30] for cortical cells derived from embryonic mice. Cell death induced by this peptide is also due to apoptosis and occurs in a concentration-dependent manner [10] with a maximal effect at around 80 μM [30]. Cell death caused by this peptide was measured using a standard MTT (3, [4,5-dimethylthiazol-2yl]-2,5 diphenyltetrazolium bromide, Sigma) assay and has proved to be an effective method for quantifying the degree of cell loss.

The first insight into the specificity of this peptide's neurotoxicity was the demonstration that cellular expression of protease-sensitive PrP (PrPC) is a requirement [30]. Büeler et al [31] were the first to generate mice that have no expression of a full prion protein transcript (*Prnp*$^{0/0}$ mice). These mice, devoid of PrPC, develop normally and apparently have little behavioral abnormalities [31], although a recent study [32] suggests they may have altered circadian rhythms. *Prnp*$^{0/0}$ mice are resistant to infection with scrapie [33].

We prepared cell cultures from the cortex of *Prnp*$^{0/0}$ mice [31] and treated them with concentrations of PrP 106–126 up to 200 μM over a period of 10 days [30]. After that time measurements made indicated that there was no loss of cells as compared to controls. However, wild-type mouse cells thus treated showed a reduction to 65% of the control value. This result clearly demonstrated that PrPC expression is necessary for the toxicity of PrP 106–126.

The PrP 106–126 peptide is also toxic to cerebellar granule cells [34]. At 80 μM PrP 106–126 is more toxic to cerebellar cells than to cortical cells, reflecting a greater sensitivity of these cells to the peptide. Nevertheless, *Prnp*$^{0/0}$ cerebellar cells are also resistant to PrP 106–126 toxicity [34].

PrPC is expressed by astrocytes [35] as well as by neurons [36]; therefore, we began investigating the effect of this peptide on other cell types in our culture system [37]. The discovery that PrP 106–126 induces proliferation in microglia [37] prompted us to investigate whether these cells play a role in the neurotoxic effect of PrP 106–126. Treatment of mixed cell cultures from the cerebellum with L-leucine-methyl ester (LLME) reduces the number of microglia in cell culture. LLME treatment does not significantly affect the survival of other cells in the culture. However, this treatment effectively abolishes the toxicity of PrP 106–126 (fig 2). Co-culturing cerebellar cells with additional microglia increases the number of cells killed by PrP 106–126. These results clearly suggest that microglia are involved in the neurotoxic effects of PrP106–126 (fig 2).

Low concentrations (eg, 20 μM) of PrP 106–126 are not toxic to wild-type cerebellar cells in culture. However, when wild-type cerebellar cells are co-cultured with additional microglia 20 μM PrP 106–126 becomes toxic to the cerebellar neurones. This toxicity increases in relation to the number of microglia added. Co-culturing cerebellar cells with astrocytes does not have the same effect. This suggests that astrocytes are not directly involved in the neurotoxic mechanism of PrP 106–126.

Neurotoxic enhancement of PrP 106–126 was also shown by us for microglia derived from *Prnp*$^{0/0}$ mice. Additional *Prnp*$^{0/0}$ microglia could also restore the neurotoxicity of PrP 106–126 to LLME-treated wild-type cerebellar cells (fig 2). However, these microglia were less effective than wild-type microglia. This result suggests that the role of microglia in this neurotoxic mechanism is at least partially independent of PrPC expression by microglia.

If destruction of neurons by PrP 106–126 were purely an indirect effect of the peptide on microglia, one would expect *Prnp*$^{0/0}$ neurons to be similarly destroyed by the presence of PrP 106–126 and wild-type microglia. *Prnp*$^{0/0}$ cortical cells co-cultured with wild-type microglia continued to be resistant to the effects of PrP 106–126, even at 200 μM. This result again emphasizes the importance of PrPC expression to neurotoxic susceptibility. Therefore, micro-

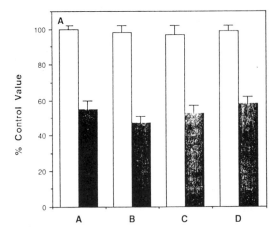

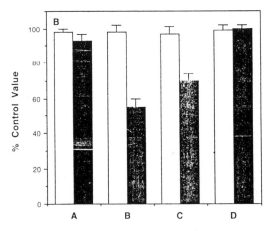

Fig 2. Neurotoxic effects of a prion protein fragment (PrP 106–126). The effect of 80 μM PrP 106–126 on wild type cerebellar cultures was determined by assessing the relative survival of treated cells as compared to control cultures not treated with the peptide and expressed as a percentage of control survival. Graph A: Cultures not treated with LLME (Sigma-Aldrich, Deisenhofen, Germany); graph B: cultures treated with LLME; open bars: control; solid bars: treatment with PrP 106–126; no co-culture (A), co-culture with normal microglia (B), with *Prnp[0/0]* microglia (C), with normal astro-cytes (D). This relative survival was determined by an MTT (Sigma) assay carried out following 10 days treatment with the peptide. PrP 106–126 treatment produced significant cell loss (Student's *t*-test, $P < 0.05$). LLME treatment of cerebellar cultures did not alter cell survival, but blocked the toxic effect of the peptide. Co-culturing LLME-treated cerebellar cells with astrocytes from normal mice did not alter survival of control cultures or restore the toxicity of PrP 106–126. Co-culturing untreated cerebellar cells with normal microglia significantly enhanced the toxicity of PrP 106–126. However, co-culturing cerebellar cells with microglia from *Prnp[0/0]* mice did not have this effect. Shown are mean survival (%) ± standard error. Methods: Cell cultures were prepared by dissociating cerebella from 6-day-old mice produced by interbreeding C57Bl/6J and 129/sv(ev) mice. The cerebella were dissociated in Hanks' medium (Biochrom, Berlin, Germany) containing 0.5% trypsin (Biochrom) and plated at $1-2 \times 10^6$ cells/cm² in 12 or 24 well trays coated with poly-D-lysine (50 μg/mL, Sigma-Aldrich, Deisenhofen, Germany). Cultures were maintained in Dulbecco's minimal essential medium (Biochrom) supplemented with 10% fetal calf serum 2 mM glutamine and 1% antibiotics (penicillin, streptomycin) and kept at 37 °C with 10% CO_2 for 10 days. Microglia were prepared from dissociating cerebral cortices [40] of newborn normal or *Prnp[0/0]* mice. Six or seven cortices were trypsinized in 0.5% trypsin and plated in 75-cm² culture flasks (Falcon Becton-Dickinson, Heidelberg, Germany). The cultures were maintained at 37 °C with 10% CO_2 for 14 days until glial cultures were confluent. Microglia were isolated from these cultures by shaking cultures at 260 rpm for 2 h. Astrocytes were prepared from the same glia preparations after most microglia had been removed by shaking. Further trypsination of adherent cells and preplating for 30 min to remove remaining microglia gave pure astrocyte cultures. Peptides used in these experiments were PrP 106–126 with the sequence KTNMKHMAGAAAAGAVVG-GLG, and as a control Pr P106–126 scrambled which consists of the same amino acids as PrP 106–126, but in a random order NGAKALMGGHGATKVMVGAAA. Peptides were synthesized on a Milligan 9050 peptide synthesizer and added to cultures at 2-day intervals from initial plating day until the 8th day. The peptide concentration was maintained at 80 μM. Cell survival was determined on day 10. MTT was diluted to 200 μM in Hanks' solution (Biochrom) and added to cultures for 2 h at 37 °C. The MTT formazan product was released from cells by addition of dimethylsulphoxide (Sigma) and measured at 570 nm in an Ultrospec III spectrophotometer (Pharmacia Biotech, Freiburg, Germany). Relative survival in comparison to untreated controls could then be determined. LLME was applied to cultures 1 day after plating, at 5 mM. Two hours after application the medium was removed, and fresh medium with or without peptides was applied. Cerebellar cells were fixed on coverslips for co-culturing and microglia or astrocytes plated at 10^5 cells/cm² were prepared in 24-well cloning trays. Coverslips were then transferred to the cloning trays for experiments. Before MTT assays, the coverslips were removed and placed in fresh trays.

glia play a necessary but insufficient role in the neurotoxic effect of the peptide. Similarly, PrP^C expression is also necessary, but at least in a mixed primary cell culture system, insufficient.

The use of antioxidants such as vitamin E and N-acetyl-cysteine allowed us to demonstrate that oxidative stress is also involved in the toxic mechanism of PrP 106–126 [34]. Both these agents blocked the neurotoxicity of the peptide on wild-type cerebellar cultures. The implication of this result is that the peptide PrP 106–126 induces the generation of oxidative substances by cells present in the mixed cell culture. Alternatively or in addition there may be a lowering of neuronal resistance to the effects of reactive oxygen species (ROS) or inhibition of defensive enzymes.

Microglia already implicated in the toxic mechanism are a major source of ROS when activated [38]. It has been shown that β-amyloid stimulates microglia to release tumor necrosis factor α (TNFα) and superoxide radicals [39]. Substances released from activated microglia such as ROS are often short-lived in solution and hard to measure. However, superoxide and nitric oxide, which both can be released from microglia and have destructive effects, can react to form a more stable compound, nitrite [40]. Microglia isolated from the cerebral cortex of wild-type mice were exposed to PrP 106–126 for 4 days and the supernatant was assayed for nitrite. First, the supernatant was treated with hydroxyammonium chloride to react all superoxide present to nitrite. Measurement of nitrite production showed that PrP 106–126 stimulates microglia to a high degree. In the presence of PrP 106–126, nitrite production was greatest from wild-type mouse microglia, but microglia from $Prnp^{0/0}$ mice could also be stimulated to release significant levels. This indicates that microglia release ROS upon activation by PrP 106–126. The role of microglia within the toxic mechanism of PrP 106–126 is thus likely to be one that induces oxidative stress on neurons by the release of substances that are themselves toxic to neurons.

PrP 106–126 stimulates microglia to release oxidative substances regardless of whether they express PrP^C. Activation of $Prnp^{0/0}$ microglia to release ROS, albeit to a lesser extent, suggests that they should also be able to kill neurons. As indicated earlier, $Prnp^{0/0}$ microglia can restore the toxic effect of PrP 106–126 to wild-type cerebellar cell cultures treated beforehand with LLME to remove their resident microglia. This indicates that these $Prnp^{0/0}$ microglia activated by PrP 106–126 can also be neurodestructive, suggesting that PrP^C expression by neurons makes them more sensitive to oxidative stress. We investigated this by subjecting cerebellar cultures from wild-type and $Prnp^{0/0}$ mice to oxidative stress from an alternative source of oxygen radicals. Xanthine oxidase produces oxygen radicals in the presence of xanthine. By replacing PrP 106–126-activated microglia with xanthine (ie, using microglia-reduced cultures), we showed that, in fact, $Prnp^{0/0}$ cerebellar cells are more sensitive to oxidative stress than wild-type cells (table I).

Cerebellar cells treated to reduce their microglial content are resistant to PrP 106–126 toxicity, but when a source of oxidative stress is added to the culture at the same time as PrP 106–126, an enhanced cell destruction is seen, greater than that induced by oxidative stress alone. Xanthine oxidase toxicity to wild-type cells was thus enhanced by the addition of 80 μM PrP 106–126 (table I). The toxicity of free radicals produced by xanthine oxidase on $Prnp^{0/0}$ cerebellar cells is not enhanced by PrP 106 126. This result therefore ties together the two previous findings. PrP 106–126 is neurotoxic as a consequence of at least two factors. First, oxidative stress is induced in the culture by the effect of PrP 106–126 on microglia. Second, PrP 106–126 is neurotoxic only to PrP^C-expressing cells. This second effect of PrP 106–126

Table I. The role of oxidative stress in PrP 106–126 neurotoxicity.

Treatment	Mouse strain	
	Normal	PrnP$^{0/0}$
Control	100 ± 4	100 ± 2
Xanthine oxidase	90 ± 5	75 ± 4
Xanthine oxidase with 80 μM PrP 106–126 scrambled	92 ± 5	77 ± 2
Xanthine oxidase with 80 μM PrP 106–126	65 ± 5	79 ± 3

Cultures of normal or *Prnp$^{0/0}$* cerebellar cells were treated with LLME to destroy microglia. The effect of 80 μM PrP 106–126 on these cell cultures was then determined in the presence of a known oxygen radical producer, xanthine oxidase. A concentration of xanthine oxidase was used which had a slight effect on the survival of normal LLME-treated cerebellar cells as compared to controls. The effect of the same concentration of xanthine oxidase on *Prnp$^{0/0}$* cerebellar cells is significantly greater (Student's *t*-test, $P < 0.05$), suggesting they are less resistant to oxidative stress. PrP 106–126 (80 μM) did not kill LLME-treated cerebellar cells (see fig 2); however, it greatly enhanced the effect of xanthine oxidase on LLME-treated cerebellar cells from normal mice, but not those from *Prnp$^{0/0}$* mice. The scrambled peptide applied at 80 μM together with xanthine oxidase does not have this effect. This suggests that PrP 106–126 treatment greatly enhances susceptibility to oxidative stress only to those cells expressing PrP^C. Methods: Cerebellar cultures, LLME treatment and survival assay were as described in figure 2. Cultures treated with xanthine oxidase were exposed to 5 μU of xanthine oxidase (Boehringer) in the presence of 5 μM xanthine for 24 h. Xanthine oxidase treated cultures were also co-exposed to 80 μM PrP 106–126 for the same time interval after which MTT survival assays were performed. Shown are means and standard errors.

results in reduced cellular resistance to the oxidative substances released by the activated microglia.

These results suggest that a significant role in prion diseases may be played by microglia. Mice infected with scrapie also show significant activation of microglia [7]. There is also evidence that this microglial activation precedes neuronal death. At present there is some evidence for cytokine production in scrapie [41], but there is no evidence as yet for the increased generation of ROS. Nevertheless, it is highly likely that if microglia are activated in prion disease there would be increased production of ROS by those microglia. As in many neurodegenerative diseases, prevention of neuronal oxidative stress may be one way to reduce or prevent neuronal loss caused by PrP^{Sc}.

Acknowledgments

The authors thank C Weissmann, University of Zurich, for providing *Prnp$^{0/0}$* mice. We thank R Fischer, W Dröse, B Maruschak and B Lage for technical assistance as well as C Bunker for editing the manuscript. This study was supported by research grants from the Bundesministerium für Bildung, Wissenschaft, Forschung und Technologie, the Deutsche Forschungsgemeinschaft, the Wilhelm-Sander-Stiftung and the Fritz-Thyssen-Stiftung.

References

1 Searle J, Kerr JFR, Bishop CJ. Necrosis and apoptosis: distinct modes of cell death with fundamentally different significance. *Pathol Annu* 1982;17(2):229-59
2 Kerr JFR, Wyllie AH, Currie AR. Apoptosis: a basic biological phenomenon with wide-ranging implications in tissue kinetics. *Br J Cancer* 1972;26:239-57
3 Buja LM, Eigenbrodt ML, Eigenbrodt EH. Apoptosis and necrosis. Basic types and mechanisms of cell death. *Arch Pathol Lab Med* 1993;117:1208-14

4   Fairbairn DW, Carnahan KG, Thwaits RN, Grigsby RV, Holyoak GR, ONeill KL. Detection of apoptosis indu-ced DNA cleavage in scrapie-infected sheep brain. *FEMS Microbiol Lett* 1994;115:341-6

5   Hogan RN, Baringer JR, Prusiner SB. Progressive retinal degeneration in scrapie-infected hamsters: a light and electron microscopical analysis. *Lab Invest* 1981;44:34-42

6   Eikelenboom P, Rozemuller JM, Kraal G et al. Cerebral amyloid plaques in Alzheimer's disease but not in scra-pie-affected mice are closely associated with a local inflammatory process. *Virchows Arch [Cell Pathol]* 1991;60:329-36

7   Williams AE, Lawson LJ, Perry VH, Fraser H. Characterization of the microglial response in murine scrapie. *Neuropathol Appl Neurol* 1994;20:47-55

8   Migheli A, Cavalla P, Marino S, Schiffer D. A study of apoptosis in normal and pathologic nervous tissue after in situ end-labeling of DNA strand breaks. *J Neuropathol Exp Neurol* 1994;53:606-16

9   Müller WEG, Ushijima H, Schroder HC et al. Cytoprotective effect of NMDA receptor antagonists on prion protein (PrionSc)-induced toxicity in rat cortical cell cultures. *Eur J Pharmacol* 1993;246:261-7

10 Forloni G, Angeretti N, Chiesa R et al. Neurotoxicity of a prion protein fragment. *Nature* 1993;362:543-6

11 Giese A, Groschup MH, Hess B, Kretzschmar HA. Neuronal cell death in scrapie-infected mice is due to apop-tosis. *Brain Pathol* 1995;5:213-21

12 Gold R, Schmied M, Rothe G et al. Detection of DNA fragmentation in apoptosis: application of in situ nick translation to cell culture systems and tissue sections. *J Histochem Cytochem* 1993;41:1023-30

13 Gavrieli Y, Sherman Y, Ben-Sasson SA. Identification of programmed cell death in situ via specific labeling of nuclear DNA fragmentation. *J Cell Biol* 1992;119:493-501

14 Bruce ME, McBride PA, Jeffrey M, Scott JR. PrP in pathology and pathogenesis in scrapie-infected mice. *Mol Neurobiol* 1994;8:105-12

15 Fraser H. Diversity in the neuropathology of scrapie-like diseases in animals. *Br Med Bull* 1993;49:792-809

16 Bruce ME. Scrapie strain variation and mutation. *Br Med Bull* 1993;49:822-38

17 Bruce ME, McConnell I, Fraser H, Dickinson AG. The disease characteristics of different strains of scrapie in Sinc congenic mouse lines: implications for the nature of the agent and host control of pathogenesis. *J Gen Virol* 1991;72:595-603

18 Buyukmihci NC, Goehring-Harmon F, Marsh RF. Photoreceptor degeneration during infection with various strains of the scrapie agent in hamsters. *Exp Neurol* 1987;97:201-6

19 Buyukmihci NC, Goehring-Harmon F, Marsh RF. Photoreceptor degeneration in experimental transmissible mink encephalopathy of hamsters. *Exp Neurol* 1987;96:727-31

20 Kozlowski PB, Moretz RC, Carp RI, Wisniewski HM. Retinal damage in scrapie mice. *Acta Neuropathol (Berl)* 1982;56:9-12

21 Hogan RN, Kingsbury DT, Baringer JR, Prusiner SB. Retinal degeneration in experimental Creutzfeldt-Jakob disease. *Lab Invest* 1983;49:708-15

22 Sloviter RS, Dean E, Neubort S. Electron microscopic analysis of adrenalectomy-induced hippocampal granule cell degeneration in the rat: apoptosis in the adult central nervous system. *J Comp Neurol* 1993;330:337-51

23 Wyllie AH, Morris RG, Smith AL, Dunlop D. Chromatin cleavage in apoptosis: association with condensed chromatin morphology and dependence on macromolecular synthesis. *J Pathol* 1984;142:67-77

24 McBride PA, Bruce ME, Fraser H. Immunostaining of scrapie cerebral amyloid plaques with antisera raised to scrapie-associated fibrils (SAF). *Neuropathol Appl Neurobiol* 1988;14:325-36

25 Bruce ME, Dickinson AG, Fraser H. Cerebral amyloidosis in scrapie in the mouse: effect of agent strain and mouse genotype. *Neuropathol Appl Neurobiol* 1976;2:471-8

26 Scott JR, Fraser H. Degenerative hippocampal pathology in mice infected with scrapie. *Acta Neuropathol (Berl)* 1984;65:62-8.

27 Bursch W, Paffe S, Putz B, Barthel G, Schulte-Hermann R. Determination of the length of the histological stages of apoptosis in normal liver and in altered hepatic foci of rats. *Carcinogenesis* 1990;11:847-53

28 Thompson CB. Apoptosis in the pathogenesis and treatment of disease. *Science* 1995;267:1456-62

29 Lassmann H, Bancher C, Breitschopf H et al. Cell death in Alzheimer's disease evaluated by DNA fragmenta-tion in situ. *Acta Neuropathol (Berl)* 1995;89:35-41

30 Brown DR, Herms J, Kretzschmar HA. Mouse cortical cells lacking cellular PrP survive in culture with a neuro-toxic PrP fragment. *Neuroreport* 1994;5:2057-60

31 Büeler H, Fischer M, Lang Y et al. Normal development and behaviour of mice lacking the neuronal cell-surface PrP protein. *Nature* 1992;356:577-82

32 Tobler I, Gaus SE, Deboer T et al. Altered circadian activity rhythms and sleep in mice devoid of prion protein. *Nature* 1996;380:639-42

33 Büeler H, Aguzzi A, Sailer A et al. Mice devoid of PrP are resistant to scrapie. *Cell* 1993;73:1339-48

34 Brown DR, Schmidt B, Kretzschmar HA. Role of microglia and host prion protein in neurotoxicity of prion protein fragment. *Nature* 1996;380:345-7

35 Moser M, Colello RJ, Pott U, Oesch B. Developmental expression of the prion protein gene in glial cells. *Neuron* 1995;14:509-17

36 Kretzschmar HA, Prusiner SB, Stowring LE, DeArmond SJ. Scrapie prion proteins are synthesized in neurons. *Am J Pathol* 1986;122:1-5

37 Brown DR, Schmidt B, Kretzschmar HA. A neurotoxic prion protein fragment enhances proliferation of microglia but not astrocytes in culture. *Glia* 1996 (in press)

38 Giulian D. Ameboid microglia as effectors of inflammation in the central nervous system. *J Neurosci Res* 1987;18:155-71

39 Meda L, Cassatella MA, Szendrei GI et al. Activation of microglial cells by β-amyloid protein and interferon-gamma. *Nature* 1995;374:647-50

40 Beckman JS, Beckman TW, Chen J, Marshall PA, Freeman BA. Apparent hydroxyl radical production by peroxynitrite: Implications for endothelial injury from nitric oxide and superoxide. *Proc Natl Acad Sci USA* 1990;87:1620-4

41 Williams AE, van Dam AM, Man-A-Hing WKH, Berkenbosch F, Eikelenboom P, Fraser H. Cytokines, prostaglandins and lipocorin-1 are present in the brains of scrapie-infected mice. *Brain Res* 1994;654:200-6

Transmissible Subacute Spongiform Encephalopathies:
Prion Diseases
L Court, B Dodet, eds
© 1996, Elsevier, Paris

Mechanisms of scrapie-induced neuronal cell death

Janet R Fraser[1], William G Halliday[2], Deborah Brown[1], Pavel V Belichenko[3], Martin Jeffrey[2]

[1]Neuropathogenesis Unit, Institute for Animal Health, Ogston Building, West Mains Road, Edinburgh EH9 3JF; [2]CVL Lasswade Laboratory, Bush Estate, Penicuik EH26 0SA, UK; [3]Brain Research Institute, Russian Academy of Medical Science, per Obuka, 5, Moscow 103064, Russian federation

Summary – The extent of neuronal loss resulting from scrapie infection is not known. In some murine models, severe loss develops in precise neuroanatomical areas with specific scrapie strain and mouse genotype combinations, including dorsal lateral geniculate nucleus (dLGN) neurons, hippocampal pyramidal cells and photoreceptor cells in the retina. Morphometric techniques have revealed substantial loss in the dLGN, a major retinal projection area, following intraocular infection with Me7 scrapie. Ultrastructural studies of the dLGN and hippocampus reveal rare cells whose morphology suggests an apoptotic mechanism of neuron death. There is growing evidence that apoptosis is the mechanism of neuron loss in scrapie, and we confirm here the recent demonstration of labelled apoptotic cells in the retina of mice with a primary scrapie retinopathy. Studies following intracerebral infection with Me7 scrapie of a different mouse strain reveal a major loss of CA1 hippocampal pyramidal cells. Golgi staining shows loss of dendritic spines, aberrant dendritic morphology and vacuolation preceding neuronal death, which may be mediated by the accumulation of abnormal prion protein.

Neuronal loss in the spongiform encephalopathies

The pathological indices of the transmissible spongiform encephalopathies are frequently quoted as vacuolar degeneration, glial hypertrophy and neuronal loss. However, although neuronal loss can be confidently identified in particular neuroanatomical locations, its full extent is not known. The first evidence for an undetected neuronal loss in the spongiform encephalopathies came from a morphometric analysis of the vestibular nucleus in bovine spongiform encephalopathy (BSE)-affected cattle, which revealed a mean 50% loss of neurons in this nucleus [1]. This led to studies in mice infected by the intraocular route with Me7 scrapie; this route targets infectivity and subsequent pathology to the central projections of the optic nerve, including the dorsal lateral geniculate nucleus (dLGN) [2, 3]. In this model, which has an incubation period of around 240 days, the first significant loss of dLGN neurons was found in two out of five mice at 150 days, and in terminal mice the number of neurons fell from a mean of around 22,000 to less than 2,000 [4, 5]. Despite this massive cell loss, 'dying' cells could not be identified by light microscopy, and in a substantial electrophysiological study of this nucleus, no abnormalities of neuron function were found [6]. This suggests that the cells were being cleared rapidly from the neuropil. At an ultrastructural level, cells were very rarely seen with irregular cytoplasmic margins, vacuolation of the endoplasmic reticulum and condensed nuclear chromatin in the dLGN and also in the hippocampus of intracerebrally

Key words: scrapie / neuronal loss / apoptosis

Me7-infected mice (fig 1). These changes are characteristic of apoptotic cells, or cells which are undergoing 'programmed cell death' [7]. This mechanism of cell death would account for the apparent absence of 'dying' cells, as cells undergoing this process can be cleared in a matter of hours. However, as the rate of cell loss in the dLGN is around 200–300 cells per day, visualizing these cells in sections using apoptotic markers is clearly difficult.

Scrapie-induced retinopathy

One model of scrapie-induced neuron loss has a rate of between 20,000 and 30,000 cells per day: the loss of photoreceptor cells in the retina in certain strains of mice infected intra-cerebrally with 79A scrapie [8, 9]. In this model, which has an incubation period of around 170 days, photoreceptor cells are lost from around 120 days post-infection, and in the terminal mouse retina, the photoreceptor layer, normally 10–12 cells thick (fig 2a), is reduced to one to two cells (fig 2b). Electroretinography reveals a reduction of b-wave amplitude coincident with the onset of cell loss, which, coupled with ultrastructural studies, suggests a deficit in disc membranogenesis [10].

Evidence for apoptosis of photoreceptors

We looked for evidence of apoptosis in a retinopathy model using the Apoptag in situ kit (Oncor), which detects digoxigenin-labelled genomic DNA in thin sections of fixed tissue. C3H/LaDk mice, which have the *rd* gene for retinal dystrophy, were included as a positive control as the photoreceptors are progressively lost through apoptosis during the 4 weeks after birth. Four VM/Dk-*Sinc^s7* mice were intracerebrally infected with 20 μL of a 1% brain homo-genate from a mouse terminally affected with Me7 scrapie; a second group of four mice were injected with normal brain homogenate. The eyes from terminal VM/Dk-*Sinc^s7* mice, normal brain-injected controls and 11-day-old C3H/LaDk mice were fixed in Davidson's solution, embedded in paraffin and 6-mm sections cut. The kit was used as directed, and labelled cells were visualized with diaminobenzidine. A small but consistent number of cells were positively labelled in the VM/Dk-*Sinc^s7* photoreceptor layer (fig 3a); no labelling was seen in the retina from the normal brain-injected control mouse (fig 3b). There was substantial labelling in the photoreceptor layer in the *rd* mouse used as a positive control (fig 3c). This result confirms a similar observation by Giese et al [11], and adds to the accruing evidence for an apoptotic mechanism of neuron death operating in the spongiform encephalopathies [12, 13, 14]. Neuronal autophagy, which can be a feature of apoptotic cell death, has been reported in experi-mental murine models of Creutzfeldt-Jakob disease [15], and we have shown that autophagic vacuoles are a conspicuous finding at ultrastructural level in mice infected with BSE [16].

Morphological changes in dying neurons

In a third model of scrapie-induced neuron loss, pathological changes preceding cell death can be identified. In the F_1 cross between C57Bl/FaBtDk and VM/Dk mice infected intracerebrally with Me7 scrapie, a substantial loss of hippocampal pyramidal neurons is found in terminal mice [17]. In this model, with an incubation period of around 210 days, Golgi labelling of 70 μm sections from terminal mice reveals stunted, distorted dendrites, vacuoles within the dendrites and a profound loss of dendritic spines (fig 4a) compared to normal brain-injected controls (fig 4b). Studies are in progress to define the relationship between spine loss, abnormal prion protein deposition and other indices of pathology.

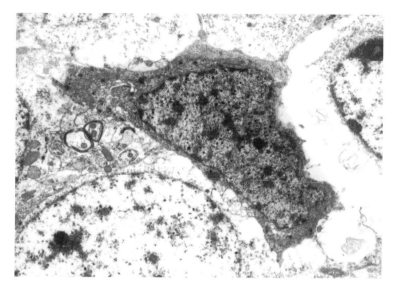

Fig 1. Electron micrograph of a neuron in CA1 of the hippocampus of a terminal mouse which had been intracerebrally infected with Me7 scrapie, showing features consistent with apoptosis: irregular cytoplasmic margins, vacuolation of the endoplasmic reticulum and condensed nuclear chromatin ($\times$ 112,500).

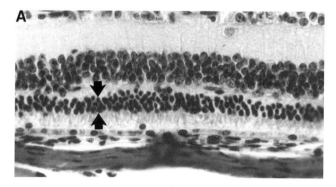

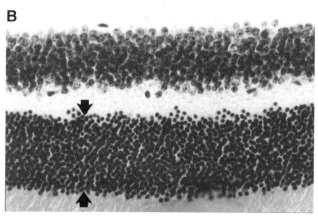

Fig 2. Photoreceptor cell loss in the retina of a mouse terminally-affected with 79A scrapie (**A**), compared with an age-matched normal brain-injected control (**B**). The arrows show the photoreceptor layer (haematoxylin and eosin, $\times$ 500).

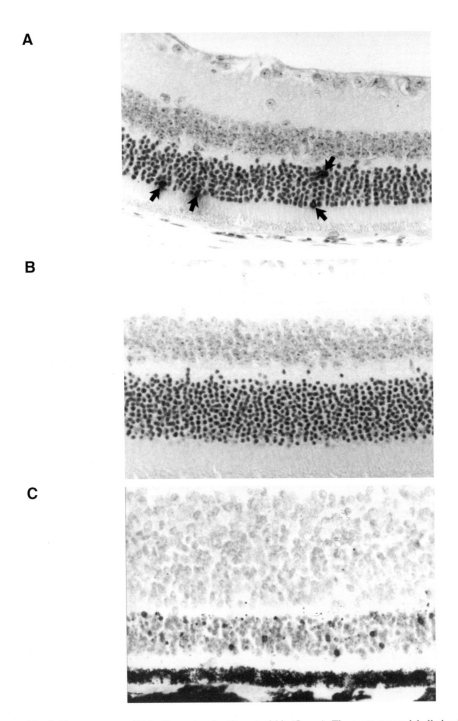

Fig 3. Photoreceptor cell labelling using the 'Apoptag' kit (Oncor). There are sparse labelled cells (arrows) in the terminal VM/Dk-*Sinc^{s7}* mouse with scrapie-induced retinopathy (**A**), no labelling in the normal brain-injected control (**B**) and substantial cell labelling in an 11-day-old C3H/LaDk mouse with retinal dystrophy (**C**) (methyl green counterstain, × 500).

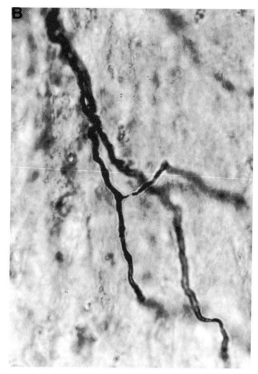

Fig 4. Golgi staining of CA1 pyramidal cell dendrites in a terminal Me7 scrapie-infected mouse (**A**), and in a normal brain-injected control (**B**), showing profound loss of dendritic spines (× 1,250).

Conclusion

In conclusion, we have shown that substantial neuronal loss occurs within the transmissible spongiform encephalopathies and can remain undetected unless revealed by morphometric techniques. There is now considerable evidence from several sources, including our own studies, implicating an apoptotic mechanism in the death of these cells.

References

1 Jeffrey M, Halliday WG, Goodsir CM. A morphometric and immunohistochemical study of the vestibular complex in bovine spongiform encephalopathy. *Acta Neuropathol (Berl)* 1992;84:651-7
2 Fraser H, Dickinson AG. Targeting of scrapie lesions and spread of agent via the retino-tectal projection. *Brain Res* 1985;346:32-41
3 Scott JR, Davies D, Fraser H. Scrapie in the central nervous system: neuroanatomical spread of infection and control of pathogenesis. *J Gen Virol* 1992;7:1637-44
4 Fraser JR, Jeffrey M, Halliday WG, Fowler N, Goodsir C, Brown D. Early loss of neurons and axon terminals in scrapie-affected mice revealed by morphometry and immunocytochemistry. *Mol Chem Neuropathol* 1995;24:245-9
5 Jeffrey M, Fraser JR, Halliday WG, Fowler N, Goodsir CM, Brown DA. Early unsuspected neuron and axon terminal loss in scrapie-infected mice revealed by morphometry and immunocytochemistry. *Neuropathol Appl Neurobiol* 1995;21:41-9
6 Black CJ. An electrophysiological investigation of normal and scrapie-infected dorsal lateral geniculate neurones in vitro. PhD thesis, Faculty of Medicine, University of Edinburgh, Edinburgh, UK, 1995

7 Kerr JFR, Wyllie AH, Currie AR. Apoptosis: a basic biological phenomenon with wide-ranging implications in tissue kinetics. *Br J Cancer* 1972;26:239-57

8 Foster JD, Fraser H, Bruce ME. Retinopathy in mice with experimental scrapie. *Neuropathol Appl Neurobiol* 1986;12:185-96

9 Foster JD, Davies D, Fraser H. Primary retinopathy in scrapie in mice deprived of light. *Neurosci Lett* 1986;72:111-4

10 Curtis R, Fraser H, Foster JD, Scott JR. The correlation of electroretinographic and histopathological findings in the eyes of mice infected with the 79A strain of scrapie. *Neuropathol Appl Neurobiol* 1989;15;75-89

11 Giese A, Groschaup MH, Hess B, Kretzschmar A. Neuronal cell death in scrapie-infected mice is due to apoptosis. *Br Pathol* 1995;5:213-21

12 Forloni et al. Neurotoxicity of a prion protein fragment. *Nature* 1993;362:543-6

13 Fairbairn DW, Carnahan KG, Thwaits RN, Grigsby RV, Holyoak GR, O'Neill KL. Detection of apoptosis induced DNA cleavage in scrapie-infected sheep brain. *FEMS Microbiol Lett* 1994;115:341-6

14 Lucassen PJ, Williams A, Chung WCJ, Fraser H. Detection of apoptosis in murine scrapie. *Neurosci Lett* 1995;198:185-8

15 Boellaard JW, Scholte W, Tateishi J. Neuronal autophagy in experimental Creutzfeldt-Jakob disease. *Acta Neuropathol (Berl)* 1989;78:410-8

16 Jeffrey M, Scott JR, Williams A, Fraser H. Ultrastructural features of spongiform encephalopathy transmitted to mice from three species of Bovidae. *Acta Neuropathol (Berl)* 1992;84:559-69

17 Scott JR, Fraser H. Degenerative hippocampal pathology in mice infected with scrapie. *Acta Neuropathol (Berl)* 1984;65:62-8

Transmissible Subacute Spongiform Encephalopathies:
Prion Diseases
L Court, B Dodet, eds
© 1996, Elsevier, Paris

Neuronal death in mice infected with the bovine spongiform encephalopathy agent

Olivier Robain[1], Corinne Lasmézas[2], Jean-Guy Fournier[3], Jean-Philippe Deslys[2]

[1]U29 and [3]U153 INSERM, Hôpital Saint Vincent de Paul, 74 avenue Denfert-Rochereau, Paris 75674 Cedex 14; [2]Service de Neurovirologie, Commissariat à l'Énergie atomique, BSV/DMR/CRSSA, 60-68 avenue Division Leclerc, BP 6, 92265 Fontenay-aux-Roses, France

Summary – Neuronal death has been studied in mice inoculated intracerebrally with the bovine spongiform encephalopathy (BSE) agent. After the primary passage of BSE, the main histological change consisted of selective neuronal death associated with mild gliosis and only a very few vacuoles of spongiosis. Terminal transferase uridine-triphosphate nick end labeling showed characteristic, selective labeling of the nuclei of degenerating neurons. Electron microscopy showed condensation and marginalization of the chromatin at the nuclear margin that are typical of apoptosis. After the second and third passages of BSE, neuronal death was associated with significant gliosis and spongiosis. These findings emphasize the major role of neuronal death in the development of BSE.

Introduction

Scrapie in sheep, Creutzfeldt-Jakob disease in humans and bovine spongiform encephalopathy (BSE) are members of a group of transmissible encephalopathies. They were originally identified by the distribution and morphology of the lesions in the central nervous system (CNS), that is, a vacuolar aspect of the neuropil and neuronal perikaryon in conjunction with glial changes. A decrease in neuronal counts has also been described and is particularly evident at the terminal stage of the disease. In some circumstances, neuronal degeneration may be the most prominent if not the only histological change. In the present study, we looked at the occurrence of neuronal death in the brain of animals that were infected with the BSE agent.

Materials and methods

Animals

C57 BL mice were injected intracerebrally with 0.02 mL of 25% brain homogenate prepared from cows with BSE.

Histology

A histological study was done in three mice after primary inoculation, in three other mice after secondary ($n = 2$) and tertiary ($n = 1$) passages, and in three normal uninjected mice. After the mice were killed by decapitation, one cerebral hemisphere was used for biochemical study, while the other, including the cerebellum, was fixed for 24 h in modified Karnovsky's solution (1% paraformaldehyde–1% glutaraldehyde in 0.15 M phosphate buffer at pH 7.4). Coronal sections (200-μm thick) of the whole hemisphere and sagittal sectioning of the cerebellum were made using a vibratome. Small areas of the

Key words: *bovine spongiform encephalopathy / neuronal death / apoptosis / transmission*

hippocampus and cerebellum were selected. After 1 h of post–fixation in 2% osmic acid, they were stained *en bloc* with uranyl acetate and embedded in Araldite. Semi-thin sections (1-μm thick) were stained with toluidine blue. Ultra-thin sections comprising the CA1 pyramidal area of the hippocampus or the cerebellar vermis were stained with uranyl acetate and lead citrate. Sections were then examined under a Philips CM10 electron microscope.

Terminal transferase uridine-triphosphate nick end labeling (TUNEL) was performed for the detection of DNA fragmentation: the semi-thin sections were incubated for 15 min at 37 °C in the transferase solution (75 μ/mL, Boehringer Manheim, Germany) containing 200 mM potassium cacodylate (pH 6.5), 25 mM Tris HCl, 10 mM β-mercaptoethanol, 10 mM $CaCl_2$, 50 mg/mL bovine serum albumin (BSA) and 0.1 nM Bio-16-dUTP (Enzo Diagnostics, New York, NY, USA). After rinsing in distilled water, sections were incubated in 0.1 M phosphate buffer saline (PBS) pH 7.4 containing 3% BSA for 15 min, then 1 h more with the antibiotin antibody diluted 1/100. After washing in PBS, the sections were incubated for 1 h with goat anti-rabbit immunoglobulin G coupled to colloidal gold (5 nm diameter) diluted 1/50. After washing in PBS and distilled water, sections were incubated in a silver enhancement solution (Biocell kit) for 10 min. Finally, they were examined without mounting medium.

Results

After primary inoculation, all the animals developed a neurological disease and displayed hindlimb paresis and ataxia. The incubation period after primary inoculation was variable and extended from 368 to 719 days. After the second and third passages, the incubation time was shorter (167 ± 2 days, $n = 3$), reflecting the adaptation of the BSE strain to mouse species.

Histological study

Histological changes were observed in several areas of the brain of infected mice, but were more apparent in the cerebellum; therefore, we describe only changes in this area.

After the primary passage, many scattered degenerating neurons were visible. The degeneration was conspicuous in the Purkinje cell layer and affected about 30% of the Purkinje cells. In semi-thin sections stained with toluidine blue, degenerated neurons appeared shrunken and dark (fig 1A) with an irregular outline (fig 1B). Neuronal lesions were rare in the internal granular layer (< 1%); however, few dark, shrunken granule cells were visible (fig 1B). Mild gliosis of Bergman's cells, with a swollen appearance of glial nuclei, was also visible (fig 1B).

After primary inoculation, very few vacuoles of spongiosis were encountered (less than one in each field at a magnification of × 10).

A similar pattern of degeneration was also observed in Purkinje cells after the second and third passages. Reactive astrocytes were either dispersed in the Purkinje cell layer or clustered to replace degenerating neurons.

Gliosis made up of protoplasmic astrocytes was present within the Purkinje cell layer. Vacuoles of spongiosis were visible both in the internal granular and Purkinje cell layers.

The TUNEL method shows characteristic, selective labeling of the nuclei of degenerating Purkinje cells (fig 2). There is also a non specific staining of cellular debris. Labeled cells were only occasionally found in the internal granular layer.

Electron microscopic examination of degenerating neurons revealed two types of alterations. The first, most frequent aspect of neuronal changes involved the nucleus and consisted of dark aggregates made up of clumps of chromatin located at the inner nuclear margin. In these degenerating neurons, the cytoplasm appeared normal, although more electron dense; however, the organelles had a normal morphological appearance (fig 3).

A second aspect visible by electron microscope affected a smaller percentage of neurons. In this case, the nucleus appeared unaffected, but major alterations occurred in the cytoplasm, ie,

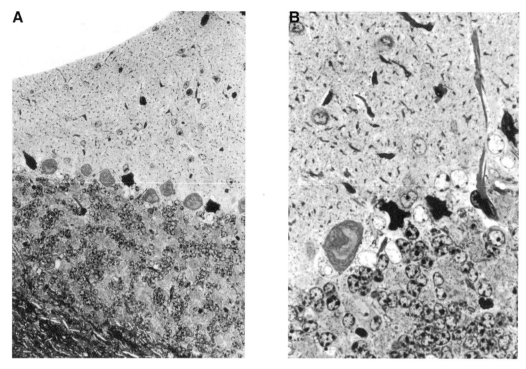

Fig 1. Semi-thin sections of the cerebellar vermis showing selective neuronal death of Purkinje cells after toluidine blue staining. A. Three degenerated Purkinje cells with dark staining both of the cytoplasm and nucleus are intermingled with six Purkinje cells of normal appearance (× 300). B. Degenerated Purkinje cells display an irregular outline. In addition, two dark cells are visible in the internal granular layer and probably reflect degenerating granule neurons. Glial cells intermingled within the Purkinje cell layer are also observed; they have a clear nucleus lightly lined with chromatin (× 750).

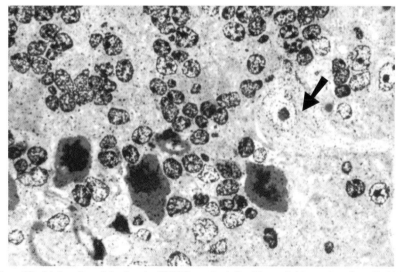

Fig 2. Labeling of Purkinje cell nuclei using the TUNEL method. On this semi-thin section, three Purkinje cells show a strong dot accumulation on their nucleus as compared to normal Purkinje cells (arrow) (× 750).

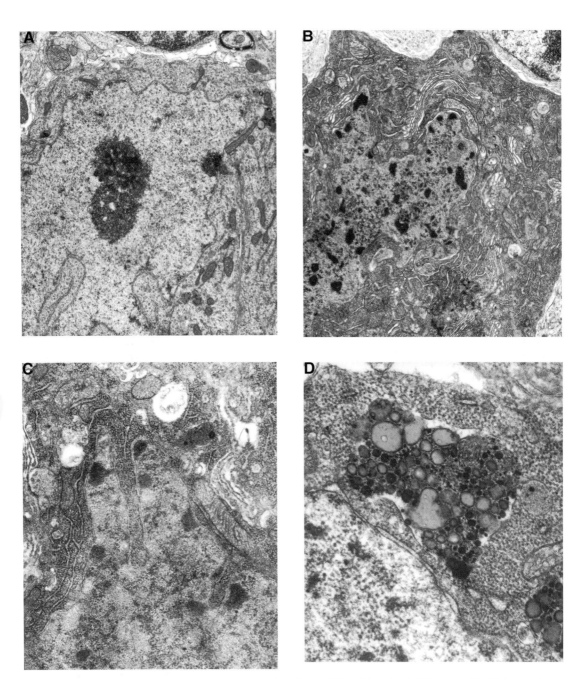

Fig 3. Ultrastructural aspect of normal (**A**) and degenerating (**B** and **C**) Purkinje cells. **A.** This normal Purkinje cell shows a clear nucleus with dispersed chromatin and an infolded nuclear membrane. The cytoplasm is electrolucent and contains normal organelles ($\times$ 8,900). **B.** A degenerated Purkinje cell with dark nucleus and cytoplasm. The chromatin is condensed and accumulated at the periphery of the nucleus. Morphologically normal mitochondria and Golgi apparatus are present in the dark cytoplasm ($\times$ 8,900). **C.** Dark aggregates of chromatin are located at the periphery along the nuclear membrane ($\times$ 21,000). **D.** Accumulation of round, clear and dark bodies in the cytoplasm of a Purkinje cell. Note the normal aspect of the nucleus ($\times$ 39,000).

vacuoles containing dense bodies or amorphous electron-dense material. Few typical apoptotic bodies were found in sections examined by electron microscopy (fig 4).

Vacuoles of spongiosis lying within dendritic processes both in the internal granular and Purkinje cell layers were visible in mice observed after the second and third passages, but only rarely in mice examined after the primary passage. Astrocytes filled with glial fibrils were more frequently visible in the Purkinje cell layer in mice analyzed after the second and third passages.

Discussion

The principal observation of this preliminary study is the occurrence of neuronal cell death without concomitant spongiosis in the brain of BSE-infected mice. In the cerebellum, Purkinje cells are more vulnerable, but other neuronal cell populations in various areas of the brain are also susceptible (Robain et al, in preparation).

Neuronal loss has already been observed in several types of transmissible spongiform encephalopathies and data were usually obtained by morphometric methods [1–6].

Our observations of DNA fragmentation by in situ end labeling (the TUNEL method) strongly suggest that neuronal loss results from apoptosis of these cells. When examined by electron microscopy, two types of cellular alterations were observed in degenerating neurons. Nuclear alterations (with clumps of chromatin at the nuclear margin) with moderate modifications of the cytoplasm were frequently observed. Such changes have been classified as apoptosis type 1 by Clarke [7]. The mechanisms involved in this type of apoptotic process are nuclear events such as endonuclease activation. In a few degenerating neurons, the nucleus appears normal, but the cytoplasm contains autophagic vacuoles. These cytoplasmic events have been identified as apoptosis type 2.

Interestingly, neuronal death was detected after primary transmission without spongiosis.

After the second and third passages, histological changes showed the classic pattern with spongiosis, gliosis and neuronal degeneration.

These data suggest that adaptation of the agent to the host is related to the development of vacuolar changes and gliosis, but that it is not necessary for neuronal death.

Acknowledgements

This work was supported by grant 94005 (Ministry of Health, France).

References

1   Gambetti P, Parchi P, Petersen RB, Chen SG, Lugaresi E. Fatal familial insomnia and familial Creutzfeldt-Jakob disease: clinical, pathological and molecular features. *Brain Pathol* 1995;5:43-51
2   Ghetti B, Dlouhy SR, Giaccone G, Bugiani O, Frangioni B, Farlow MR, Tagliavini F. Gerstmann-Sträussler-Scheinker disease and the Indiana kindred. *Brain Pathol* 1995;5:61-76
3   Jeffrey M, Halliday WG, Goodsir CM. A morphometric and immunohistochemical study of the vestibular complex in bovine spongiform encephalopathy. *Acta Neuropathol* 1992;84:651-7
4   Jeffrey M, Fraser JR, Halliday WG, Fowler N, Goodsir CM, Brown DA. Early unsuspected neuron and axon terminal loss in scrapie-infected mice revealed by morphometry and immunocytochemistry. *Neuropathol Appl Neurobiol* 1995;21:41-9
5   Kretschmar HA, Brown DR, Giese A, Herms J. Gliosis and cell death in prion disease. *Neuropathol Appl Neurobiol* 1996;22:8
6   Wells GAH, Wilesmith JW. The neuropathology and epidemiology of bovine spongiform encephalopathy. *Brain Pathol* 1995;5:91-103
7   Clarke PGH. Development of cell death: morphological diversity and multiple mechanisms. *Anat Embryol (Berl)* 1990;181:195-213

PATHOGENESIS

Transmissible Subacute Spongiform Encephalopathies:
Prion Diseases
L Court, B Dodet, eds
© 1996, Elsevier, Paris

Distinct subcellular localizations of PrP^C detected by immunoelectron microscopy: functional implications

Jean-Guy Fournier[1], Françoise Escaig-Haye[1], Thierry Billette de Villemeur[2], Olivier Robain[3]

[1]*Inserm U153, Institut de Myologie, Hôpital Pitié-Salpêtrière, 47 boulevard de l'Hôpital, 75651 Paris cedex 13; [2]Service de Neuropédiatrie, Hôpital Trousseau, 75012 Paris; [3]Inserm U29, Hôpital St-Vincent-de-Paul 75014, Paris, France*

Summary – Cellular protease-sensitive prion protein (PrP^C) is a membrane sialoglycoprotein synthesized in the central nervous system and extraneural tissues. Its abnormal disease-specific isoform (PrP^{Sc}) is found in the brains of subjects with transmissible spongiform encephalopathies. One major unanswered question concerns the physiological role of PrP^C, depending largely on the limited knowledge of its in vivo subcellular localization. Using a sensitive post-embedding electron microscopy technique, we reported that in hamster brain, synaptic *boutons* constituted the submicroscopic site where PrP^C was preferentially observed in conjunction with synaptophysin, indicating a possible role for PrP^C in neurotransmission. Outside the brain, PrP^C-expressing cells were seen in several hamster tissues, where the immunogold labeling was associated with plasma membrane, secretory granules and the Golgi apparatus. These distinct PrP^C subcellular sites, also observed in some human tissues, suggest a cell-specific role for PrP^C.

Introduction

Transmissible spongiform encephalopathies or prion diseases include several neurodegenerative disorders in animals (scrapie and bovine spongiform encephalopathy) and humans (kuru, Creutzfeldt-Jakob disease, Gertsmann-Sträussler-Scheinker disease). In humans, they are characterized by a long preclinical incubation period and occur in sporadic, iatrogenic and genetic forms. Biochemically, the hallmark is the accumulation in the brain tissue of a pathological disease-specific prion protein (PrP^{Sc}) [1, 2]. Although PrP^{Sc} copurifies with the transmissible agent, the infectivity of the protein itself remains to be demonstrated. PrP^{Sc} is produced by a post-translational modification of a normal cellular protease-sensitive prion protein isoform (PrP^C), encoded by a chromosomal single copy gene [3], the expression of which is not modified during the course of infection [4, 5]. PrP^C is a glycoprotein attached to membranes by a glycosyl-phosphatidylinositol anchor. From in vitro experiments, it has been proposed that a possible membrane site would be the neuron plasma membrane [6, 7]. However, the membrane target in vivo is unclear and the physiologic role of PrP^C remains unknown [8]. In order to gain insight into the function of PrP^C, we examined the subcellular

Key words: cellular prion protein / ultrastructural detection

site of PrP^C in the hamster brain and in various normal hamster and human tissues, using a sensitive immunogold ultrastructural method.

Materials and methods

Tissue fixation and embedding

Syrian hamsters were anesthetized with Nembutal and perfused transcardially with 4% paraformalde-hyde–0.1% glutaraldehyde in buffer (CaCl$_2$ 0.02 mM, sodium phosphate 0.1 M, pH 7.4). Several organs were removed and small fragments (1 mm^3) were processed for electron microscopy. For brains, 200-μm thick sections were cut with a vibratome, then the hippocampus was dissected and three distinct regions (CA1, CA3 and the dentate gyrus) were isolated and processed for embedding. Following serial dehydration in ethanol at low temperatures, sections were embedded in Lowicryl K4M methacrylate resin as previously described [9, 10].

Antibodies

The immunodetection study was conducted with a specific polyclonal anti-PrP antibody (Br-1) directed against the protease-resistant prion protein, PrP 27–30, isolated from the hamster brain inoculated with the 263K scrapie agent [2]. A monoclonal antibody against PrP 27–30 (3F4) [8] and two other rabbit polyclonal anti-PrP antibodies were used; one raised against PrP 27–30 (RO73) [11] and the other (P38) against the synthetic peptide (encompassing the amino acids 88–107).

Immunocytochemistry

Light microscopy
Semi-thin sections (1 μm) were deposited on a clean glass microscope slide covered with aminopropyl-silane. The sections were incubated for 30 min in phosphate buffered saline (PBS)–3% bovine serum albumin (BSA) PBS–3% BSA, drained a few seconds and incubated with the primary antibody (the time and conditions adapted for each antibody used were similar to those used for electron microscopy). After a rapid wash in PBS, the sections were incubated in the secondary antibody conjugated with 5-nm gold particles (Janssen, Amersham) diluted 1/50 in PBS–3% BSA. Following a further wash in PBS (5 min) and water (five times for 2 min each), the gold signal was visualized after silver intensification (Biocell silver-enhancing kit).

Electron microscopy
Thin sections (80 nm) were cut and retrieved on 600–mesh acetone-treated naked gold grids. After incubation in a blocking solution (0.1 M PBS, 1% BSA, pH 7) for 20 min, sections were placed on the surface of a drop of the blocking solution containing the primary antibody diluted 1/50–1/300 for 1 h at room temperature. With the RO73 antibody (1/200 dilution), polyclonal synthetic anti-PrP anti-body (1/25 dilution) and 3F4 monoclonal anti-PrP antibody (1/2 dilution), the sections were incubated overnight at 4 °C. The grids were then rinsed twice for 10 min in PBS buffer containing 0.05% Tween 20 and were kept for 1 h in contact with goat anti-rabbit or anti-mouse serum complexed with 10-nm gold particles (1/50 dilution) in PBS–1% BSA solution. Subsequently, the grids were rinsed in PBS–Tween 20 (0.05%) and distilled water twice for 10 min each. Sections were then stained with uranyl acetate for 20 min, rinsed twice for 5 min in distilled water and observed under a Phillips CM 10 electron microscope at 60 kV. The specificity of the immunogold signal was controlled by incubation in pre-immune serum or by omission of the first antibody. Tissue known to be negative for PrP^C (liver) was also included in the control. In addition, positive control under our conditions was assessed using scrapie-infected hamster brain in which it is known that amyloid plaques occur beneath the ependyma (fig 1a).

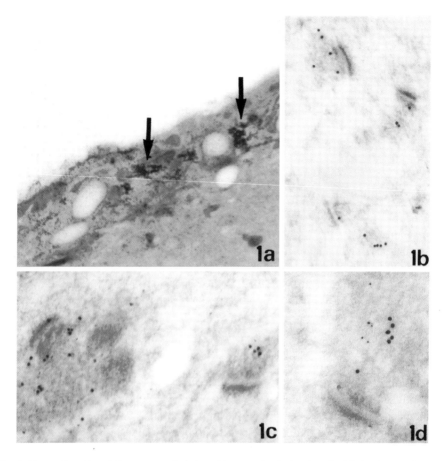

Fig 1. Semi-thin section (1 μm) from a scrapie-infected hamster brain incubated with the Br-1 anti-PrP antibody.
a. Visualization of amyloid plaques (arrows) in the subependymal zone after silver enhancement ($\times$ 800).
b. Immunoreaction for PrP^C over synaptic structures in the hamster hippocampus dentate gyrus ($\times$ 40,000).
c. Reactivity over the presynaptic region with a monoclonal anti-synaptophysin antibody ($\times$ 45,000). **d.** Double labeling showing co-localization of PrP^C (10-nm gold particles) and synaptophyin (5-nm gold particles) ($\times$ 95,000).

Results

Hamster brain

After careful examination of about 50 sections from each hippocampal region, an immunogold signal specific to PrP^C was found to be localized in the dentate formation, consistently associated with synaptic complexes present in the molecular layer. The detection of PrP was obtained with Br-1 and RO73 polyclonal anti-PrP antibodies, but at a higher frequency with the former (fig 1b).

Labeling was observed on the electron-dense zone reflecting the presence of synaptic vesicles. This region reacted with an anti-synaptophysin antibody (fig 1c). Double labeling of the same section showed co-localization of PrP^C and synaptophysin (fig 1d).

Extraneural tissues

The analysis of PrP^C in several non infected prion tissues has been performed on semi-thin sections. A specific signal was observed only in tissues in which cells strongly express PrP^C in specific structures.

This is particularly the case in human (fig 2a) and hamster stomach tissues (fig 2b) in which an immunopositive reaction obtained after silver enhancement was observed in epithelial mucous gland cells. On the luminal side of the gland, reactivity was sometimes observed; however, only when positive cells were present in the gland (fig 2a).

Submicroscopic localization of PrP^C

Three major subcellular sites were found to be immunoreactive with the anti-PrP antibodies used.

Secretory granules
The silver signal observed on semi-thin sections from human and hamster stomachs corresponds to the presence of PrP in secretory granules (figs 3 and 4). Labeling is concentrated on the clear zone of the granules in human tissue while, in the hamster, it appears more homogeneous. In hamster intestine and lung tissue (fig 5), labeling is also observed on secretory granules.

Golgi apparatus
Epitopes recognized by anti-PrP antibodies and associated with the Golgi apparatus were found to be present in human stomach and hamster intestine cells (fig 6).

Plasma membrane
The location of PrP^C at the plasma membrane was seen in human cells from kidney, lymph node and stomach (fig 7). In hamster tissues, plasma membrane was found to be immunoreactive for PrP^C in stomach epithelial and kidney proximal tubule cells (fig 8).

Discussion

We observed the subcellular localization of PrP^C in hamster brain and in hamster and human extraneuronal tissues with a sensitive post-embedding immunogold-electron microscopic protocol allowing the preservation of the antigen and thus a high sensitivity of detection [12]. Our observations confirm that PrP^C is widely distributed throughout the body.

In hamster brain, PrP^C is known to be highly concentrated in the hippocampus [8, 13]. Probing this brain formation with two anti-PrP antibodies, we observed that the dentate gyrus was the privileged region where antibodies reacted,. particularly in synapses localized in the molecular layer. Based upon ultrastructural criteria, strengthened by co-localization with synaptophysin, it can be assumed that PrP^C is present in synaptic *boutons* [14, 15]. Further analysis would be required to establish whether PrP^C is associated with the synaptic vesicle membrane in order to specify the role that PrP^C might play in neurotransmission. The accumulation of PrP^C at the synaptic level is in agreement with its detection in the neuropil of the rodent brain [16] as well as the fact that PrP^C is transported by rapid anterograde axonal flow [17]. It is important to note that, in other neurodegenerative diseases such as Alzheimer's and Huntington's, it has also been stated that the protein of interest could have a synaptic localization in the central nervous system [18, 19]; thus, the synapse may be considered the main target in the pathogenesis of these diseases [20, 21].

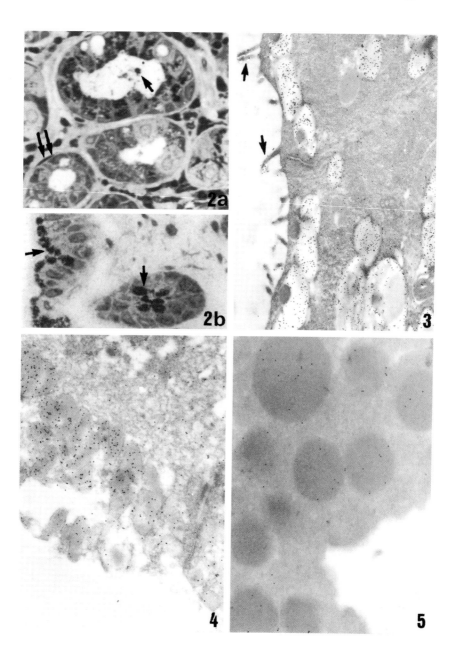

Fig 2. a. Semi-thin section of human stomach incubated with the Br-1 anti-PrP antibody. Silver staining specific to PrP^C associated with mucous gland cells. Silver deposition on the luminal side (arrow). Negative gland for PrP^C (double arrows). **b.** Semi-thin section of the hamster stomach incubated with the RO73 anti-PrP antibody. Silver label specific to PrP^C associated with apical epithelium and gland cells (arrows) (a and b: × 800).
Fig 3. PrP immunogold labeling over human stomach secretory granules and microvilli (arrows) (× 15,000).
Fig 4. Hamster stomach secretory granules immunopositive for PrP^C (× 20,000).
Fig 5. Hamster lung ultra-thin section incubated with the 3F4 monoclonal anti-PrP antibody. Non ciliated cell showing PrP^C immunogold labeling over secretory granules (× 20,000).

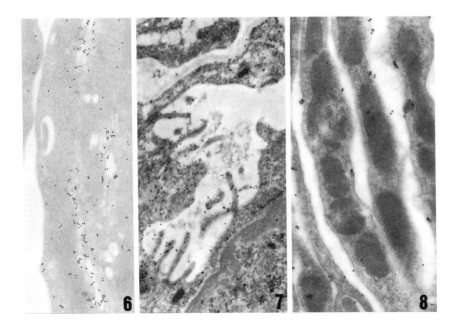

Fig 6. Golgi apparatus of a hamster intestinal cell reacting with the RO73 anti-PrP antibody ($\times$ 25,000).
Fig 7. Endothelial cell plasma membrane in human stomach showing PrP^C immunogold signal (Br-1 antibody) ($\times$ 25,000).
Fig 8. Immunogold detection of PrP^C over the plasma membrane of hamster kidney proximal tubule cells (Br-1 antibody) ($\times$ 30,000).

Outside the brain and depending on the tissue, some or a very few cells express PrP^C. Three major subcellular sites of PrP^C were observed, in agreement with the known secretory pathway of PrP^C: Golgi apparatus, secretory vesicles and granules, and plasma membrane [1, 22]. The latter site, considered to be the unique localization of PrP^C, has only been observed in certain cells such as those of the hamster kidney and the human spleen as well as human parietal and endothelial cells. The most striking finding of the present study is the presence of PrP^C in the secretory granules, especially well observed in the hamster and human digestive tracts. However, we are as yet unable to state whether PrP^C is associated with the granule membrane or is a secretion product. The fact that PrP has been detected in microvilli of epithelial cells and on the luminal side suggests that PrP^C is both a constituent of the granule membrane and a secretory product. It is known that PrP^C can be present under these two forms [23], and in vitro and in vivo PrP^C has been detected in extracellular fluid [6, 7, 24]. The morphological aspect of the PrP^C secretory function was reported in the non ciliated cells of the hamster lung [8], which we have been able to confirm.

Taken together, the present data argue for a cell-specific role of PrP^C. The different subcellular distributions suggest that PrP^C is involved in a post-Golgi vesicular membrane trafficking leading, according to the cell type, to the plasma membrane or the secretory organelle membrane, possibly adopting an exocrine secretory form.

One interesting point is the presence of PrP^C along the digestive tract, in view of the fact that epithelial cells of that tissue are in direct contact with the infectious agent during oral infection. To date, no information exists concerning the synthesis of the pathological form of PrP (PrP^Sc) in the gastrointestinal tissue. If the agent resides in epithelial cells, in the absence of PrP^Sc formation, it can be assumed that those cells are defective in cofactors required for the conversion of PrP^C into PrP^Sc (eg, the X element) [25].

The implication of the digestive tract in the pathogenesis remains to be elucidated. The presence of PrP^C at the digestive tract epithelial cell surface could promote the penetration of the agent into the organism, while the secreted form of PrP^C within the lumen could prevent its penetration by a molecular interaction with the PrP^Sc component of the agent.

Acknowledgments

We wish to thank Drs P Brown, R Rubenstein, S Prusiner and JL Laplanche for the gift of Br-1, 3F4, RO73 and P38 anti-PrP antibodies, respectively. The present work was supported by the Ministère de l'enseignement supérieur et de la recherche (ACC-SV- 10) and Assistance public (AOA 94005).

References

1 Prusiner SB. Molecular biology of prion diseases. *Science* 1991;252:1515-22
2 Brown P, Coker-Vann M, Pomeroy K et al. Diagnosis of Creutzfeldt-Jakob disease by Western blot identification of marker protein in human brain tissue. *N Engl J Med* 1986;314:547-51
3 Oesch B, Westaway D, Walchli M et al. A cellular gene encodes Scrapie PrP 27–30 protein. *Cell* 1985;40: 735-46
4 Kretzchmar HA, Prusiner SB, Stowring LE, DeArmond SJ. Scrapie prion proteins are synthesized in neurons. *Am J Pathol* 1986;122:1-5
5 Brown HR, Goller N, Rudelli G, Wolfe GC, Wisniewski HM, Robakis NK. The mRNA encoding the scrapie agent protein is present in a variety of non-neuronal cells. *Acta Neuropathol* 1990;80:1-6
6 Caughey B, Raymond GJ. The scrapie-associated form of PrP is made from a cell surface precursor that is both protease and phospholipase sensitive. *J Biol Chem* 1991;266:8217-23
7 Stahl N, Borchelt DR, Prusiner SB. Differential release of cellular and scrapie prion protein from cellular membranes by phosphatidylinositol-specific phospholipase C. *Biochemistry* 1990;29:5405-12
8 Bendheim PE, Brown HR, Rudelli RD. Nearly ubiquitous tissue distribution of the scrapie agent precursor protein. *Neurology* 1992;42:149-56
9 Escaig-Haye F, Grigoriev V, Peranzi G, Lestienne P, Fournier JG. Analysis of human mitochondria transcripts using electron microscopic in situ hybridization. *J Cell Sci* 1991;100:851-62
10 Escaig-Haye F, Grigoriev V, Sharova I, Rudneva V, Buckrinskaya A, Fournier JG. Ultrastructural localization of HIV1 RNA and core proteins. Simultaneous visualization using double immunogold labeling after in situ hybridization and immunocytochemistry. *J Submicrosc Cytol Pathol* 1992;24:437-43
11 Taraboulos A, Serban D, Prusiner SB. Scrapie prion proteins accumulate in the cytoplasm of persistently infected cultured cells. *J Cell Biol* 1990;110:2117-32
12 Carlemalm E, Villiger W, Hobot J, Acetarin J, Kellenberger E. Low temperature embedding with Lowicryl resins: two new formulations and some applications. *J Microsc* 1985;140:55-72
13 DeArmond S, Mobley W, DeMott D, Barry R, Beckstead J, Prusiner SB. Changes in the localization of brain prion proteins during Scrapie infection. *Neurology* 1987;87:1271-80
14 Kitamoto T, Shin RW, Doh-Ura K et al. Abnormal isoform of prion proteins accumulates in the synaptic structures of the central nervous system in patients with Creutzfeldt-Jakob disease. *Am J Pathol* 1992;140:1285-94
15 Fournier JG, Escaig-Haye F, Billette de Villemeur T, Robain O. Ultrastructural localization of cellular prion protein (PrP^C) in synaptic boutons of normal hamster hippocampus. *CR Acad Sci Paris Life Sci* 1995;318: 339-44
16 Taraboulos A, Jendroska K, Serban D, Yang S, DeArmond S, Prusiner SB. Regional mapping of prion proteins in brains. *Proc Natl Acad Sci USA* 1992;89:7620-4
17 Borchelt DR, Koliatsos VE, Guarnieri M, Pardo CA, Sisodia S, Price DL. Rapid anterograde axonal transport of the cellular prion glycoprotein in the peripheral and central nervous systems. *J Biol Chem* 1994;269:14711-4
18 Schubert W, Prior R, Weidemann A et al. Localization of Alzheimer beta-A4 amyloid precursor protein at central and peripheral synaptic sites. *Brain Res* 1991;563:184-94
19 Sharp A, Loev S, Schilling G et al. Widespread expression of Huntington's disease gene (IT15) protein product. *Neuron* 1995;14:1065-74

20 Masliah E, Terry R. The role of synaptic proteins in the pathogenesis of disorders of the central nervous system. *Brain Pathol* 1993;3:77-85

21 Clinton J, Forsyth C, Royston MC, Roberts GW. Synaptic degeneration is the primary neuropathological feature in prion disease: a preliminary study. *Neuroreport* 1993;4:65-8

22 Borchelt DR, Taraboulos A, Prusiner SB. Evidence for synthesis of scrapie prion protein in the endocytic pathway. *J Biol Chem* 1992;267:16188-99

23 Hay B, Prusiner SB, Lingappa VR. Evidence for a secretory form of the cellular prion protein. *Biochemistry* 1987;26:8110-15

24 Tagliavini F, Prelli F, Porro M, Salmona M, Bugiani O, Frangione B. A soluble form of prion protein in human cerebrospinal fluid: implications for prion-related encephalopathies. *Biochem Biophys Res Commun* 1992;184: 1398-404

25 Telling GC, Scott M, Masrianni J et al. Prion propagation in mice expressing human and chimeric PrP transgenes implicates the interaction of cellular PrP with another protein. *Cell* 1995;83:79-90

Transmissible Subacute Spongiform Encephalopathies:
Prion Diseases
L Court, B Dodet, eds
© 1996, Elsevier, Paris

Subcellular localization and toxicity of pre-amyloid and fibrillar prion protein accumulations in murine scrapie

Martin Jeffrey[1], Caroline M Goodsir[1], Moira E Bruce[1], Patricia A McBride[2], Janice R Fraser[2]

[1]*Lasswade Veterinary Laboratory, Bush Estate, Penicuik, Midlothian, EH26 OSA;* [2]*Institute for Animal Health, AFRC and MRC Neuropathogenesis Unit, West Mains Road, Edinburgh EH9 3J, Scotland, UK*

Summary – Ultrastructural immunolocalization of prion protein (PrP) in 87V and Me7 murine scrapie shows that, irrespective of the pattern of PrP accumulation viewed by light microscopy, disease-specific PrP accumulates in association with the plasmalemma of neurons, diffusing from the neuronal cell surface into the extracellular space around cell processes prior to aggregation and fibril assembly. The initial events occur without the involvement of glial cells. Plaques are formed by the localized release of pre-amyloid PrP from the plasmalemma of dendrites. Several plaques may form along the length of a single dendrite. Both pre-amyloid PrP released from the surface of scrapie-infected cells and PrP included into fibrillar amyloid contains N-terminal PrP amino acid residues, suggesting that truncation of PrP is not a prerequisite for fibril formation. In many brain areas cells releasing PrP and the adjacent neuropil into which PrP is released are morphologically normal. However, aggregation of PrP into fibrils is invariably associated with subcellular pathology and glial cell activation. In addition, a highly specific axon terminal degeneration is seen in some brain regions showing pre-amyloid PrP accumulations. These results suggest that intracellular accumulation of PrP in cells is not initally damaging to neurons, but that there are two different mechanisms by which excess extracellular PrP may damage brain tissue: a) a selective, site-specific tissue degeneration; or b) non specific degeneration in response to amyloid fibril formation.

Introduction

Transmissible spongiform encephalopathies are characterized by the accumulation of a host-coded cell surface sialoglycoprotein designated prion protein (PrP) [1–4]. This protein, which accumulates in an abnormal partially protease-resistant form [2, 3] in the brains of infected animals, is essential to the development of disease [5], but its precise role in the pathogenesis of infection is not definitively established [6–10]. In order to further characterize the role of PrP in the pathogenesis of scrapie, we have studied its subcellular localization in the 87V and Me7 strains of murine scrapie.

Materials and methods

Details of the methods used have previously been published [11]. Briefly, terminally ill VM mice infected intracerebrally with the cloned 87V or the Me7 strains of scrapie were killed and perfused with a paraformaldehyde and gluteraldehyde solution. Samples of brain were post-fixed in osmium tetroxide, dehydrated and embedded in Araldite. Sections (1 μm) were cut and etched with saturated sodium

---

Key words: scrapie / ultrastructure / prion protein

ethoxide or potassium methoxide 18 crown bAq dimethyl sulfoxide (DMSO). Sections were de-osmicated, treated with formic acid and then subjected to peroxidase anti-peroxidase immunohisto-chemical labelling using 1B3 or 1A8 anti-PrP antiserum.

Serial and semi-serial (1-μm and 80-nm) sections were taken from blocks with appropriate patterns of PrP immunolabelling so that specific cells or PrP deposits seen in 1-μm thick stained sections could be identified in the electron microscope. For electron microscopy, 80-nm sections were etched in sodium periodate, or in potassium methoxide before being treated with hydrogen peroxide in methanol and then formic acid. Primary antibody (1A8 or 1B3) or pre-bleed antiserum was then applied for 1 h. After rinsing, sections were incubated with Auroprobe 1 and reacted with silver for 3 and 5 min. Grids were post-fixed with 2.5% gluteraldehyde and counterstained with uranyl acetate and lead citrate. In addition to immunostaining with Auroprobe 1, serial (65-nm) sections were also labeled with Extravidin 10-nm gold.

Immunolabelling for both light and electron microscopy was also carried out using antibodies raised to N-terminal peptides. For electron microscopy, pre-embedding methods of tissue preparation and anti-body labelling were necessary [12].

Results and discussion

Subcellular localization of PrP

The 87V strain of murine scrapie is a distinct biologically cloned strain [13] in which light microscopical PrP accumulation is found in restricted brain areas [13, 14]. The form of PrP accumulation occurs variously as frequent classical and subependymal plaques, or granular neuropil deposits or in a distinctive perineuronal pattern (fig 1a). These PrP accumulations are considered to be disease-specific because no immunohistochemical reaction occurs at these sites in control mice. The Me7 strain shows a diffuse granular pattern of disease-specific PrP accumulation throughout all brain areas (although some brain areas are predictably more severely affected) [13, 15]. A small number of plaques are also found.

Patterns of PrP accumulation similar to that seen by light microscopy were located under the electron microscope (Fig1b). Areas of neuropil in which primitive plaques (see below) or granular or perineuronal PrP accumulations are found often show no pathological changes [16]. In such areas PrP is present at the plasmalemma of neurons and neurites and in the extra-cellular spaces between different types of processes including axons, dendrites, axon terminals, glial cell processes and myelinated axons. Similar findings were observed in the diffuse type of labelling pattern present in Me7 scrapie. These findings show that PrP is released at the plasmalemma of neurons and major neurites with subsequent diffusion of pre-amyloid PrP between adjacent neurites and glial processes [16, 17]. The occasional appearance of sparse amyloid filaments between cell processes in the absence of other morphological changes in areas of intense PrP accumulation in the extracellular space suggests that PrP may assemble into fibrils without interaction with other cell types. In areas of intense disease-specific PrP accumulation, microglia and astrocytes showed intralysosomal PrP accumulation, suggesting that these cells are involved in phagocytosis of excess PrP. The similar subcellular patterns of PrP and filament localization – irrespective of the light micro-scopical forms of PrP accumulation – in both Me7 and 87V scrapie indicate that there is a common sequence involved in the release, local diffusion and fibrillar aggregation of PrP.

Amyloid plaques

Classical plaques composed of a central core of bundles of PrP-immunoreactive amyloid fibrils were found not infrequently in 87V scrapie, but very rarely in Me7. PrP was also found at the

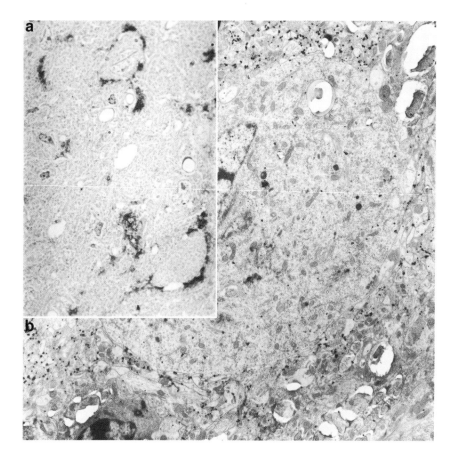

Fig 1. Disease-specific PrP accumulation around neurons in the lateral hypothalamus of 87V scrapie-infected mouse brains. **a)** Thick section (1 μm) labelled for PrP by peroxidase anti-peroxidase method showing PrP accumulation around neuronal perikarya. **b)** Electron micrograph showing the soma and surrounding neuropil of a neuron shown in figure 1a. Although there is perikaryonal immunolabelling by the silver-enhanced 1-nm gold labeled-particles, there is neither significant accumulation of filaments, nor significant subcellular pathology (the apparent vacuolation is an artifact related to the effects of the pretreatment regimes on myelin).

plaque periphery, in the absence of fibrils, at the plasmalemma of cell processes and in the associated extracellular spaces [16]. Other plaques, which were identified by light microscopical immunocytochemistry, had few or no recognizable amyloid fibrils when examined by electron microscopy and were termed primitive plaques. PrP could be demonstrated in a non fibrillar form at the plasmalemma and in the extracellular spaces between neurites of such plaques (fig 2). PrP was closely associated with the plasmalemma of occasional dendrites passing towards the centre of primitive plaques [18]. These results suggest that plaques are formed around one or more PrP-releasing dendrites (and more than one plaque may form along the length of a single dendrite). PrP accumulates in the extracellular spaces adjacent to such processes prior to its spontaneous aggregation into fibrils and subsequent maturation into classical plaques [18].

Scrapie-specific PrP is operationally defined by its partial resistance to protease treatment. No currently available antibody is able to distinguish between scrapie-specific PrP and normal PrP. The 33–35-kDa disease-specific form of PrP is partially resistant to protease digestion whereas the normal form of PrP can be completely digested. Protease K digestion of the murine disease-specific form of PrP produces diverse forms of low molecular mass PrP, some of which are N-terminally truncated at amino acid residue 49 or 57 within the octapeptide repeat segment [19]. Using synthetic peptide antibodies to the N-terminus of PrP (which are not present in the truncated disease-specific PrP) and antibodies to the protease-resistant fraction of PrP, plaques and pre-amyloid deposits in the brains of mice experimentally infected with the 87V strain of scrapie were immunolabelled for examination by light and electron microscopy. Classical fibrillar amyloid deposits in plaques (fig 3) as well as pre-amyloid deposits were both immunolabelled by antibodies to the N-terminus of PrP and to the protease-

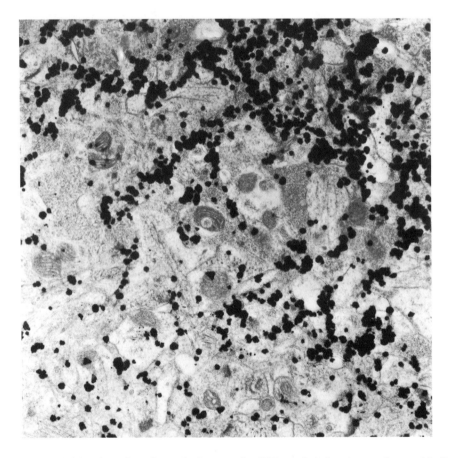

Fig 2. Detail of a primitive plaque from the cerebral cortex of an 87V scrapie-infected mouse immunolabelled using the 1-nm immunogold/silver technique. The plaque does not have any visible fibrils; the neuropil is essentially normal. This small field was representative of the whole plaque. Although the large size of the silver grains prevents accurate interpretation of their location, the gold/silver deposits tend to be arranged around the periphery of a range of different kinds of cell processes.

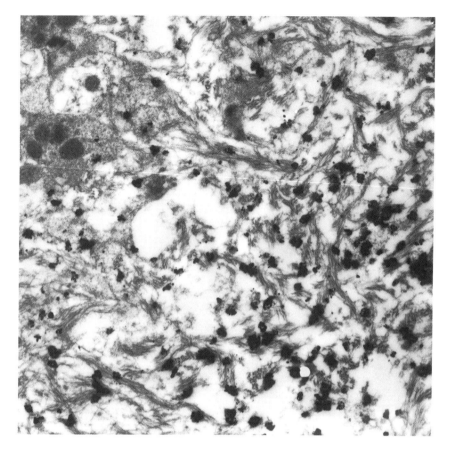

Fig 3. Part of a classical plaque obtained from the cerebral cortex of an 87V scrapie-infected mouse. The plaque was immunolabelled by a pre-embedding technique and reacted with an antibody recognizing N-terminal amino acids 15–40 of the PrP molecule. Bundles of fibrillar amyloid are immunolabelled with gold/silver complexes indicating that N-terminal amino acids are present in the amyloid fibrils.

resistant core of the PrP molecule [12]. This suggests that both N-terminal and core amino acid residues are present in disease-specific PrP released from scrapie-infected cells in vivo. The results also suggest that N-terminal truncation of PrP is not a prerequisite for subsequent aggregation of PrP into amyloid fibrils [12]. It also leaves open the possibility that the conformational states which lead to partial protease resistance and associated N-terminal truncation of the scrapie-specific forms of PrP protein do not occur until after the release of PrP by scrapie-infected cells.

In vivo toxicity of PrP

Initial PrP accumulation at the plasmalemma around neurons, in early plaques or in areas of diffuse or localized PrP accumulation was generally not associated with morphological changes of either the neuron or the dendrite releasing the PrP or in the adjacent neuropil in which excess PrP accumulated. However, accumulation of pre-amyloid PrP in some brain areas was associated with specific degeneration of dendritic spines and axon terminals (fig 4).

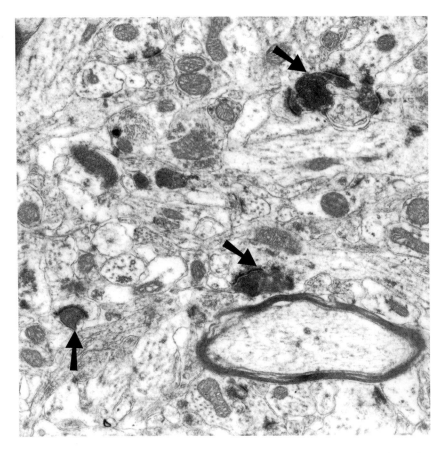

Fig 4. Selective axon terminal degeneration (arrowheads) in the hippocampus of a mouse with Me7 scrapie. This area of the hippocampus had moderate pre-amyloid PrP accumulation (not shown).

Initial PrP aggregation into fibrils was also associated with tissue damage in both Me7 and 87V plaques and diffuse accumulations. Tissue damage associated with fibrillogenesis was confined to the region of the neuropil containing nascent fibrils. As natural sheep scrapie and bovine spongiform encephalopathy have only extremely sparse plaques and fibrillar PrP deposits (personal observations), it is unlikely that neuropil damage associated with fibrillar PrP would be the causative neurological signs in these diseases. We conclude that pre-amyloid PrP release and accumulation are not invariably toxic either to the neuron releasing PrP or to the neuropil into which it is released. However, axon terminal degeneration and dendritic spine loss in specific neuroanatomical areas may be indicative of PrP toxicity and is possibly the main cause of neurological dysfunction in the strains of murine scrapie that we have examined.

Conclusion

These immunogold electron microscopy studies have shown that, although there are different patterns of PrP accumulation when viewed by light microscopy, the patterns of subcellular PrP accumulation are essentially similar, at least in the two mouse models of disease studied.

The subcellular localization of PrP has not yet shown any link between PrP accumulation and vacuolation or the presence of dystrophic neurites or tubulovesicular bodies. PrP release from infected cells was not associated with morphological changes in the releasing cell, but both extracellular pre-amyloid and fibrillar PrP were associated with degenerative changes of the neuropil.

References

1 Bolton DC, McKinley MP, Prusiner SB. Identification of a protein that co-purifies with the scrapie prion. *Science* 1982;218:1309-11

2 Hope J, Morton LJD, Farquhar CF, Multhaup G, Beyreuther K, Kimberlin RH. The major protein of scrapie-associated fibrils (SAF) has the same size, charge distribution and N-terminal protein sequence as predicted for the normal brain protein (PrP). *EMBO J* 1986;5:2591-7

3 McKinley MP, Bolton DC, Prusiner SB. A protease resistant protein is a structural component of the scrapie prion. *Cell* 1983;35:57-62

4 Oesch B, Westaway D, Walchi M et al. A cellular gene encodes scrapie PrP 27–30 protein. *Cell* 1985;40:735-46

5 Bueler H, Aguzzi A, Sailer A et al. Mice devoid of PrP are resistant to scrapie. *Cell* 1993;73:1339-47

6 Bruce ME, McConnell I, Fraser H, Dickinson AG. The disease characteristics of different strains of scrapie in *Sinc* congenic mouse lines: implications for the nature of the agent and host control of pathogenesis. *J Gen Virol* 1991;72:595-603

7 Prusiner SB. Prion biology. In: Prusiner SB, Collinge J, Powell J, Anderton B, eds. *Prion Diseases of Humans and Animals*. London: Ellis Horwood, 1992;533-67

8 Prusiner SB. Transgenetic investigations of prion diseases of humans and animals. *Philos Trans R Soc Lond B* 1993;239-54

9 Rohwer RG. The scrapie agent: a virus by any other name. In: Chesebro BW, ed. *Current Topics in Microbiology and Immunology: Transmissible Spongiform Encephalopathies*. 172. Berlin: Springer Verlag, 1991;195-232

10 Somerville RA, Bendheim PE, Bolton DC. The transmissible agent causing scrapie must contain more than protein. *Med Virol* 1991;1:131-44

11 Jeffrey M, Goodsir CM. Immunohistochemistry of resinated tissues for light and electron microscopy. In: Baker HF, Ridley RM, eds. *Methods in Molecular Medicine. Prion Diseases*. Totowa: Humana Press Inc, 1996: 301-12

12 Jeffrey M, Goodsir CM, Fowler N, Hope J, Bruce ME, McBride PA. Ultrastructural immunolocalisation of synthetic prion protein antibodies in 87V murine scrapie. *Neurodegeneration* 1996;5:101-9

13 Bruce ME, McBride PA, Farquhar CF. Precise targeting of the pathology of the sialoglycoprotein PrP, and vacuolar degeneration in mouse scrapie. *Neurosci Lett* 1989;102:1-6

14 Jeffrey M, Goodsir CM, Bruce ME, McBride PA, Scott JR, Halliday WG. Infection specific prion protein (PrP) accumulates on neuronal plasmalemma in scrapie infected mice. *Neurosci Lett* 1992;147:106-9

15 Bruce ME, McBride PA, Jeffrey M, Rozemuller JM, Eikelenboom P. PrP in scrapie and B/A4 in Alzheimer's disease show similar patterns of deposition in the brain. In: Corain B, Iqbal K, Nicolini M, Winblad B, Wisniewski H, Zatta P, eds. *Alzheimer's Disease: Advances in Clinical and Basic Research*. London: John Wiley and Sons Ltd, 1993:481-7

16 Jeffrey M, Goodsir CM, Bruce ME, McBride PA, Scott JR, Halliday WG. Correlative light and electron microscopy studies of PrP localisation in 87V scrapie. *Brain Res* 1994;656:329-43

17 Jeffrey M, Goodsir CM, Bruce ME, McBride PA, Fowler N, Scott JR. Murine scrapie-infected neurons in vivo release excess PrP into the extracellular space. *Neurosci Lett* 1994;174:39-42

18 Jeffrey M, Goodsir CM, Bruce ME, McBride PA, Farquhar C. Morphogenesis of amyloid plaques in 87V murine scrapie. *Neuropathol Appl Neurobiol* 1994;20:535-42

19 Hope J, Multhaup G, Reekie LJD, Kimberlin RH, Beyreuther K. Molecular pathology of scrapie-associated fibril protein (PrP) in mouse brain affected by the Me7 strain of scrapie. *Eur J Biochem* 1988;172:271-7

Transmissible Subacute Spongiform Encephalopathies:
Prion Diseases
L Court, B Dodet, eds
© 1996, Elsevier, Paris

Immunogold electron microscopy studies of tubulovesicular structures: they are not composed of prion protein (PrP)

Pawel P Liberski[1], Martin Jeffrey[2], Caroline Goodsir[2]

[1]*Laboratory of Electron Microscopy and Neuropathology, Laboratories of Tumor Biology, Chair of Oncology, Medical Academy Lodz, Paderewski Street 4, PL93-509 Lodz, Poland;* [2]*Central Veterinary Laboratory, Lasswade Veterinary Laboratory, Bush Estate, Penicuik, Edinburgh, Scotland*

Summary – Tubulovesicular structures (TVS) are the disease-specific particles found by thin-section electron microscopy in all of the transmissible spongiform encephalopathies. We used the immunogold methods (both 10-nm immunogold and the 1-nm immunogold silver-enhanced methods) for ultrastructural localization of prion protein (PrP). In all scrapie models examined (263K and 22CH in hamsters and 87V and Me7 in mice), TVS-containing processes were readily detected and neither these processes nor TVS themselves were decorated with gold particles. Even when amyloid plaques were observed in close contact with TVS-containing neuronal processes, the plaques were decorated with gold particles, while the processes remained unstained. TVS located in areas adjacent to plaques in the 87V model and in areas of diffuse PrP immunolabelling in Me7 were also unlabelled with anti-PrP sera.

Introduction

Tubulovesicular structures (TVS) are disease-specific particles found by thin-section electron microscopy in all transmissible spongiform encephalopathies (TSE) or prion diseases, including natural Creutzfeldt-Jakob disease (CJD) and Gerstmann-Sträussler-Scheinker disease [1–3]. TSE are caused by a still incompletely characterized infectious agent variously referred to as prion, slow, unconventional virus, or virino [4, 5]. It has even been suggested that this agent is a single-stranded (ss) DNA virus of unusual structure, called 'nemavirus', which is visualized as TVS (designated 'tubulofilamentous particles') by thin-section electron microscopy or as 'thick tubules' when the touch preparation method was applied [6–8]. Furthermore, it was proposed that TVS are composed of an abnormal isoform of prion protein (PrP) [6–8]. Using thin-section electron microscopy and immunogold techniques, we report here that TVS are not recognized by anti-PrP antisera and thus are not composed of PrP. Moreover, we report that 'tubulofilamentous particles' are merely non-specific findings related neither to TSE nor to the TVS as they were observed in brains of control animals and represent swollen microtubules.

Key words: immunogold electron microscopy / tubulovesicular structures / scrapie / Creutzfeldt-Jakob disease

Materials and methods

Animals

Outbred, 6-week-old golden Syrian hamsters were inoculated intracerebrally with 0.05 mL of a 10% brain suspension of the 263K and 22CH strains of scrapie (kindly supplied by R Kimberlin, SARDAS, Edinburgh, Scotland, UK, and R Carp, IBR, Staten Island, New York, USA, respectively), perfused with 100 mL of 1.25% glutaraldehyde and 1% paraformaldehyde prepared in cacodylate buffer (pH 7.4) followed by 50 mL of 5% glutaraldehyde and 4% paraformaldehyde and routinely prepared for transmission electron microscopy. Mouse brains infected with 87V or Me7 strains of scrapie were also available for this study. As additional control tissues, rats were injected intracisternally with $MgCl_2$ (in a 'scrapie-free' laboratory) to develop mercury encephalopathy and processed analogously as hamsters.

Immunogold procedures

The immunogold methods used for ultrastructural localization of PrP are as previously reported [9–10]. Both 10-nm immunogold and the 1-nm immunogold silver-enhanced (IGSS) methods were applied. Briefly, these methods are as follows: 65–80-nm sections were taken from blocks previously identified as containing, on immunolabeled 1-mm thick sections, accumulations of PrP either in the form of amyloid plaques or as other forms. Most sections originated from cerebral cortex, thalamus or hippocampus, but occasional samples taken from the cerebellum, midbrain or medulla were also examined. Sections placed on 400-mesh nickel grids were etched with sodium periodate for 60 min or in potassium methoxide 18 crown bAq dimethyl sulfoxide (DMSO) for 15 min. Endogenous peroxidase was blocked and sections deosmicated with 3% hydrogen peroxide for 3 min. Primary antibody (1B3 or 1A8 kindly supplied by J Hope, MRC and AFRC Neuropathogenesis Unit, Edinburgh, Scotland, UK) was then applied diluted 1:100 or 1:400, respectively, in incubation buffer for 1 h. After rinsing, sections were incubated with extravidin 10-nm colloidal gold diluted 1:10 in incubation buffer for 1 h. Sections prepared for the IGSS method were stained with 1-nm gold and silver enhanced. Grids were post-fixed with 2.5% glutaraldehyde in phosphate-buffered saline and counterstained with uranyl acetate and lead citrate.

Results

In an 87V murine model, PrP-conjugated gold particles decorated typical stellate amyloid plaques and the cell surface of numerous dendrites [9, 10]. In both the 263K and 22CH hamster models, anti-PrP antibodies readily decorated primitive plaques composed of randomly oriented amyloid fibrils at the subependymal areas as well as fibrils floating within ventricles. In these models neither stellate plaques nor PrP-immunodecorated dendrites were observed. In all models TVS-containing processes were readily detected and neither these processes nor TVS themselves were decorated with gold particles (figs 1–3). Even if amyloid plaques were observed in close contact with TVS-containing neuronal processes, the plaques were decorated with gold particles, while the processes remained unlabeled (fig 3). TVS located in areas adjacent to plaques in the 87V model (fig 1) and in areas of diffuse PrP immunolabelling in the Me7 model (fig 2) were also unlabelled with anti-PrP antisera. In both hamsters and rats, whether inoculated with 263K or 22CH scrapie agents, sham-inoculated or $MgCl_2$-intoxicated, thick 'tubulofilamentous particles' were easily found, particularly in longitudinally cut myelinated axons. These particles were indistinguishable from swollen microtubules and, as reported by Narang et al [6–8], occasionally showed cross striations. 'Tubulofilamentous particles' were particularly easily found in those brain areas which were suboptimally fixed.

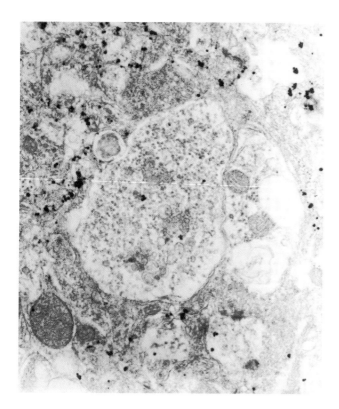

Fig 1. Process containing TVS adjacent to a plaque in an 87V mouse scrapie model. Although immunoreactivity to PrP is present in the extracellular space and also on loosely aggregated amyloid fibrils, no significant immunogold accumulation is present on the TVS-containing process (IGSS method and 1B3 antibody; original magnification × 30,000).

Discussion

Using the immunogold technique we were unable to label TVS with anti-PrP antibodies. As this technique proved to be sensitive enough to immunolabel not only amyloid plaques but also pre-amyloid accumulations of PrP [9, 10], we strongly believe that the absence of labeling reflects the real structure of TVS which are not composed of PrP. That TVS are PrP-negative may have several important implications for hypotheses about their nature. Principally, it falsifies the suggestion put forward by Narang et al [6–8] that TVS are cross sections of 'thick tubules' or 'tubulofilamentous particles' visualized by touch preparations of scrapie-affected mouse and hamster brains. These 'thick tubules' were further claimed to represent an ultra-structural correlate of the 'nemavirus' which, in turn, was proposed to be an elusive scrapie agent. Indeed, the presence of these 'tubulofilamentous particles' in normal mouse and hamster brains suggests that they are not scrapie-specific and represent merely swollen microtubules probably resulting from less than optimal fixation [11]. In fact, Narang et al's figure 2 [12] clearly shows numerous areas of myelin sheath dissolution which are regarded as stigmata of suboptimal fixation. Furthermore, Chasey [13], in a recent comment on the Narang papers, suggested that 'thick tubules' may represent microtubular doublets; again these are entirely normal subcellular components. The same arguments apply to the spurious relationship between 'tubulofilamentous particles', TVS and scrapie-associated fibrils (SAF). Narang et al have claimed that TVS, visualized as 'thick tubules' [7, 8] in touch preparations, undergo transformation into SAF following treatment with sodium dodecyl sulfate (SDS). Grids which

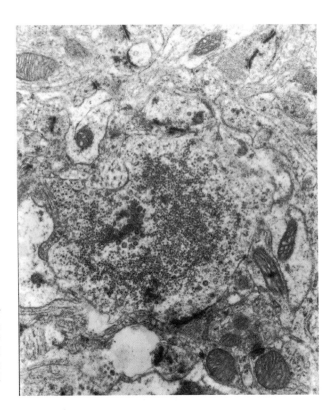

Fig 2. Process containing TVS in an area of diffuse PrP immunolabeling in a Me7 mouse scrapie model. No immunoreactivity to PrP is present in the process. The 10-nm gold particles are not visible at the magnification shown (10-nm immunogold method and 1A8 antibody).

were not treated with SDS did not show SAF. Grids treated with proteinase K (PK) alone did not show SAF, while the addition of DNase or mung bean nuclease produced SAF [6]. This sequence of enzymatic digestion and SDS treatment corresponds to that already reported by McKinley et al [14], who demonstrated that SDS treatment and proteolytic digestion are mandatory for the formation of prion rods which, like SAF, are composed of PrP. As inhibitors of proteolytic enzymes were not used by Narang et al prior to SDS treatment, SAF were formed, even in the apparent absence of PK; similar data have already been reported by Beekes et al [15]. Such an interpretation is further substantiated by Narang et al's immunogold localization of PrP along SAF [6–8], which also merely recapitulates already published data. Obviously, more work is necessary to characterize TVS molecularly.

The nature and significance of TVS is unknown. In hamsters inoculated with the 263K strain of scrapie, TVS were initially observed as early as 3 weeks after intracerebral inoculation, but their number increased only with the onset of disease 9 to 10 weeks after inoculation [1]. Noteworthy, vacuolation and astrocytosis were detected 8 weeks post-inoculation and, thus, followed the appearance of TVS. In Webster-Swiss mice, TVS were found 12 and 16 weeks after intracerebral inoculation and intradermal (footpads) inoculation, respectively [16]. In NIH Swiss mice infected with the Fujisaki strain of CJD, TVS were first seen 13 weeks after intracerebral inoculation and their numbers had increased dramatically 18 weeks after inoculation, when the first signs of clinical disease were noted [1]. Vacuolation and astrocytosis were detected at the same time, and the increase in the number of processes containing TVS paralleled the increasing intensity of vacuolation and astrocytosis. Thus, TVS appeared

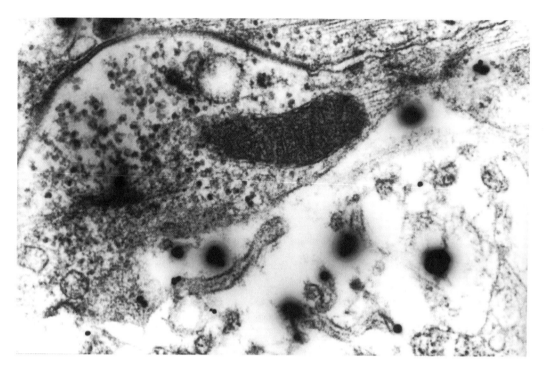

Fig 3. Process containing TVS adjacent to a plaque in a 263K hamster model. Note the PrP immunoreactivity on amyloid fibrils but not in the process (IGSS method and 1B3 antibody; original magnification × 30,000).

early in the incubation period, preceding the onset of clinical disease. Furthermore, in scrapie-infected hamsters, TVS preceded the appearance of other neuropathological changes. The approximately 1,000-fold lower infectivity titer of the Fujisaki strain of CJD, compared to the 263K strain of scrapie, may have accounted for the delayed appearance of TVS in experimental CJD. The apparent correlation between the number of neuronal processes containing TVS and infectivity titer may explain why in cell cultures infected with the scrapie agent, TVS could not be found and why their number in experimental scrapie in sheep was so low [17]. Collectively, it was suggested that TVS may represent the infectious agent or aggregate of the disease rather than its pathological product. The latter hypothesis was, however, advocated by Gibson and Doughty [17] who interpreted TVS as breakdown products of microtubules. The predominant theory of the scrapie agent is now the 'prion hypothesis' and its derivatives which implies that a conformationally altered abnormal isoform (PrP^{Sc}) of normal cellular membrane glycoprotein (PrP^{C}) is the agent and its accumulation merely mimicks replication [5]. The nucleation hypothesis formulated by Gajdusek suggests that not the monomer but the aggregate ('crystal') of PrP^{Sc} is the agent ('infectious amyloid') [4]. Indeed, recent in vitro studies elegantly showed the protein-protein (PrP^{Sc}–PrP^{C}) interactions which may underlie de novo formation of the agent [18, 19]. However, de novo generation of the infectivity was neither addressed nor proven in these experiments, and thus, they did not discriminate between two possibilities: PrP^{Sc} as the scrapie agent versus PrP being merely the amyloidogenic protein. If PrP is the agent, TVS being not composed of this protein could not be an ultrastructural correlate of it. If, however, it is found that TVS is more than merely a useful ultra-structural marker for the whole group of TSE, it may suggest that PrP and the agent are two separate entities. The analogies to molecular pathogenesis of another brain amyloidosis, Alzheimer's disease, are so striking [20] that it is tempting to suggest that PrP plays the role of the amyloidogenic protein in TSE, the processing of which is triggered by a still elusive agent of which TVS may be an ultrastructural correlate.

Acknowledgements

This paper is a part of the European Community concerted action 'Prion Diseases: From Neuropathology to Pathobiology and Molecular Biology' (Project leader: Prof Herbert Budka, Vienna, Austria). PPL is supported by grants from the KBN and the Maria Sklodowska-Curie Foundation while in Poland and by the British Council fellowship while in the United Kingdom.

References

1 Liberski PP, Yanagihara R, Gibbs CJ Jr, Gajdusek DC. Appearance of tubulovesicular structures in Creutzfeldt-Jakob disease and scrapie precedes the onset of clinical disease. *Acta Neuropathol (Berl)* 1989;79:349-54
2 Liberski PP, Budka H, Sluga E, Barcikowska M, Kwiecinski H. Tubulovesicular structures in Creutzfeldt-Jakob disease. *Acta Neuropathol (Berl)* 1992;84:238-43
3 Liberski PP, Budka H. Tubulovesicular structures in Gerstmann-Sträussler-Scheinker disease. *Acta Neuropathol (Berl)* 1994;88:491-2
4 Gajdusek DC. Infectious amyloids: subacute spongiform encephalopathies as transmissible cerebral amyloidoses. In: Fields B, Knipe DM, Howley PM, eds. *Fields Virology*, 3rd edn. Philadelphia, New York: Lippincott-Raven Publishers, 1995;2851-901
5 Liberski PP. Prions, β-sheets and transmissible dementias: is there still something missing? *Acta Neuropathol (Berl)* 1995;90:113-25
6 Narang HK. Evidence that scrapie-associated tubulofilamentous particles contain a single stranded DNA. *Intervirology* 1993;36:1-10
7 Narang HK, Asher DM, Gajdusek DC. Tubulofilaments in negatively stained scrapie-infected brains: relationships to scrapie-associated fibrils. *Proc Natl Acad Sci USA* 1987;84:7730-4
8 Narang HK, Asher DM, Gajdusek DC. Evidence that DNA is present in abnormal tubulofilamentous structures found in scrapie. *Proc Natl Acad Sci USA* 1988;85:3375-9
9 Jeffrey M, Goodsir C, Bruce ME, McBride PA, Farquhar C. Morphogenesis of amyloid plaque in 87V murine scrapie. *Neuropathol Appl Neurobiol* 1994;20:535-42
10 Jeffrey M, Goodsir C, Bruce ME, McBride PA, Halliday WG. Correlative light and electron microscopy studies of PrP localization in 87V scrapie. *Brain Res* 1994;656:329-43
11 Liberski PP. "Tubulofilamentous particles" are not scrapie-specific and are unrelated to tubulovesicular structures. *Acta Neurobiol Exp (Warsz)* 1995;55:149-54
12 Narang HK. Scrapie-associated tubulofilamentous particles in scrapie hamsters. *Intervirology* 1992;34:105-11
13 Chasey D. Comment on the paper of HK Narang "Evidence that scrapie-associated tubulofilamentous particles contain a single-stranded DNA". *Intervirology* 1994;37:106
14 McKinley MP, Meyer RK, Kenaga L et al. Scrapie prion rod formation in vitro requires both detergent extraction and limited proteolysis. *J Virol* 1991;65:1340-51
15 Beekes M, Oberdinieck U, Malchow M, Haupt M, Diringer H. Evidence for the formation of scrapie-associated fibrils (SAF) in vivo: prior addition of proteinase K as well as inhibitors of endogenous protease activity with subsequent detergent extraction results in isolation of SAF from scrapie infected brains. Presented at the IXth International Congress of Virology, Glasgow, Scotland, UK, 8–13 August 1993
16 Narang HK. A chronological study of experimental scrapie in mice. *Virus Res* 1991;9:293-306
17 Gibson PH, Doughty LA. An electron microscopic study of inclusion bodies in synaptic terminals of scrapie-infected animals. *Acta Neuropathol (Berl)* 1989;77:420-5
18 Bessen RA, Kocisko DA, Raymond GJ, Nandan S, Lansbury PT, Caughey B. Non genetic propagation of strain-specific properties of scrapie prion protein. *Nature* 1995;375:698-700
19 Kocisko DA, Come JH, Priola SA et al. Cell free formation of protease-resistant prion protein. *Nature* 1994; 370:471-4
20 Liberski PP. Transmissible cerebral amyloidoses as a model for Alzheimer's disease: an ultrastructural perspective. *Mol Neurobiol* 1994;8:67-77

Transmissible Subacute Spongiform Encephalopathies:
Prion Diseases
L Court, B Dodet, eds
© 1996, Elsevier, Paris

Pathogenesis of scrapie in hamsters after oral and intraperitoneal infection

Michael Beekes, Elizabeth Baldauf, Heino Diringer

Robert Koch Institut, Bundesinstitut für Infektionskrankheiten und nicht übertragbare Krankheiten, FG 123, Nordufer 20, 13353 Berlin, Germany

Summary – Transmissible spongiform encephalopathies (TSE) are associated with distinct pathognomonic markers, such as infectivity and TSE-specific amyloid protein (also referred to as prion protein). An assessment of the quantitative relationship between the agent titer and the amount of pathological protein in the central nervous system (CNS) of hamsters orally infected with scrapie revealed a parallel accumulation of both markers and a relatively constant ratio of about 10^6 molecules of the protein per infectious unit during the course of infection. On this basis, the spread of infection after oral uptake of agent was analyzed by tracing the accumulation of TSE-specific amyloid protein in the CNS. The pathogenetic process first appeared in the spinal cord between vertebrae T4 and T9. Subsequently, the accumulation process showed an anterograde spread to the brain and a retrograde spread to the lumbar spinal cord. Additional findings indicated a possible alternative route of access to the brain. Preliminary results of a further study addressing the pathogenesis after intraperitoneal administration of the agent strongly suggest that infection enters the CNS not only via the spinal cord, but also reaches the brain directly from the periphery via the medulla oblongata and possibly the pons.

Study design and results

This article presents some results of our studies on the pathogenesis of scrapie in hamsters ofter oral and intraperitoneal infection. Using a new purification method for the extraction of transmissible spongiform encephalopathy (TSE)-specific amyloid protein with an established yield of about 70% [1] and a Western blot assay for the quantification of this protein, we initially assessed the quantitative relationship between infectivity and TSE-specific amyloid protein in the central nervous system (CNS) of hamsters orally infected with scrapie [2].

Three to five hamsters were sacrificed at different times between 77 and 133 days post-infection (dpi). The brain and cervical spinal cord were removed and the infectivity in these tissues was quantified by bioassay. The TSE-specific amyloid protein was extracted and quantified in the Western blot. The concentration of the protein in the tissue homogenate was calculated from the blotting results and compared to the infectivity found in the homogenate. Figure 1a and b shows the results of this analysis and demonstrates a parallel accumulation of agent and TSE-specific amyloid protein in the cervical spinal cord and in the brain. During the recorded course of infection there was a relatively constant ratio of about 10^6 molecules of the pathological protein per infectious unit in the tissue homogenates. Therefore, we considered

Key words: scrapie / transmissible spongiform encephalopathies (TSE) / TSE-specific amyloid protein

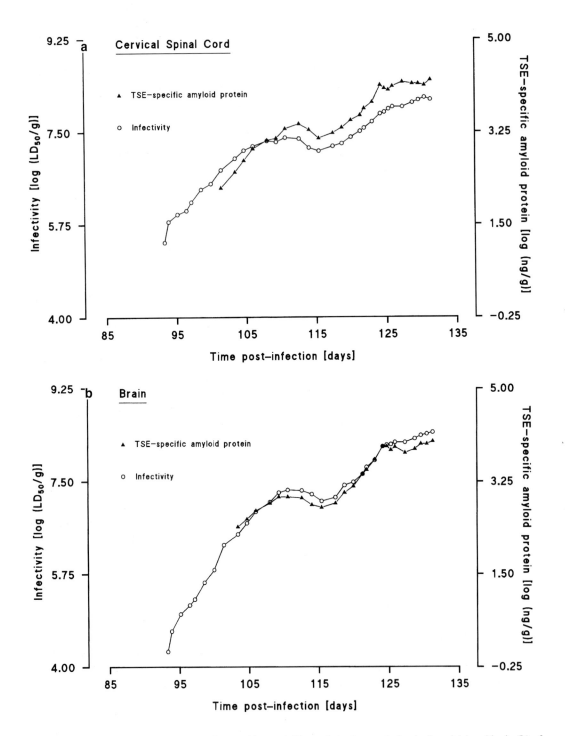

Fig 1. Accumulation of infectivity and TSE-specific amyloid protein in the cervical spinal cord (**a**) and brain (**b**) of hamsters orally infected with scrapie. The graphs were constructed using rolling means for the *x*- and *y*-values. Mean error of infectivity assay: ± 0.4 log; mean error of protein determination: ± 0.2 log.

the TSE-specific amyloid protein as a reliable marker for infectivity which could be used for tracing the spread of infection in our animal model.

On this basis we performed an analysis of the spread of infection in the CNS after oral uptake of the agent by monitoring the accumulation of the pathological protein [2]. Again, hamsters were infected orally with scrapie and sacrificed at different times after infection. The brain and spinal cord were removed and the spinal cord was dissected into seven defined segments as shown in figure 2. There were two cervical segments between the cervical vertebrae C1 and C7, four thoracic segments (between T1 and T13) and one lumbar segment (L1–L3). The spinal cord segments representing about 20–50 mg of tissue and the entire brain (about 1 g of tissue) were homogenized. The TSE-specific amyloid protein was subsequently extracted from the entire spinal cord samples and from 1/20th of the brain homogenate, representing 50 mg of tissue.

The pathological protein was visualized and quantified in the Western blot, and the amount of TSE-specific amyloid protein in the tissue homogenate was calculated from the blotting results.

Up to 89 dpi we were not able to detect the pathological protein in the brain or anywhere in the spinal cord. When we analyzed animals sacrificed at more than 89 dpi, accumulation of TSE-specific amyloid protein showed the pattern presented in figure 2. The pathogenetic process was first detected in the thoracic spinal cord between vertebrae T4 and T9. Subsequently, it exhibited an anterograde spread via segment T1–T3 and via the cervical spinal cord to the brain and a retrograde spread in the lumbar direction (fig 3).

Our experimental design allowed detection of TSE-specific amyloid protein in the brain only after the pathological protein had appeared in all spinal cord segments between vertebrae C1 and T9. However, as we analyzed only 1/20th of the homogenate of an entire brain, the amount of TSE-specific amyloid protein in the brain in the early stages of the infection may have been underestimated. In addition, TSE-specific amyloid protein in localized foci might have been diluted below the detection limit.

Figure 4 shows the relationship between the increase in the total amount of the pathological protein in the CNS and its local accumulation in the different spinal cord segments and in the brain. At early stages of the infection, when the total amount of TSE-specific amyloid protein in the CNS is still very low, only the accumulation in T4–T9 contributes to the total amount of the pathological protein. Then the other spinal cord segments start one after the other to show accumulation of TSE-specific amyloid protein in anterograde and retrograde directions. Accumulation in the brain was first detected when the total amount of the pathological protein in the CNS was about 40 ng. However, extrapolation of the curve for the brain in figure 4 (dashed line) shows that the onset of accumulation in the brain probably could have been detected together with the onset in T4–T9 if the entire brain had been analyzed instead of only 50 mg of brain tissue. This observation indicates the possibility of an alternative route of spread of infection to the brain.

In order to investigate this route of access more precisely, we performed a detailed analysis of the spread of infection in the spinal cord and brain after intraperitoneal administration of agent. The design of the study was similar to that of the previous experiments, with the difference that the brain as well as the spinal cord was divided into segments. The medulla oblongata, pons, cerebellum, cortex and remaining stem (together with the diencephalon) were analyzed separately. This approach revealed an onset of accumulation of the pathological protein as shown in figure 5a–c (Baldauf, preliminary unpublished results). In some animals the first accumulation of TSE-specific amyloid protein was detected in the medulla oblongata

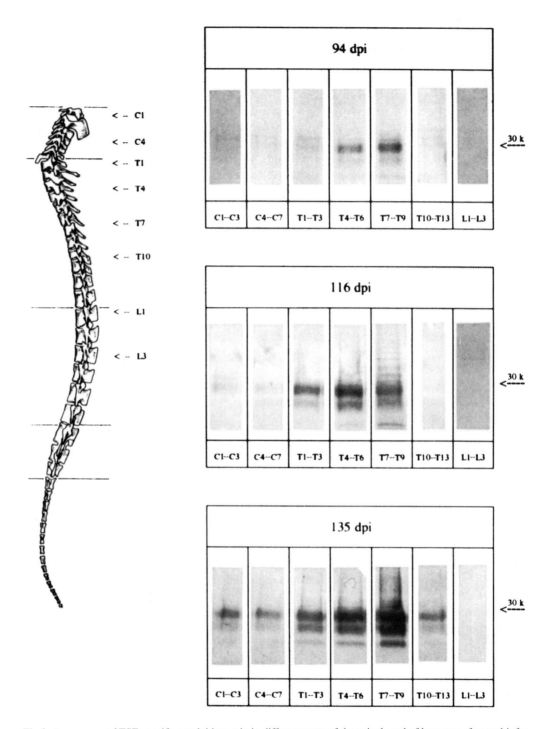

Fig 2. Appearance of TSE-specific amyloid protein in different parts of the spinal cord of hamsters after oral infection with scrapie: Western blot detection with (mAb) 3F4 at 94, 116 and 135 dpi; 20–40 mg of tissue for the different spinal cord samples between C1 and L3.

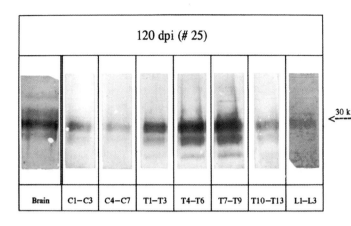

Fig 3. Appearance of TSE-specific amyloid protein in different parts of the CNS of hamsters after oral infection with scrapie: Western blot detection with mAb 3F4 at 120 dpi; 50 mg of tissue for the brain sample and 20–40 mg of tissue for the spinal cord samples between C1 and L3.

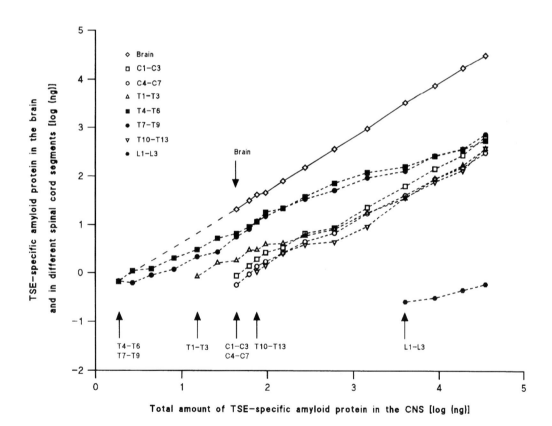

Fig 4. Sequential and quantitative progression of the accumulation of TSE-specific amyloid protein in the CNS of hamsters orally infected with scrapie. Arrows indicate the relative sequence of first detection in the different tissues. The graph was constructed using rolling means for the *x*- and *y*-values. Mean error of protein determination: ± 0.2 log.

a   56 dpi (# 24)

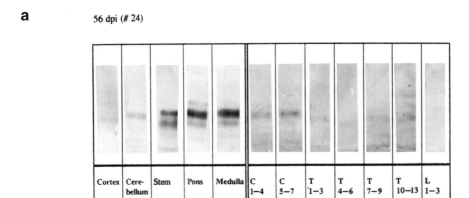

b   64 dpi (# 36)

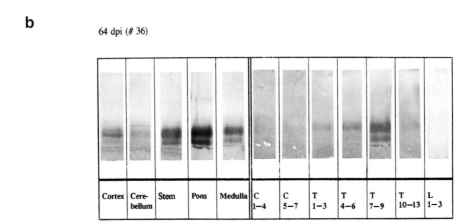

c   74 dpi (# 45)

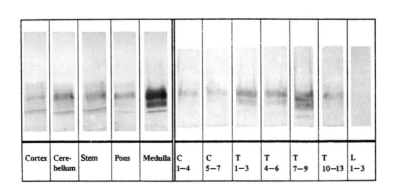

Fig 5. Appearance of TSE-specific amyloid protein in different parts of the spinal cord and of the brain of hamsters after intraperitoneal infection with scrapie: Western blot detection with mAb 3F4 at 56 (**a**), 64 (**b**) and 74 (**c**) dpi; 20–50 mg of tissue for the different spinal cord, medulla and pons samples; up to 100 mg for samples of the cerebellum, cortex and remaining stem (together with the diencephalon).

and in the pons (fig 5a). Other animals showed a simultaneous onset of accumulation in these brain areas and in the thoracic spinal cord (fig 5b, c). According to these preliminary findings the infection does not enter the CNS via the spinal cord alone, but also reaches the brain directly from the periphery via the medulla oblongata and possibly the pons. However, an overall evaluation of the data available so far suggests that the medulla oblongata, not the pons, is the actual point of entry. Examples showing the pons as the first site of detection probably reflect difficulties in achieving a precise separation of pons and medulla oblongata rather than the true pathogenetic situation.

Our findings for the accumulation of TSE-specific amyloid protein in the spinal cord of hamsters orally or intraperitoneally challenged with scrapie are entirely consistent with observations made by Kimberlin and Walker [3–8] in pathogenesis studies on the spread of infectivity after intraperitoneal or intragastric infection of small rodents. Our data extend the pathogenetic concept introduced by these authors to scrapie in orally infected hamsters and show the existence of an alternative route of spread of infection from the periphery to the brain, other than via the spinal cord. A direct route of access to the brain as demonstrated in our model system was already suggested earlier [5] and is also in agreement with immunohistochemical findings reported more recently [9, 10].

Acknowledgments

The technical assistance of S Lichy and H Wohlert is greatly appreciated. This project was supported by EU Grant B102-CT93-0248 and by the Hertie-Stiftung.

References

1 Beekes M, Baldauf E, Cassens S et al. Western blot mapping of disease-specific amyloid in various animal species and humans with transmissible spongiform encephalopathies using a high-yield purification method. *J Gen Virol* 1995;76:2567-76
2 Beekes M, Baldauf E, Diringer H. Sequential appearance and accumulation of pathognomonic markers in the central nervous system of hamsters orally infected with scrapie. *J Gen Virol* 1996;77:1925-34
3 Kimberlin RH, Walker CA. Pathogenesis of mouse scrapie: dynamics of agent replication in spleen, spinal cord and brain after infection by different routes. *J Comp Pathol* 1979;89:551-62
4 Kimberlin RH, Walker CA. Pathogenesis of mouse scrapie: evidence for neural spread of infection to the CNS. *J Gen Virol* 1980;61:183-7
5 Kimberlin RH, Walker CA. Pathogenesis of mouse scrapie: patterns of agent replication in different parts of the CNS following intraperitoneal infection. *J R Soc Med* 1982;75:618-24
6 Kimberlin RH, Walker CA. Invasion of the CNS by scrapie agent and its spread to different parts of the brain. In: Court LA, Cathala F, eds. *Virus Non Conventionnels et Affections du Système Nerveux Central.* Paris: Masson, 1983;17-33
7 Kimberlin RH, Walker CA. Pathogenesis of scrapie (strain 263K) in hamsters infected intracerebrally, intraperitoneally or intraocularly. *J Gen Virol* 1986;67:255-63
8 Kimberlin RH, Walker CA. Pathogenesis of scrapie in mice after intragastric infection. *Virus Res* 1989;12: 213-20
9 Muramoto T, Kitamoto T, Tateishi J, Goto I. Accumulation of abnormal prion protein in mice infected with Creutzfeldt-Jakob disease via intraperitoneal route: a sequential study. *Am J Pathol* 1993;143:1470-9
10 van Keulen LJM, Schreuder BEC, Meloen RH et al. Immunohistochemical detection and localization of prion protein in brain tissue of sheep with natural scrapie. *Vet Pathol* 1995;32:299-308

Transmissible Subacute Spongiform Encephalopathies:
Prion Diseases
L Court, B Dodet, eds
© 1996, Elsevier, Paris

Scrapie in severe combined immunodeficient mice

Corinne Ida Lasmézas[1], Jean-Yves Cesbron[2], Jean-Philippe Deslys[1],
Rémi Demaimay[1], Karim Adjou[1], Catherine Lemaire[2], Jean-Pierre Decavel[2],
Dominique Dormont[1]

[1]Service de Neurovirologie, Commissariat à l'Energie Atomique, DSV/DRM/CRSSA, 60–68, avenue
Division Leclerc, BP 6, 92265 Fontenay-aux-Roses cedex; [2]Laboratoire de Pathogénie Expérimentale,
Institut Pasteur de Lille, INSERM U 415, 1, rue du Professeur A-Calmette, BP 245, 59019 Lille cedex,
France

Summary – Knowledge of the pathogenesis of transmissible spongiform encephalopathies (TSE) is still incomplete. Infectivity and accumulation of PrP-res (the protease-resistance pathological isoform of the host-encoded prion protein [PrP]) are found in the organs of the lymphoreticular system (LRS) before the involvement of the brain. Moreover, early splenectomy of intraperitoneally infected mice prolongs the incubation period, and severe combined immunodeficient (SCID) mice cannot be infected via the peripheral route with the Creutzfeldt-Jakob disease agent or low doses of the scrapie agent. We questioned the requirement of an LRS replication phase in the pathogenesis of the disease. SCID, immunologically reconstituted SCID and CB17 control mice were intraperitoneally inoculated with the C506M3 scrapie strain. Immunological reconstitution restored the full susceptibility to scrapie infection, while only 33% of the SCID mice developed the disease. The incubation period was significantly prolonged in reconstituted mice in comparison to the control group. PrP-res levels at the terminal stage of disease were identical in the brains of the animals from the three groups. However, no PrP-res was detectable in the spleen of the diseased SCID mice, and the PrP-res level was four times lower in the spleen of reconstituted SCID as compared to CB17 control mice. Moreover, immunoglobulin quantification indicated that reconstitution was incomplete; therefore, the number of functional cells of the immune system seems to correlate with the number of target cells of the scrapie agent. This study shows the role of the immune system as a primary target of the scrapie agent and that a primary replication step of the agent in the LRS favours neuroinvasion and hence, the development of clinical disease. On the contrary, the fact that a small proportion of the SCID mice also exhibited scrapie within the same period of time as the control mice, and without PrP-res detection in the spleen, strongly suggests that the scrapie agent can spread directly to the nervous system from the peritoneum, probably through autonomic visceral nerve fibres. From this point of view, the role of the peripheral replication step would not involve processing of the agent, but would merely consist of increasing the probability for the agent to reach the central nervous system.

Introduction

Scrapie is one of the transmissible spongiform encephalopathies [TSE] [1] occurring as Creutzfeldt-Jakob disease (CJD) in humans [2, 3] and bovine spongiform encephalopathy in animals [4]. The nature of the causative agent is still controversial. A modified, partially protease-resistant isoform of the host encoded prion protein (PrP), named PrP-res, is associated with infectivity in experimental models [5, 6]. PrP-res accumulates in the brains and organs

___

Key words: *severe combined immunodeficient mouse / scrapie / immune system / pathogenesis*

of the lymphoreticular system (LRS) of infected animals [7–9]. Infectivity and PrP-res accumulation in the spleen are detectable long before the involvement of the central nervous system (CNS) [10–15]. Scrapie incubation time is unaffected by thymectomy [14], and no specific immunological change could be detected during the time course of the disease [16–18]. Studies involving splenectomy [14], whole body irradiation [19] and spleen fractionation [20] have shown that cells sustaining replication within the LRS are mitotically quiescent and located in the stromal fraction of the spleen. They seem to be restricted to cells with the morphological characteristics and the antigen trapping capacity of follicular dendritic cells [21–23]. There is strong evidence showing that the pathway of neuroinvasion from the sites of early infection and replication in LRS tissues, such as the spleen, involves visceral autonomic (probably sympathetic) fibres that form part of the splanchnic nerve complex. This pathway targets the infection to the mid-thoracic spinal cord whence it spreads to the brain [24]. However, the results of splenectomy suggest that, following intraperitoneal (ip) infection of Syrian hamsters with the 263K strain of the scrapie agent, neuroinvasion may occur directly via prevertebral ganglia or nerve endings in the peritoneal wall without the need for substantial agent replication in the LRS [25]. The possibility of a direct infection of nerve endings has also been raised from the splenectomy data obtained after intragastric infection of mice [26]. Severe combined immunodeficient (SCID) mice, CB17 mouse mutants [27] which lack functional follicular dendritic cells in addition to lymphocytes [28], did not develop clinical signs after ip administration of the CJD strain Fukuoka-1, although they did after infection via the intracerebral (ic) route [21]. Neither did SCID mice develop scrapie after inoculation of the Me7 scrapie strain by scarification, although this route was highly effective in immunocompetent mice [29]. The spleens of ip and ic inoculated animals with the CJD strain did not contain any detectable PrP-res [21]. Similarly, SCID mouse spleens collected during the first 90 days after ip infection with the the Me7 strain contained neither infectivity nor PrP-res [30]. These results strongly suggest that an LRS infection phase, and presumably some replication, is necessary prior to neuroinvasion and the development of clinical disease when infection is achieved by a peripheral route. We therefore investigated the role of the immune system in the early pathogenesis of scrapie after a peripheral inoculation by means of the SCID model and its immunological reconstitution.

Materials and methods

Three-week-old SCID mice were reconstituted with unfractionated CB17 splenic cells. The CB17 spleens were gently homogenised. The single-cell suspensions were injected ip into SCID mice, with a ratio of one spleen per SCID mouse, in a scrapie-free laboratory. Scrapie infection was achieved 5 weeks later.

The experimental C506M3 scrapie strain was kindly provided by P Brown (National Institutes of Health, Bethesda, MD, USA) and titrated in our laboratory. SCID, reconstituted SCID and CB17 control mice were inoculated ip at 2 months of age with 100 μL of a 0.2% brain homogenate of a C57BL/6 mouse terminally infected with the disease, corresponding to 1.6×10^5 ic 50% lethal dose (LD$_{50}$) of the scrapie agent. To control the absence of contamination during the time course of the experiments, three SCID and three reconstituted SCID mice were sham infected with 100 μL of a 0.2% C57BL/6 mouse brain homogenate. The mice were reared in specific pathogen-free conditions. They were sacrificed just before death, and the spleens and brains were immediately frozen in liquid nitrogen and were stored at -80 °C for PrP analysis. The incubation period was defined as the interval between injection and sacrifice [31].

PrP-res was purified by a scrapie-associated fibril protocol previously described [32]. Proteinase K was used at 10 μg/mL for 1 h at 37 °C. Samples equivalent to 2.5 mg of brain or 4 mg of spleen tissue (which represented 1/5 to 1/10 of the total spleen weight) were loaded in each lane, run on a 12% poly-acrylamide gel and transferred onto a nitrocellulose membrane (Schleicher and Schuell, Ecquevilly, France). Immunoblotting was performed with a polyclonal anti-mouse PrP antibody kindly provided by RJ Kascsak (NYS Institute for Basic Research in Developmental Disabilities, Staten Island, NY, USA) and a peroxidase-conjugated secondary antibody (Southern Biotechnology, Birmingham, AL, USA). The blots were developed with a chemiluminescence enhancement kit (Amersham, Les Ulis, France), quantified with a radioimager (Bio-Rad, Ivry-sur-Seine, France) and visualised on autoradiographic films. The sensitivity of detection is 10 μg of scrapie-infected mouse brain.

Three of the 12 SCID survivors, as well as the unique survivor in the reconstituted group, were sacrificed at 500 days post-inoculation (dpi) to compare their immunoglobulin concentrations with those of the diseased SCID, reconstituted SCID and CB17 mice. Spleen lysates were spun for 15 min at 13,000 g. The supernatants were serially diluted in phosphate-buffered saline-bovine serum albumin (0.005%) and loaded onto a nitrocellulose membrane by a slot blot apparatus (Bio-Rad). Mouse immunoglobulin (Ig) was detected with a peroxidase-conjugated antibody (Southern Biotechnology). Quantification was performed as described for the Western blot procedure.

Results

All CB17 mice developed the disease following infection. In addition, all but one reconstituted SCID mouse developed scrapie, however with a statistically significant delay of 42 days compared with that of the CB17 mice (table I). A significant proportion of the non reconstituted SCID mice also infected by the ip route developed scrapie. In this case, the transmission rate was 33%, and none of the survivors exhibited any clinical signs of disease up to 500 dpi. However, the incubation time for the infected SCID group was not statistically different from that for the CB17 control group, and some mice developed the disease even earlier than the CB17 mice.

In the spleens of those 33% non reconstituted SCID mice who developed the disease, no PrP-res was detected at the terminal stage of the disease (fig 1). On the other hand, PrP-res was readily detected in the spleens of the CB17 mice and, at a four-fold lower level, in the spleens of the reconstituted SCID mice (fig 1). In contrast, the brains of the diseased non reconstituted SCID, reconstituted SCID and CB17 mice exhibited similar amounts of PrP-res (fig 1). None of the sham-infected mice developed any neurological signs nor exhibited PrP-res in either the brain or the spleen.

The difference in the transmission rates between reconstituted (93%) and non-reconstituted (33%) SCID mice, and the fact that PrP-res was detectable only in spleens in the former of these two groups strongly suggest that a primary replication step of the agent in the LRS

Table I. Transmission data of mice inoculated ip with 100 μL of 0.2% brain homogenate of a mouse at the terminal stage of scrapie, strain C506M3.

	CB17	Reconstituted SCID	SCID
No of dead mice (total no of mice)	19 (19)	13 (14)	6 (18)
Percentage of transmission	100	93	33
Incubation period in days (mean ± SEM)	324 ± 11	366 ± 12[1]	344 ± 45[2]

[1]Statistically significantly different from incubation period of CB17 mice; t-test: $P = 0.0184$; Mann and Whitney: $P = 0.028$. [2]Not statistically different from incubation period of CB17 mice. SCID: severe combined immunodeficiency.

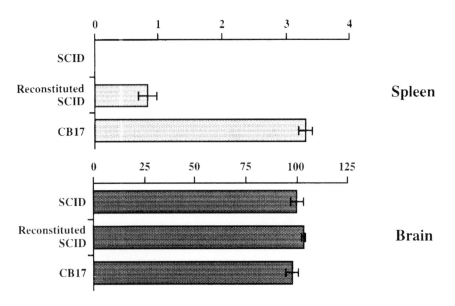

Fig 1. Western blot detection of PrP-res in the spleens and brains of mice at the terminal stage of scrapie. Each mouse group corresponds to a triplicate. PrP-res levels shown in severe combined immunodeficiency (SCID) mice correspond to those of the 33% of SCID mice that developed scrapie. PrP-res levels are expressed as the percentages of PrP-res in the brains of C57BL/6 mice at the terminal stage of scrapie (positive control).

favours neuroinvasion. The results of many other experiments have led to the concept of limited scrapie replication sites in both the LRS and the CNS [33–35]. The smaller amounts of PrP-res detected in the spleens of reconstituted mice and the delay in the onset of the disease compared with those in the CB17 control animals may be related to the incomplete reconstitution of these mice with some subset of immune cells, as suggested by the lower levels of Ig found in the reconstituted animals (fig 2). Under these circumstances, the peripheral scrapie replication rate seems to depend directly on the number of target cells of the LRS.

Although most of the non-reconstituted SCID mice did not become affected by the ip route, 33% developed scrapie within the same period of time as the CB17 controls, suggesting that the severely immunologically depleted LRS organs of the animals were not involved in the pathogenesis of the disease. This discrepancy between our results and those of other studies with no peripheral transmission in SCID mice [21, 29] may be accounted for by the use of a different strain of agent and/or different doses. Indeed, it has recently been shown that, although SCID mice resist ip infection with 10^3–10^4 ic LD_{50} of the Me7 strain, higher doses can produce disease (KL Brown, this volume). Nevertheless, the absence of detection of PrP-res in the spleens of infected SCID mice is in agreement with other assessments that SCID mouse spleens do not support TSE agent replication [21, 30]. It is very unlikely that a 'leaky' SCID phenotype [36], which leads to the development of oligoclonal B and T cells, accounts for the pathogenesis of the disease in scrapie-infected SCID mice. Indeed, as in surviving animals, the diseased SCID mice had very low Ig levels, indicating the homogeneity of the SCID group (fig 2).

The possibility of a persistent peripheral infection, with neuroinvasion occurring only in 33% of the SCID mice, cannot be discounted [37]. However, this hypothesis implies that the

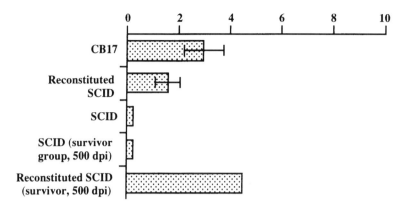

Fig 2. Slot blot quantification of total immunoglobulins (Ig) in spleen cell lysates of scrapie-infected severe combined immunodeficiency (SCID), reconstituted SCID and CB17 mice, and comparison with the SCID survivor group as well as the unique reconstituted SCID survivor. Each group was tested in triplicate. Arbitrary units were used to express the relative levels of Ig. Standard deviations of the two SCID groups represented were too small to appear in the figure. dpi: days post-inoculation

immune system of SCID mice would be able to support TSE agent replication, which is not in accordance with current data [21, 30] and with our results showing the absence of detectable PrP-res in the spleens and the profound depletion of the immune systems of the diseased SCID mice. Therefore, the most likely explanation for the transmissibility of the disease to the SCID mice is that peripheral scrapie infection can occur independently of the LRS, probably via a direct neural spread from the peritoneum, as has been previously proposed by Kimberlin and Walker [25, 26].

The transmission rate of the disease would reflect the probability for the agent to reach the nervous system. We observed a relative variability in the incubation periods of SCID mice. One explanation could be that the time necessary to reach the CNS along neuroanatomical pathways would be variable, depending, for example, on the type of nerve cell encountered. The fact that the majority of the SCID mice did not develop scrapie after ip injection strongly suggests the occurrence of clearance of the inoculum in the peritoneum [38]. In the other groups, replication in LRS organs would have prevented elimination of the agent from the organism. Besides, the prolonged but homogeneous incubation periods in the reconstituted mice suggest that the time before neuroinvasion depends on the number of peripheral target cells.

Therefore, the LRS role would be first to trap the agent in immunocompetent, non dividing cells such as follicular dendritic cells, and, second, by allowing continuous replication of the agent, to permit the agent to reach its nervous targets. An important point would be to determine the requirement of normal PrP for the peripheral phase of infection and for the transport of the agent to the CNS, as its importance for the development of disease and pathological lesions in the brain has already been demonstrated [39, 40].

The understanding of the initial pathway of infection in TSE is of major importance because no efficient treatment to prevent infection in cases of an accidental contamination by TSE agents is yet available. This study suggests that early inhibition of agent replication, by allowing the apparently natural clearance of the contaminating agent, may prevent spread to the nervous system and clinical disease in cases of peripheral accidental contamination.

Acknowledgments

The Laboratoire de Pathogénie Expérimentale (Institut Pasteur, Lille) is supported by grants from the Institut Pasteur de Lille, the Centre National de la Recherche Scientifique, the Institut National de la Santé et de la Recherche Médicale and the Ministère de l'Enseignement Supérieur et de la Recherche (MESR ACCSV no 10). TSE research in the 'Service de Neurovirologie' is made possible by grants from DRET and MESR ACCSV no 10 (Paris, France).

References

1 Cuillé J, Chelle PL. La maladie dite tremblante du mouton est-elle inoculable? *C R Acad Sci* 1936;203:1552-4
2 Creutzfeldt HG. Uber eine eigenartige herdformige erkrankung des zentralnervensystems. *Z Neurol U Psychiatr* 1920;57:1-18
3 Jakob A. Uber eine eigenartige Erkrankung des Zentral-nervensystems mit bemerkenswertem anatomischem Befunde (spastische pseudosklerotische Encephalomyelopathie mit disseminierten Degenerationsherden). *Dtsch Z Nervenheilk* 1921;70:132-46
4 Wells GAH, Scott AC, Johnson CT et al. A novel progressive spongiform encephalopathy in cattle. *Vet Rec* 1987;121:419-20
5 Prusiner SB. Novel proteinaceous infectious particles cause scrapie. *Science* 1982;216:136-44
6 Bolton DC, McKinley MP, Prusiner SB. Identification of a protein that purifies with the scrapie prion. *Science* 1982;218:1309-11
7 Doi S, Ito M, Shinagawa M, Sato G, Isomura M, Goto H. Western blot detection of scrapie-associated fibril protein in tissues outside the central nervous system from preclinical scrapie-infected mice. *J Gen Virol* 1988;69: 955-60
8 Kitamoto T, Mohri S, Tateishi J. Organ distribution of proteinase-resistant prion protein in humans and mice with Creutzfeldt-Jakob disease. *J Gen Virol* 1989;70:3371-9
9 Rubenstein R, Merz PA, Kascsak RJ et al. Scrapie-infected spleens: analysis of infectivity, scrapie-associated fibrils, and protease-resistant proteins. *J Infect Dis* 1991;164:29-35
10 Fraser H, Dickinson AG. Pathogenesis of scrapie in the mouse: the role of the spleen. *Nature* 1970;226:462-3
11 Eklund CM, Kennedy RC, Hadlow WJ. Pathogenesis of scrapie virus infection in the mouse. *J Infect Dis* 1967;117:15-22
12 Kuroda Y, Gibbs CJ, Amyx HL, Gajdusek DC. Creutzfeldt-Jakob disease in mice: persistent viremia and preferential replication of virus in low-density lymphocytes. *Infect Immun* 1983;41:154-61
13 Kimberlin RH, Walker CA. Pathogenesis of mouse scrapie: dynamics of agent replication in spleen, spinal cord and brain after infection by different routes. *J Comp Pathol* 1979;89:551-62
14 Fraser H, Dickinson AG. Studies of the lymphoreticular system in the pathogenesis of scrapie: the role of spleen and thymus. *J Comp Pathol* 1978;88:563-73
15 Kimberlin RH, Walker CA. Incubation periods in six models of intraperitoneally injected scrapie depend mainly on the dynamics of agent replication within the nervous system and not the lymphoreticular system. *J Gen Virol* 1988;69:2953-60
16 Dormont D, Herodin F, Laupretre ND et al. Biochemical and immunological events during unconventional virus infections. 1. GFAP, IgG, IgM, complement evolutive patterns. 2. Effects of modifications of reticuloendothelial system on scrapie incubation period. In: Court LA, Dormont D, Brown P, Kingsbury DT, eds. *Unconventional Virus Diseases of the Central Nervous System (Paris, 2–6 December 1986)*. Fontenay-aux Roses, France: CEA Diffusion, 1989:271-87
17 Kingsbury DT, Smeltzer DA, Gibbs CJJ, Gajdusek DC. Evidence for normal cell-mediated immunity in scrapie-infected mice. *Infect Immun* 1981;32:1176-80
18 Kasper KC, Stites DP, Bowman KA, Panitch H, Prusiner SB. Immunological studies of scrapie infection. *J Neuroimmunol* 1982;3:187-201
19 Fraser H, Farquhar CF. Ionising radiation has no influence on scrapie incubation period in mice. *Vet Microbiol* 1987;13:211-23

20 Clarke MC, Kimberlin RH. Pathogenesis of mouse scrapie: distribution of agent in the pulp and stroma of infect-ed spleens. *Vet Microbiol* 1984;9:215-25

21 Kitamoto T, Muramoto T, Mohri S, Doh-Ura K, Tateishi J. Abnormal isoform of prion protein accumulates in follicular dendritic cells in mice with Creutzfeldt-Jakob disease. *J Virol* 1991;65:6292-5

22 McBride PA, Eikelenboom P, Kraal G, Fraser H, Bruce ME. PrP protein is associated with follicular dendritic cells of spleens and lymph nodes in uninfected and scrapie-infected mice. *J Pathol* 1992;168:413-8

23 Muramoto T, Kitamoto T, Tateishi J, Goto I. The sequential development of abnormal prion protein accumula-tion in mice with Creutzfeldt-Jakob disease. *Am J Pathol* 1992;140:1411-20

24 Kimberlin RH, Walker CA. Pathogenesis of mouse scrapie: evidence for neural spread of infection to the CNS. *J Gen Virol* 1980;51:183-7

25 Kimberlin RH, Walker CA. Pathogenesis of scrapie (strain 263K) in hamsters infected intracerebrally, intra-peritoneally or intraocularly. *J Gen Virol* 1986;67:255-63

26 Kimberlin RH, Walker CA. Pathogenesis of scrapie in mice after intragastric infection. *Virus Res* 1989;12: 213-20

27 Bosma GC, Caster RP, Bosma MJ. A severe combined immunodeficiency mutation in the mouse. *Nature* 1983;301:527-30

28 Kapasi ZF, Burton GF, Shultz LD, Tew JG, Szakal AK. Induction of functional follicular dendritic cell devel-opment in severe combined immunodeficiency mice. *J Immunol* 1993;150:2648-58

29 Taylor DM, McConnell I, Fraser H. Scrapie infection can be established readily through skin scarification in immunocompetent but not immunodeficient mice. *J Gen Virol* 1996;77:1595-9

30 O'Rourke KI, Hulf TP, Leathers CW, Robinson MM, Gorham JR. SCID mouse spleen does not support scrapie agent replication. *J Gen Virol* 1994;75:1511-4

31 Dickinson AG, Meikle VMH, Fraser HG. Identification of a gene which controls the incubation period of some strains of scrapie agent in mice. *J Comp Pathol* 1968;78:293-9

32 Kascsak RJ, Rubenstein R, Merz PA, Carp RL, Wisniewski HM, Diringer H. Biochemical differences among scrapie-associated fibrils support the biological diversity of scrapie agents. *J Gen Virol* 1985;66:1715-22

33 Dickinson AG, Fraser H, Meikle VMH, Outram GW. Competition between different scrapie agents in mice. *Nature New Biol* 1972;237:244-5

34 Dickinson AG, Fraser H, McConnell I, Outram GW, Sales DI, Taylor DM. Extraneural competition between different scrapie agents leading to loss of infectivity. *Nature* 1975;253:6

35 Kimberlin RH, Walker CA. Competition between strains of scrapie depends on the blocking agent being infec-tious. *Intervirology* 1985;23:75-81

36 Bosma MJ, Carroll AM. The SCID mouse mutant: definition, characterization, and potential uses. *Annu Rev Immunol* 1991;9:323-50

37 Collis SC, Kimberlin RH. Long-term persistence of scrapie infection in mouse spleens in the absence of clinical disease. *FEMS Microbiol Lett* 1985;29:111-4

38 Dickinson AG, Fraser H, Outram GW. Scrapie incubation time can exceed natural lifespan. *Nature* 1975;256:732-3

39 Büeler H, Aguzzi A, Sailer A et al. Mice devoid of PrP are resistant to scrapie. *Cell* 1993;73:1339-47

40 Brandner S, Isenmann S, Raeber A et al. Normal host prion protein necessary for scrapie-induced neurotoxicity. *Nature* 1996;379:339-43

Transmissible Subacute Spongiform Encephalopathies:
Prion Diseases
L Court, B Dodet, eds

Scrapie in immunodeficient mice

Karen L Brown, Karen Stewart, Moira E Bruce, Hugh Fraser

BBSRC and MRC Neuropathogenesis Unit, Institute for Animal Health, West Mains Road, Edinburgh, Scotland, UK

Summary – Infection of severe combined immunodeficient (SCID) mice with high dose Me7 scrapie by intracerebral (ic), intraperitoneal (ip) or two subcutaneous (sc) routes, produces central nervous system (CNS) disease. Bioassay of spleens from SCID mice infected with 10% Me7 scrapie brain homogenate by ip, ic or abdominal sc injection showed traces or low levels of infectivity in the spleen. However, sc injection beneath the skin of the neck failed to infect the spleen. This suggests that following high-dose infection, peripherally infected SCID mice develop CNS disease without agent replication in spleen. CB20 bone marrow reconstitution of SCID mice resulted in the regeneration of a normal lymphoid architecture in the spleen. Spleens from these reconstituted mice, infected intracerebrally with 10% Me7 inocula contained high levels of infectivity. These results suggest that the ability to replicate the agent in the spleen or lymphoid tissue depends on the restoration of normal lymphoid structure and in particular the presence of differentiated follicular dendritic cells.

Introduction

Although scrapie is a degenerative disease only of the central nervous system (CNS), the replication of infectivity in the lymphoreticular system can precede that in the CNS by many weeks or months. There is strong evidence that the target cell permitting agent replication in the CNS is the neuron [1], but the candidate cells for peripheral replication are not yet defined [2]. The use of whole body ionizing radiation before or after infection has no effect on the incubation period or replication of the scrapie agent in the spleen in several scrapie models [3, 4]. The major conclusion from these findings is that scrapie agent replication outside the CNS depends on nondividing, long-lived cells [5], and their identity as follicular dendritic cells (FDC) has been proposed [2, 3]. The presence of cells with the morphology and location of FDC in the spleen and lymph nodes of mice, which immunolabel with anti-prion protein (PrP) and entrap immune complexes provides further support for this suggestion [6]. The availability of the severe combined immunodeficient (SCID) mouse, which shows resistance to peripheral infection with a murine strain of Creutzfeldt-Jakob disease [7], provides further opportunities to investigate the role of the immune system in the peripheral pathogenesis of scrapie. These mice have an autosomal recessive mutation that prevents the formation of functional B and T lymphocytes. In addition, SCID mice possess nonfunctional FDC, a defect which can be reversed after reconstitution with isogenic bone marrow [8].

Key words: severe combined immunodeficiency / follicular dendritic cells / scrapie

Materials and methods

Me7 infection of SCID mice and spleen bioassay

SCID mice were infected with a 10% Me7 brain homogenate by intracerebral (ic), intraperitoneal (ip) and two subcutaneous (sc) routes (into the scruff of the neck or into the skin of the ventral abdomen). Injections were carried out using a 25-gauge needle and a volume of 0.02 mL. Two animals from each route of injection were sacrificed 50 days post-infection (dpi) and spleens removed for bioassay. Spleens from individual animals were prepared as 10% w/v homogenates in physiological saline and a 0.02-mL volume injected ic into groups of 12 recipient C3H mice. The remaining SCID mice and the bioassay mice were scored weekly to determine the incubation period of the clinical disease according to previously established criteria [9].

Bone marrow reconstitution and scrapie infection

Bone marrow was prepared from femurs and tibias of adult CB20 mice. Cells were suspended in Iscove's Dulbecco's medium (Gibco BRL) and a 0.1-mL volume was injected intravenously (iv) into the tail vein. When the iv injection was unsuccessful, 0.1 mL was injected ip, or the injection was termed perivenous and the volume less than 0.1 mL. SCID mice receiving bone marrow were maintained under standard specific pathogen-free conditions. Five weeks post-reconstitution, mice were infected ic with a 10% Me7 brain homogenate. At 5, 10, 15 and 20 weeks post-injection, the animals were sacrificed and their spleens harvested aseptically. Half of the spleens were taken for bioassay and the remainder used for immunocytochemistry.

Histology/immunocytochemistry

Spleens from scrapie-infected or uninfected SCID and bone marrow-reconstituted SCID mice were immunolabelled with anti-PrP antibody (1B3) [6], an anti-FDC monoclonal antibody (mAb) (designated FDC-M1) [10]. Cryostat sections of spleen, briefly fixed in acetone, were used with the FDC-M1 mAb, whereas periodate–lysine–paraformaldehyde (PLP)-fixed, paraffin-embedded sections were used with the anti-PrP antibody. Immunolabelling was carried out using the avidin–biotin complex (ABC) technique or the standard peroxidase–anti-peroxidase technique.

Results

Me7 infection of SCID mice

All SCID mice became affected with scrapie, using the ic, ip and both sc routes of infection. This result contrasts with the failure of SCID mice to develop CNS disease following ip challenge with lower doses of Me7 scrapie (1 or 0.1% dilutions) [11]. There was inconsistency in the presence of the scrapie agent in SCID mouse spleens (table I). Abdominal sc injection resulted in a low level of the agent in one of the two spleens assayed while the agent was absent in both spleens after sc injection at the neck. Some difficulty was encountered with the administration of the sc injections into the abdomen and it is probable that one such injection in one mouse was inadvertently ip. On the other hand, ip injection produced consistent but low levels of the agent in both spleens with a wide variation in the length of incubation periods between individual animals. After ic injection only traces of infectivity were found in the SCID spleen with a very low incidence of scrapie in assay mice.

Bioassay of bone marrow-reconstituted SCID spleens

Infectivity in the spleens of reconstituted SCID mice was assessed 5, 10, 15 and 20 weeks after ic infection (table II). High levels were present in spleens of mice receiving bone marrow compared with unreconstituted SCID infected by the ic route (see table I). Spleens from SCID

Table I. Infectivity bioassay of spleens from SCID mice infected with a 10% dilution of scrapie brain by one of four routes of injection and harvested 50 dpi. Bioassay carried out by ic injection of 10% spleen homogenate into C3H mice.

Route of infection	Incubation period and incidence (x/y) in SCID mean ± SE	Bioassay spleen number	Scrapie incidence in assay mice	Individual incubation periods in assay mice or mean ± SE (days)
Intracerebral	155 ± 3 (11/11)	1	2/12	260, 489
		2	3/12	289, 556, 568
Intraperitoneal	256 ± 4 (9/9)	1	11/11	236 ± 14
		2	11/12	263 ± 12
Subcutaneous				
Neck	229 ± 12 (11/11)	1	0/12	–
		2	0/12	–
Abdomen	247 ± 13 (7/7)	1	0/11	–
		2	9/11	312 ± 21

Table II. Infectivity bioassay of half spleens from SCID mice reconstituted with CB20 bone marrow and infected ic with a 10^{-1} dilution of Me7 35 days following reconstitution. Bioassay carried out by ic injection of C3H mice.

Spleen bioassay (wpi)	Incidence	Incubation period (days) mean ± SE	Route of bone marrow injection
5	11/11	170 ± 2[a]	iv
5	12/12	179 ± 2	ip
10	12/12	166 ± 2	iv
10	12/12	179 ± 3[b]	iv
10	11/11	213 ± 7	Perivenous
15	12/12	178 ± 2	Perivenous
15	11/11	198 ± 4	Perivenous
15	12/12	179 ± 2	ip
15	12/12	182 ± 1	ip
20	11/11	176 ± 2	ip
20	12/12	173 ± 2	ip
20	12/12	247 ± 6	iv
20	12/12	181 ± 1	iv

[a]The corresponding histology for this spleen is depicted in figure 5; [b]the corresponding histology for this spleen is depicted in figure 6.

mice receiving bone marrow iv (*n* = 5) contained high levels of infectivity, with one exception where the level of infectivity present at 140 dpi was lower than in the other spleens bioassayed. Spleens from SCID mice receiving bone marrow ip (*n* = 5) also contained high levels of infectivity and of the three spleens from mice which received perivenous injections, two contained low levels of infectivity, whereas one spleen contained slightly higher levels.

Spleen morphology

Immunolabelling with the FDC-M1 mAb in the infected and uninfected SCID spleens identified some very immature and undifferentiated cells in some spleens although there was variation in the presence and intensity of this labelling between animals and even within single spleens (fig 1). Immunolabelling with the anti-PrP antibody in both infected and uninfected SCID mice showed weak labelling of possible immature FDC in some spleens, similar to that obtained with the FDC-M1 mAb (fig 2). In contrast, labelling the CB20 spleen with the FDC mAb (fig 3) and the anti-PrP antibody revealed a normal structure (fig 4). SCID mice receiving bone marrow iv had spleens which were morphologically almost indistinguishable from those of normal mice. In addition, there was strong immunolabelling with the antibody to PrP (fig 5) and with the FDC mAb (fig 6), thus identifying the restoration of normal FDC structure. Ip injection of bone marrow also produced an effective restoration of lymphoid structure in the spleen of SCID examined. Inadvertent perivenous bone marrow injection achieved variable levels of nevertheless good reconstitution between animals.

Discussion

In this study we have shown that the development of scrapie in peripherally infected SCID mice is dose-dependent. Previously, peripheral infection of SCID mice with 1 or 0.1% Me7

Fig 1. Cryostat section of uninfected SCID spleen immunolabelled with the FDC-specific antibody (FDC-M1) with no labelling.

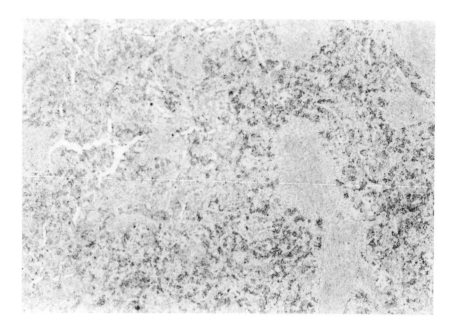

Fig 2. Paraffin section of periodate-lysine-paraformaldehyde-fixed spleen from a SCID mouse at the terminal stage of the disease. Immunolabelling with antibody to PrP (1B3) with no evidence of PrP labelling.

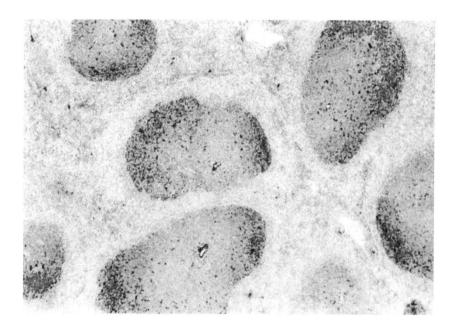

Fig 3. Cryostat section of uninfected CB20 spleen immunolabelled with FDC-M1 showing labelling of FDC in the white pulp.

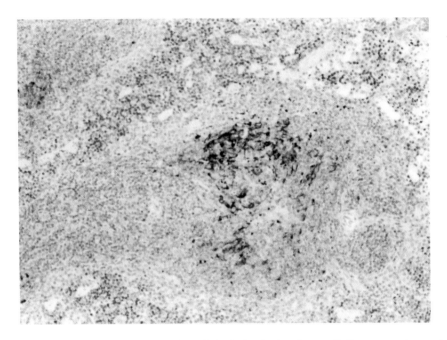

Fig 4. Paraffin section of periodate-lysine-paraformaldehyde-fixed spleen from a CB20 mouse at the terminal stage of the disease, immunolabelled with 1B3, showing PrP labelling in the white pulp.

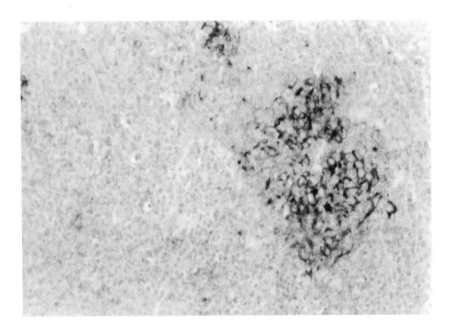

Fig 5. Paraffin section of periodate-lysine-paraformaldehyde fixed spleen from a CB20 bone marrow reconstituted SCID mouse 10 weeks post-Me7 infection immunolabelled with 1B3. Strong PrP labelling is present in the white pulp in contrast with the unreconstituted SCID spleen.

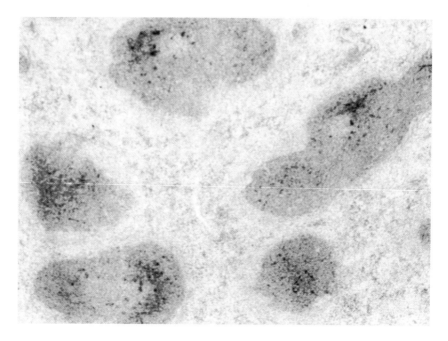

Fig 6. Cryostat section from a CB20 bone-marrow-reconstituted SCID spleen 5 weeks post-Me7 infection immuno-labelled with FDC-M1. Immunolabelling shows development of FDC in the white pulp in contrast with the lack of labelling and lymphoid structure in the unreconstituted SCID mice spleen.

failed to produce disease during a 600-day observation period [11]. One possible explanation is that peripheral nerve infection occurs at high dose, with direct entry to the CNS without replication in the spleen [12]. Bioassay of spleens indicated that while spleens taken from SCID mice injected by the ip and sc (abdomen) routes contained low levels of infectivity, the ic and most sc routes produced only trace levels or no splenic infectivity. It is clear that replication is not occurring in these SCID mouse spleens, although it seems that residual inoculum remaining in the peritoneal cavity becomes associated with the spleen. These results contrast with those of a similar study in which high dose infection of SCID mice did not result in disease or infectivity in the spleen, although in this study spleen bioassay was carried out using centrifuged supernatants [13].

In contrast, bioassay of spleens taken from the ic-infected bone-marrow reconstituted SCID mice at 5 to 20 weeks, in which reconstitution was iv or ip produced 100% incidence of disease in assay mice with incubation periods consistent with replication to high titre. The morphology of these spleens, in terms of FDC structure and lymphoid development, was almost indistinguishable from those of the normal mice examined. In the bioassay of three spleens for which reconstitution was perivenous there was variability in the amounts of infectivity present which correlated to the success of bone marrow reconstitution in these spleens. These results show that the amount of infectivity present in spleen is related to the success of the reconstitution, in terms of normal lymphoid structure and in the structure and presence of the FDC. Although this does not eliminate other interpretations it supports other data suggesting that FDC are permissive for scrapie replication. The SCID mouse model is being used in further experiments with the aim of confirming or excluding the FDC as the specific cell type responsible for replication of scrapie in the spleen.

Acknowledgments

The authors would like to acknowledge the gift of the mAb to FDC supplied by M Koscoe of the Basel Institute of Immunology and the gift of the mAb THY-1 from G Kraal of the Free University of Amsterdam. We would also like to acknowledge D Drummond for technical assistance.

References

1 Fraser H, Dickinson AG. Targeting of scrapie lesions and agent via the retino-tectal projection. *Brain Res* 1985;346:32-41

2 Fraser H, Bruce ME, Davies D, Farquhar CF, McBride PA. The lymphoreticular system in the pathogenesis of scrapie. In: Prusiner SB, Collinge J, Powell J, Anderton B, eds. *Prion Diseases of Humans and Animals.* Chichester: Ellis Horwood, 1992;308-17

3 Fraser H, Farquhar CF. Ionising radiation has no influence on scrapie incubation period in mice. *Vet Microbiol* 1987;13:211-23

4 Fraser H, Farquhar CF, McConnell I, Davies D. The scrapie disease process is unaffected by ionising radiation. In: Iqbal K, Wisniewski HM, Winblad B, eds. *Alzheimer's Disease and Related Disorders.* Chichester: Ellis Horwood, 1989;653-8

5 Fraser H, Davies D, McConnell I, Farquhar CF. Are radiation-resistant, post-mitotic, long-lived (RRPMLL) cells involved in scrapie replication? In: Court LA, Dormont D, Brown P, Kingsbury D, eds. *Unconventional Virus Diseases of the Central Nervous System.* Fontenay-aux-Roses: Commissariat à l'Energie Atomique, 1989;563-74

6 McBride PA, Eikelenboom P, Kraal G, Fraser H, Bruce ME. PrP protein is associated with follicular dendritic cells of spleens and lymph nodes in uninfected and scrapie-infected mice. *J Pathol* 1992;168:413-8

7 Kitamoto T, Muramoto T, Mohri S, Doh-Ura K, Tateishi J. Abnormal isoform of prion protein accumulates in follicular dendritic cells in mice with Creutzfeldt-Jakob disease. *J Virol* 1991;65:292-5

8 Kapasi ZF, Burton GF, Schulz LD, Tew JG, Szakal AK. Induction of functional follicular dendritic cell development in severe combined immunodeficient mice. *J Immunol* 1993;150:2648-58

9 Bruce ME, McConnell I, Fraser H, Dickinson AG. The disease characteristic of different strains of scrapie in *Sinc* congenic mouse lines: implications for the nature of the agent and host control of pathogenesis. *J Gen Virol* 1991;72:595-603

10 Koscoe MH, Pflugfelder E, Gray D. Follicular dendritic cell-dependent adhesion and proliferation of B cells in vitro. *J Immunol* 1992;148:2331-40

11 Fraser H, Brown KL, Stewart K, McConnell I, McBride P, Williams A. Replication of scrapie in spleens of SCID mice follows reconstitution with wild type bone marrow. *J Gen Virol* 1996;77:1935-40

12 Kimberlin RH, Walker CA. Pathogenesis of experimental scrapie. In: Bock G, Marsh J, eds. *Novel Infectious Agents and the Central Nervous System.* Ciba Foundation Symposium 135. Chichester: Wiley, 1988;37-53

13 O'Rourke KI, Huff TP, Leather CW, Robinson MM, Gorham JR. SCID mouse spleen does not support scrapie agent replication. *J Gen Virol* 1994;75:1511-4

Transmissible Subacute Spongiform Encephalopathies:
Prion Diseases
L Court, B Dodet, eds
© 1996, Elsevier, Paris

Cytokine immunoreactivity in the pathogenesis of murine scrapie: its relationship to PrPSc, gliosis, vacuolation and neuronal apoptosis

Alun Williams[1], Anne-Marie van Dam[2], Paul J Lucassen[3,4], Diane Ritchie[1]

[1]Institute for Animal Health, BBSRC & MRC Neuropathogenesis Unit, Ogston Building, West Mains Road, Edinburgh, EH9 3JF, UK; [2]Research Institute Neurosciences, Free University, Department of Pharmacology, Faculty of Medicine, van der Boechorststraat 7, 1081 BT, Amsterdam; [3]Graduate School Neurosciences Amsterdam, Netherlands Institute for Brain Research, Meibergdreef 33, 1105 AZ Amsterdam; [4]Division of Medical Pharmacology, Leiden/Amsterdam Centre for Drug Research, PO Box 9503, 2300 RA Leiden, The Netherlands

Summary – The relationship between prion protein (PrP) deposition, cytokine immunoreactivity, glial activation and vacuolation was investigated in the hippocampus and thalamus of a murine scrapie model. Within an affected area, PrP deposition preceded the onset of cytokine immunoreactivity and glial activation. These changes occurred early in the course of disease while vacuolation occurred later. A study of the relationship between these events and the occurrence of apoptosis showed that the onset of cytokine production by activated glia preceded the apoptotic loss of CA1-CA2 hippocampal neurons. These data suggest that the infection of neurons which leads to abnormal PrP metabolism also provides a stimulus for cytokine production by activated glia. Whether these in turn affect neuronal responses to infection and the course of disease remains to be determined.

The neuropathology of the transmissible spongiform encephalopathies (TSE or 'prion diseases') is generally described as a triumvirate of vacuolation of neurons and neuropil, a marked gliosis and neuronal loss [1; see also Fraser et al, this publication]. Accumulation of disease-specific prion protein (PrPSc) also occurs in affected areas of the brain [2–4]. The gliosis consists of both an astrocytosis and a microglial activation [5–10].

Activated glia in scrapie upregulate many factors including the cytokines interleukin IL-1, IL-6 and tumour necrosis factor (TNF) α [11–13]. These cytokines exert various effects in the central nervous system (CNS) (see ref [14] for review). In Alzheimer's disease, which shows many similar pathological features to the TSE [15], glia showing immunoreactivity for cytokines such as IL-1 and IL-6 have been detected in association with the amyloid (βA4) plaques [16], and IL-1 has been shown to increase the expression of the amyloid precursor protein (APP) mRNA [17, 18].

Previous studies have reported increased expression of mRNA for the cytokines IL-1, IL-6 and TNFα in mice showing clinical signs of TSE [11, 12]. Increased immunoreactivity of IL-1, IL-6, TNFα and related factors such as the prostaglandins E_2 (PGE$_2$) and $F_{2\alpha}$, and lipocortin-1 has also been reported in scrapie-affected mice [13]. However, although the intra-

Key words: scrapie / cytokine / apoptosis / glia

ocular injection of TNFα has been demonstrated to produce myelin ballooning in the optic nerve similar to that observed in the white matter lesions of the panencephalopathic form of Creutzfeldt-Jakob disease (CJD) [19], the role of cytokines in the neuropathogenesis of TSE is not known.

In this paper, we present our investigations of cytokine immunoreactivity in the CNS during the development of scrapie. We investigated hippocampal and thalamic cytokine immuno-reactivity during the incubation period of the 301V/VM scrapie model, and compared tem-poral aspects of this immunoreactivity with those of prion protein (PrP) deposition, glial activa-tion, vacuolation and neuronal loss. Mice, inoculated intracerebrally with 20 μL of a 1% brain homogenate from a clinically affected donor mouse, were studied at 30-day intervals through-out the incubation period (mean, 117 ± 1 days). The mice were perfusion-fixed using Bouin's fixative (for cytokine immunoreactivity, glial fibrillary acidic protein (GFAP) staining for astrocytes and F4/80 staining for microglia), paraformaldehyde-periodate-lysine (PLP; for PrP immunocytochemistry and GFAP), PLP containing 0.05% glutaraldehyde (PLP-G) for F4/80 or 4% paraformaldehyde (PFA) for in situ end-labelling studies. Other mice were sacrificed by cervical dislocation and their brains immersion-fixed in formol saline (for routine histo-pathological examination). Bouin's fixed brains were examined as 50 μm-vibratome sections; those fixed using PLP, PFA or formal saline as 5-μm wax sections and PLP-G fixed brains as 10 μm cryostat sections. These methods of preservation were chosen to maximize immuno-reactivity of the various factors examined (eg, cytokines by Bouin's fixation, F4/80 by PLP-G) and to provide good glial morphology to compare with the cytokine staining. Immuno-reactivity for all factors was detected using a PAP method; in addition, an ABC method was used to detect F4/80 staining. Control tissue was obtained from mice inoculated with a homo-genate of normal brain. Sections stained with preadsorbed antibody (IL-1β, IL-6, TNFα, PGE$_2$), normal serum or without applying a primary antibody acted as controls for the staining protocols. Cytokine and PGE$_2$ immunoreactivity was abolished or markedly reduced after preadsorption of the primary antibody with 1 μM of the appropriate recombinant protein; no immunoreactivity for these factors, PrP, GFAP or F4/80, was detected in the absence of primary antibody or when normal serum was used.

There was no increase in glial staining or detectable induction of cytokine immunoreactivity at 30 days post-inoculation (dpi) and vacuolation was also absent. However, PrP deposition was present in the CA1-CA2 pyramidal neuron layer in the hippocampus and in the dorsal thalamic nuclei in 75% of the mice examined (table I). It was present as small foci of granular deposits in the neuropil and perineuronal staining.

The amount of PrP immunoreactivity was greater at 60 dpi and was present in all mice examined; both granular and perineuronal staining were observed and some immunoreactivity was associated with glia. There were more immunoreactive foci in the hippocampus (partic-ularly the CA1-CA2 region) and thalamus than at 30 dpi and the intensity of staining was greater. Scrapie-induced cytokine and PGE$_2$ immunoreactivity were also detected at 60 dpi. In all mice, IL-1β and TNFα staining were detected in perivascular cells in areas of the thalamus that also showed PrP immunoreactivity. Glial PGE$_2$ staining was also observed in both the thalamus and hippocampus of 50% of the mice examined. GFAP and F4/80 staining revealed activated astrocytes and microglia, respectively, confined to those areas of the thalamus and hippocampus showing PrP deposition and cytokine/PGE$_2$ immunoreactivity. Vacuolation was still absent.

Table I. Summary of results in scrapie-infected mice.

| | Days post-inoculation | | | | |
	30	60	90	105	115
PrP	±	+	++	+++	++++
IL-1	–	+	++	ND	+++
IL-6	–	–	+	ND	+++
TNFα	–	+	++	ND	+++
PGE$_2$	–	±	++	ND	+++
Lipocortin-1	–	–	+	ND	+++
F4/80	–	±	++	+++	++++
GFAP	–	±	++	+++	++++
Vacuolation	–	–	±	+	++
ISEL*	–	–	±	+++	+

–: no increase in immunoreactivity/in situ end-labelling (ISEL)/vacuolation compared to control mice; ±: mild increase in immunoreactivity/sparse ISEL-positive cells/low grade vacuolation in a proportion of mice; +: mild–moderate increase in immunoreactivity/(ISEL)/moderate vacuolation in all mice; ++: moderate increase in immunoreactivity/moderate vacuolation; +++: marked increase in immunoreactivity/ISEL; ++++: severe increases in immunoreactivity; ND: not done; *: described in hippocampus only.

At 90 dpi, PrP deposition was more diffusely distributed throughout the hippocampus and thalamus, although the intensity of this staining was comparable to that at 60 dpi. Within the hippocampus, more PrP immunoreactivity was present in the CA1-CA2 regions than elsewhere. Glial cytokine immunoreactivity (IL-1β, IL-6, TNFα) was present in all mice examined, in those areas showing PrP deposition. PGE$_2$ staining was stronger than at 60 dpi. Glial activation was more prominent than at 60 dpi and was also confined to those areas showing PrP deposition and cytokine immunoreactivity. Comparisons of the morphology of GFAP, F4/80 and cytokine staining indicated that both astrocytes and microglia were immunoreactive for IL-1, IL-6 and TNFα, but that only astrocytes were immunoreactive for PGE$_2$. Low grade vacuolation was present in the hippocampal CA1-CA2 regions and in the thalamus of 67% of the mice studied, occurring specifically in those areas showing PrP staining.

PrP deposition was most widespread and most intense in mice sacrificed when showing clinical signs of disease (111–114 dpi). Within the hippocampus, it was still most prominent in the CA1-CA2 region. Glial cytokine, PGE$_2$ and lipocortin immunoreactivity were stronger than at 90 days, but largely confined to those areas showing PrP deposition. Glial activation in these areas was also more pronounced and the intensity of vacuolation greater. Examination of haematoxylin-eosin stained sections also revealed a marked loss of CA1 pyramidal neurons in all mice examined at term. This loss of neurons was almost total in some mice, but had not been obvious on close visual inspection at 90 dpi.

This observation prompted a study of hippocampal neuronal loss using an in situ end-labelling (ISEL) technique [20]. Previous studies using this technique had shown that neuronal apoptosis could be demonstrated in clinically affected animals of various murine scrapie models, including the hippocampus of the 301V/VM model, and in some areas (but not the hippocampus) late in the incubation period of the 79A/C57BL model [21,22]. ISEL was therefore performed on PFA-fixed sections from the above series. In addition, mice sacrificed at 105 dpi were examined using both ISEL and electron microscopy. These investigations revealed the presence of low numbers of labelled apoptotic cells in clinically affected mice but many ISEL-labelled cells in the CA1-CA2 region of the hippocampus at 105 dpi, at a time when moderate neuronal loss was detectable on heamatoxylin-eosin sections. There were no significant differences in ISEL staining in the CA1-CA2 region of the hippocampus between scrapie-infected and control mice at 30 and 60 dpi, but sparse, individual labelled cells were present at 90 dpi. The ultrastructural studies of mice at 105 dpi confirmed the neuronal apoptosis.

These studies therefore suggest that within affected areas of the brain, PrP deposition precedes cytokine immunoreactivity and glial activation which in turn precedes neuronal loss and vacuolation. This scrapie-induced cytokine production and glial activation occurred early in the course of disease and before clinical signs were apparent (the initial, non specific signs of illness were observed from 95 dpi). These data therefore confirm and extend previous studies of cytokine expression in the CNS of a Chandler/SWRj murine scrapie model with a 10-week clinical course, where the onset of detectable increases in IL-1 and TNFα mRNA coincided with the onset of glial activation (increased mRNAs for GFAP and a murine analogue of α1-antichymotrypsin) and the onset of clinical symptoms [12].

Our findings support the concept that the infection of neurons by the scrapie agent which leads to PrP^{Sc} accumulation also provides a stimulus for glial cytokine production early in the course of disease. This glial cytokine production may act as a further signal for glial activation [23, 24] and may also contribute to the development of other pathological lesions in scrapie. In this respect, it has been proposed that TNFα may be responsible for the white matter lesions observed in certain forms of CJD [19].

Recent in vitro studies have suggested that the neurotoxicity of the PrP peptide, human PrP 106–126, is mediated via microglia [25; see also Kretzschmar et al, this publication], although direct neurotoxicity may also occur [26]. The data presented here demonstrate that accumulation of PrP^{Sc}, glial activation and induction of cytokine synthesis occur early in the course of disease and before most (if not all) of the neuronal apoptosis. While it is tempting to speculate that these data provide in vivo evidence to support the concept that the neuronal loss in scrapie is mediated by glial products (in a manner analogous to that proposed for Alzheimer's disease [16]), the question whether these cytokines affect the PrP metabolism and the neuronal response to infection by the TSE agent clearly requires further investigation.

References

1　Masters CL, Richardson EP Jr. Subacute spongiform encephalopathy (Creutzfeldt-Jakob disease): the nature and progression of spongiform change. *Brain* 1978;101:333-44

2　DeArmond SJ, Mobley WC, DeMott DL, Barry RA, Beckstead JH, Prusiner SB. Changes in the localization of brain prion proteins during scrapie infection. *Neurology* 1987;37:1271-80

3　Bruce ME, McBride PA, Farquhar CF. Precise targeting of the pathology of the sialoglycoprotein, PrP, and vacuolar degeneration in mouse scrapie. *Neurosci Lett* 1989;102:1-6

4　Bell JE, Ironside JW. Neuropathology of spongiform encephalopathies in humans. *Br Med Bull* 1993;49:738-77

5　Fraser H. The pathology of natural and experimental scrapie. In: Kimberlin RH, ed. *Slow Virus Diseases of Animals and Man.* Amsterdam & Oxford: North Holland Publishing Company, 1976;267-305

6　Jendroska K, Heinzel FP, Torchia M et al. Proteinase-resistant prion protein accumulation in Syrian hamster brain correlates with regional pathology and scrapie infectivity. *Neurology* 1991;41:1482-90

7　Barcikowska M, Liberski PP, Boellaard JW, Brown P, Gajdusek DC. Microglia is a component of the prion protein amyloid plaque in the Gerstmann-Sträussler-Scheinker syndrome. *Acta Neuropathol (Berl)* 1993;85: 623-7

8　Sasaki A, Hirato J, Nakazato, Y. Immunohistochemical study of microglia in the Creutzfeldt-Jakob diseased brain. *Acta Neuropathol (Berl)* 1993;86:337-44

9 Williams AE, Lawson L, Perry VH, Fraser H. Characterization of the microglial response in murine scrapie. *Neuropathol Appl Neurobiol* 1994;20:47-55

10 Mulheisen H, Gehrmann J, Meyermann R. Reactive microglia in Creutzfeldt- Jakob disease. *Neuropathol Appl Neurobiol* 1995;21:505-17

11 Liberski PP, Nerurkar VR, Yanagihara R, Gajdusek DC. Tumor necrosis factor: cytokine-mediated myelin vacuolation in experimental Creutzfeldt-Jakob disease. *Proc VIIIth Int Congr Virol* Berlin, Germany, 1990; Abstract P69-015:421

12 Campbell IL, Eddelston M, Kemper P, Oldstone MBA, Hobbs MV. Activation of cerebral cytokine gene expression and its correlation with onset of reactive astrocyte and acute-phase response gene expression in scrapie. *J Virol* 1994;68:2383-7

13 Williams AE, Van Dam AM, Man-A-Hing WKH, Berkenbosch F, Eikelenboom P, Fraser H. Cytokines, prostaglandins and lipocortin-1 are present in the brains of scrapie-infected mice. *Brain Res* 1994;654:200-6

14 Rothwell NJ, Hopkins SJ. Cytokines and the nervous system. II. Actions and mechanisms of action. *Trends Neurosci* 1995;18:130-6

15 Bruce ME, McBride PA, Jeffrey M, Rozemuller JM, Eikelenboom P. PrP in scrapie and βA4 in Alzheimer's disease show many similar patterns of deposition in the brain. In: Corain B, Iqbal K, Nicolini M, Winblad B, Wisniewski H, Zatta P, eds. *Alzheimer's Disease: Advances in Clinical and Basic Research.* New York: John Wiley & Sons Ltd, 1993;481-7

16 Eikelenboom P, Zhan SS, van Gool WA, Allsop D. Inflammatory mechanisms in Alzheimer's disease. *Trends Neurosci* 1994;15:447-50

17 Goldgaber D, Harris HW, Hla T et al. Interleukin 1 regulates synthesis of amyloid β-protein precursor mRNA in human endothelial cells. *Proc Natl Acad Sci USA* 1989;86:7606-10

18 Gray CW, Patel AJ. Regulation of β-amyloid precursor protein isoform mRNAs by transforming growth factor-β1 and interleukin-1β in astrocytes. *Mol Brain Res* 1993;19:251-6

19 Liberski PP, Yanagihara R, Nerurkar VR, Gajdusek DC. Tumour necrosis factor-α produces Creutzfeldt-Jakob disease-like lesions in vivo. *Neurodegeneration* 1993;2:215-25

20 Lucassen PJ, Chung WCJ, Vermeulen JP, van Lookeren Campagne M, van Dierendonck JH, Swaab DF. Microwave enhanced in situ end labeling of fragmented DNA: parametric studies in relation to post mortem delay and fixation of rat and human brain. *J Histochem Cytochem* 1995;43:1163-71

21 Giese A, Groschup MH, Hess B, Kretzschmar HA. Neuronal death in scrapie-infected mice is due to apoptosis. *Brain Pathol* 1995;5:213-21

22 Lucassen PJ, Williams A, Chung WCJ, Fraser H. Detection of apoptosis in murine scrapie. *Neurosci Lett* 1995;198:1-4

23 Giulian D, Woodward J, Young DG, Krebs JF, Lachman LB. Interleukin-1 injected into mammalian brain stimulates astrogliosis and neovascularisation. *J Neurosci* 1988;8:2485-90

24 Dickson DW, Lee SC, Mattiace LA, Yen SC, Brosnan C. Microglia and cytokines in neurological disease, with special reference to AIDS and Alzheimer's disease. *Glia* 1993;7:75-83

25 Brown DR, Schmidt B, Kretzschmar HA. Role of microglia and host prion protein in neurotoxicity of a prion protein fragment. *Nature* 1996;380:345-7

26 Hope J, Shearman MS, Baxter HC, Chong A, Kelly SM, Price NC. Cytotoxicity of prion protein peptide (PrP106-126) differs in mechanism from the cytotoxic activity of the Alzheimer's disease amyloid peptide, Aβ 25-35. *Neurodegeneration* 1996;5:1-11

Transmissible Subacute Spongiform Encephalopathies:
Prion Diseases
L Court, B Dodet, eds
© 1996, Elsevier, Paris

Telencephalic brain grafts in the study of scrapie pathogenesis

Adriano Aguzzi[1], Thomas Blättler[1], Michael Klein[1], Alex Räber[2], Charles Weissmann[2], Sebastian Brandner[1]

[1]*Department of Pathology, Institute of Neuropathology;* [2]*Institute of Molecular Biology, University of Zurich, CH-8091 Zurich, Switzerland*

Summary – It is still unknown whether central nervous system (CNS) neuropathology in spongiform encephalopathies is due to neurotoxicity of PrP^{Sc}, acute depletion of PrP^C, or to some other mechanism. To address this question, we grafted neural tissue from Tg*20* transgenic mice which overexpress PrP^C into the brain of scrapie-resistant PrP-deficient mice, and inoculated it with scrapie prions. Infected grafts developed all histopathological changes of scrapie to an extreme degree, and contained high amounts of PrP-res and infectivity, while neighboring cells remained unaffected. The life span of infected mice was not reduced, and no neurological symptoms developed. Surprisingly, graft-borne PrP migrated from the transplants in amounts sufficient to induce the formation of plaque-like deposits in the host brain. Even 16 months after infection, no pathological changes were seen in the PrP-deficient tissue immediately adjacent to the grafts and to such deposits. Therefore, the availability of endogenous PrP^C to the infectious agent in vivo rather than deposition of PrP-res correlates directly with scrapie neurotoxicity. In a second set of experiments, we addressed the question of the spread of the infectious agent from peripheral sites to the CNS of an infected organism. We found that prion infectivity administered to the eye does not reach the CNS of *Prnp$^{0/0}$* mice. To detect prion spreading, we transplanted embryonic neuroectodermal tissue from Tg*20* mice overexpressing PrP to the caudoputamen of *Prnp$^{0/0}$* recipient mice. Following intracerebral inoculation with mouse prions, grafts developed localized, severe spongiform encephalopathy. However, neither histopathological changes, production of proteinase-resistant PrP^{Sc}, nor replication of prion infectivity were detected in the grafts of mice inoculated intraocularly. Humoral immunity to PrP developed in several animals soon after Tg*20* grafting, independently of the route of prion inoculation; anti-PrP titers did not exert any influence on the course of the disease after intracerebral inoculation. Introduction of a *lck-Prnp* transgene abolished the immune response, but did not reconstitute the spread of infectivity via the intraocular route. These results uncover a novel role for PrP^C in the pathogenesis of transmissible spongiform encephalopathies, ie, support of infectious spreading within the CNS of an infected host.

Introduction

Two still unresolved questions in the field of prion diseases have captured much of the attention of researchers. The first question relates to the actual nature of the infectious agent, which has been named prion and replicates in the central nervous system (CNS) and in certain additional tissues of infected animals and humans. The second question, however, is no less intriguing: by which mechanisms can prions bring about the damage to the CNS which is characteristic of almost all transmissible spongiform encephalopathies? This problem raises many additional related questions. Is the damage related to the actual replication of the prion? Or is

---

Key words: transgenic mice / brain grafts / intraocular inoculation / prion spread

the spongiform encephalopathy the result of accumulation of toxic metabolites within or around neurons?

One prime candidate for the latter hypothesis of toxicity would certainly be disease-specific prion protein (PrP^{Sc}), the pathologically changed isoform of the normal prion protein, PrP^{C}. If so, would PrP^{Sc} be toxic only if it was generated within cells, or would it damage nervous cells if it acted from without or were internalized? And anyway, since prions appear to replicate, or at least to accumulate, in the organs of the lymphoreticular system, such as spleen, lymph nodes and Peyer's plaques of the intestine, why is that we do not observe immune deficiencies or structural pathologies of the latter organs after infection with prions? In other words, is susceptibility to prion toxicity a unique property of neural tissue, or is it rather the result of the 100-fold higher levels of PrP^{Sc} accumulation seen in brains of terminally scrapie-sick mice as compared to lymphoreticular organs?

The complexity of these questions, along with our limited understanding of the nature of the infectious agent, suggests that it may be very difficult to devise suitable systems to address them experimentally. However, the recent generation of genetic in vivo model systems, such as transgenic mice expressing, at various levels, normal and mutated forms of the prion protein, as well as knockout mice which bear hetero- or homozygous ablations of the *Prnp* gene which encodes the prion protein, has now opened new, promising avenues of investigation.

Along these lines, we have taken advantage of various strains of transgenic mice (constructed in the laboratory of C Weissmann) and have asked whether the neurografting technique could be used to address the question of prion neurotoxicity. Here, we present a characterization of biological properties such as tissue growth, proliferation and differentiation in neuroepithelial grafts. Special emphasis is laid on the development of the blood-brain barrier (BBB) after grafting. We then discuss how embryonic telencephalic grafting was applied to the study of scrapie pathogenesis.

Biological characteristics of mouse neuroectodermal grafts

Neural grafting has often been used to address questions related to developmental neurobiology [1–4]. Several studies investigated the establishment of neuronal organization within grafts and interactions with the host CNS [2, 5, 6]. More recently, grafting studies were aimed at questions related to neural plasticity. For example, it was asked whether and to what extent undifferentiated progenitor cells can integrate and take part in the formation of the host CNS [7–9]. Other studies were undertaken to address questions related to the tumorigenic potential of various oncogenes by grafting retrovirally transduced cells into the rodent CNS [10–12].

In the field of neurodegenerative disorders, grafting studies have been mainly aimed at reconstituting certain pathways or particular functions after surgical or toxic lesions to selected functional systems [1, 13–15]. In these models, an artificial lesion leads to degeneration of specific neuronal systems. Grafting of neural tissue or genetically engineered cells aims at functional repair of induced lesions [13]. Many such experiments were carried out in the rat system, which is well suited for developmental studies and allows stereotaxic surgical interventions with appropriate accuracy. However, with the advent of transgenic techniques, it has become possible to study in more detail the role played by single molecules during development and in pathological processes in mice [16, 17]. The generation of knockout mice by targeted deletion of genes of interest [18] has further broadened our insight into the molecular mechanisms of neural development and the pathogenesis of CNS diseases. A number of transgenic and knockout mice have provided valuable models for neurodegenerative diseases [17, 19–21]. Others, however, show early postnatal [22–24] or even embryonic [25, 26] lethal phenotypes which can be difficult to interpret. Although these models provide evidence for a crucial role of the respective gene products during development and hint at an important role of these factors for the determination of cell fates during differentiation [25–27], they do not allow us to study the role these factors play in secondary pathologic processes such as neuro-degeneration. In an effort to overcome this problem, we have employed transplantation approaches for neural tissue derived from such mouse embryos. Using grafting techniques, it has been possible to study neural tissue of mice with premature lethal genotypes at time points exceeding by far the life span of the mutant mice [28–30].

The grafting procedure is relatively simple [28, 29]. Embryos are harvested from timed pregnant dams at defined stages at mid-gestation. Graft tissue can be radiolabeled for later identification by autoradiography [5, 28] and injected into the caudoputamen or lateral ventricles of recipient mice using a stereotaxic frame [10, 28]. If histocompatible strains of mice are used, signs of graft rejection, such as lymphocytic infiltration and tissue necrosis, remain an exceptional finding and are detected in less than 5% of neural grafts [28].

In an effort to determine the optimal time for embryonic tissue preparation and transplanta-tion, we compared the final size of grafts resulting from tissue harvested at various embryonic stages. We found murine telencephalic tissue from embryonic day (E) 12.5 to reliably diffe-rentiate into large neural grafts that are suitable for detailed graft analysis. Tissue harvested at earlier embryonic stages often resulted in grafts containing non-neural tissue portions, since it was difficult to clearly separate mesenchymal tissue from the neural anlage at E 9.5–E 11.5. Such tissue portions induce permanent BBB leakage after grafting and were thus considered unfavorable. In contrast, neural tissue harvested at later embryonic stages (E 13.5–E16.5) was easily separated from the meninges. However, proliferation and growth potential were markedly reduced, resulting in smaller transplants that were only partially accessible to thorough examination. Moreover, when tissue was harvested and transplanted at E 12.5, the total number of neural grafts was higher than with tissue harvested at other embryonic stages [28].

Graft cell proliferation, as determined with immunocytochemical detection of incorporated 5-bromo-2'-deoxyuridine [31], showed that proliferation indices decreased sharply from initially 35% of grafted cells to around 5% during the second week after transplantation and to less than 1% after more than 7 weeks [28]. At the same time, differentiation of grafted cells proceeds to the terminal postmitotic state. Thus, mature neuroepithelial grafts contain neurons with myelinated processes and a dense synaptic network, glia (astrocytes, oligodendrocytes and microglia) and blood vessels 4 weeks after grafting (projected age of grafted tissue approximately P20) [28, 29]. Taken together, these findings indicate that embryonic neuro-epithelial tissue grafted into an adult host brain follows a program of maturation and differen-tiation similar to the in vivo time course [28].

Blood-brain barrier (BBB) and brain grafts

The BBB maintains the homeostatic environment in the brain by preventing blood-borne compounds from free entry into the CNS parenchyma. The barrier is formed by tight junctions in the vascular endothelia which are probably induced by astrocytes [32, 33]. A number of pathological CNS processes, such as inflammation, demyelination, tumor or degeneration, can induce breakdown of the BBB. In turn, BBB leakage might induce CNS dysfunction caused

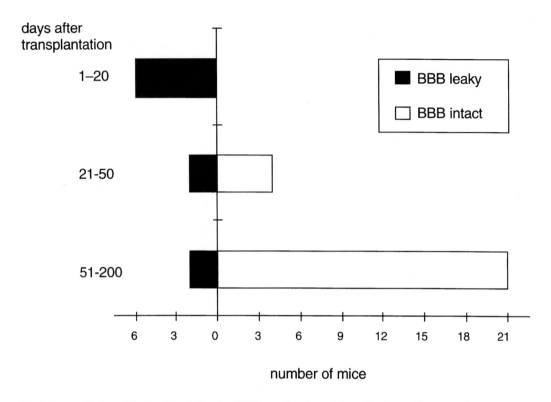

Fig 1. Reconstitution of the blood-brain barrier (BBB) as a function of time after the grafting procedure.

by blood-borne neurotoxic compounds normally excluded from the brain parenchyma [34, 35].

Various investigators have reported controversial findings on the post-transplantation status of the BBB in rodents. An early, yet most valuable study has suggested that the type of donor tissue determines the characteristics and BBB properties of graft supplying vessels [36]. According to this hypothesis, neural grafts would be expected to induce BBB properties in the supplying blood vessels. In fact, several authors described complete BBB reconstitution after neural grafting to the CNS. Some even found no residual BBB leakage as early as 1 week after grafting [37–40]. Other studies, however, have claimed the BBB to remain permanently disrupted after neural grafting to the CNS [41, 42]. Our group carried out studies in the model described using four independent marker molecules to detect damage to the BBB [43]. The results obtained with various techniques were surprisingly consistent, with minor variations owing to variable sensitivity of individual techniques rather than differing findings (fig 1). Magnetic resonance imaging using a contrast agent in vivo indicates, in agreement with the histological data, that in our paradigm, ie, grafting of tissue fragments as opposed to single cell suspensions, the BBB is reconstituted in 67% of all grafts after 3 weeks, and in 90% of the grafts 7 weeks after grafting [43]. These findings are particularly important with respect to the exploitation of neurografting techniques in neurodegenerative diseases. They indicate that the grafting procedure as such does not induce permanent BBB leakage that might expose the

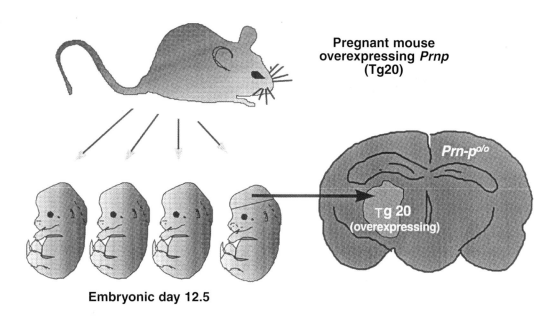

Pregnant mouse overexpressing *Prnp* (Tg20)

Prn-p$^{o/o}$

Tg 20 (overexpressing)

Embryonic day 12.5

Fig 2. Use of the neurografting technique for the study of mouse scrapie. The neuroectodermal anlage from a PrP^C-overexpressing mouse is implanted into the brain of a *Prnp*-deficient knockout mouse.

grafted tissue to a nonphysiological environment, and suggest that the genotype of the grafted tissue determines the BBB properties of the graft. Thus, a pathologic condition affecting exclusively the graft can result in secondary BBB disruption.

Neurografts in prion research

Having established that a neurograft from a donor which would have undergone early lethality may be kept alive in a nonaffected surrounding, we decided to apply this technique to the study of mouse scrapie. *Prnp$^{0/0}$* mice, which are devoid of PrP^C, resistant to scrapie and do not propagate prions [19, 44]. Because these mice show normal development and behavior [45, 46], it has been argued that scrapie pathology may come about because PrP^{Sc} deposition is neurotoxic [47], rather than by depletion of cellular PrP^C. In the latter case, one might expect that the lack of PrP^C would result in embryonic or perinatal lethality, especially since PrP^C is encoded by a unique gene for which no related family members have been found.

   To address the question of neurotoxicity, we exposed brain tissue of *Prnp$^{0/0}$* mice to a continuous source of PrP^{Sc}. For the reasons exposed earlier, we thought that a convenient way to do so would be to graft embryonic telencephalic tissue from transgenic mice overexpressing PrP into various structures (in most experiments we chose the caudoputamen or the lateral ventricles or the forebrain) of *Prnp$^{0/0}$* mice and to inoculate it with scrapie prions (fig 2). The donors, Tg20 mice [48], contain two arrays of 30 to 50 gene copies encoding PrP, overexpress PrP^C by five- to eight-fold and show incubation times of around 60 days as compared to 160 days for CD-1 wild-type mice. In the terminal stage of scrapie, both mouse strains exhibit similar prion titers, while PrP^{Sc} levels in Tg20 are 25–50% lower than those in CD-1 brain.

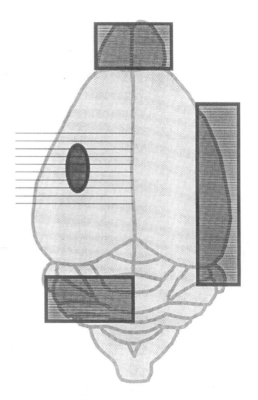

- **Conventional histology**

- **Histoblots**

- **Western blot analysis**

- **Titration of infectivity (indicator mice)**

Fig 3. Analysis of grafted brains was performed using various methods on the grafts and on the shaded regions of the host brain (designated as 'frontal', 'parietal contralateral' and 'cerebellum').

The recipient mice were monitored clinically for the development of scrapie symptoms. In addition, we performed histological analyses of the grafts and their surroundings (conventional histology, immunohistochemistry, in situ hybridization), and we determined the content of PrP^C and PrP^{Sc} by Western blotting and histoblotting. Since to date it would not be appropriate to equate the amount of PrP^{Sc} with infectivity, we decided to independently determine the amount of infectivity present in the graft and in regions of the brain at various distances from the graft by bioassay titration, using Tg*20* recipient mice. All these procedures are summarized in figure 3.

We observed that all mice remained free of scrapie symptoms for at least 70 weeks; this exceeds by at least seven-fold the survival time of scrapie-infected Tg*20* mice. Therefore, the presence of a continuous source of PrP^{Sc} and of scrapie prions does not exert any clinically detectable adverse effects on the physiological functions of a mouse devoid of PrP^C.

On the other hand, histological analysis revealed that Tg*20* and wild-type grafts developed characteristic histopathological features of scrapie 70 and 160 days post-inoculation (dpi), respectively, reflecting the incubation time of scrapie in the respective donor animals [19, 48]. Uninfected or mock-infected Tg*20* grafts occasionally showed mild gliosis but never spongiosis. Therefore, high expression of PrP by itself did not induce neurodegeneration in grafts.

Early stages of the disease in the graft (70–140 dpi) were characterized by spongiosis and gliosis (fig 4). In addition, we observed reduced synaptophysin immunoreactivity, which

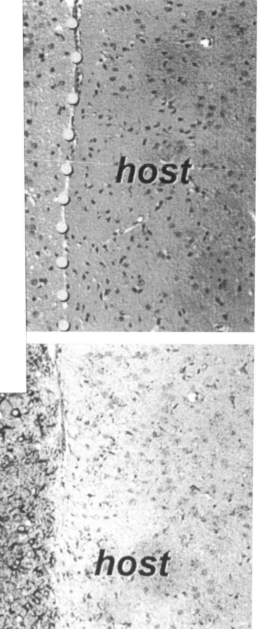

Fig 4. Typical histological appearance of a PrP^C-expressing graft (left side of the figure) at the interface to the knock-out host brain (right side). Note the spongiform microcystic changes (upper panel) and the brisk astrocytic reaction evidenced by immunocytochemical staining for glial fibrillary acidic protein (GFAP, lower panel).

we take as a sign of damage to the neuronal trees and subsequent decrease in the density of synaptic junctions. These changes are similar to those found in terminally sick Tg20 mice, and since they occurred at graft ages similar to the life expectancy of scrapie-exposed donor mice, we concluded that neurografted PrPC-expressing tissue indeed constitutes a realistic model for the scrapie encephalopathy. Intermediate stage grafts (140–280 dpi) showed status spongiosus with dramatic ballooning and loss of neurons, gliosis and stripping of neuronal processes. At late stages (280–480 dpi), cellular density increased from 900 to over 4,000 cells/mm². Astrocytes constituted the main cell population and synaptophysin immunoreactivity was almost completely absent. Intriguingly, the grafts underwent progressive disruption of the BBB during the course of the disease.

Although grafts had extensive contact with the recipient $Prnp^{o/o}$ brain, histopathological changes never extended into the host tissue, even at the latest stages. Wild-type mice engrafted with Tg20 tissue showed severe histopathological alterations in the graft and milder changes in the recipient brain, in good accordance with the general observation that the level of available PrPC determines the speed of onset and often also the extent of pathology. Surprisingly, histoblot analysis [49] of non-inoculated, engrafted $Prnp^{o/o}$ brain revealed that PrP immunoreactivity extended to the white matter of the recipient brains. In infected grafts, PrPSc was detected in both grafts and recipient brain, where it formed fine granules along white matter tracts and even in the contralateral hemisphere. Furthermore, immunohistochemistry revealed PrP deposits in the host hippocampus and occasionally in the parietal cortex of all animals harboring PrP-expressing grafts. Up to 35 clusters of PrP deposits per section appeared late in infection, each consisting of 25–120 globules 2–4 μm in diameter closely associated with astrocytic processes; no deposits were observed in inoculated or mock inoculated non-engrafted $Prnp^{o/o}$ brains or in $Prnp^{o/o}$ brains engrafted with $Prnp^{o/o}$ tissue. It is unlikely that such deposits were produced locally by Tg20 cells emigrating from the graft, since the graft borders were always sharply demarcated. PCR analysis of host brain regions containing PrP

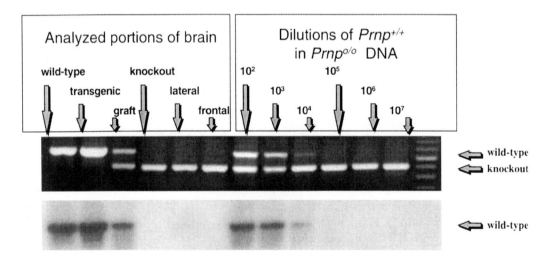

Fig 5. PCR analysis of grafts and surrounding host brain. The transgenic amplification product is identical to the product of the wild-type gene, and displays a slower electrophoretic mobility than the product of the knockout allele. No transgenic DNA was detected in the regions of the host brain in which PrP deposits were detected.

Table I. Determination of scrapie prion infectivity in the grafts and in regions of the host brain adjacent to, or distant from, the graft. The presence of prion infectivity correlates with the presence of histologically detectable PrP deposits.

Source of infectivity	Days after inoculation	Transmission	Incubation time of recipients
Standard prion inoculum			
RML, 10-1	–	4/4	58, 59, 62, 67
RML, 10-3	–	2/2	65, 65
RML, 10-5	–	4/4	83, 84, 84, 95
RML, 10-7	–	1/3	109, > 217, > 217
RML, 10-9	–	0/4	> 217, > 217, > 217, > 217
RML, 10-11	–	0/3	> 217, > 217, > 217
Mouse tissue			
Graft region	245	2/2	75, 75
Contralateral frontal cortex	336	0/2	> 170, > 170
Contralateral frontal cortex	350	0/2	> 170, > 170
Contralateral frontal cortex	428	0/2	> 170, > 170
Contralateral frontal cortex	454	0/2	> 170, > 170
Contralateral parietal cortex	285	2/2	103/121
Cerebellum	285	0/3	> 170, > 170, > 170
Spleen	285	0/1	> 170

deposits failed to reveal PrP-encoding DNA (fig 5, detection limit $< 1:10^4$) and graft-derived cells were never detected distant from the graft by in situ hybridization or autoradiography of brain engrafted with ^{3}H-thymidine-labeled tissue [29]. We conclude that graft-derived PrP^C and PrP^{Sc} are transferred from the graft to distant areas of the host brain.

It could be argued that pathology does not spread to the surrounding areas because there may not be sufficient physiological connections between the graft and the surrounding tissue. To address this potential issue, we grafted Tg20 PrP^C-overexpressing tissue in wild-type mice and determined the effects of intracerebral prion inoculations. This procedure resulted in histopathologically verifiable scrapie in both graft and host tissue and in clinical scrapie of the host mouse without any modulation of the time course of the disease. However, the extent of pathology was more pronounced in the graft than in the host, once again reinforcing the general observation that the availability of PrP^C, rather than the deposition of PrP^{Sc}, is the rate-limiting step and also the major pathogenetic determinant in the development of scrapie.

To determine infectivity, various portions of brains containing grafts with severe histopathological lesions were inoculated into Tg20 indicator mice. Samples of graft led to terminal scrapie after 74 days, indicating a titer of approximately 5.7 logLD$_{50}$ (50% lethal dose) units per mL 10% homogenate (table I). While the frontal brain and cerebellum, in which no deposits were detected, did not show infectivity, about 1.5 logLD$_{50}$ units per mL 10% homogenate were detected in the contralateral hemisphere. Infectivity was not due to residual inoculum, because within 4 days after inoculation no infectivity could be detected in recipient brain [19]. Thus, infectious prions moved from the grafts to some regions of the PrP-deficient host brain without causing pathological changes or clinical disease. The distribution of PrP^{Sc} in the white matter tracts of the host brain suggests diffusion within the extracellular space [50] rather than axonal transport (S Brandner and A Aguzzi, unpublished results).

Why was no scrapie pathology observed in PrP^C-deficient tissue, even in regions adjoining the graft which contained high levels of PrP^{Sc}, and was clearly leaking such material? Perhaps

PrP^{Sc} is inherently nontoxic and PrP^{Sc} plaques found in spongiform encephalopathies are an epiphenomenon rather than a cause of neuronal damage. Indeed, the extent of PrP deposition in the brains of humans succumbing to prion diseases with similar clinical presentation is extremely variable [51, 52].

Alternatively, PrP^{Sc} is toxic only when it is formed and accumulated within the cell, but not when it is presented from without. Finally, it may be that PrP^{Sc} is pathogenic when presented from without, but only to cells expressing PrP^C, either because it initiates conversion of PrP^C to PrP^{Sc} at the cell surface and/or because it is internalized by way of association with PrP^C, which is endocytosed efficiently [53]. Along with previous transgenic studies [54, 55] showing delayed onset of disease in *Prnp$^{o/+}$* mice despite massive accumulation of PrP^{Sc} and reports of typical scrapie histopathology in fatal familial insomnia-inoculated mouse brains devoid of detectable levels of PrP^{Sc} [56], the present data imply that not deposition of PrP^{Sc} but rather the availability of PrP^C for some intracellular process elicited by the infectious agent is directly linked to spongiosis, gliosis and neuronal death.

Acknowledgments

The authors wish to thank K Stepinski, R von Rotz and R Chase for secretarial assistance. The work described is being supported by the Canton of Zurich, by grants from the *Schweizerischer Nationalfonds* and *Nationales Forschungsprogramm* NFP38 to A Aguzzi and C Weissmann, of the European Union and the *Bundesamt für Gesundheitswesen* to A Aguzzi, of the Human Frontier Science Program to C Weissmann and by a postdoctoral fellowship of the Ernst Hadorn Foundation to S Brandner.

References

1 Fisher LJ, Gage FH. Grafting in the mammalian central nervous system. *Physiol Rev* 1993;73:583-616
2 Kromer LF, Bjorklund A, Stenevi U. Intracephalic embryonic neural implants in the adult rat brain. I. Growth and mature organization of brainstem, cerebellar, and hippocampal implants. *J Comp Neurol* 1983;218:433-59
3 O'Leary DD, Stanfield BB. Selective elimination of axons extended by developing cortical neurons is dependent on regional locale: experiments utilizing fetal cortical transplants. *J Neurosci* 1989;9:2230-46
4 Renfranz PJ, Cunningham MG, McKay RD. Region-specific differentiation of the hippocampal stem cell line HiB5 upon implantation into the developing mammalian brain. *Cell* 1991;66:713-29
5 Jaeger CB, Lund RD. Transplantation of embryonic occipital cortex to the tectal region of newborn rats: a light microscopic study of organization and connectivity of the transplants. *J Comp Neurol* 1980;194:571-97
6 Lund RD, Hauschka SD. Transplanted neural tissue develops connections with host rat brain. *Science* 1976;193:582-4
7 Brüstle O. Targeted introduction of neurons into the embryonic brain. *Neuron* 1995;15:1275-85
8 Campbell C, Olsson M, Björklund A. Regional incorporation and site-specific differentiation of striatal precursors transplanted to the embryonic forebrain ventricle. *Neuron* 1995;15(1259):1273
9 Gage FH, Coates PW, Palmer TD et al. Survival and differentiation of adult neuronal progenitor cells transplanted to the adult brain. *Proc Natl Acad Sci USA* 1995;92:11879-83
10 Aguzzi A, Kleihues P, Heckl K, Wiestler OD. Cell type-specific tumor induction in neural transplants by retrovirus-mediated oncogene transfer. *Oncogene* 1991;6:113-8
11 Brüstle O, Aguzzi A, Talarico D, Basilico C, Kleihues P, Wiestler OD. Angiogenic activity of the K-fgf/hst oncogene in neural transplants. *Oncogene* 1992;7:1177-83
12 Wiestler OD, Brüstle O, Eibl RH, Radner H, Aguzzi A, Kleihues P. Retrovirus-mediated oncogene transfer into neural transplants. *Brain Pathol* 1992;2:47-59
13 Dunnett SB. Neural transplantation in animal models of dementia. *Eur J Neurosci* 1990;2:567-87
14 Dunnett SB, Björklund A, Stenevi U, Iversen SD. Grafts of embryonic substantia nigra reinnervating the ventrolateral striatum ameliorate sensorimotor impairments and akinesia in rats with 6-OHDA lesion of the nigrostriatal pathway. *Brain Res* 1981;229:209-17
15 Lindvall O. Prospects of transplantation in human neurodegenerative diseases. *Trends Neurosci* 1991;14:376-84
16 Aguzzi A, Brandner S, Sure U, Ruedi D, Isenmann S. Transgenic and knock-out mice: models of neurological disease. *Brain Pathol* 1994;4:3-20

17 Aguzzi A, Brandner S, Isenmann S, Steinbach J, Sure U. Transgenic and gene disruption techniques in the study of neurocarcinogenesis. *Glia* 1995;15:348-64

18 Thomas KR, Capecchi MR. Site-directed mutagenesis by gene targeting in mouse embryo-derived stem cells. *Cell* 1987;51:503-12

19 Büeler HR, Aguzzi A, Sailer A et al. Mice devoid of PrP are resistant to scrapie. *Cell* 1993;73:1339-47

20 Games D, Adams D, Alessandrini R et al. Alzheimer-type neuropathology in transgenic mice overexpressing V717F beta-amyloid precursor protein. *Nature* 1995;373:523-7

21 LaFerla FM, Tinkle BT, Bieberich CJ, Haudenschild CC, Jay G. The Alzheimer's A beta peptide induces neuro-degeneration and apoptotic cell death in transgenic mice. *Nat Genet* 1995;9:21-30

22 Klein R, Smeyne RJ, Wurst W et al. Targeted disruption of the trkB neurotrophin receptor gene results in nervous system lesions and neonatal death. *Cell* 1993;75:113-22

23 Magyar JP, Bartsch U, Wang ZQ et al. Degeneration of neural cells in the central nervous system of mice deficient in the gene for the adhesion molecule on glia, the beta 2 subunit of murine Na, K-ATPase. *J Cell Biol* 1994;127:835-45

24 Smeyne RJ, Klein R, Schnapp A et al. Severe sensory and sympathetic neuropathies in mice carrying a disrupted Trk/NGF receptor gene. *Nature* 1994;368:246-9

25 Bladt F, Riethmacher D, Isenmann S, Aguzzi A, Birchmeier C. Essential role for the c-met receptor in the migration of myogenic precursor cells into the limb bud [see comments]. *Nature* 1995;376:768-71

26 Meyer D, Birchmeier C. Multiple essential functions of neuregulin in development [see comments]. *Nature* 1995;378:386-90

27 Gassmann M, Casagranda F, Orioli D et al. Aberrant neural and cardiac development in mice lacking the ErbB4 neuregulin receptor. *Nature* 1995;378:390-4

28 Isenmann S, Brandner S, Sure U, Aguzzi A. Telencephalic transplants in mice: characterization of growth and differentiation patterns. *Neuropathol Appl Neurobiol* 1996;21:108-17

29 Isenmann S, Molthagen M, Brandner S et al. Neural transplants of AMOG/b2-deficient mice survive but do not express the β2-subunit of the Na,K-ATPase. *Glia* 1995;15:377-88

30 Isenmann S, Brandner S, Kühne G, Boner J, Aguzzi A. Comparative in vivo and pathological analysis of the blood-brain barrier in mouse telencephalic transplants. *Neuropathol Appl Neurobiol* 1996;21:118-28

31 Gratzner HG. Monoclonal antibody to 5-bromo- and 5-iododeoxyuridine: a new reagent for detection of DNA replication. *Science* 1982;218:474-5

32 Janzer RC, Raff MC. Astrocytes induce blood-brain barrier properties in endothelial cells. *Nature* 1987;325: 253-57

33 Risau W, Wolburg H. Development of the blood-brain barrier [see comments]. *Trends Neurosci* 1990;13:174-8

34 Rosenstein JM, Brightman MW. Circumventing the blood-brain barrier with autonomic ganglion transplants. *Science* 1983;221:879-81

35 Svendgaard NA, Bjorklund A, Hardebo JE, Stenevi U. Axonal degeneration associated with a defective blood-brain barrier in cerebral implants. *Nature* 1975;255:334-6

36 Stewart PA, Wiley MJ. Developing nervous tissue induces formation of blood-brain barrier characteristics in invading endothelial cells: a study using quail-chick transplantation chimeras. *Dev Biol* 1981;84:183-92

37 Bertram KJ, Shipley MT, Ennis M, Sanberg PR, Norman AB. Permeability of the blood-brain barrier within rat intrastriatal transplants assessed by simultaneous systemic injection of horseradish peroxidase and Evans blue dye. *Exp Neurol* 1994;127:245-52

38 Broadwell RD, Charlton HM, Ebert P, Hickey WF, Villegas JC, Wolf AL. Angiogenesis and the blood-brain barrier in solid and dissociated cell grafts within the CNS. *Prog Brain Res* 1990;82:95-101

39 Broadwell RD, Charlton HM, Ganong WF, Salcman M, Sofroniew M. Allografts of CNS tissue possess a blood-brain barrier. I. Grafts of medial preoptic area in hypogonadal mice. *Exp Neurol* 1989;105:135-51

40 Swenson RS, Shaw P, Alones V, Kozlowski G, Zimmer J, Castro AJ. Neocortical transplants grafted into the newborn rat brain demonstrate a blood-brain barrier to macromolecules. *Neurosci Lett* 1989;100:35-9

41 Rosenstein JM. Neocortical transplants in the mammalian brain lack a blood-brain barrier to macromolecules. *Science* 1987;235:772-4

42 Rosenstein JM, Brightman MW. Alterations of the blood-brain barrier after transplantation of autonomic ganglia into the mammalian central nervous system. *J Comp Neurol* 1986;250:339-51

43 Isenmann S, Brandner S, Kühne G, Boner J, Aguzzi A. Comparative in vivo and pathological analysis of the blood-brain barrier in mouse telencephalic transplants. *Neuropathol Appl Neurobiol* 1996;21:118-28

44 Sailer A, Büeler H, Fischer M, Aguzzi A, Weissmann C. No propagation of prions in mice devoid of PrP. *Cell* 1994;77:967-8

45 Büeler HR, Fischer M, Lang Y et al. Normal development and behaviour of mice lacking the neuronal cell-surface PrP protein. *Nature* 1992;356:577-82

46 Manson JC, Clarke AR, Hooper ML, Aitchison L, McConnell I, Hope J. 129/Ola mice carrying a null mutation in *PrP* that abolishes mRNA production are developmentally normal. *Mol Neurobiol* 1994;8:121-7

47 Forloni G, Angeretti N, Chiesa R et al. Neurotoxicity of a prion protein fragment. *Nature* 1993;362:543-6

48 Fischer M, Rülicke T, Raeber A et al. Prion protein with amino terminal deletions restoring susceptibilty of *PrP* knockout mice to scrapie. *EMBO J* 1996;15:1255-64

49 Taraboulos A, Jendroska K, Serban D, Yang SL, DeArmond SJ, Prusiner SB. Regional mapping of prion proteins in brain. *Proc Natl Acad Sci USA* 1992;89:7620-4

50 Jeffrey M, Goodsir CM, Bruce ME, McBride PA, Fowler N, Scott JR. Murine scrapie-infected neurons in vivo release excess prion protein into the extracellular space. *Neurosci Lett* 1994;174:39-42

51 Collinge J, Owen F, Poulter M et al. Prion dementia without characteristic pathology. *Lancet* 1990;336:7-9

52 Hayward PA, Bell JE, Ironside JW. Prion protein immunocytochemistry: reliable protocols for the investigation of Creutzfeldt-Jakob disease. *Neuropathol Appl Neurobiol* 1994;20:375-83

53 Shyng SL, Huber MT, Harris DA. A prion protein cycles between the cell surface and an endocytic compartment in cultured neuroblastoma cells. *J Biol Chem* 1993;268:15922-8

54 Büeler H, Raeber A, Sailer A, Fischer M, Aguzzi A, Weissmann C. High prion and PrPSc levels but delayed onset of disease in scrapie-inoculated mice heterozygous for a disrupted *PrP* gene. *Mol Med* 1994;1(1):19-30

55 Manson JC, Clarke AR, McBride PA, McConnell I, Hope J. *PrP* gene dosage determines the timing but not the final intensity or distribution of lesions in scrapie pathology. *Neurodegeneration* 1994;3:331-40

56 Collinge J, Palmer MS, Sidle KCL et al. Transmission of fatal familial insomnia to laboratory animals. *Lancet* 1995;346:569-70

Transmissible Subacute Spongiform Encephalopathies:
Prion Diseases
L Court, B Dodet, eds
© 1996, Elsevier, Paris

Effect of amphotericin B on 263K and 139H strains of scrapie in Armenian hamsters

Maurizio Pocchiari[1], Loredana Ingrosso[1], Franco Cardone[1], Anna Ladogana[1], Claudia Eleni[2], Umberto Agrimi[2]

[1]Laboratory of Virology, Viale Regina Elena, 299, 00161 Rome; [2]Istituto Zooprofilattico Sperimentale delle Regioni Lazio e Toscana, via Appia Nuova, 1411, 00178 Rome, Italy

Summary – Scrapie is a transmissible disease of the central nervous system occurring naturally in sheep and goats. We have previously reported that treatment with the polyene antibiotic amphotericin B (AmB) prolongs the incubation period of hamsters infected intracerebrally or intraperitoneally with the 263K strain of scrapie. In this paper, we show that AmB also prolongs the incubation period of Armenian hamsters infected with the Syrian hamster-adapted 263K strain of scrapie (first passage). However, AmB therapy was ineffective in Armenian hamsters infected with 263K from passage in Armenian hamsters (second passage). There was no anti-scrapie effect of AmB in Armenian hamsters infected with the first or second passage of 139H scrapie strain.

Introduction

Spongiform encephalopathies are neurodegenerative diseases of humans and animals characterized by the formation of a partially protease-resistant protein (PrP-res or PrP 27–30) which derives from a conformational modification of an endogenous cellular protease-sensitive isoform (PrP-sen or PrP^C). In the brains of infected individuals, PrP-res aggregates and accumulates as amyloid fibrils [1].

In experimental animal models, sulphated polyanions [2–7], Congo red [8] and the polyene antibiotic amphotericin B (AmB) [9–15] have given encouraging results. These drugs prolong or sometimes even prevent the appearance of the disease by delaying the formation of PrP-res and/or inhibiting scrapie replication [5, 6, 13, 16–19].

AmB differs from polyanions in two ways: first, because its beneficial effect is retained, even when the scrapie agent is given intracerebrally. Second, because the effect of AmB is strain specific, ie, there are some strains of scrapie (263K, C506) or Creutzfeldt-Jakob disease (CJD) (KFu) which are susceptible to the treatment and others (139H, 139A, Me7 and 87V) which are not [13, 14, 20, 21]. Moreover, AmB is able to dissociate in vivo the replication of the scrapie agent from PrP-res accumulation [13].

We have previously shown [21] that AmB treatment is ineffective with the 139H strain, regardless of the route of infection (intracerebral or intraperitoneal) and the species of hamster (Syrian, Armenian or Chinese). This result strengthens the notion that the 139H strain of scrapie is resistant, per se, to AmB treatment. On the other hand, the beneficial effect of AmB on

Key words: scrapie / amphotericin B / scrapie strains

the 263K strain of scrapie is maintained during the first passage from Syrian to Armenian hamsters, but not to Chinese hamsters [21].

In this paper, we further investigate the effect of AmB therapy in scrapie-infected hamsters.

Materials and methods

Infection of animals

Weanling outbred golden Syrian hamsters (*Mesocricetus auratus*) and inbred Armenian (*Cricetus cricetus*) hamsters were purchased from Charles River, Calco, Como, Italy. The hamsters were injected with the 263K strain [22] or the 139H strain [23] of scrapie. The inocula were obtained from 10% (w/v) phosphate-buffered saline suspensions of pooled brains from clinically affected animals. The brain suspensions were centrifuged at 500 g for 10 min; the supernatants were collected and stored at –70 °C in 5 mL aliquots. Before use, the samples were thawed and vigorously shaken by vortexing. The animals were either injected intracerebrally into the left hemisphere (0.05 mL) or intraperitoneally into the lower left quadrants of the abdomen (0.1 mL) using sterile syringes with 26-gauge needles. The animals were housed in stainless (Syrian hamsters) or plastic cages (Armenian hamsters) with water and food ad libitum, observed 5 days per week, and scored for the presence of clinical signs.

Amphotericin B treatment

AmB (Sigma, catalogue no A9528) was freshly prepared for each administration by dissolving the powder (50 mg) in 10 mL of sterile distilled water. AmB was then diluted appropriately with 5% glucose, and 0.5 mL (1 mg of AmB/hamster, equivalent to approximately 10 mg/kg body weight) was injected intraperitoneally once a week alternatively in the lower right and left quadrants of the abdomen. AmB treatment was started the same day as inoculation and maintained for 16 (intracerebrally scrapie-infected animals) or 20 weeks (intraperitoneally scrapie-infected animals). Controls were untreated scrapie-inoculated hamsters. There was no difference in the mean incubation periods of scrapie-infected hamsters receiving an injection of saline solution and untreated hamsters [10].

Histopathological procedures

At the late stage of clinical signs of scrapie, Syrian hamsters were sacrificed with chloroform and their brains dissected and fixed in 10% buffered neutral formalin. Paraffin sections (7 μm) were hematoxylin–eosin-stained and scored for the intensity of vacuolar degeneration on a scale of 0 to 5 in nine defined areas: medulla, cerebellar cortex, superior colliculus, hypothalamus, thalamus, hippocampus, septum and posterior and anterior cerebral cortices, as previously reported [24, 25]. Coded sections were scored by two independent observers. After decoding, lesion profiles were constructed from the average score in each area.

Results

The first intracerebral passage of 263K from Syrian to Armenian hamsters gave an incubation period of 166.4 ± 6.6 days (mean ± SD, $n = 9$) and 165.5 ± 8.8 days ($n = 20$) in two separate experiments, respectively. In these two experiments, AmB treatment prolonged the mean incubation periods by 43.1 (incubation period, 209.5 ± 2.4 days, $n = 4$) and 43.6 (incubation period, 209.1 ± 7.2 days, $n = 14$) days, respectively. As expected, the second intracerebral passage of 263K into Armenian hamsters (ie, brain homogenate from scrapie-infected Armenian to non-infected Armenian hamsters) produced scrapie disease in only 115.7 ± 4.9 days ($n = 15$). However, AmB treatment of these animals did not produce any beneficial effect (incubation period, 118.5 ± 5.6 days, $n = 12$; fig 1).

The second intracerebral passage of the 139H strain of scrapie into Armenian hamsters also resulted in a shorter incubation period (117.6 ± 4.6 days, $n = 15$) than that observed at the first passage (152.0 ± 0.9 days, $n = 10$). However, contrary to results with the 263K strain, AmB treatment was ineffective after both the first and second intracerebral passage of the 139H strain in Armenian hamsters (152.5 ± 2.8 days, $n = 8$ and 117.2 ± 4.9 days, $n = 11$, respectively; fig 1).

The effect of AmB treatment was also tested in the first intraperitoneal passage of 263K and 139H strains from Syrian to Armenian hamsters. As expected, the anti-scrapie effect of AmB was observed in animals infected with the 263K strain (310.5 ± 75.2 days, $n = 12$ versus 234.1 ± 19.4 days, $n = 14$; $P < 0.001$; 32.6% of controls), but not with the 139H strain (195.0 ± 8.8 days, $n = 8$ versus 194.6 ± 10.2 days, $n = 15$; 0.2% of controls) of scrapie.

The histological examination of the brains of untreated and AmB-treated Syrian hamsters sacrificed at the clinical stage of scrapie disease revealed that the characteristic lesion profile is maintained during AmB treatment in both the AmB-sensitive (263K) and AmB-resistant (139H) strain of scrapie (fig 2).

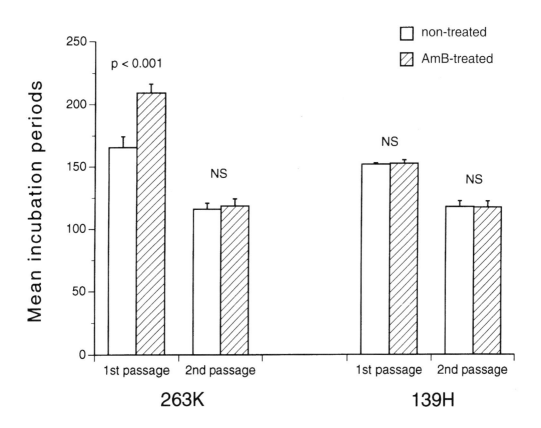

Fig 1. Effect of AmB in Armenian hamsters intracerebrally infected with the first or second passage of 263K and 139H strains of scrapie. Bars represent the standard deviations. The difference between the mean incubation periods of untreated (non-treated) and AmB-treated animals was tested with two-tailed Student's *t* test. NS: not significant.

Discussion

The extended incubation period seen at the first passage from Syrian to Armenian hamsters of the 263K or 139H scrapie strains is referred to as the 'species barrier' effect [26] and depends on the differences in the amino-acid sequence of PrP-sen between Syrian and Armenian hamsters [27]. At the second passage (Armenian to Armenian), the incubation periods for both scrapie strains were reduced. This result is not unexpected since subsequent passages of a scrapie isolate in a new host species always results in a shortened incubation period [27–30]. In this second passage, however, the 263K strain lost its sensitivity to AmB, indicating that a new mutant strain had emerged. We propose to call this new strain 263KAr. It is unlikely that the lack of an anti-scrapie effect of AmB on 263KAr depends on the new host genotype since in the first passage, the sensitivity to AmB was retained. Furthermore, it is unlikely that during the passage from Syrian to Armenian hamsters we selected a different strain from a mixture which was present in the original inoculum since the 263K strain of scrapie has been carefully cloned [22, 23]. It must be defined whether this strain would retain its phenotypic properties (incubation period, lesion profile and AmB sensitivity) when back-passage (from Armenian to Syrian hamsters) occurs.

At present, it is much more difficult to recognize whether a new strain has emerged from the 139H strain following the passage from Syrian to Armenian hamsters since this strain retains its resistance to AmB treatment. Further work is necessary to establish that.

Interestingly, the lesion profiles of 263K or 139H scrapie-infected Syrian hamsters are not modified by therapy with AmB. Thus, AmB therapy does not affect the characteristics of the scrapie strain. This is in agreement with our previous observation that the anti-scrapie effect of AmB is directed towards the interaction between the host PrP and the infectious agent rather than towards the scrapie agent alone [13]. It remains to be investigated whether the drug could affect the appearance of the lesions in the various areas of the central nervous system.

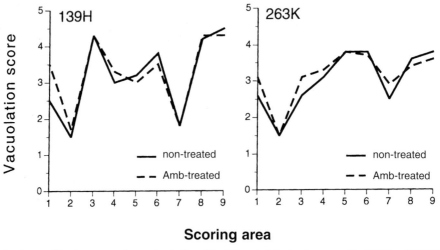

Scoring area

Fig 2. Lesion profiles in untreated (non-treated) and AmB-treated Syrian hamsters for 139H and 263K strains of scrapie. 1: medulla; 2: cerebellar cortex; 3: superior colliculus; 4: hypothalamus; 5: thalamus; 6: hippocampus; 7: septum; 8 and 9: posterior and anterior cerebral cortex, respectively.

Acknowledgments

We thank A Garozzo for editorial assistance. We also thank F Varano for his invaluable help in the organization of the animal facility and M Bonanno for assistance in animal care. This work was partially supported by the National Registry of Creutzfeldt-Jakob disease, financed by the Italian Ministry of Health-Instituto Superiore di Sanità and by a grant from CNR no 95.01681.CT04.

References

1 Pocchiari M. Prions and related neurological diseases. *Mol Aspects Med* 1994;15:195-291
2 Ehlers B, Diringer H. Dextran sulphate 500 delays and prevents mouse scrapie by impairment of agent replication in spleen. *J Gen Virol* 1984;65:1325-30
3 Ehlers B, Rudolph R, Diringer H. The reticuloendothelial system in scrapie pathogenesis. *J Gen Virol* 1984;65: 423-8
4 Farquhar CF, Dickinson AG. Prolongation of scrapie incubation period by an injection of dextran sulphate 500 within the month before or after infection. *J Gen Virol* 1986;67:463-73
5 Kimberlin RH, Walker CA. Suppression of scrapie infection in mice by heteropolyanion 23, dextran sulfate, and some other polyanions. *Antimicrob Agents Chemother* 1986;30:409-13
6 Diringer H, Ehlers B. Chemoprophylaxis of scrapie in mice. *J Gen Virol* 1991;72:457-60
7 Ladogana A, Casaccia P, Ingrosso L et al. Sulphate polyanions prolong the incubation period of scrapie-infected hamsters. *J Gen Virol* 1992;73:661-5
8 Ingrosso L, Ladogana A, Pocchiari M. Congo red prolongs incubation period of scrapie-infected hamsters. *J Virol* 1995;69:506-8
9 Amyx H, Salazar AM, Gajdusek DC, Gibbs CJ Jr. Chemotherapeutic trials in experimental slow virus disease. *Neurology* 1984;34(suppl 1):149
10 Pocchiari M, Schmittinger S, Masullo C. Amphotericin B delays the incubation period of scrapie in intracerebrally inoculated hamsters. *J Gen Virol* 1987;68:219-23
11 Pocchiari M, Casaccia P, Ladogana A. Amphotericin B. A novel class of antiscrapie drug. *J Infect Dis* 1989;160: 795-802
12 Casaccia P, Ladogana A, Xi YG et al. Measurement of the concentration of amphotericin B in brain tissue of scrapie-infected hamsters with a simple and sensitive method. *Antimicrob Agents Chemother* 1991;35:1486-8
13 Xi YG, Ingrosso L, Ladogana A, Masullo C, Pocchiari M. Amphotericin B treatment dissociates in vivo replication of the scrapie agent from PrP accumulation. *Nature* 1992;356:598-601
14 Demaimay R, Adjou K, Lasmézas C et al. Pharmacological studies of a new derivative of amphotericin B, MS-8209, in mouse and hamster scrapie. *J Gen Virol* 1994;75:2499-503
15 McKenzie D, Kaczkowski J, Marsh R, Aiken J. Amphotericin B delays both scrapie agent replication and PrP-res accumulation early in infection. *J Virol* 1994;68:7534-6
16 Caughey B, Race RE. Potent inhibition of scrapie associated PrP accumulation by Congo red. *J Neurochem* 1992;59:768-71
17 Caughey B, Raymond GJ. Sulfated polyanion inhibition of scrapie-associated PrP accumulation in cultured cells. *J Virol* 1993;67:643-50
18 Caughey B, Ernst D, Race RE. Congo red inhibition of scrapie agent replication. *J Virol* 1993;67:6270-2
19 Gabizon R, Meiner Z, Halimi M, Ben-Sasson SA. Heparin-like molecules bind differentially to prion-proteins and change their intracellular metabolic fate. *J Cell Physiol* 1993;157:319-25
20 Carp RI. Scrapie, unconventional infectious agent. In: Specter S, ed. *Neuropathogenic Viruses and Immunity.* New York: Plenum Press, 1992;111-36
21 Pocchiari M, Ingrosso L, Ladogana A. Effect of Amphotericin B on different experimental strains of spongiform encephalopathy agents. In: CJ Gibbs Jr, ed. *Bovine Spongiform Encephalopathy: the BSE Dilemma.* New York: Springer Verlag, 1996;271-81
22 Kimberlin RH, Walker C. Characteristics of a short incubation model of scrapie in the golden hamster. *J Gen Virol* 1977;34:295-304
23 Kimberlin RH, Walker CA, Fraser H. The genomic identity of different strains of mouse scrapie is expressed in hamsters and preserved on reisolation in mice. *J Gen Virol* 1989;70:2017-25
24 Fraser H, Dickinson, AG. The sequential development of the brain lesion of scrapie in three strains of mice. *J Comp Pathol* 1968;78:301-11
25 Fraser H, Dickinson AG. Scrapie in mice. Agent-strain differences in the distribution and intensity of gray matter vacuolation. *J Comp Pathol* 1973;83:29-40
26 Dickinson AG. Scrapie in sheep and goats. *Front Biol* 1976;44:209-41

27 Lowenstein DH, Butler DA, Westaway D, McKinley MP, DeArmond SJ, Prusiner SB. Three hamster species with different scrapie incubation times and neuropathological features encode distinct prion proteins. *Mol Cell Biol* 1990;10:1153-63
28 Hecker R, Taraboulos A, Scott M et al. Replication of distinct scrapie prion isolates is region specific in brains of transgenic mice and hamsters. *Genes Dev* 1992;6:1213-28
29 Bruce ME. Scrapie strain variation and mutation. *Br Med Bull* 1993;49:822-38
30 Dickinson AG, Meikle VMH. Host-genotype and agent effects in scrapie incubation: change in allelic interaction with different strains of agent. *Mol Gen Genet* 1971;112:73-9

Transmissible Subacute Spongiform Encephalopathies:
Prion Diseases
L Court, B Dodet, eds
© 1996, Elsevier, Paris

Polyene antibiotics in experimental transmissible spongiform encephalopathies

Rémi Demaimay[1], Karim T Adjou[1], Vincent Beringue[1], Séverine Demart[1], François Lamoury[1], Corinne Lasmézas[1], Jean-Philippe Deslys[1], Michel Seman[2], Dominique Dormont[1]

[1]*Service de Neurovirologie, CEA/DRM/SSA, BP 6, 92265 Fontenay-aux-Roses Cedex;* [2]*Laboratoire d'Immunodifférenciation, Institut J-Monod, Université Denis-Diderot, Paris, France*

Summary – Amphotericin B (AmB) has been reported to increase the survival time of 263K-infected hamsters. However, this effect is limited by the acute toxicity of the drug. Use of a less toxic AmB derivative, MS-8209, has made it possible to detect a significant effect in a murine model. We then explored two features of scrapie pathogenesis: the link between infectivity and protease-resistant prion protein (PrP-res) accumulation in the brain, and the effects of polyene antibiotics administration during the late stages of the disease. The results reported here show that infectivity was linked to PrP-res accumulation in a 263K-infected hamster model and that high doses of AmB and MS-8209 prolong survival time in mice treated at late stages of the disease.

Introduction

Several experimental treatments have been administered to Creutzfeldt-Jakob disease (CJD) affected humans and rodents experimentally infected with transmissible spongiform encephalopathy (TSE) agents. Among the drugs used, three classes prolong the survival time in experimental models: polyanions [1, 2], Congo red [3] and amphotericin B (AmB) [4, 5]. AmB is a widely used antifungal drug, with a broad range of activity [6], utilized for the last 30 years to treat systemic mycoses [7]. AmB treatment prolongs survival time of 263K-infected hamsters and induces a delay in protease-resistant prion protein (PrP-res) accumulation in the brain. Two distinct features emerge from experiments done in experimental scrapie: i) AmB effects seem to be restricted to the Syrian golden hamster infected with the 263K scrapie strain and ii) an increase in survival time is only observed when animals are treated early after experimental infection. Restriction to host and scrapie strain could only be due to the low dose of AmB used. Indeed, the first report of an AmB effect, not yet reproduced, was evidenced in mice and monkeys inoculated with the CJD agent [5]. Extensive research on AmB derivatives led to the synthesis of molecules with lower toxicity [6], such as MS-8209 [8]. Its effect was similar to those of AmB in intracerebrally (ic) or intraperitoneally (ip) 263K-infected hamsters, and a 10-mg/kg regimen increased survival time by 100% (ic inoculation [9]). In treated hamsters, a delay in PrP-res accumulation was observed with both drugs, concomitantly with a decrease

---

Key words: prion / experimental scrapie / amphotericin B / MS-8209 / prion protein / glial fibrillary acidic protein

in glial fibrillary acidic protein (GFAP, specific to astrocytes) gene hyperexpression [9]. In C506M3-infected mice, treatment with MS-8209 and AmB prolongs survival time [10]. This delay is related to the dose used and the length of the treatment and correlated to a slowdown in PrP-res accumulation and *GFAP* gene expression. As we could postulate that the published inefficacy of AmB administered in the late stages of the disease could have the same origin as the host restriction, we tested AmB and its derivative during the late stages of the murine scrapie infection. Furthermore, as action on TSE agent multiplication in 263K-infected hamsters is still controversial [11, 12], we evaluated infectious titer in the brain of animals treated with high doses of polyene antibiotics.

Materials and methods

Animals and inoculations

We used female golden hamsters (René Janvier, France) and C57BL/6 mice (Iffa-Credo, France). scrapie strains 263K (4.4×10^8 50% lethal dose [LD_{50}] per gram of brain, H Fraser, UK) and C506M3 (7.9×10^8 LD_{50} per gram of brain, DC Gajdusek, USA) were obtained from the brain homogenate of infected animals. This brain homogenate (1% w/v) was injected into the right cerebral hemisphere (20 or 50 μL, respectively, for mice and hamsters). Infectious titers were estimated either by measuring the incubation period in recipient mice [13], or by the end-point dilution method (Reed and Muench).

Drugs and treatments

AmB (Fungizone®, Squibb, Paris, France) and MS-8209 (*N*-methyl glucamine salt of 1-deoxy-1-amino-4,6,*O*-benzylidine-D-fructosyl-AmB, Mayoly-Spindler, Chatou, France) were suspended in a 5% glucose (w/v) sterile solution. All animals were treated 6 days a week by the intraperitoneal route. After sacrifice by cervical column disruption, the central nervous system (including the cerebellum) was dissected, rapidly frozen in liquid nitrogen and kept at –80 °C until use. The animals were regularly examined for determination of the onset of neurological symptoms.

Protein analysis [10]

Briefly, PrP-res was isolated by the scrapie-associated fibrils procedure following proteinase K digestion. The prion protein (PrP) was visualized by immunoblot using polyclonal anti-PrP (007JB) or monoclonal anti-PrP (3F4, RJ Kascasck, New York, NY, USA) antibodies. Immunodetection was carried out with the Enhanced Chemiluminescence kit (ECL, Amersham) and signals were quantified with a Radio-Imager (Bio-Rad).

Gene expression [14]

Total RNA was size-fractionated and blotted onto a nylon membrane. Blots were hybridized with α-^{32}P dCTP-labeled probes. Probes were as follows: i) a glyceraldehyde-3-phosphate-dehydrogenase (GAPDH) probe (P Fort, Université de Montpellier, France); ii) a GFAP probe (NJ Cowan, New York University, NY, USA); iii) a PrP probe (D Westaway, University of San Fransisco, CA, USA). After stringent washes, signals were quantified with a Radio-Imager and normalized in comparison with the glyceraldehyde-3-phosphate-deshydrogenase (GAPDH) signals.

Results

Infectivity studies in treated hamsters

This experiment was done in order to verify a possible dissociation between PrP-res accumulation and TSE agent multiplication in the brain after treatment as published elsewhere [11]. PrP-res accumulation was lower than in untreated infected animals (data not shown) after long-term treatment. The difference is only due to a delay in PrP-res accumulation kinetics while the same amounts of PrP-res can be observed at the terminal stage of the disease. The greatest difference in PrP-res accumulation was observed between treated and untreated hamsters at 40 days post-infection (dpi) (fig 1) as previously reported [11]. We then titrated the infectivity at 40 dpi by the end-point dilution method in recipient hamsters. At this time, the TSE agent load was significantly lower in the groups treated with 2.5 mg/kg AmB and 10 or 25 mg/kg MS-8209 versus solvent-treated and untreated groups (fig 1). The major difference between the two previous experiments reported in the literature was the use for inoculum of crude brain homogenates [12] or purified PrP-res [11]. In our experiment, the same difference in infectious titers has been observed between treated and control groups (solvent-treated and untreated) when we used both preparations as inoculum. Moreover, purification by SAF did not lead to a loss in the infectious titer (data not shown).

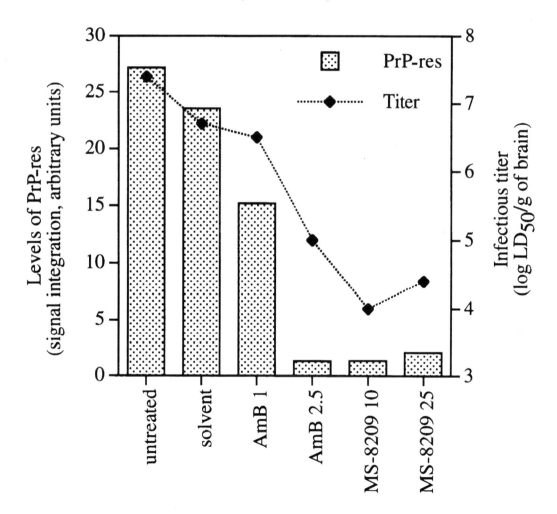

Fig 1. Effect of polyene antibiotic administration (drug, dose in mg/kg) on PrP-res accumulation and infectious titer in the brain of 263K-infected hamsters. The animals were sacrificed 40 dpi. PrP-res amounts were estimated by immunoblotting with 3F4 antibody and infectious titers were calculated using the end-point dilution method.

Preclinical treatment in mouse scrapie

In a preliminary experiment, intracerebrally infected mice treated with 2.5 mg/kg of MS-8209 or AmB from 90 to 120 dpi had lower PrP-res and GFAP mRNA levels in comparison with controls at the end of the treatment, whereas, 30 days after the end of the treatment (150 dpi), PrP-res accumulation was lower in AmB-treated groups and *GFAP* gene expression was similar in all groups (data not shown). Surprisingly, survival time was prolonged (data not shown). Mice were then treated from the commencement of precise events in the course of the disease – the beginning of detectable PrP-res accumulation (80 dpi), the appearance of histo-pathological lesions (110 dpi) and the prodromic phase (140 dpi) – until death. For treatment beginning 80 dpi, PrP-res levels were analyzed 110 dpi; they were similar in the three groups (untreated, solvent-treated, AmB-treated groups at 1 mg/kg) and lower for the others. PrP-res levels were inversely proportional to the dose of drug administered (fig 2). All treated groups had less hyperexpression of the *GFAP* gene in comparison with controls, except for mice treated with 1 mg/kg AmB. These results are in accordance with those obtained after a treatment from 90 to 120 dpi. For treatment beginning 110 dpi, analyses were done at 140 dpi; except in the untreated group, PrP-res levels were similar in all groups (approximate-ly half that of the untreated group). All of the groups had similar levels of *GFAP* gene expres-sion. Untreated and solvent-treated mice developed clinical signs in the same range of incuba-tion time. During treatment beginning 140 dpi, in all treated groups animals died within a few minutes after injection, suggesting stress hypersensitivity; these animals were excluded from the statistical analysis. All treated groups had a prolonged incubation time compared with the solvent-treated group when treatment was started 80 and 110 dpi ($P < 0.01$, fig 2A). When the treatment started 140 dpi, a significant increase in survival time was only observed for mice treated with 25 mg/kg of MS-8209 (fig 2A). It is noteworthy that we obtained an effect with a 2.5-mg/kg treatment starting 80 dpi, comparable to that of long-term treatment beginning on the day of inoculation at the same dose (39 vs 44 days [10]).

Discussion

Polyene antibiotics delayed both TSE agent multiplication and PrP-res accumulation [4, 12]. This is probably related to host mechanisms because of the lack of effect when the drugs were mixed with the scrapie inoculum [15]. Thus, PrP-res accumulation is linked to TSE agent multiplication even after treatment with polyene antibiotic. We then investigated the effects of treatments at the late stages of the disease, when pathology in the brain is independent of TSE agent multiplication in peripheral organs [16]; they led to reduced PrP-res accumulation and increased the animal survival time. In contrast, HPA23, which is the only polyanion able to prolong survival time of intracerebrally infected mice, did not affect survival time in the same model when used during the late stages of the disease, despite its ability to cross the blood–brain barrier [17].

Effects obtained in the early stages of the experimental disease are of another nature from those observed at the late stages. When treatment began close to inoculation, the effects were similar to those observed with polyanions: survival times increased and PrP-res levels were reduced even when treatment was stopped early in the experimental disease [18], as observed with a diluted inoculum. Thus, polyene antibiotics seem to interfere directly or indirectly with the infectious process which was reported to be linked to protease-sensitive prion protein (PrP^C) expression [19, 20]. This hypothesis is confirmed by the differences observed in infec-

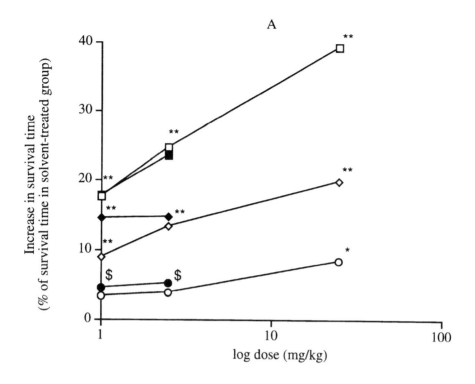

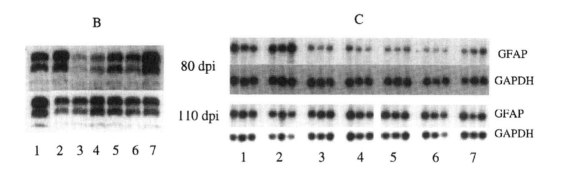

Fig 2. Effects of preclinical treatment with polyene antibiotics. **A.** Survival time after three different treatments beginning 80 (squares), 110 (diamonds) or 140 dpi (circles) (solid symbols: AmB-treated; open symbols: MS-8209-treated). ******Values significantly different from the solvent-treated group ($P < 0.01$); *significantly different from the solvent-treated group ($P < 0.05$); $significantly different from the untreated group ($P < 0.05$); unmarked: not significantly different from the untreated and solvent-treated groups. **B.** PrP-res accumulation and **C.** *GFAP* gene expression, 30 days after beginning of the treatment 80 and 110 dpi. Lane 1: untreated mice; lane 2: solvent-treated mice; lane 3: 25 mg/kg of MS-8209; lane 4: 2.5 mg/kg of MS-8209; lane 5: 1 mg/kg of MS-8209; lane 6: 2.5 mg/kg of AmB; lane 7: 1 mg/kg of AmB.

tious titers in treated hamsters. When treatments were given during preclinical stages of the disease, we observed a delay in PrP-res accumulation and lower infectious titers (data not shown); this effect was observed only when treatment started before the appearance of neurological injuries. Moreover, while survival time was increased, *GFAP* gene expression and PrP-res accumulation were not affected by the drug, and there was no relationship between the dose of AmB and the survival time. Consequently, two observations should be noted: i) the dissociation between PrP-res accumulation and survival time at a later stage and ii) the similar effects of long-term administration and treatment beginning 80 dpi. Thus, we cannot exclude multiple effects of polyene antibiotics during the disease – on the infectious process and on the neuropathological process. AmB could inhibit PrP-res-mediated neurotoxicity [21, 22] directly at the neuron level or indirectly at the glial cell levels. It seems unlikely that: i) AmB has a direct neuroprotective effect because pharmacological data have demonstrated its neurotoxicity [23] and ii) AmB has an inhibitory effect on glial cells, whereas it has an immunostimmulatory effect on peripheral macrophages [24, 25]. Another explanation is that polyene antibiotics limit PrP-res-mediated neurotoxicity via their effects on PrP-res accumulation which is the main determinant of neuronal death. Indeed, there is a marked effect on PrP-res accumulation with these antibiotics: an effect that is not related to *Prnp* gene underexpression (*Prnp* gene expression is similar after treatment with MS-8209 or AmB in vivo; unpublished data). Experiments performed by Caughey and Raymond [26] in ScMNB cells showed that AmB, unlike polyanions, did not alter PrP-res production. As we consider neuroblastomal cells infected with the scrapie agent as a model of the infected neurons, the AmB effect on PrP-res accumulation in neurons seems to be indirect. We must therefore characterize the effects of such molecules on PrP-res-mediated cell toxicity.

As we hypothetized that AmB interferes with the infectious process, prolongation of survival time reflects only the inability of the TSE agent to infect new cells. We may expect that PrP-res accumulation topography could be modified after treatment. Nevertheless, global evaluation of the PrP-res level did not permit us to predict subtle modifications that may be important in neurodegeneration. It will be of a great interest to correlate survival time and modification of PrP-res accumulation and *GFAP* gene hyperexpression topographies in precise areas of the brain.

Acknowledgments

Thanks to K Cherifi for the drug process. Special thanks to R Rioux and JC Mascaro for animal care. This work was made possible by a grant from DRET and the ACCSV of the French Ministry of Research.

References

1 Ehlers B, Diringer H. Dextran sulphate 500 delays and prevents mouse scrapie by impairment of agent replication in spleen. *J Gen Virol* 1984;65:1325-30
2 Kimberlin RH, Walker CA. The antiviral compound HPA-23 can prevent scrapie when administered at the time of infection. *Arch Virol* 1983;78:9-18
3 Ingrosso L, Ladogana A, Pocchiari M. Congo red prolongs the incubation period in scrapie-infected hamsters. *J Virol* 1995;69:506-8
4 Pocchiari M, Schmittinger S, Ladogana A, Masullo C. Effects of amphotericin B in intracerebrally scrapie inoculated hamster. In: Court LA, Dormont D, Brown P, Kingsbury DT, eds. *Unconventional Virus Diseases of the Central Nervous System*. Paris: CEA Diffusion, 1986:1:314-23
5 Amyx H, Salazar AM, Gajdusek DC, Gibbs CJ. Chemotherapeutic trials in experimental slow virus disease. *Neurology* 1984;34:149
6 Bolard J. How do the macrolide antibiotics affect the cellular membranes properties? *Biochim Biophys Acta* 1986;864:257-304

7 Joly V, Yeni P. Amphotéricine B. In: Carbon C, Régnier B, Saimot AG, Vildé JL, Yeni P, eds. *Médicaments anti-infectieux.* Paris: Médecines-Sciences, 1994:455-61

8 Saint-Julien L, Joly V, Seman M, Carbon C, Yeni P. Activity of MS-8209, a nonester amphotericin B derivative, in treatment of experimental systemic mycoses. *Antimicrob Agents Chemother* 1992;36:2722-8

9 Adjou KT, Demaimay R, Lasmézas C, Deslys JP, Seman M, Dormont D. MS-8209, a new amphotericin-B derivative, provides enhanced efficacy in delaying hamster scrapie. *Antimicrob Agents Chemother* 1995;39:2810-2

10 Demaimay R, Adjou K, Lasmézas C et al. Pharmacological studies of a new derivative of amphotericin B, MS-8209, in mouse and hamster scrapie. *J Gen Virol* 1994;75:2499-503

11 Xi YG, Ingrosso L, Ladogana A, Masullo C, Pocchiari M. Amphotericin B treatment dissociates in vivo replication of the scrapie agent from PrP accumulation. *Nature* 1992;356:598-601

12 McKenzie D, Kaczkowski J, Marsh R, Aiken J. Amphotericin B delays both scrapie agent replication and PrP-res accumulation early in infection. *J Virol* 1994;68:7534-6

13 Prusiner SB, Cochran P, Groth DF, Downey DE, Bowman KA, Martinez HM. Measurement of the scrapie agent using an incubation time interval assay. *Ann Neurol* 1982;11:353-8

14 Lazarini F, Deslys JP, Dormont D. Regulation of the glial fibrillary acidic protein, β actin and prion protein mRNAs during brain development in mouse. *Mol Brain Res* 1991;10:343-6

15 Carp RI. Unconventional infectious agent. In: Specter S, ed. *Neuropathogenic Viruses and Immunity.* New York: Plenum Press, 1992:111-36

16 Kimberlin RH, Walker CA. Pathogenesis of mouse scrapie: dynamics of agent replication in spleen, spinal cord and brain after infection by different routes. *J Comp Pathol* 1979;89:551-62

17 Dormont D, Yeramian P, Lambert P et al. In vitro and in vivo antiviral effects of HPA23. In: Court LA, Dormont D, Brown P, Kingsbury DT, eds. *Unconventional Virus Diseases of the Central Nervous System.* Paris: CEA Diffusion 1986;1:324-37

18 Adjou KT, Demaimay R, Lasmézas C, Deslys JP, Seman M, Dormont D. Differential effects of a new amphotericin B derivative, MS-8209, on mouse BSE and scrapie: implications for the mechanism of action of polyene antibiotics. *Res Virol* 1996;147:213-8

19 Büeler H, Aguzzi A, Sailer A et al. Mice devoid of *PrP* are resistant to scrapie. *Cell* 1993;73:1339-47

20 Manson JC, Clarke AR, McBride PA, McConnell I, Hope J. *PrP* gene dosage determines the timing but not the final intensity or distribution of lesions in scrapie pathology. *Neurodegeneration* 1994;3:331-40

21 Müller WEG, Ushijima H, Schröder HC et al. Cytoprotective effect of NMDA receptor antagonists on prion protein (prionSc)-induced toxicity in rat cortical cell cultures. *Eur J Pharmacol* 1993;246:261-7

22 Brandner S, Isenmann S, Raeber A et al. Normal host prion protein necessary for scrapie-induced neurotoxicity. *Nature* 1996;379:339-43

23 Walker RW, Rosenblum MK. Amphotericin B-associated leukoencephalopathy. *Neurology* 1992;42:2005-10

24 Tokuda Y, Tsuji M, Kimura S, Abe S, Yamaguchi H. Augmentation of murine tumor necrosis factor production by amphotericin B in vitro and in vivo. *Antimicrob Agents Chemother* 1993;37:2228-30

25 Cleary JD, Chapman SW, Nolan RL. Pharmacologic modulation of interleukin-1 expression by amphotericin B-stimulated human mononuclear cells. *Antimicrob Agents Chemother* 1992;36:977-81

26 Caughey B, Raymond GJ. Sulfated polyanion inhibition of scrapie-associated PrP accumulation in cultured cells. *J Virol* 1993;67:643-50

Transmissible Subacute Spongiform Encephalopathies:
Prion Diseases
L Court, B Dodet, eds
© 1996, Elsevier, Paris

Altered properties of prion proteins in Creutzfeldt-Jakob disease patients heterozygous for the E200K mutation*

Ruth Gabizon[1], Glenn Telling[2], Zeev Meiner[1], Michele Halimi[1], Irit Kahana[1], Stanley B Prusiner[2,3]

[1]Department of Neurology, Hadassah University Hospital, Jerusalem, Israel; [2]Departments of Neurology, and [3]Biochemistry and Biophysics, University of California, San Francisco, CA 94143-0518, USA

Summary – Creutzfeldt-Jakob disease (CJD) in Libyan Jews is an autosomal dominant disorder linked to a mutation at codon 200 of the prion protein gene (*PRNP*), resulting in the substitution of lysine (K) by glutamate (E). Using monospecific antisera to synthetic prion protein (PrP) peptides with either E or K at position 200, we showed that, in heterozygous CJD patients of Libyan origin, protease-resistant PrP is composed largely if not entirely of mutant PrP. However, both wild-type (wt) and mutant PrP remain insoluble in the presence of nondenaturing detergents, another hallmark of the disease-specific isoform (PrPSc). Mutant PrP was found to have a different glycosylation than wt PrP, and to be more stable. Our findings indicate that, although only the mutant PrP becomes protease resistant, both PrP proteins in the brains of patients who died of an inherited prion disease underwent changes in their properties.

Introduction

Inherited prion diseases constitute a group of fatal neurodegenerative illnesses which are vertically transmitted as autosomal dominant disorders [1]. They are caused by mutations in the human prion protein (PrP) gene, designated *PRNP*, which is located in the short arm of chromosome 20. Three groups of inherited prion diseases are recognized based on the clinical manifestations of these diseases: familial Creutzfeldt-Jakob disease (CJD), Gerstmann-Sträussler-Scheinker (GSS) disease, and fatal familial insomnia (FFI) [2–4].

In Libyan Jews suffering from CJD, a point mutation results in the substitution of K for E at residue 200 of PrP and is referred to as PrP(E200K) [5, 6]. This mutation was found in all definitely affected individuals and yields a maximum load score of 4.85 [7]. The same mutation is also present in other clusters of the disease [8–10].

Many lines of evidence argue that the abnormal disease-specific isoform of the prion protein (PrPSc) or CJD prion protein (PrPCJD) is necessary for both the transmission and pathogenesis of these diseases [11]. The fundamental event underlying prion propagation features a

*Adapted from *Nature Med* 1996;2:59-64

Key words: Creutzfeldt-Jakob disease / E200K mutation

conformational change in PrP [12]. The protease-sensitive prion protein (PrP^C) is thought to adopt a structure composed of four α-helices, whereas in PrP^{Sc} some of this helical structure is lost and a large amount of β-sheet is formed [13]. The pathological PrP mutations may destabilize one or more α-helices of PrP^C. Studies with transgenic (Tg) mice led to the proposal that prion propagation involves the formation of a complex between PrP^{Sc} and the homotypic substrate PrP^C to form a heterodimeric intermediate which is transformed into two molecules of PrP^{Sc} [14, 15].

Most patients with inherited prion diseases are heterozygous and produce, we presume, both wild-type (wt) and mutant PrP^C. Whether mutant PrP^{Sc} recruits both wt and mutant, or only mutant PrP^C to form PrP^{Sc} has not yet been resolved. It has been shown by protein sequencing that in GSS disease with mutations in one of the codons 102, 145, 198 and 217, amyloid plaques were composed mostly of mutant PrP [16–18].

Since the E200K mutation is found in the most common inherited prion disease, we prepared monospecific antisera to synthetic peptides with either E or K at residue 200. Using these antisera, we found that protease-resistant PrP in CJD patients with the E200K mutation is composed primarily if not exclusively of mutant PrP. However, the protein originating from the wt allele, although not protease-resistant, remains insoluble in the presence of nondenaturing detergent, and thus displays one of the hallmarks of PrP^{Sc} rather than PrP^C [19, 20]. In addition, our results suggest that mutant PrP^C is more stable than wt PrP, and therefore accumulates in carriers of the mutation [20].

Results

In order to investigate the status of mutant and wt PrP in heterozygous CJD patients carrying the E200K mutation, we prepared rabbit antisera to peptides with sequences around codon 200 of the *PRNP* gene, with either K or E at residue 200. Two of these antisera, 1E (raised against keyhole limpet hemocyanin (KLH) coupled to peptide GENFTETDVKMMERVVEQM) and 4K (raised against KLH coupled to peptide KQHTVTTTTKGENFTKTDVKMMER) reacted specifically with wt PrP and PrP(E200K), respectively, both by immunoprecipitation and by Western blotting [20]. In all the experiments described below, the CJD brain samples used were from individuals heterozygous for the E200K mutation, while control brain was from a homozygous wild-type (wt) individual.

In addition to its protease resistance, PrP^{Sc}, in contrast to PrP^C, is insoluble in detergents [19]. In order to assess these properties in wt and mutant PrP present in the brain of a heterozygous CJD patient, we solubilized a crude membrane preparation of control and CJD(E200K) brain in a buffered solution containing 2% Sarkosyl followed by centrifugation of the lysate at 20,000 g for 1 h. The pellets and supernatants were digested in the presence and absence of 50 μg/mL proteinase K for 30 min at 37 °C.

Figure 1 is an immunoblot of undigested supernatant fractions from control and CJD(E200K) brains as well as pellet from CJD(E200K) brain in the presence and absence of proteinase K (PK). Three antibodies were used, 3F4 monoclonal antibody (mAb) and the two monospecific antisera. The 3F4 mAb presumably shows the relative concentration of PrP in the samples regardless of the allelic origin of the protein. As expected, the 1E antiserum reacted strongly with the supernatant of control brain and also with the undigested pellet of the CJD brain, but not with the protease-resistant PrP present in the CJD brain pellet. The 4K antiserum did not react with the supernatant of the control brain and reacted weakly with the CJD supernatant, since it did not react with wt PrP. In contrast, 4K reacted strongly with both the PK-digested and undigested CJD pellets. Surprisingly, in the heterozygous CJD brain, wt PrP, as probed by antiserum 1E, was mainly present in the pellet and not in the supernatant, in contrast to the expected location of PrP^C. We considered the possibility that wt PrP, even if not participating in the disease process, may be affected by the neurodegenerative state of the brain. This does not seem to be the case, however, since the control sample in figure 1 is from the brain of an individual suffering from Alzheimer's disease.

In immunoblots, the 4K antiserum reacted primarily with the lower bands of mutant PrP, which represent partial glycosylated protein (fig 1). This is probably due to the fact that the antibody epitope is masked by the sugar residues attached to the N residue at position 197. In contrast to the to the 4K antiserum, the 1E antiserum reacted with fully glycosylated PrP in the control brain.

The relative molecular mass (M_r) of PrP^C in cultured fibroblasts from patients homozygous for the E200K mutation was lower than that of wt PrP^C derived from the fibroblasts of control

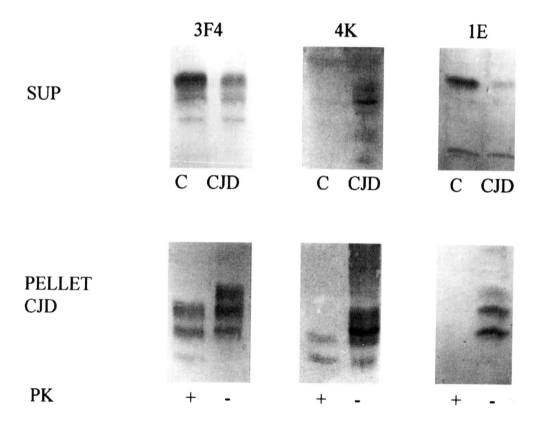

Fig 1. Immunoblot of fractions from control and Creutzfeldt-Jakob disease (CJD) brains with 1E and 4K antisera. One hundred mL of brain microsomes from a wild-type control and a E200K heterozygous CJD brain were diluted to 2 mL with TNE (10 mM Tris pH 7.5, 100 mM NaCl, 1 mM EDTA) with 2% Sarkosyl and centrifuged at 20,000 *g* for 1 h at 4 °C. The pellet of the CJD sample was resuspended in TNE with 2% Sarkosyl and digested in the presence and absence of 50 μg/mL proteinase K (PK) for 30 min at 37 °C. The supernatants (SUP) were concentrated by methanol precipitation and the four samples were immunoblotted with three different antibodies, 3F4 mAb and the 1E and 4K antisera. Molecular mass markers were 53, 34.9, 28.7, 20.5 kDa.

individuals (fig 2). To determine whether the reduced M_r of PrP(E200K) is caused by a different glycosylation pattern of PrP or to an abnormal truncation of the protein, we cultured fibroblasts from normal controls and CJD patients heterozygous and homozygous for the codon-200 mutation in the presence of tunicamycin, which inhibits the Asn-linked glycosylation of proteins. PrP from all three sources synthesized in the presence of tunicamycin had similar M_r. The difference in M_r between wt and mutant PrP can therefore be attributed to differences in glycosylation of PrP^C (E200K).

To compare the relative stabilities of wt and mutant PrP, we cultured cells derived from normal individuals and patients homozygous for the codon-200 mutation for increasing periods of time in the presence of tunicamycin (fig 3). Since PrP was produced only in its unglycosylated form under these conditions, we could compare the rate of disappearance of fully glycosylated wt and mutant PrP^C. Fully glycosylated species disappeared more rapidly in the case of wt than in mutant PrP, indicating that the degradation of PrP^C (E200K) is slower than for the wt protein.

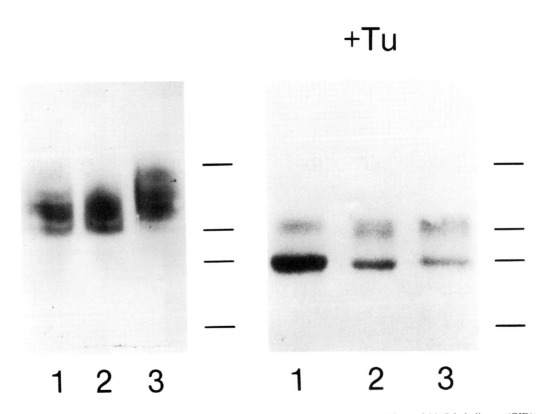

Fig 2. Immunoblot analysis of cultured human fibroblasts from control subjects and Creutzfeldt-Jakob disease (CJD) patients. Cultured fibroblasts derived from skin biopsies of wild-type controls and CJD patients were grown either in the absence (left panel) or presence (right panel) of tunicamycin (Tu). Cells were derived from a normal control individual (lanes 3), a patient heterozygous for the PrP(E200K) mutation (lanes 1) and a patient homozygous for the PrP(E200K) mutation (lanes 2). The cells were extracted with TNE and 1% NP40. The supernatant of a low-speed spin was concentrated by methanol precipitation, followed by immunoblotting of the precipitates with 3F4 mAb. Molecular mass markers were 45, 31, 21.5 kDa.

A.

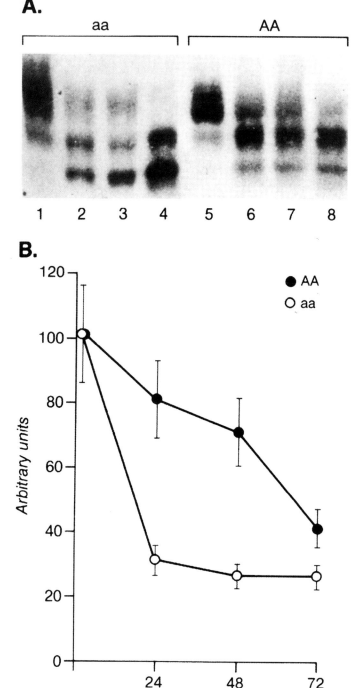

B.

Fig 3. The turnover of PrP^C-(E200K) is slower in cultured cells than that of wild-type PrP^C. A. Fibroblasts from a normal individual or from a CJD patient homozygous for the PrP(E200K) mutation were cultured in the presence of tunicamycin for 0, 3, 24 and 72 h, respectively. After the designated time points, the cells were extracted as explained in the previous figures and the methanol-precipitated extracts were immunoblotted with 3F4 mAb. Lanes 1–4: fibroblasts from a normal control individual; lanes 5–8: fibroblasts from an individual homozygous for the PrP(E200K) mutation. B. Each of the upper bands in A which represent mature, fully glycosylated PrP^C was scanned, and the scan values were plotted against hours of incubation in the presence of tunicamycin. Open circles indicate the control individual and solid circles, the homozygous E200K CJD patient. Error bars are standard deviations for the means of four experiments.

Discussion

We produced monospecific antisera which enabled us to distinguish wt from mutant PrP. Using these antisera, we were able to show that, in heterozygous E200K CJD patients, only mutant PrP becomes protease-resistant during the disease. Using the 3F4 mAb, we found that the concentration of total PrP in the cells of patients carrying the mutation was considerably higher than that in control subjects, apparently due to a slower degradation rate of the E200K mutant PrP. In addition, the M_r of mutant PrP was lower than that of wt PrP, probably due to a different glycosylation profile.

In experimental scrapie as well as in sporadic CJD, the distinct properties of PrPSc include partial protease resistance and insolubility in nondenaturing detergents [21].

As opposed to mutant PrP(E200K), wt PrP in the brains of heterozygous CJD patients was not protease-resistant (fig 1). However, as opposed to normal PrPC, wt PrP in these patients' brains was insoluble in the presence of nondenaturing detergents.

Our findings suggest a mechanism in which PrPC (E200K), as compared to wt PrPC, accumulates in the cells of heterozygous individuals due to a slower degradation rate. PrPC(E200K) is not produced as protease-resistant [21] and has to be transformed into PrPSc. This transformation is most likely dependent on some as yet unknown mechanism which is probably favored by the E200K mutation or the different degree of glycosylation also caused by the mutation. Lower glycosylation of PrP increases the rate by which PrP becomes protease-resistant, since tunicamycin-treated ScN2a produce PrPSc more rapidly than untreated cells [22]. It is also possible that spontaneous formation of PrPSc is a stochastic process that depends on the concentration of PrPC. It has been shown that the level of PrPC expression in Tg mice is inversely related to the incubation time in both infectious and genetic forms of the disease [23].

Development of treatments for CJD and specifically for genetic CJD depends on a better understanding of the pathogenic mechanism that causes the disease. To achieve this goal, it is crucial to investigate the biochemical properties and metabolic pathways of wt and mutant PrP in heterozygous patients, as well as the interaction between these two proteins. The use of our new antibodies, which can distinguish wt from mutant PrP(E200K), will facilitate this research.

References

1 Prusiner SB. Inherited prion diseases. *Proc Natl Acad Sci USA* 1994;91:4611-4
2 Masters CL, Gajdusek DC, Gibbs CJ Jr. Creutzfeldt-Jakob disease virus isolations from the Gerstmann-Sträussler syndrome. *Brain* 1981;104:559-88
3 Hsiao K, Baker HF, Crow TJ et al. Linkage of the prion protein missense variant to Gerstmann-Sträussler syndrome. *Nature* 1989;338:342-5
4 Goldfarb LG, Petersen RB, Tabaton M et al. Fatal familial insomnia and familial Creutzfeldt-Jakob disease: different prion proteins determined by a DNA polymorphism. *Science* 1992;258:806-8
5 Goldfarb LG, Korczyn AD, Brown P, Chapman J, Gajdusek CD. Mutation in codon 200 of scrapie amyloid precursor gene linked to Creutzfeldt-Jakob disease in Sephardic Jews of Libyan and non-Libyan origin. *Lancet* 1990;336:514-5
6 Hsiao K, Meiner Z, Kahana E et al. Mutation of the prion protein in Libyan Jews with Creutzfeldt-Jakob disease. *N Engl J Med* 1991;324:1091-7
7 Gabizon R, Rosenmann H, Meiner Z et al. Mutation and polymorphism of the prion protein gene in Libyan Jews with Creutzfeldt-Jakob disease. *Am J Hum Genet* 1993;53:828-35
8 Goldfarb LG, Mitrova E, Brown P, Tob BH, Gajdusek DC. Mutation in codon 200 of scrapie amyloid protein gene in two clusters of Creuzfeldt-Jakob disease in Slovakia. *Lancet* 1990;336:637-8
9 Brown P, Galves S, Goldfarb LG et al. Familial Creutzfeldt-Jakob disease in Chile is associated with the codon 200 mutation of the *Prnp* amyloid precursor gene on chromosome 20. *J Neurol Sci* 1992;112:65-7

10 Bertoni JM, Brown P, Goldfarb L, Gajdusek D, Omaha NE. Familial Creutzfeldt-Jakob disease with the *PRNP* codon 200lys mutation and supranuclear palsy but without myoclonus or periodic EEG complexes. *Neurology* 1992;42(4, suppl 3):350

11 Prusiner SB. Molecular biology of prion diseases. *Science* 1991;252:1515-22

12 Safar J, Roller PP, Gajdusek DC, Gibbs CJ Jr. Conformational transitions, dissociation, and unfolding of scrapie amyloid (prion) protein. *J Biol Chem* 1993;268:20276-84

13 Pan KM, Balwin M, Nguyen J et al. Conversion of alpha helices into beta sheets features in the formation of the scrapie prion proteins. *Proc Natl Acad Sci USA* 1993;90:10962-6

14 Scott M, Foster D, Mirenda C et al. Transgenic mice expressing hamster prion protein produce species-specific scrapie infectivity and amyloid plaques. *Cell* 1989;59:847-57

15 Prusiner SB, Scott M, Foster D et al. Transgenic studies implicate interactions between homologous PrP isoforms in scrapie prion replication. *Cell* 1990;63:673-86

16 Kitamoto T, Iizuka R, Tateishi J. An amber mutation of prion protein in Gerstmann-Sträussler syndrome with mutant prion plaques. *Biochem Biophys Res Commun* 1993;192:525-31

17 Tagliavini F, Prelli F, Porro M et al. Amyloid fibrils in Gerstmann-Sträussler-Scheinker disease (Indiana and Swedish kindreds) express only PrP peptides encoded by the mutant allele. *Cell* 1994;79:695-703

18 Kitamoto T, Yamaguchi K, Doh-Ura K, Tateshi J. A prion protein missense variant is integrated in kuru plaque cores in Gerstmann-Sträussler syndrome. *Neurology* 1991;41:306-10

19 Meyer RK, McKinley MP, Bowman KA, Braunfeld MB, Barry RA, Prusiner SB. Separation and properties of cellular and scrapie prion proteins. *Proc Natl Acad Sci USA* 1986;83:2310-4

20 Gabizon R, Telling G, Meiner Z, Halimi M, Kahana I, Prusiner SB. Insoluble wild type and protease resistant mutant PrP were found in the brains of patients who died of inherited prion disease. *Nature Med* 1996;2:59-64

21 Meiner Z, Halimi M, Polakiewicz RD, Prusiner SB, Gabizon R. Presence of prion protein in peripheral tissues of Libyan Jews with Creutzfeldt-Jakob disease. *Neurology* 1992;42:1355-60

22 Tarabolous A, Rogers M, Borchelt DR et al. Acquisition of protease resistance by prion proteins in scrapie-infected cells does not require asparagine-linked glycosylation. *Proc Natl Acad Sci* 1990;87:1355-60

23 Telling GC, Scott M, Mastrianni J et al. Prion propagation in mice expressing human and chimeric PrP transgenes implicates the interaction of cellular PrP with another protein. *Cell* 1995;83:79-90

MOLECULAR BIOLOGY AND TRANSGENIC MODELS

Transmissible Subacute Spongiform Encephalopathies:
Prion Diseases
L Court, B Dodet, eds
© 1996, Elsevier, Paris

Prion biology and diseases

Stanley B Prusiner*

Departments of Neurology and of Biochemistry and Biophysics, HSE-781, University of California, San Francisco, CA 94143-0518, USA

Summary – Prions cause a group of human and animal neurodegenerative diseases which are now classified together because their etiology and pathogenesis involve modification of the prion protein (PrP). Prion diseases are manifest as infectious, genetic and sporadic disorders. The human disorders include kuru, Creutzfeldt-Jakob disease (CJD), Gerstmann-Sträussler-Scheinker disease and fatal familial insomnia. Prion diseases can be transmitted among mammals by the infectious particle designated 'prion'. Despite intensive searches over the past three decades, no nucleic acid has been found within prions; yet, a modified isoform of the host-encoded PrP designated PrP^{Sc} is essential for infectivity. In fact, considerable experimental data argue that prions are composed exclusively of PrP^{Sc}. Studies of the prion diseases have taken on new significance with the recent report of ten cases of variant CJD in teenagers and young adults. All of these cases have been reported from Great Britain where over 160,000 cattle have died of bovine spongiform encephalopathy, or 'mad cow disease' as it is often called. It now seems possible that bovine prions have been passed to humans through the consumption of tainted beef products. The study of prion biology and diseases seems to be a new and emerging area of biomedical investigation. While prion biology has its roots in virology, neurology and neuropathology, its relationships to the disciplines of molecular and cell biology as well as protein chemistry have become evident only recently. Certainly, the possibility that learning how prions multiply and cause disease will open up new vistas in biochemistry and genetics seems likely.

Introduction

Prions cause a group of human and animal neurodegenerative diseases which are now classified together because their etiology and pathogenesis involve modification of the prion protein (PrP) [1]. Prion diseases are manifest as infectious, genetic and sporadic disorders. These diseases can be transmitted among mammals by the infectious particle designated 'prion' [2]. Despite intensive searches over the past three decades, no nucleic acid has been found within prions [3–6]; yet, a modified isoform of the host-encoded PrP designated PrP^{Sc} is essential for infectivity [1, 7–10]. In fact, considerable experimental data argue that prions are composed exclusively of PrP^{Sc}. Earlier terms used to describe prion diseases include: transmissible encephalopathies, spongiform encephalopathies and slow virus diseases [11–13]. Human prion disorders include: kuru, Creutzfeldt-Jakob disease (CJD), Gerstmann-Sträussler-Scheinker (GSS) disease and fatal familial insomnia (FFI).

Key words: prion / mad cows / bovine spongiform encephalopathy / scrapie / Creutzfeld-Jakob disease / protein conformation

Studies of prion diseases have taken on new significance with the recent report of ten cases of atypical CJD in three teenagers and seven young adults [14–16]. All of these cases have been reported from Great Britain where over 160,000 cattle have died of bovine spongiform encephalopathy (BSE) or 'mad cow disease' as it is often called [17]. It now seems possible that bovine prions have been passed to humans through the consumption of tainted beef products. How many cases of atypical CJD caused by bovine prions will occur in the years ahead is unknown. Until more time passes, we shall be unable to assess the magnitude of this potential problem. Whatever its size, the need for more knowledge about the biology of prions and the pathogenesis of prion diseases would seem to be substantial.

Experimental transmissions of CJD and kuru from humans to apes ushered in an exciting period of neurological research which focused first on kuru – a disease of epidemic proportions centered in the Fore region of the New Guinea highlands [18] – and subsequently on CJD and GSS disease [19, 20]. Upon transmission of kuru to apes in 1966, the possibility that kuru was transmitted amongst the New Guinea natives by ritualistic cannibalism was no longer remote [21]. Such transmission studies were stimulated by the insightful suggestion of Hadlow that the neuropathology of kuru resembled scrapie, and thus that the disease might be transmissible after intracerebral inoculation of brain extracts into apes following a prolonged incubation time [22].

Scrapie of sheep and goats possesses an equally fascinating history. For many years, British scientists had argued about whether natural scrapie was a genetic or an infectious disease [23]. Scrapie, like kuru and CJD, produced death of the host without any sign of an immune response to a "foreign infectious agent". Debate about scrapie was heightened in 1966, when Tikvah Alper and colleagues reported the extraordinary resistance of the scrapie agent to inactivation by ionizing and ultraviolet (UV) irradiation [3, 4]. Speculation about the composition of the scrapie agent increased over the following decade.

Origin of the prion concept

Once an effective protocol was developed for preparation of partially purified fractions of scrapie agent from hamster brain, investigators could demonstrate that procedures modifying or hydrolyzing proteins diminish scrapie infectivity [2, 24]. At the same time, tests done in search of a scrapie-specific nucleic acid could not demonstrate any dependence of infectivity on a polynucleotide [2], in agreement with earlier studies reporting the extreme resistance of infectivity to UV irradiation at 254 nm [4]. Based on these findings, the term 'prion' was introduced to distinguish the *pro*teinaceous *in*fectious particles that cause scrapie, CJD, GSS disease, and kuru from both viroids and viruses [2]. It now seems likely that the scrapie agent may be composed only of a protein that adopts an abnormal conformation [25–28].

Prion proteins

After it was established that scrapie prion infectivity in partially purified fractions depended upon protein [24], the search for a scrapie-specific protein intensified. While the insolubility of scrapie infectivity made purification problematic, we took advantage of this property along with its relative resistance to degradation by proteases to extend the degree of purification. In subcellular fractions from Syrian hamster (SHa) brain enriched for scrapie infectivity, a protease-resistant prion protein of 27 to 30 kDa, designated PrP 27–30, was identified that was absent from controls [29–31]. Radioiodination of partially purified fractions revealed a protein unique to preparations from scrapie-infected brains [29, 30]. This protein was later named PrP with an apparent relative molecular mass of 27–30 or more succinctly, as PrP 27–30 [31]. The existence of this protein was rapidly confirmed [32].

Purification of PrP 27–30 to homogeneity allowed determination of its NH_2-terminal amino acid sequence [33]. This permitted the synthesis of an isocoding mixture of oligonucleotides that was subsequently used to identify PrP cDNA clones [34–37]. It was found that PrP is encoded by a chromosomal gene and not by a nucleic acid within the infectious scrapie prion particle [34, 36]. Levels of PrP mRNA remain unchanged throughout the course of scrapie infection – an observation which led to the identification of the normal *Prnp* gene product, a protein of 33 to 35 kDa, designated PrPC [34, 36]. PrPC is protease-sensitive, while PrP 27–30 is the protease-resistant core of a 33 to 35 kDa disease-specific protein, designated PrPSc [36, 38]. At the same time, it was found that brains of normal and scrapie-infected hamsters express similar levels of PrP mRNA and a protease-sensitive prion protein designated PrPC [36]. PrPC or a subset of PrP molecules is the substrate for PrPSc.

The discovery of PrP 27–30 in fractions enriched for scrapie infectivity was accompanied by the identification of rod-shaped amyloid particles [7, 30]. Solubilization of PrP 27–30 into liposomes with retention of infectivity [39] demonstrated that large PrP polymers are not required for infectivity and permitted the immunoaffinity copurification of PrPSc and infectivity [40, 41]. Of note, the amyloid plaques in prion diseases contain PrP, as determined by immunoreactivity and amino acid sequencing [42–46].

Prnp *gene structure, organization and expression*

The entire open reading frame (ORF) of all known mammalian and avian *Prnp* genes resides within a single exon [34, 47–49], which eliminates the possibility that PrPSc arises from alternative RNA splicing [34, 49, 50]. The two exons of the SHa *Prnp* gene are separated by a 10-kilobase (kb) intron: exon 1 encodes a portion of the 5' untranslated leader sequence, while exon 2 encodes the ORF and 3' untranslated region [34]. The mouse (Mo) and sheep *Prnp* genes contain three exons with exon 3 analogous to exon 2 of the hamster [50–52]. The promoters of both the SHa- and Mo*Prnp* genes contain multiple copies of G–C rich repeats and are devoid of TATA boxes. These G–C nonamers represent a motif that may function as a canonical binding site for the transcription factor Sp1 [53]. Although PrP mRNA is constitutively expressed in the brains of adult animals [36, 37], it is highly regulated during development. In the septum, levels of PrP mRNA and choline acetyltransferase increase in parallel during development [54]. In other brain regions, *Prnp* gene expression occurs at an earlier age. In situ, hybridization studies show that the highest levels of PrP mRNA occur in neurons [55]. It was shown that there are < 0.002 *Prnp* gene sequences per 50% infectious dose (ID_{50}) unit in purified prion fractions, indicating that a gene encoding PrPSc is not a component of the infectious prion particle [36]. This is a major feature which distinguishes prions from viruses – including those retroviruses that carry cellular oncogenes – and from satellite viruses that derive their coat proteins from other viruses previously infecting plant cells. Purified fractions enriched for prion infectivity were analyzed for a scrapie-specific nucleic acid, but no such nucleic acid was found [56]. If such a molecule exists, then its size is 80 nucleotides (nt) or less [6, 57].

Cell biology of PrPSc *formation*

In scrapie-infected cells, PrPC molecules destined to become PrPSc exit to the cell surface prior to conversion into PrPSc [58–61]. Like other glycosyl-phosphatidyl inositol (GPI)-anchored proteins, PrP[C] appears to re-enter the cell through a subcellular compartment bounded by cholesterol-rich, detergent-insoluble membranes [62] which might be caveolae or early endosomes [63, 64]. Within this cholesterol-rich, non acidic compartment, GPI-anchored PrP[C] can be either converted into PrP[Sc] or partially degraded [62]. PrP[Sc] is trimmed at the N-terminus in an acidic compartment in scrapie-infected cultured cells, to form PrP 27–30 [61, 65]. In contrast, N-terminal trimming of PrP[Sc] is minimal in brain, where little PrP 27–30 is found [66].

Structures of PrP isoforms

In a search for a post-translational chemical modification that might explain the differences in the properties of these two PrP isoforms, PrP[Sc] was analyzed by mass spectrometry and gas phase sequencing. The amino acid sequence was the same as that deduced from the translated ORF of the *Prnp* gene, and no modifications that might differentiate PrP[C] from PrP[Sc] were found [25]. These findings forced consideration of the possibility that conformation distinguishes the two PrP isoforms.

Molecular modeling of PrP isoforms

By comparing the amino acid sequences of 11 mammalian and one avian prion proteins, structural analyses using a neural network algorithm [67, 68] predicted four α-helical regions [69, 70] (fig 1). Interestingly, the four putative α-helical domains of PrP [69] showed both strong helix preference in the α/α class prediction and strong β-sheet preference in the β/β class prediction. These results are consistent with the hypothesis that these domains undergo conformational changes from α-helices to β-sheets during the formation of PrP[Sc] [71].

Spectroscopy of PrP isoforms

PrP[C] and PrP[Sc] were purified using non denaturing procedures [26, 72]. Fourier transform infrared (FTIR) spectroscopy and circular dichroism (CD) demonstrated that PrP[C] has a high α-helix content (42%) and no β-sheet (3%) [26]. In contrast, the β-sheet content of PrP[Sc] was 43% and the α-helix 30% as measured by FTIR [26] and CD spectroscopy [73]. As determined in earlier studies, N-terminally truncated PrP[Sc] derived by limited proteolysis and designated PrP 27–30 has an even higher β-sheet content (54%) and a lower α-helix (21%) content [74–76]. Denaturation of PrP 27–30 under conditions which reduced scrapie infectivity resulted in concomitant diminution of β-sheet content [75, 76]. Since PrP[Sc] seems to be the only component of the 'infectious' prion particle, it is likely that a conformational transition is a fundamental event in the propagation of prions.

   We and others have suggested that the conversion of PrP[C] into PrP[Sc] may proceed through a metastable or partially unfolded intermediate designated PrP* [77, 78]. Intermediates in the refolding of PrP 27–30 from a denatured state have been identified by fluorescence and CD spectroscopy [79]. How PrP[C] unfolds and refolds into PrP[Sc] is unknown, but the profound change in protein structure that occurs during this process is likely to be associated with a large activation barrier [78]. Whether a catalyst such as a molecular chaperone mediates this process is unknown. Some features of PrP[Sc] formation seem to mimic an enzymatic reaction: PrP[C] is the substrate and PrP[Sc] is the product; the reaction rate depends upon the concentration of substrate, PrP[C], which is manifestly inverse with respect to the incubation time; PrP[Sc] is an allosteric effector which regulates the conversion of PrP[C] into PrP[Sc]; pseudo substrates such as heterotypic PrP[C] molecules can slow the conversion through competition. Besides being the product of the conversion reaction and acting as a ligand, it has been suggested that PrP[Sc] func-

A

B

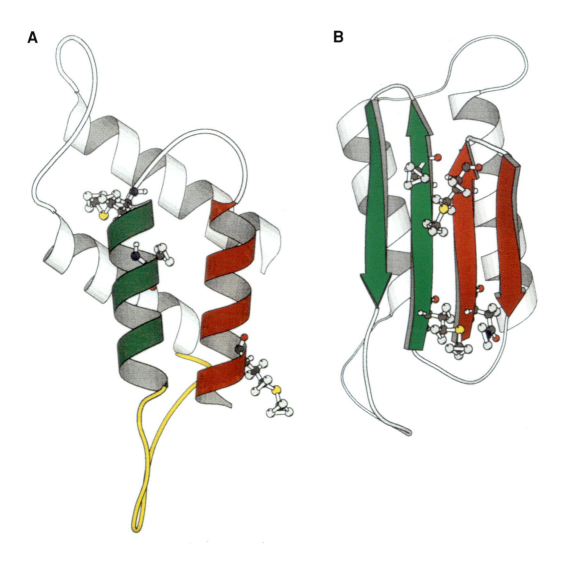

Fig 1. Plausible models for the tertiary structures of PrP^{Sc} and PrP^C. **A.** The proposed three-dimensional structure of PrP^C [70]. Helix 1 is shown in red, while helix 2 is in green. We think that helices 1 and 2 are converted into β-sheet structure during the formation of PrP^{Sc}. The H_2–H_3 loop corresponding to the S_{2b}-H_3 loop in PrP^{Sc} is shown in yellow. Four residues (Asn[108], Met[112], Met[129] and Ala[133]) implicated in the species barrier as noted above. They are shown in the ball and stick model. **B.** The proposed three-dimensional structure of PrP^{Sc}. This structure was chosen from the six penultimate models of PrP^{Sc} because it appeared to correlate best with genetic data on residues involved in the species barrier. It contains a 4-strand mixed β-sheet with two α-helices packed against one face of the sheet. Strands 1a and 1b (in red) correspond to helix 1 in PrP^C, while strands 2a and 2b (in green) correspond to helix 2. Helices 3 and 4 in this model remain unchanged from the PrP^C model [70]. Four residues (Asn[108], Met[112], Met[129] and Ala[133]) implicated in the species barrier [105] are shown in the ball and stick model. They cluster on the solvent accessible surface of the β-sheet which might provide a plausible interface for the PrP^{Sc}–PrP^C interaction. The S_{2b}–H_3 loop connecting the β-sheet and helix 3 is implicated in the species barrier and is shown in yellow. This conformationally flexible loop could come into contact with PrP^C during the formation of the PrP^{Sc}–PrP^C complex. Therefore, the specific molecular recognition during prion replication might involve both the β-sheet as the primary binding site and the S_{2b}–H_3 loop as an additional site for interaction.

tions as a chaperone-like protein in mediating the conversion [80]. Though unlikely, this additional role has not been ruled out by the available data. Changes in the inducibility of heat shock proteins (Hsp), some of which function as molecular chaperones, as well as their subcellular distribution, have been found in scrapie-infected cells [81, 82]. Whether the Hsp function as chaperones in the conversion of PrP^C into PrP^{Sc} or accumulate in response to PrP^{Sc}, which is perceived as a misfolded protein, remains to be determined.

Experiments with immunoprecipitated [^{35}S] Met-labeled SHaPrP^C and a more than 50-fold excess of purified SHaPrP 27–30 have been interpreted as showing that PrP^C becomes protease-resistant under these conditions [27]. To date, there is no evidence that the properties of PrP^C bound to PrP^{Sc} have been changed into those of PrP^{Sc}. Only reisolation of PrP^C after disruption of the PrP^C–PrP^{Sc} complexes will allow determination as to whether the properties of these PrP^C molecules have been altered or not. Our attempts to disrupt such PrP^C–PrP^{Sc} complexes formed in the presence of excess PrP^{Sc} have been unsuccessful using non denaturing procedures [28]. Using a similar protocol, we have shown that synthetic peptides can induce protease resistance in PrP^C, but the resulting PrP^C/peptide complexes have not caused disease after injection into experimental animals, to date [28].

Transgenetics and gene targeting

Species barrier

To test the hypothesis that differences in *Prnp* gene sequences might be responsible for the species barrier, transgenic (Tg) mice expressing SHa*Prnp* were constructed [83, 84]. The *Prnp* genes of Syrian hamsters and mice encode proteins differing at 16 positions. Inoculation of Tg(SHa*Prnp*) mice with SHa prions demonstrated abrogation of the species barrier resulting in abbreviated incubation times due to a non stochastic process [83, 84]. The length of the incubation time after inoculation with SHa prions was inversely proportional to the level of SHaPrP^C in the brains of Tg(SHa*Prnp*) mice [83]. Bioassays of brain extracts from clinically ill Tg(SHa*Prnp*) mice inoculated with Mo prions revealed that only Mo prions but no SHa prions were produced. Conversely, inoculation of Tg(SHa*Prnp*) mice with SHa prions only led to the synthesis of SHa prions.

Modeling GSS disease

The codon-102 point mutation found in GSS disease patients was introduced into the Mo*Prnp* gene and Tg(Mo*Prnp*-P^{101}L)H mice were created expressing high (H) levels of the mutant transgene product. Five lines of Tg(Mo*Prnp*-P^{101}L)H mice developed central nervous system (CNS) degeneration indistinguishable from experimental murine scrapie with neuropathology consisting of widespread spongiform morphology and astrocytic gliosis [85] and PrP amyloid plaques [86]. Brain extracts prepared from spontaneously ill Tg(Mo*Prnp*-P^{101}L)H mice transmitted CNS degeneration to 40% of Tg^{196} mice and some SHa [86]. The Tg^{196} mice express low levels of the mutant transgene product MoPrP^C (P^{101}L), but do not develop spontaneous disease. Much higher rates of transmission of prions from ill Tg(Mo*Prnp*-P^{101}L)H mice to Tg^{196} /*Prnp*$^{0/0}$ mice have been found (G Telling and SB Prusiner, in preparation). These studies as well as transmission of prions from patients who died of GSS disease to apes and monkeys [20, 87], non-Tg mice [88] and mouse/human (Hu) chimeric Tg(MHu2M-P^{102}L) mice [89] argue persuasively that prions are devoid of nucleic acid.

PrP-deficient mice

Ablation of the *Prnp* gene *(Prnp$^{0/0}$)* in mice did not affect the development of these animals [90, 91], but one study of *Prnp$^{0/0}$* mice reports loss of Purkinje cells in the cerebellum after 70 weeks of age [92]. In fact, they are healthy at almost 2 years of age. *Prnp$^{0/0}$* mice are resistant to prions [8] and do not propagate scrapie infectivity [9, 93, 94]. One group of investigators described low levels of prion infectivity in *Prnp$^{0/0}$* mice in an initial report [8], but that was attributed to contamination of pooled samples in a subsequent communication in which the absence of infectivity in individual brains was reported [95].

Hippocampal slices from *Prnp$^{0/0}$* mice have been reported to show abnormalities in gamma-aminobutyric acid (GABA)-mediated synaptic transmission and diminished long-term potentiation (LTP) [96–98]. However, our attempts to reproduce these findings were unsuccessful [99].

Transmission of human prions to transgenic mice

Because our initial transgenetic studies had shown that the 'species barrier' between mice and SHa for the transmission of prions can be abrogated by expression of a SHa*Prnp* transgene in mice [84], Tg mice expressing the human gene designated Hu*PRNP* were constructed. These Tg(Hu*PRNP*)FVB mice inoculated with Hu prions failed to develop CNS dysfunction more frequently than non-Tg controls [100]. Faced with this apparent dichotomy, we constructed mice expressing a chimeric Hu/Mo *PRNP* transgene designated MHu2M. Earlier studies had shown that chimeric SHa/Mo*Prnp* transgenes supported the transmission of either Mo or SHa prions [101, 102]. Hu*PRNP* differs from mouse PrP at 28 of 254 positions [103], while chimeric Hu/MoPrP differs at nine residues. We found that mice expressing the MHu2M transgene are susceptible to human prions and exhibit abbreviated incubation times [100].

When Tg(Hu*PRNP*) mice were crossed with *Prnp$^{0/0}$* mice, their offspring were rendered susceptible to Hu prions, unexpectedly. These findings suggested that Tg(Hu*PRNP*)FVB mice were resistant to Hu prions because MoPrP^C inhibited the conversion of HuPrP^C into PrP^{Sc}; once MoPrP^C was removed by gene ablation, the inhibition was abolished [89]. While earlier studies argued that PrP^C forms a complex with PrP^{Sc} during the formation of nascent PrP^{Sc} [83], these findings suggested that PrP^C also binds to another macromolecule during the conversion process. We provisionally designated this second macromolecule 'protein X'. Like the binding of PrP^C to PrP^{Sc} which is most efficient when the two isoforms have the same sequence [83], the binding of PrP^C to protein X seems to exhibit the highest affinity when these two proteins are from the same species. Whether protein X is a chaperone involved in catalyzing the conformational changes that feature in the formation of PrP^{Sc} [26] remains to be established.

Binding sites on PrP^C

Based largely on the foregoing studies with chimeric transgenes, we surmise that H1 and H2 are the region of PrP^C that binds to PrP^{Sc} during the formation of nascent PrP^{Sc}. This conclusion is supported by correlations between species-specific amino acid substitutions and transmission studies across species [104, 105].

Molecular modeling studies suggest that PrP^C is a four-helix-bundle protein: the four putative helices are designated H1, H2, H3, and H4 [70]. A comparison of amino acid substitutions among species in and adjacent to H3 and H4 suggests that the binding of PrP^C to protein X may occur through interactions at the C-terminal end of H4. Such a model can explain why chimeric MHu2MPrP^C but not HuPrP^C is converted into PrP^{Sc} in the presence of MoPrP^C and why HuPrP^C is converted into PrP^{Sc} in the absence of MoPrP^C. The proposed model is also consistent with the observation that the Hu and MHu2M prions exhibit similar incubation times in Tg(MHu2M) mice [89]. Further support is derived from studies on the transmission of prions from inherited prion diseases to mice expressing chimeric *PrP* transgenes. Tg(MHu2M) mice were resistant to Hu prions from a patient with GSS disease who carried the P102L mutation, but susceptible to prions from patients with familial CJD who harbor the E200K mutation; however, Tg(MHu2M-P^{101}L) mice were susceptible to GSS disease prions. These findings demonstrate that a single amino acid mismatch at codon 102 near H1 prolongs the incubation time; whereas a mismatch at codon 200 in the loop between H3 and H4 does not.

The residue 129 lies at the N-terminus of H2; this polymorphic residue is either M or V in humans. Prions from a patient homozygous for M129 produced an incubation time of ~170 d in Tg(Hu*PRNP*-M^{129})440/*Prnp$^{0/0}$* mice expressing Hu*PRNP* encoding M129 on the null background; the same inocula gave much longer incubation times in Tg(Hu*PRNP*-V^{129})152/*Prnp$^{0/0}$* mice expressing Hu*PRNP* with a V at codon 129 [89]. Molecular modeling studies predict that residue 129 lies at the N-terminus of H2 within the putative PrP^C–PrP^{Sc} interface. These findings demonstrate that a single amino acid mismatch at codon 129, like codon 102 in the region of helices 1 and 2, prolongs the incubation time, in contrast to a mismatch at codon 200.

Binding of PrP^C to PrP^{Sc} and synthetic PrP peptides

Once studies of mice expressing SHa*Prnp* transgenes indicated that PrP^C and PrP^{Sc} form a complex during the formation of nascent PrP^{Sc} [83], we attempted to demonstrate the PrP^{Sc} formation through such complexes by mixing purified fractions containing equimolar amounts of the two isoforms [106]. Unable to demonstrate conversion of PrP^C into PrP^{Sc} in these mixtures, we pursued the interactions of synthetic PrP peptides since we had found that such peptides corresponding to regions of putative secondary structure display conformational pluralism [69, 107]. In contrast to our earlier findings, other investigators were able to demonstrate an interaction between PrP^{Sc} and PrP^C by mixing a 50-fold excess of PrP^{Sc} with labeled PrP^C [27].

Having found that large, synthetic PrP peptides encompassing the first two putative α-helical regions mimic more of the structural features displayed by the two PrP isoforms than smaller peptides [108, 109], we asked if such peptides could alter the properties of PrP^C. Up to 70% of the PrP^C was complexed when a peptide of 56 amino acids corresponding to PrP residues 90–145 was used [28]. Unexpectedly, the peptide in a β-sheet conformation did not bind PrP^C; whereas, the random coil did. The PrP^C–peptide complex was resistant to proteolytic digestion, displayed a high β-sheet content and sedimented at 100,000 g for 1 h. The ultrastructure of the complexes was that of fibrous aggregates. Addition of 2% sarkosyl disrupted the complex and rendered PrP^C sensitive to protease digestion. Complex formation could be prevented by including an α-PrP monoclonal antibody (mAb) in the PrP^C–peptide mixture. When the pathologic A^{117}V mutation was substituted in the peptide, the concentration of peptide required for complex formation could be reduced by a factor of ~10. Our findings in concert with transgenetic investigations argue that PrP^C interacts with PrP^{Sc} through a domain which contains the first two putative α-helices.

Although other investigators have reported the in vitro formation of PrP^{Sc} by mixing a large excess of PrP^{Sc} with [^{35}S]PrP^C, their conclusions assume that protease-resistant PrP^C is equivalent to PrP^{Sc} [27]. Interestingly, the binding of PrP^C to PrP^{Sc} was found to be dependent upon the same residues [110] that render Tg(MHu2M) mice susceptible to SHa prions [102] and it seems to be strain dependent [111]. Though we were able to confirm the binding of PrPC to PrPSc in the presence of a large excess of PrPSc, we were unable to reproduce renaturation of PrPSc from guanidine hydrochloride (Gdn HCl) as judged by restoration of protease resistance [28].

Attempts to separate [^{35}S]PrPC from PrPSc under conditions in which scrapie infectivity is preserved were unsuccessful, using a variety of detergents, α-PrP mAb, detergent-lipid-complexes (DLPC) and synthetic peptides. Until such conditions are identified, we cannot determine whether PrPC has been converted into PrPSc or is only tightly bound. The experiments with PrP peptides which bind to PrPC and render it protease-resistant argue that the latter possibility is more likely to be correct since sarkosyl disrupted the PrPC–peptide complex and made PrPC sensitive to protease. Since the PrP peptides used in our studies have not exhibited prion infectivity, it will be possible to determine if the PrPC–peptide complexes that mimic many of the features of PrPSc are capable of transmitting prion disease to inoculated animals.

Monomers or dimers?

Whether PrPC and PrPSc exist as monomers, dimers or oligomers is unknown. Molecular modeling studies suggest that PrPC may form dimers through a solvent accessible hydrophobic surface of H1 [70]. The dimerization of PrPC provides an explanation for the results of genetic studies using (PrP-A × PrP-B)F$_1$ mice inoculated with 22-A prions [112, 113]. These F$_1$ mice exhibit incubation periods exceeding those of both parents. Ionizing radiation inactivation studies suggest a target size of 55 kDa for the infectious prion particle suggesting that PrPSc is also a dimer [3, 114]. It remains to be established whether the PrPC–PrPSc complex is a heterodimeric or heterotetrameric structure (fig 2B).

On the binding of PrPC to protein X

Although MoPrnp and HuPRNP differ at 28 residues, only nine or fewer amino acids between codons 96 and 167 feature in the species barrier for the transmission of Hu prions into mice as demonstrated by the susceptibility of Tg(MHu2M) mice to Hu prions [100]. Breaching the species barrier by substitution of residues in H1 and H2 is analogous to the results of studies with chimeric SHa/MoPrnp transgenes, as noted above. That Tg(HuPRNP) mice crossed with gene-targeted mice in which the MoPrnp gene has been disrupted (Prnp$^{0/0}$) rendered the mice susceptible to Hu prions, which in combination with the susceptibility of Tg(MHu2M)/FVB mice to Hu prions, argues that MoPrPC inhibits the conversion of HuPrPC, but not that of MHu2MPrPC into PrPSc [89]. Perhaps the most conservative explanation for these findings is the existence of a macromolecule other than PrPSc that binds PrPC.

If Mo protein X has a higher affinity for MoPrPC than it does for HuPrPC, then MoPrPC would inhibit the conversion of HuPrPC into HuPrPSc. When MoPrPC production was eliminated by gene targeting, then HuPrPC is converted into HuPrPSc in murine cells expressing protein X. If the binding site of PrPC to protein X occurs through the N- or C-terminal region of PrPC, then chimeric PrP molecules such as MHu2M would be likely to compete effectively with MoPrPC for binding to protein X. The experimental data argue that this is the case (table I).

Since truncation of the N-terminus of PrP in cultured cells still permitted the formation of PrPSc-like molecules as indicated by their acquisition of protease resistance [115], it seems

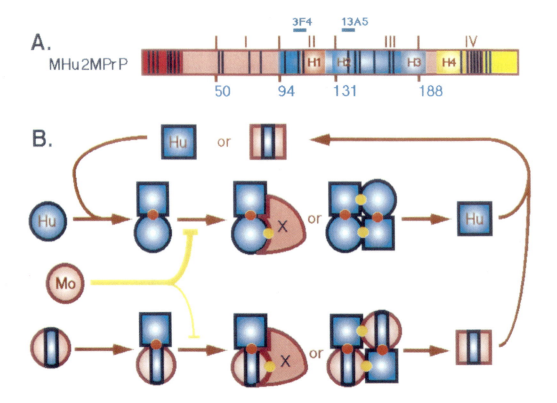

Fig 2. The chimeric Hu/Mo *PRNP* transgene and a model for the formation of PrPSc. **A.** Schematic representation of the open reading frame (ORF) of the prion protein gene showing the 28 differences between the Hu and Mo versions (vertical blue lines). The ORF of Hu*PRNP* encoded a protein of 253 amino acids and that of Mo*Prnp* of 254 residues. The four putative α-helical regions of PrPC are designated H1 (red), H2 (green), H3 (light blue) and H4 (yellow). In segments II and III, DNA fragment from the Hu*PRNP* gene (cyan rectangle) was substituted for Mo*Prnp* (mauve rectangle). The signal peptide at the N-terminus is denoted by the purple rectangle. The signaling sequence at the C-terminus which is removed upon GPI anchor addition is denoted by the yellow rectangle. **B.** A possible model for the formation of PrPSc. PrPC is represented by circles and PrPSc by squares. HuPrP is cyan and MoPrP is mauve. The site for binding of PrPSc to PrPC is denoted by the orange disk and the site for binding of PrPC to protein X by the yellow disk. When present, MoPrPC inhibits the formation of HuPrPSc more than it inhibits the production of MHu2MPrPSc as denoted by the yellow effector lines. An alternative model for the formation of PrPSc is shown where protein X is PrPSc. If this is the case, then the yellow disk designates the binding of the C-terminal domain of PrPC to PrPSc. While a tetrameric intermediate is shown, a dimeric intermediate is also possible where the same molecule of PrPSc binds a molecule of PrPC at two sites.

more likely that the site at which PrPC binds to protein X is at the C-terminal end of PrPC. A comparison of amino acid substitutions in and adjacent to H3 and H4 shows sufficient variation between Hu and Mo as well as between SHa and Mo to account for the results of studies with chimeric transgenes. These sequence comparisons suggest that the binding of PrPC to protein X occurs through interactions at the C-terminal end of H4 (fig 2A). PrP residue 215 is a reasonable candidate for participation in the binding of PrPC to protein X. In HuPrP, this residue is an I, while in MoPrP it is a V, and in SHaPrP, a T. The side-chain of residue 215 is predicted to lie on the external surface of the fourth putative helix (H4). In addition to residue

Table I. Transmission of Hu prions to Tg(MHu2*MPRNP*)mice.

Inoculum[a]	Incubation time mean d ± SEM (n/no)	
(A) Tg(MHu2*M*)FVB mice inoculated with sporadic or infectious CJD		
sCJD(RG)	238 ± 3	(8/8)[b]
sCJD(EC)	218 ± 5	(7/7)[b]
iCJD(364)	232 ± 3	(9/9)[b]
iCJD(364)[c]	231 ± 6	(9/9)
sCJD(MA)	222 ± 1	(4/4)
(B) Tg(MHu2*M*)*Prnp⁰/⁰* mice inoculated with sporadic or infectious CJD		
sCJD(RC)	207 ± 4	(8/10)
sCJD(RG)	191 ± 3	(10/10)
iCJD(364)	192 ± 6	(8/8)
iCJD(364)[c]	211 ± 5	(8/9)
sCJD(MA)	180 ± 5	(8/8)
sCJD(RO)	217 ± 2	(9/9)
(C) Tg(MHu2*M*)*Prnp⁰/⁰* mice inoculated with inherited GSS or CJD		
GSS(JJ,P102L)	> 310	(0/10)
fCJD(LJ1,E200K)	170 ± 2	(10/10)
fCJD(CA,E200K)	180 ± 9	(9/9)
fCJD(FH,E200K)	> 290	(0/7)

[a]All samples were 10% (w/v) brain homogenates unless otherwise noted that were diluted 1:10 prior to inoculation. sCJD is sporadic CJD, iCJD is iatrogenic CJD, GSS disease involves the codon-102 mutation and fCJD is familial CJD with the codon-200 mutation. Patients' initials referring to inocula in table IB are given in parentheses. If the *PRNP* gene of the patient carried a mutation, then the mutation is noted after the patients' initials. [b]Transmissions previously reported [100]. [c]This is a second inoculum prepared from a different brain region of iatrogenic CJD patient 364.

215, amino acids at positions 219 and 220 may participate in the binding of PrP^C to protein X. It seems unlikely that residues at 228 or 230 are involved in the binding to protein X since they are adjacent to the GPI anchor which is attached to an S residue at 231. The proposed model can explain why chimeric MHu2MPrP^C but not HuPrP^C is converted into PrP^{Sc} in the presence of MoPrP^C and why HuPrP^C is converted into PrP^{Sc} in the absence of MoPrP^C. This model is also consistent with the observation that Hu and MHu2M prions exhibit similar incubation times in Tg(MHu2*M*) mice if the concentration of protein X is not rate limiting.

Does PrP^{Sc} bind protein X?

The transmission of Hu and MHu2M prions into Tg(MHu2*M*)/FVB mice yielded similar incubation times. This finding argues that the differences in the Hu or Mo sequences of PrP^{Sc} at the N- and C-termini have little effect on the transmission of prions to Tg(MHu2*M*) mice (tables I and II). Conversely, the region of PrP containing H1 and H2 clearly governs prion transmission to Tg(MHu2*M*)/*Prnp⁰/⁰* mice: inoculation of Hu or MHu2M prions produced disease but Mo prions did not.

SB Prusiner

Table II. Serial transmission of chimeric Hu/Mo prions in Tg(MHu2M) mice.

Recipient mouse line	Inoculum[a]	Incubation times mean d ± SEM (n/no)	
		Illness	Death
(A) Chimeric prions produced in Tg(MHu2M) mice inoculated with CJD prions			
Tg(MHu2M)5378/FVB	MHu2M(sCJD)[b]	220 ± 3 (7/7)[c]	226 ± 1 (5)
Non Tg5378/FVB	MHu2M(sCJD)[b]	> 400	
Tg(MHu2M)5378/FVB	MHu2M(sCJD)[d]	226 ± 3 (9/9)	228 ± 1 (6)
Non Tg5378/FVB	MHu2M(sCJD)[d]	> 400	
Tg(MHu2M)5378/Prnp[0/0]	MHu2M(sCJD)[b]	189 ± 4 (8/8)	192 ± 1 (4)
Tg(MHu2M)5378/Prnp[0/0]	MHu2M(sCJD)[d]	183 ± 5 (7/7)	190 ± 3 (4)
(B) Mouse prions produced in Tg(MHu2M) or non-Tg mice inoculated with RML prions			
Tg(MHu2M)5378/FVB	Mo(RML)	178 ± 3 (19/19)	203 ± 2 (14)[e]
NonTg5378/FVB	Mo(RML)	127 ± 2 (18/18)	156 ± 2 (5)
Tg(MHu2M)5378/FVB	MHu2M(RML)[f]	185 ± 1 (7/7)	211 ± 1 (3)
Tg(MHu2M)5378/FVB	MHu2M(RML)[g]	189 ± 2 (7/7)	211 ± 9 (3)
Non-Tg5378/FVB	MHu2M(RML)[g]	134 ± 3 (5/5)	ND
Tg(MHu2M)5378/Prnp[0/0]	Mo(RML)	> 400	
Tg(MHu2M)5378/Prnp[0/0]	MHu2M(RML)[f]	> 360	
Tg(MHu2M)5378/Prnp[0/0]	MHu2M(RML)[g]	> 360	

[a]Notations in parentheses indicate inoculum used in initial passage in Tg(MHu2M) mice. [b]Mice were inoculated with chimeric prions generated in the brain of a Tg(MHu2M)5378/FVB mouse that had been inoculated with a brain homogenate prepared from patient EC who died of sporadic CJD. [c]Number of mice developing CNS illness divided by the number inoculated are given in parentheses. [d]Mice were inoculated with chimeric prions generated in the brain of a second Tg(MHu2M)5378/FVB mouse that had been inoculated with a brain homogenate prepared from patient EC who died of sporadic CJD. [e]Data from [100]. [f]Mice were inoculated with Mo prions generated in the brain of a Tg(MHu2M)5378/FVB mouse that had been inoculated with Rocky Mountain Lab (RML) Mo prions. [g]Mice were inoculated with Mo prions generated in the brain of a second Tg(MHu2M)5378/FVB mouse that had been inoculated with RML Mo prions.

Although the lack of binding of PrPSc to protein X is what we expect for the product of an 'enzyme-substrate' reaction, this situation can be construed as an argument favoring the hypothesis that PrPSc is protein X. If PrPSc binds to both sites on PrPC, then the infrequent transmission of Hu prions to Tg(HuPRNP)/FVB mice suggests the HuPrPSc has a higher affinity at the C-only terminal binding site for MoPrPC than for HuPrPC. This conclusion contrasts with the observation that homotypic interactions between PrPC and PrPSc have produced the shortest incubation times when the same amino acid sequence is shared by the two PrP isoforms.

While most cases of Hu prion disease are not readily transmitted to Tg(HuPRNP)/FVB, an exception has been noted [100]. Undoubtedly, other such cases will be found since it has been reported that a group of GSS disease cases harboring the P102L mutation transmit CNS degeneration to non-Tg mice in less than 400 days while a second group of cases with the same genotype does not [116]. One explanation for these cases of prion disease which are unusual with respect to their transmission characteristics is that they represent different strains of prions. Upon the binding of PrPSc to PrPC, the resulting conformation of PrPC allows it to bind protein X.

No evidence that amyloid formation features in PrP^{Sc} synthesis

Some investigators have suggested that scrapie agent multiplication proceeds through a crystallization process involving PrP amyloid formation [77, 117–119]. While the high β-sheet content of PrP^{Sc} is clearly a hallmark distinguishing it from PrP^C [26, 73], there is no evidence that PrP^{Sc} forms paracrystalline arrays as is the case for amyloids which also have a high β-sheet content [120, 121]. Purified infectious prions isolated from scrapie-infected SHa brains with a cocktail of protease-inhibitors contain only PrP^{Sc} which exists as amorphous aggregates; only if PrP^{Sc} molecules are exposed to detergents and limited proteolysis do they then polymerize into prion rods exhibiting the ultrastructural and tinctorial features of amyloid [26, 66]. Furthermore, dispersion of prion rods into DLPC results in a 10- to 100-fold increase in scrapie titer, and no rods could be identified in these fractions by electron microscopy [39]. Consistent with these findings is the absence or rarity of amyloid plaques in many prion diseases, as well as the inability to identify any amyloid-like polymers in cultured cells chronically synthesizing prions [66, 83].

While prion replication seems to be 'seeded' by PrP^{Sc}, it can also be initiated by specific mutations, the specificity of which has been dramatically demonstrated recently when the change from a glutamate to a lysine at codon 200 produces PrP^{CJD} and CJD in humans [122], while the same glutamate-to-lysine mutation at codon 219 is a common polymorphism occurring in ~6% of the Japanese population [123].

The discovery of two binding sites on PrP^C, one of which interacts with homotypic PrP^{Sc} and the other with protein X, argues for a high degree of specificity in the prion replication process. While SHaPrP^C or a chimeric SHa/MoPrP^C can be converted into the corresponding PrP^{Sc} molecules in Tg mice inoculated with SHa prions [83, 102], the situation with Hu prions is different [89, 100]. As described above, Tg(Hu*PRNP*)/FVB mice rarely produce HuPrP^{Sc} after inoculation with Hu prions, but after crossing Tg(Hu*PrP*) mice onto the null *Prnp*$^{0/0}$ background their offspring are rendered susceptible to Hu prions. Furthermore, Tg(MoHu2*M*)/FVB mice expressing a chimeric Hu/MoPrP^C readily produce chimeric PrP^{Sc} after inoculation with Hu prions [100]. These results argue persuasively for two different species-specific binding sites on PrP^C in the propagation of prion infectivity.

It is difficult to reconcile the foregoing data with the conjecture that crystallization is the mechanism by which prions replicate. A crystallization phenomenon producing multidimensional macroscopic ordered arrays implies a high degree of at least two-dimensional, and more commonly three-dimensional, organization in aggregates of PrP^{Sc}. It seems more likely that a complex process involving chaperones or similar molecules is responsible for transforming PrP^C into PrP^{Sc}.

Prion diversity

For many years, studies of experimental scrapie were performed exclusively with sheep and goats. The disease was first transmitted by intraocular inoculation [124] and later by intracerebral, oral, subcutaneous, intramuscular and intravenous injections of brain extracts from sheep developing scrapie. Incubation periods of 1–3 years were common, and often many of the inoculated animals failed to develop disease [125–127]. Different breeds of sheep exhibited markedly different susceptibilities to scrapie prions inoculated subcutaneously, suggesting that the genetic background might influence host permissiveness [128].

The diversity of scrapie prions was first appreciated in goats inoculated with 'hyper' and 'drowsy' isolates [129]. Subsequently, studies in mice demonstrated the existence of many scrapie 'strains' [130–133] which continues to pose a fascinating conundrum. What is the macromolecule that carries the information required for each strain to manifest a unique set of biological properties if it is not a nucleic acid?

There is good evidence for multiple 'strains' or distinct isolates of prions as defined by specific incubation times, distribution of vacuolar lesions and patterns of PrP^{Sc} accumulation [134–137]. The lengths of the incubation times have been used to distinguish prion strains inoculated into sheep, goats, mice and hamsters. Dickinson and his colleagues developed a system for 'strain typing' by which mice with genetically determined short and long incubation times were used in combination with the F_1 cross [112, 134, 138]. For example, C57BL mice exhibited short incubation times of ~150 days when inoculated with either the Me7 or Chandler isolates; VM mice inoculated with these same isolates had prolonged incubation times of ~300 days.

Scrapie incubation times

The identification of a widely available mouse strain with long incubation times accelerated studies of prion strains [139, 140]. Once molecular clones of the *Prnp* gene were available, we were able to demonstrate genetic linkage between the Mo*Prnp* and a gene modulating incubation times *(Prn-i)* [141]. Subsequently, other investigators have confirmed the genetic linkage, and one group has shown that the scrapie incubation time gene *Sinc* is also linked to *Prnp* [142, 143]. Much evidence now argues that the *Prnp*, *Prn-i* and *Sinc* genes are congruent; the term *Sinc* is no longer used [144]. The PrP sequences of NZW with short and long scrapie incubation times, respectively, differ at codons 108 (L→F) and 189 (T→V) [49].

Although the amino acid substitutions in PrP that distinguish *Prnp*^{a} from *Prnp*^{b} mice argued for the congruency of *Prnp* and *Prn-i*, experiments with *Prnp*^{a} mice expressing *Prnp*^{b} transgenes demonstrated a 'paradoxical' shortening of incubation times [50]. We had predicted that these Tg mice would exhibit a prolongation of the incubation time after inoculation with RML prions based on (*Prnp*^{a} × *Prnp*^{b})F1 mice which do exhibit long incubation times. We described those findings as 'paradoxical shortening' because we and others had thought for many years that long incubation times are dominant traits [134, 141]. From studies of congenic and transgenic mice expressing different numbers of the *a* and *b* alleles of *Prnp* (table III), we now realize that these findings were not paradoxical; indeed, they result from increased *Prnp* dosage [113]. When the RML isolate was inoculated into congenic and transgenic mice, increasing the number of copies of the *a* allele was found to be the major determinant in reducing the incubation time; however, increasing the number of copies of the *b* allele also reduced the incubation time, but not to the same extent as that seen with the *a* allele (table III).

The discovery that incubation times are controlled by the relative dosage of *Prnp*^{a} and *Prnp*^{b} alleles was foreshadowed by studies of Tg(SHa*Prnp*) mice in which the length of the incubation time after inoculation with SHa prions was inversely proportional to the transgene product, SHaPrP^C [83]. Not only does the *Prnp* gene dose determine the length of the incubation time, but also the passage history of the inoculum, particularly in *Prnp*^{b} mice [113]. The PrP^{Sc} allotype in the inoculum produced the shortest incubation times when it was the same as that of PrP^C in the host [145]. The term 'allotype' is used to describe allelic variants of PrP. The effect of the allotype barrier was small when measured in *Prnp*$^{a/a}$ mice, but was clearly demonstrable in *Prnp*$^{b/b}$ mice. Previous passage of prions in *Prnp*^{b} mice shortened the incu-

Table III. MoPrP-A expression is a major determinant of incubation times in mice inoculated with the RML scrapie prions.

Mice	Prnp genotype (copies)	Prnp transgenes (copies)	Alleles a	b	Incubation Time[a] (days ± SEM)	n
Prnp^{0/0}	0/0		0	0	> 600	4
Prnp^{+/0}	a/0		1	0	426 ± 18	9[a]
B6.I-1	b/b		0	2	360 ± 16	7[b]
B6.I-2	b/b		0	2	379 ± 8	10[b]
B6.I-3	b/b		0	2	404 ± 10	20
(B6 x B6.I-1)F1	a/b		1	1	268 ± 4	7
B6.I-1 x Tg(MoPrnp-B^{0/0})15	a/b		1	1	255 ± 7	11[c]
B6.I-1 x Tg(MoPrnp-B^{0/0})15	a/b		1	1	274 ± 3	9[d]
B6.I-1 x Tg(MoPrnp-B^{+/0})15	a/b	bbb/0	1	4	166 ± 2	11[c]
B6.I-1 x Tg(MoPrnp-B^{+/0})15	a/b	bbb/0	1	4	162 ± 3	8[d]
C57BL/6J (B6)	a/a		2	0	143 ± 4	8
B6.I-4	a/a		2	0	144 ± 5	8
non-Tg(MoPrnp-B^{0/0})15	a/a		2	0	130 ± 3	10
Tg(MoPrnp-B^{+/0})15	a/a	bbb/0	2	3	115 ± 2	18
Tg(MoPrnp-B^{+/+})15	a/a	bbb/bbb	2	6	111 ± 5	5
Tg(MoPrnp-B^{+/0})94	a/a	> 30[b]	2	> 30	75 ± 2	15[e]
Tg(MoPrnp-A^{+/0})B4053	a/a	> 30[a]	> 30	0	50 ±	16

[a]Data from [9]. [b]Data from [239]. [c]The homozygous Tg(MoPrnp-B^{+/+})15 mice were maintained as a distinct subline selected for transgene homozygosity two generations removed from the (B6 × LT/Sv)F$_2$ founder. Hemizygous Tg(MoPrnp-B^{+/0})15 mice were produced by crossing the Tg(MoPrnp-B^{+/+})15 line with B6 mice. [d]Tg(MoPrnp-B^{+/0})15 mice were maintained by repeated backcrossing to B6 mice. [e]Data from [50].

bation time by ~40% when assayed in *Prnp^b* mice, compared to those inoculated with prions passaged in *Prnp^a* mice [145].

Patterns of PrP^Sc deposition

Besides measurements of the length of the incubation time, profiles of spongiform degeneration have also been used to characterize different prion strains [135, 146]. With the development of a new procedure for in situ detection of PrP^Sc, designated histoblotting [147], it became possible to localize and quantify PrP^Sc as well as to determine whether or not 'strains' produce different, reproducible patterns of PrP^Sc accumulation [137, 148]. Histoblotting overcame two obstacles that plagued PrP^Sc detection in brain by standard immunohistochemical techniques: the presence of PrP^C and the weak antigenicity of PrP^Sc [149].

Comparisons of PrP^Sc accumulation on histoblots with histologic sections showed that PrP^Sc deposition preceded vacuolation and only those regions with PrP^Sc underwent degeneration [137]. Microdissection of individual brain regions confirmed the conclusions of the histoblot studies: those regions with high levels of PrP 27–30 had intense vacuolation [150]. Thus, we concluded that the deposition of PrP^Sc is responsible for the neuropathologic changes found in the prion diseases.

While studies with both mice and Syrian hamsters established that each isolate has a specific signature as defined by a specific pattern of PrP^Sc accumulation in the brain [113, 137, 148], comparisons must be done on an isogenic background [86, 102]. Variations in the patterns of PrP^Sc accumulation were found to be equally as great as those seen between two strains when a single strain is inoculated in mice expressing different *Prnp* genes. Based upon the initial studies which were performed in animals of a single genotype, we suggested that PrP^{Sc} synthesis occurs in specific populations of cells for a given distinct prion isolate.

Are prion strains different PrP^{Sc} conformers?

Explaining the problem of multiple distinct prion isolates might be accommodated by multiple PrP^{Sc} conformers that act as templates for the folding of de novo synthesized PrP^{Sc} molecules during prion 'replication'. Although it is clear that passage history can be responsible for the prolongation of the incubation time when prions are passed between mice expressing different *Prnp* allotypes [145] or between species [83], many scrapie strains show distinct incubation times in the same inbred host [151].

Although the proposal for multiple PrP^{Sc} conformers is rather unorthodox, we already know that PrP can assume at least two profoundly different conformations: PrP^{C} and PrP^{Sc} [26]. Of note, two different isolates from mink that had died of transmissible mink encephalopathy exhibit different sensitivities of PrP^{Sc} to proteolytic digestion, supporting the suggestion that isolate-specific information might be carried by PrP^{Sc} [152–154]. How many conformations PrP^{Sc} can assume is unknown. The molecular mass of a PrP^{Sc} homodimer is consistent with the ionizing radiation target size of 55,000 ± 9,000 Da as determined for infectious prion particles independent of their polymeric form [114]. If prions are oligomers of PrP^{Sc}, which seems likely, then this offers another level of complexity which in turn generates additional diversity.

Inherited prion diseases

The recognition that ~10% of cases are familial led to the suspicion that genetics plays a role in this disease [20, 155–163]. Like sheep scrapie, the relative contributions of genetic and infectious etiologies in the human prion diseases remained puzzling. The discovery of the *Prnp* gene and its linkage to scrapie incubation times in mice [141] raised the possibility that mutation might feature in the hereditary human prion diseases. A proline (P)→leucine (L) mutation at codon 102 was shown to be linked genetically to the development of GSS disease with a log of the odds (LOD) score exceeding 3 (fig 3) [47]. This mutation may be caused by deamination of a methylated CpG in a germline *PRNP* gene resulting in the substitution of a thymine (T) for cytosine (C). This mutation has been found in ten different families in nine different countries including the original GSS disease family [164–167].

An insert of 144 bp containing six octarepeats at codon 53 was described in patients with CJD from four families residing in southern England [168–170]. Genealogic investigations have shown that all four families are related, arguing for a single founder born more than two centuries ago. The LOD score for this extended pedigree exceeds 11. Studies from several laboratories have demonstrated that two, four, five, six, seven, eight or nine octarepeats in addition to the normal five are found in individuals with inherited CJD [169, 171–176], whereas deletion of one octarepeat has been identified in subjects without the neurologic disease [177–179].

The unusually high incidence of CJD among Israeli Jews of Libyan origin was thought to be due to the consumption of lightly cooked sheep brain or eyeballs [180–185]. Recent studies have shown that some Libyan and Tunisian Jews in families with CJD have a *PRNP* point mutation at codon 200 resulting in a Glu→Lys substitution [186–188]. The E200K mutation has also been found in Slovaks originating from Orava in North Central Czechoslovakia [186], in a cluster of familial cases in Chile [189], and in a large German family living in the United

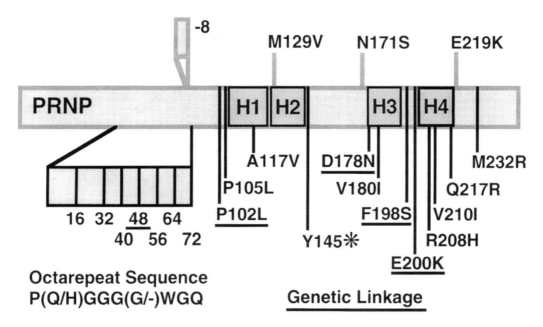

Fig 3. Human prion protein gene *PRNP*. The ORF is denoted by the large gray rectangle. Human *PRNP* wild-type coding polymorphisms are shown above the rectangle while mutations that segregate with the inherited prion diseases are depicted below. The human wild-type *PRNP* gene contains five octarepeats [P(Q/H)GGG(G/-)WGQ] from codons 51 to 91 [103]. Deletion of a single octarepeat at codon 81 or 82 is not associated with prion disease [177, 178, 240]; whether this deletion alters the phenotypic characteristics of a prion disease is unknown. There are common polymorphisms at codons 117 (Ala→Ala) and 129 (Met→Val); homozygosity for Met or Val at codon 129 appears to increase susceptibility to sporadic CJD [241]. Additional polymorphisms have been found at codons 171 (Asn→Ser) and 219 (Glu→Lys) [123, 242]. Octarepeat inserts of 16, 32, 40, 48, 56, 64, and 72 amino acids at codons 67, 75 or 83 are designated by the small rectangle below the ORF. These inserts segregate with familial CJD, and genetic linkage has been demonstrated where sufficient specimens from family members are available [169, 171–174, 179, 243, 244]. Point mutations are designated by the wild-type amino acid preceding the codon number and the mutant residue follows, eg, P^{102}L. These point mutations segregate with the inherited prion diseases and significant genetic linkage (underlined mutations) has been demonstrated where sufficient specimens from family members are available. Mutations at codons 102 (Pro→Leu), 117 (Ala→Val), 198 (Phe→Ser) and 217 (Gln→Arg) are found in patients with GSS disease [47, 164, 165, 186, 244–249]. Point mutations at codons 178 (Asp→Asn), 200 (Glu→Lys), 208 (Arg→His) and 210 (Val→Ile) are found in patients with familial CJD [187, 188, 193, 217, 250, 251]. Point mutations at codons 198 (Phe→Ser) and 217 (Gln→Arg) are found in patients with GSS disease who have PrP amyloid plaques and neurofibrillary tangles [252, 253]. Additional point mutations at codons 145 (Tyr→Stop), 105 (Pro→Leu), 180 (Val→Ile) and 232 (Met→Arg) have been recently reported [254, 255]. Single letter code for amino acids is as follows: A, Ala; D, Asp; E, Glu; F, Phe; I, Ile; K, Lys; L, Leu; M, Met; N, Asn; P, Pro; Q, Gln; R, Arg; S, Ser; T, Thr; V, Val; Y, Tyr.

States [190]. If carriers with the codon-200 mutation live long enough, they eventually develop prion disease [191, 192].

Many families with CJD have been found to have a point mutation at codon 178 resulting in an Asp→Asn substitution [193–197]. FFI which presents with insomnia is also associated with the D^{178}N mutation [198, 199]. The neuropathology in patients with FFI is restricted to selected nuclei of the thalamus. It is unclear whether all patients with the D^{178}N mutation or only a subset present with sleep disturbances. It appears that the allele with the D^{178}N mutation encodes an M at position 129 in FFI, while a V is encoded at position 129 in familial CJD [200]. The discovery that FFI is an inherited prion disease clearly widens the clinical spectrum of these disorders and raises the possibility that many other degenerative diseases of unknown etiology may be caused by prions [199, 201]. The codon-178 mutation has been linked to the development of prion disease with a LOD score exceeding 5 [202].

Bovine spongiform encephalopathy

Beginning in 1986, an epidemic of a previously unknown prion disease named bovine spongiform encephalopathy or 'mad cow' disease appeared in cattle in Great Britain [203], in which protease-resistant PrP was found in brains of ill cattle [204, 205]. It has been proposed that BSE represents a massive common source epidemic which has caused more than 160,000 cases to date [17]. Dairy cows were routinely fed meat and bone meal (MBM) prepared by rendering the offal of sheep and cattle as a nutritional supplement. Since 1988, the practice of using dietary protein supplements for domestic animals derived from rendered sheep or cattle offal has been banned. Recent statistics argue that the epidemic is now under control as a result of the 1988 food ban.

Brain extracts from cattle with BSE have transmitted the disease to mice, cattle, sheep and pigs after intracerebral inoculation [206–209]. Transmissions to mice and sheep suggest that cattle preferentially propagate a single 'strain' of prions [210]. Of particular significance is the transmission of BSE to the marmoset after intracerebral inoculation [211]. Recent cases of atypical CJD in three teenagers and seven young adults in the UK raise the possibility of transmission of BSE to humans [14–16].

Our experience with Tg(Hu*PRNP*)FVB mice expressing both Hu and MoPrP^C [89, 100] contrasts with that reported for a case of iatrogenic CJD. When Tg(Hu*PRNP*)FVB mice were inoculated with Hu prions from this iatrogenic case of CJD, the mice developed CNS dysfunction ~300 days later [212]. When this same inoculum was injected into Tg(Hu*PRNP*)*Prnp*$^{0/0}$ mice, the incubation was reduced to ~200 days which is similar to those reported by us for other Hu prions [89]. Based on the susceptibility of Tg(Hu*PRNP*)*Prnp*$^{0/0}$ mice to Hu prions from this iatrogenic CJD case and the resistance of these mice to bovine prions from cattle dying of BSE, the authors concluded that BSE prions are unlikely to be pathogenic for humans [212]. This conclusion seems contrary to recent reports describing more than ten cases of atypical CJD in teenagers and young adults who are thought to have become ill after consuming tainted beef in Great Britain as well as possibly in France and Switzerland [14, 15].

It is notable that bovine PrP is more homologous with Hu*Prnp* than is sheep PrP in the region between codons 96 and 167 which feature in the species barrier in the transmission of Hu prions into mice, as demonstrated by the susceptibility of Tg(MHu2M) mice to Hu prions [89, 100]. Whether the differences in the amino acid sequence between bovine and sheep PrP in this central domain are responsible for the apparent differential susceptibility of humans to bovine and sheep prions remains to be established. It should be noted that epidemiologic studies over nearly three decades have failed to establish convincing evidence for transmission of sheep prions to humans [213–216] and that the high incidence of CJD among Libyan Jews was initially attributed to the consumption of lightly cooked sheep brain [180]; however, subsequent studies showed that this geographic cluster of CJD is due to a mutation at codon 200 of the *PRNP* gene [122, 188, 217].

Conclusion

Prions are not viruses

The study of prions has taken several unexpected directions over the past few years. The discovery that prion diseases in humans are uniquely both genetic and infectious has greatly strengthened and extended the prion concept. To date, 18 different mutations in the human *PRNP* gene, all resulting in nonconservative substitutions, have been found to either be linked genetically to or segregate with the inherited prion diseases. Yet, the transmissible prion particle is composed largely, if not entirely, of an abnormal isoform of the prion protein designated PrPSc [1]. These findings argue that prion diseases should be considered pseudo-infections, since the particles transmitting disease appear to be devoid of a foreign nucleic acid and thus differ from all known microorganisms as well as viruses and viroids. Because much information, especially about scrapie of rodents, has been derived using experimental proto-cols adapted from virology, we continue to use terms such as infection, incubation period, transmissibility and end-point titration in studies of prion diseases.

Do prions exist in lower organisms?

In *Saccharomyces cerevisiae*, [URE2] and [URE3] mutants were described that can grow on ureidosuccinate under conditions of nitrogen repression such as glutamic acid and ammonia [218]. Mutants of [URE2] exhibit Mendelian inheritance, whereas [URE3] is cytoplasmically inherited [219]. The [URE3] phenotype can be induced by UV irradiation and by overexpression of ure2p, the gene product of *ure2*; deletion of *ure2* abolishes [URE3]. The function of Ure2p is unknown, but it has substantial homology with glutathione-S-transferase; attempts to demonstrate this enzymic activity with purified Ure2p have not been successful [220]. Whether the [URE3] protein is a post-translationally modified form of Ure2p which acts upon unmodified Ure2p to produce more of itself remains to be established.

Another possible yeast prion is the [PSI] phenotype [219]. [PSI] is a non-Mendelian inher-ited trait that can be induced by expression of the *PNM2* gene [221]. The [PSI] phenotype has been shown to require moderate levels of expression of Hsp104; ablation of the *HSP104* gene or overexpression prevented [PSI] [222]. Both [PSI] and [URE3] can be cured by exposure of the yeast to 3 mM Gdn HCl. The mechanism responsible for abolishing [PSI] and [URE3] with a low concentration of Gdn HCl is unknown. In the filamentous fungus *Podospora anserina*, the *het-s* locus controls the vegetative incompatibility; conversion from the S^s to the s state seems to be a post-translational, autocatalytic process [223].

If any of the above cited examples can be shown to function in a manner similar to prions in animals, then many new, more rapid and economical approaches to prion diseases should be forthcoming.

Common neurodegenerative diseases

The knowledge accrued from the study of prion diseases may provide an effective strategy for defining the etiologies and dissecting the molecular pathogenesis of the more common neuro-degenerative disorders such as Alzheimer's disease, Parkinson's disease and amyotrophic lateral sclerosis (ALS). Advances in the molecular genetics of Alzheimer's disease and ALS suggest that, like prion diseases, an important subset is caused by mutations that result in non conservative amino acid substitutions in proteins expressed in the CNS [224–238]. Since people at risk for inherited prion diseases can now be identified decades before neurologic dysfunction is evident, the development of an effective therapy is imperative.

Future studies

Tg mice expressing foreign or mutant *Prnp* genes now permit virtually all facets of prion diseases to be studied and have created a framework for future investigations. Furthermore, the structure and organization of the *Prnp* gene suggested that PrPSc is derived from PrPC or a precursor by a post-translational process. Studies with scrapie-infected cultured cells have provided much evidence that the conversion of PrPC to PrPSc is a post-translational process that probably occurs within a subcellular compartment bounded by cholesterol-rich membranes. The molecular mechanism of PrPSc formation remains to be elucidated, but chemical and physical studies have shown that conformations of PrPC and PrPSc are profoundly different.

The study of prion biology and diseases seems to be a new and emerging area of biomedical investigation. While prion biology has its roots in virology, neurology and neuropathology, its relationships to the disciplines of molecular and cell biology as well as protein chemistry have become evident only recently. Certainly, it is likely that learning how prions multiply and cause disease will open up new vistas in biochemistry and genetics.

Acknowledgments

We thank M Baldwin, D Borchelt, G Carlson, C Cooper, R Fletterick, M Gasset, R Gabizon, D Groth, R Koehler, L Hood, K Hsiao, Z Huang, V Lingappa, KM Pan, D Riesner, M Scott, A Serban, N Stahl, A Taraboulos, G Telling, M Torchia and D Westaway for their help in these studies. Supported by grants from the National Institutes of Health (NS14069, AG08967, AG02132, NS22786 and AG10770) and the American Health Assistance Foundation, as well as by gifts from the Sherman Fairchild Foundation and Bernard Osher Foundation.

References

1 Prusiner SB. Molecular biology of prion diseases. *Science* 1991;252:1515-22
2 Prusiner SB. Novel proteinaceous infectious particles cause scrapie. *Science* 1982;216:136-44
3 Alper T, Haig DA, Clarke MC. The exceptionally small size of the scrapie agent. *Biochem Biophys Res Commun* 1966;22:278-84
4 Alper T, Cramp WA, Haig DA, Clarke MC. Does the agent of scrapie replicate without nucleic acid? *Nature* 1967;214:764-6
5 Hunter GD. Scrapie: a prototype slow infection. *J Infect Dis* 1972;125:427-40
6 Riesner D, Kellings K, Meyer N, Mirenda C, Prusiner SB. In: Prusiner SB, Collinge J, Powell J, Anderton B, eds. *Prion Diseases of Humans and Animals.* London: Ellis Horwood, 1992;341-58
7 Prusiner SB, McKinley MP, Bowman KA et al. Scrapie prions aggregate to form amyloid-like birefringent rods. *Cell* 1983;35:349-58
8 Büeler H, Aguzzi A, Sailer A et al. Mice devoid of PrP are resistant to scrapie, *Cell* 1993;73:1339-47
9 Prusiner SB, Groth D, Serban A et al. Ablation of the prion protein (PrP) gene in mice prevents scrapie and facilitates production of anti-PrP antibodies. *Proc Natl Acad Sci USA* 1993;90:10608-12
10 Prusiner SB, Groth D, Serban A, Stahl N, Gabizon R. Attempts to restore scrapie prion infectivity after exposure to protein denaturants. *Proc Natl Acad Sci USA* 1993;90:2793-7
11 Sigurdsson B. Rida, a chronic encephalitis of sheep with general remarks on infections which develop slowly and some of their special characteristics. *Br Vet J* 1954;110:341-54
12 Gajdusek DC. Unconventional viruses and the origin and disappearance of kuru. *Science* 1977;197: 943-60
13 Gajdusek DC. Subacute spongiform virus encephalopathies caused by unconventional viruses. In: Maramorosch K, McKelvey JJ Jr, eds. *Subviral Pathogens of Plants and Animals: Viroids and Prions.* Orlando: Academic Press, 1985;483-544
14 Bateman D, Hilton D, Love S et al. Sporadic Creutzfeldt-Jakob disease in a 18-year-old in the UK. *Lancet* 1995;346:1155-6 (Lett)
15 Britton TC, Al-Sarraj S, Shaw C, Campbell T, Collinge J. Sporadic Creutzfeldt-Jakob disease in a 16-year-old in the UK. *Lancet* 1995;346:1155 (Lett)

16 Will RG, Ironside JW, Zeidler M et al. A new variant of Creutzfeldt-Jakob disease in the UK. *Lancet* 1996;347:921-5

17 Wells GAH, Wilesmith JW. The neuropathology and epidemiology of bovine spongiform encephalopathy. *Brain Pathol* 1995;5:91-103

18 Gajdusek DC, Zigas V. Degenerative disease of the central nervous system in New Guinea – The endemic occurrence of 'kuru' in the native population. *N Engl J Med* 1957;257:974-8

19 Gibbs CJ Jr, Gajdusek DC, Asher DM et al. Creutzfeldt-Jakob disease (spongiform encephalopathy): transmission to the chimpanzee. *Science* 1968;161:388-9

20 Masters CL, Gajdusek DC, Gibbs CJ Jr. Creutzfeldt-Jakob disease virus isolations from the Gerstmann-Sträussler syndrome. *Brain* 1981;104:559-88

21 Gajdusek DC, Gibbs CJ Jr, Alpers M. Experimental transmission of a kuru-like syndrome to chimpanzees. *Nature* 1966;209:794-6

22 Hadlow WJ. Scrapie and kuru. *Lancet* 1959;2:289-90

23 Parry HB. In: Oppenheimer DR, ed. *Scrapie Disease in Sheep.* New York: Academic Press, 1983;31-59

24 Prusiner SB, McKinley MP, Groth DF et al. Scrapie agent contains a hydrophobic protein. *Proc Natl Acad Sci USA* 1981;78:6675-9

25 Stahl N, Baldwin MA, Teplow DB et al. Structural analysis of the scrapie prion protein using mass spectrometry and amino acid sequencing. *Biochemistry* 1993;32:1991-2002

26 Pan KM, Baldwin M, Nguyen J et al. Conversion of α-helices into β-sheets features in the formation of the scrapie prion proteins. *Proc Natl Acad Sci USA* 1993;90:10962-6

27 Kocisko DA, Come JH, Priola SA et al. Cell-free formation of protease-resistant prion protein. *Nature* 1994;370:471-4

28 Kaneko K, Peretz D, Pan KM et al. Prion protein (PrP) synthetic peptides induce cellular PrP to acquire properties of the scrapie isoform. *Proc Natl Acad Sci USA* 1995;32:11160-4

29 Bolton DC, McKinley MP, Prusiner SB. Identification of a protein that purifies with the scrapie prion. *Science* 1982;218:1309-11

30 Prusiner SB, Bolton DC, Groth DF et al. Further purification and characterization of scrapie prions. *Biochemistry* 1982;21:6942-50

31 McKinley MP, Bolton DC, Prusiner SB. A protease-resistant protein is a structural component of the scrapie prion. *Cell* 1983;35:57-62

32 Diringer H, Gelderblom H, Hilmert H et al. Scrapie infectivity, fibrils and low molecular weight protein. *Nature* 1983;306:476-8

33 Prusiner SB, Groth DF, Bolton DC Kent SB, Hood LE. Purification and structural studies of a major scrapie prion protein. *Cell* 1984;38:127-34

34 Basler K, Oesch B, Scott M et al. Scrapie and cellular PrP isoforms are encoded by the same chromosomal gene. *Cell* 1986;46:417-28

35 Locht C, Chesebro B, Race R, Keith JM. Molecular cloning and complete sequence of prion protein cDNA from mouse brain infected with the scrapie agent. *Proc Natl Acad Sci USA* 1986;83:6372-6

36 Oesch B, Westaway D, Wälchli M et al. A cellular gene encodes scrapie PrP 27–30 protein. *Cell* 1985;40:735-46

37 Chesebro B, Race R, Wehrly K et al. Identification of scrapie prion protein-specific mRNA in scrapie-infected and uninfected brain. *Nature* 1985;315:331-3

38 Meyer RK, McKinley MP, Bowman KA, et al. Separation and properties of cellular and scrapie prion proteins. *Proc Natl Acad Sci USA* 1986;83 2310-4

39 Gabizon R, McKinley MP, Prusiner SB. Purified prion proteins and scrapie infectivity copartition into liposomes. *Proc Natl Acad Sci USA* 1987;84:4017-21

40 Gabizon R, McKinley MP, Groth DF, Prusiner SB. Immunoaffinity purification and neutralization of scrapie prion infectivity. *Proc Natl Acad Sci USA* 1988;85:6617-21

41 Gabizon R, Prusiner SB. Prion liposomes. *Biochem J* 1990;266:1-14

42 Bendheim PE, Barry RA, DeArmond SJ, Stites DP, Prusiner B. Antibodies to a scrapie prion protein. *Nature* 1984;310:418-21

43 DeArmond SJ, McKinley MP, Barry RA et al. Identification of prion amyloid filaments in scrapie-infected brain. *Cell* 1985;41:221-35

44 Kitamoto T, Tateishi J, Tashima I et al. Amyloid plaques in Creutzfeldt-Jakob disease stain with prion protein antibodies. *Ann Neurol* 1986;20:204-8

45 Roberts GW, Lofthouse R, Allsop D et al. CNS amyloid proteins in neurodegenerative diseases. *Neurology* 1988;38:1534-40

46 Tagliavini F, Prelli F, Ghisto J et al. Amyloid protein of Gerstmann-Sträussler-Scheinker disease (Indiana kindred) is an 11-kd fragment of prion protein with an N-terminal glycine at codon 58. *EMBO J* 1991;10: 513-9

47 Hsiao K, Baker HF, Crow TJ et al. Linkage of a prion protein missense variant to Gerstmann-Sträussler syndrome. *Nature* 1989;338:342-5

48 Gabriel J-M, Oesch B, Kretzschmar H, Scott M, Prusiner SB. Molecular cloning of a candidate chicken prion protein. *Proc Natl Acad Sci USA* 1992;89:9097-101

49 Westaway D, Goodman PA, Mirenda CA et al. Distinct prion proteins in short and long scrapie incubation period mice. *Cell* 1987;51:651-62

50 Westaway D, Mirenda CA, Foster D et al. Paradoxical shortening of scrapie incubation times by expression of prion protein transgenes derived from long incubation period mice. *Neuron* 1991;7:59-68

51 Westaway D, Cooper C, Turner S et al. Structure and polymorphism of the mouse prion protein gene. *Proc Natl Acad Sci USA* 1994;91:6418-22

52 Westaway D, Zuliani V, Cooper CM et al. Homozygosity for prion protein alleles encoding glutamine-171 renders sheep susceptible to natural scrapie. *Genes Dev* 1994;8:959-69

53 McKnight S, Tjian R. Transcriptional selectivity of viral genes in mammalian cells. *Cell* 1986;46:795-805

54 Mobley WC, Neve RL, Prusiner SB, McKinley MP. Nerve growth factor increases mRNA levels for the prion protein and the beta-amyloid protein precursor in developing hamster brain. *Proc Natl Acad Sci USA* 1988;85:9811-5

55 Kretzschmar HA, Prusiner SB, Stowring LE, DeArmond SJ. Scrapie prion proteins are synthesized in neurons. *Am J Pathol* 1986;122:1-5

56 Meyer N, Rosenbaum V, Schmidt B et al. Search for a putative scrapie genome in purified prion fractions reveals a paucity of nucleic acids. *J Gen Virol* 1991;72:37-49

57 Kellings K, Meyer N, Mirenda C, Prusiner SB, Riesner D. Further analysis of nucleic acids in purified scrapie prion preparations by improved return refocussing gel electrophoresis (RRGE). *J Gen Virol* 1992;73:1025-9

58 Stahl N, Borchelt DR, Hsiao K, Prusiner SB. Scrapie prion protein contains a phosphatidylinositol glycolipid. *Cell* 1987;51:229-40

59 Caughey B, Raymond GJ. The scrapie-associated form of PrP is made from a cell surface precursor that is both protease- and phospholipase-sensitive. *J Biol Chem* 1991;266:18217-23

60 Borchelt DR, Taraboulos A, Prusiner SB. Evidence for synthesis of scrapie prion proteins in the endocytic pathway. *J Biol Chem* 1992;267:16188-99

61 Taraboulos A, Raeber AJ, Borchelt DR, Serban D, Prusiner SB. Synthesis and trafficking of prion proteins in cultured cells. *Mol Biol Cell* 1992;3:851-63

62 Taraboulos A, Scott M, Semenov A et al. Cholesterol depletion and modification of COOH-terminal targeting sequence of the prion protein inhibit formation of the scrapie isoform. *J Cell Biol* 1995;129: 121-32

63 Keller GA, Siegel MW, Caras IW. Endocytosis of glycophospholipid-anchored and transmembrane forms of CD4 by different endocytic pathways. *EMBO J* 1992;11:863-74

64 Anderson RGW. Caveolae: where incoming and outgoing messengers meet. *Proc Natl Acad Sci USA* 1993; 90:10909-13

65 Caughey B, Raymond GJ, Ernst D, Race RE. N-Terminal truncation of the scrapie-associated form of PrP by lysosomal protease(s): implications regarding the site of conversion of PrP to the protease-resistant state. *J Virol* 1991;65:6597-603

66 McKinley MP, Meyer R, Kenaga L et al. Scrapie prion rod formation in vitro requires both detergent extraction and limited proteolysis. *J Virol* 1991;65:1440-9

67 Presnell SR, Cohen BI, Cohen FE. MacMatch: a tool for pattern-based protein secondary structure prediction. *Cabios* 1993;9:373-4

68 Kneller DG, Cohen FE, Langridge R. Improvement in protein secondary structure prediction by an enhanced neural network. *J Mol Biol* 1990;214:171-82

69 Gasset M, Baldwin MA, Lloyd D et al. Predicted α-helical regions of the prion protein when synthesized as peptides form amyloid. *Proc Natl Acad Sci USA* 1992;89:10940-4

70 Huang Z, Gabriel JM, Baldwin MA et al. Proposed three-dimensional structure for the cellular prion protein. *Proc Natl Acad Sci USA* 1994;91:7139-43

71 Huang Z, Prusiner SB, Cohen FE. Scrapie prions: a three-dimensional model of an infectious fragment. *Folding Design* 1996;1:13-9

72 Turk E, Teplow DB, Hood LE, Prusiner SB. Purification and properties of the cellular and scrapie hamster prion proteins. *Eur J Biochem* 1988;176:21-30

73 Safar J, Roller PP, Gajdusek DC, Gibbs CJ Jr. Conformational transitions, dissociation, and unfolding of scrapie amyloid (prion) protein. *J Biol Chem* 1993;268:20276-84

74 Caughey BW, Dong A, Bhat KS et al. Secondary structure analysis of the scrapie-associated protein PrP 27–30 in water by infrared spectroscopy. *Biochemistry* 1991;30:7672-80

75 Gasset M, Baldwin MA, Fletterick RJ, Prusiner SB. Perturbation of the secondary structure of the scrapie prion protein under conditions associated with changes in infectivity. *Proc Natl Acad Sci USA* 1993;90:1-5

76 Safar J, Roller PP, Gajdusek DC, Gibbs CJJ. Thermal-stability and conformational transitions of scrapie amyloid (prion) protein correlate with infectivity. *Protein Sci* 1993;2:2206-16

77 Jarrett JT, Lansbury PT Jr. Seeding 'one-dimensional crystallization' of amyloid: a pathogenic mechanism in Alzheimer's disease and scrapie? *Cell* 1993;73:1055-8

78 Cohen FE, Pan KM, Huang Z et al. Structural clues to prion replication. *Science* 1994;264:530-1

79 Safar J, Roller PP, Gajdusek DC, Gibbs CJ Jr. Scrapie amyloid (prion) protein has the conformational characteristics of an aggregated molten globule folding intermediate. *Biochemistry* 1994;33:8375-83

80 Liautard J-P. Prions and molecular chaperones. *Arch Virol* 1993;7(suppl):227-43

81 Kenward N, Hope J, Landon M, Mayer RJ. Expression of polyubiquitin and heat-shock protein 70 genes increases in the later stages of disease progression in scrapie-infected mouse brain. *J Neurochem* 1994;62: 1870-7

82 Tatzelt J, Zuo J, Voellmy R, et al. Scrapie prions selectively modify the stress response in neuroblastoma cells. *Proc Natl Acad Sci USA* 1995;92:2944-8

83 Prusiner SB, Scott M, Foster D et al. Transgenetic studies implicate interactions between homologous PrP isoforms in scrapie prion replication. *Cell* 1990;63:673-86

84 Scott M, Foster D, Mirenda C et al. Transgenic mice expressing hamster prion protein produce species-specific scrapie infectivity and amyloid plaques. *Cell* 1989;59:847-57

85 Hsiao KK, Scott M, Foster D et al. Spontaneous neurodegeneration in transgenic mice with mutant prion protein. *Science* 1990;250:1587-90

86 Hsiao KK, Groth D, Scott M et al. Serial transmission in rodents of neurodegeneration from transgenic mice expressing mutant prion protein. *Proc Natl Acad Sci USA* 1994;91:9126-30

87 Brown P, Gibbs CJ Jr, Rodgers-Johnson P et al. Human spongiform encephalopathy: the National Institutes of Health series of 300 cases of experimentally transmitted disease. *Ann Neurol* 1994;35:513-29

88 Tateishi J, Kitamoto T, Hoque MZ, Furukawa H. Experimental transmission of Creutzfeldt-Jakob disease and related diseases to rodents. *Neurology* 1996;46:532-7

89 Telling GC, Scott M, Mastrianni J et al. Prion propagation in mice expressing human and chimeric *PrP* transgenes implicates the interaction of cellular PrP with another protein. *Cell* 1995;83:79-90

90 Büeler H, Fischer M, Lang Y et al. Normal development and behaviour of mice lacking the neuronal cell-surface PrP protein. *Nature* 1992;356:577-82

91 Manson JC, Clarke AR, Hooper ML et al. 129/Ola mice carrying a null mutation in *PrP* that abolishes mRNA production are developmentally normal. *Mol Neurobiol* 1994;8:121-7

92 Sakaguchi S, Katamine S, Nishida N et al. Loss of cerebellar Purkinje cells in aged mice homozygous for a disrupted *PrP* gene. *Nature* 1996;380:528-31

93 Manson JC, Clarke AR, McBride PA, McConnell I, Hope J. *PrP* gene dosage determines the timing but not the final intensity or distribution of lesions in scrapie pathology. *Neurodegeneration* 1994;3:331-40

94 Sakaguchi S, Katamine S, Shigematsu K et al. Accumulation of proteinase K-resistant prion protein (PrP) is restricted by the expression level of normal PrP in mice inoculated with a mouse-adapted strain of the Creutzfeldt-Jakob disease agent. *J Virol* 1995;69:7586-92

95 Sailer A, Büeler H, Fischer M, Aguzzi A, Weissmann C. No propagation of prions in mice devoid of *PrP*. *Cell* 1994;77:967-8

96 Collinge J, Whittington MA, Sidle KC et al. Prion protein is necessary for normal synaptic function. *Nature* 1994;370:295-7

97 Whittington MA, Sidle KCL, Gowland I et al. Rescue of neurophysiological phenotype seen in *PrP* null mice by transgene encoding human prion protein. *Nature Genet* 1995;9:197-201

98 Manson JC, Hope J, Clarke AR et al. *PrP* gene dosage and long term potentiation. *Neurodegeneration* 1995;4: 113-4 (Lett)

99 Lledo P-M, Tremblay P, DeArmond SJ, Prusiner SB, Nicoll RA. Mice deficient for prion protein exhibit normal neuronal excitability and synaptic transmission in the hippocampus. *Proc Natl Acad Sci USA* 1996;93:2403-7

100 Telling GC, Scott M, Hsiao KK et al. Transmission of Creutzfeldt-Jakob disease from humans to transgenic mice expressing chimeric human–mouse prion protein. *Proc Natl Acad Sci USA* 1994;91:9936-40

101 Scott MR, Köhler R, Foster D, Prusiner SB. Chimeric prion protein expression in cultured cells and transgenic mice. *Protein Sci* 1992;1:986-97

102 Scott M, Groth D, Foster D et al. Propagation of prions with artificial properties in transgenic mice expressing chimeric *PrP* genes. *Cell* 1993;73:979-88

103 Kretzschmar HA, Stowring LE, Westaway D et al. Molecular cloning of a human prion protein cDNA. *DNA* 1986;5:315-24

104 Lowenstein DH, Butler DA, Westaway D et al. Three hamster species with different scrapie incubation times and neuropathological features encode distinct prion proteins. *Mol Cell Biol* 1990;10:1153-63

105 Schätzl HM, Da Costa M, Taylor L, Cohen FE, Prusiner SB. Prion protein gene variation among primates. *J Mol Biol* 1995;245:362-74

106 Raeber AJ, Borchelt DR, Scott M, Prusiner SB. Attempts to convert the cellular prion protein into the scrapie isoform in cell-free systems. *J Virol* 1992;66:6155-63

107 Nguyen J, Baldwin MA, Cohen FE, Prusiner SB. Prion protein peptides induce α-helix to β-sheet conformational transitions. *Biochemistry* 1995;34:4186-92

108 Zhang H, Kaneko K, Nguyen JT et al. Conformational transitions in peptides containing two putative α-helices of the prion protein. *J Mol Biol* 1995;250:514-26

109 Nguyen JT, Inouye H, Baldwin MA et al. X-ray diffraction of scrapie prion rods and PrP peptides. *J Mol Biol* 1995;252:412-22

110 Kocisko DA, Priola SA, Raymond GJ et al. Species specificity in the cell-free conversion of prion protein to protease-resistant forms: a model for the scrapie species barrier. *Proc Natl Acad Sci USA* 1995;92: 3923-27

111 Bessen RA, Kocisko DA, Raymond GJ et al. Non-genetic propagation of strain-specific properties of scrapie prion protein. *Nature* 1995;375:698-700

112 Dickinson AG, Meikle VMH. Host-genotype and agent effects in scrapie incubation: change in allelic interaction with different strains of agent. *Mol Gen Genet* 1971;112:73-9

113 Carlson GA, Ebeling C, Yang SL et al. Prion isolate specified allotypic interactions between the cellular and scrapie prion proteins in congenic and transgenic mice. *Proc Natl Acad Sci USA* 1994;91:5690-4

114 Bellinger-Kawahara CG, Kempner E, Groth DF, Gabizon R, Prusiner SB. Scrapie prion liposomes and rods exhibit target sizes of 55,000 Da. *Virology* 1988;164:537-41

115 Rogers M, Yehiely F, Scott M, Prusiner SB. Conversion of truncated and elongated prion proteins into the scrapie isoform in cultured cells. *Proc Natl Acad Sci USA* 1993;90:3182-6

116 Tateishi J, Kitamoto T. Inherited prion diseases and transmission to rodents. *Brain Pathol* 1995;5:53-9

117 Gajdusek DC. Transmissible and non-transmissible amyloidoses: autocatalytic post-translational conversion of host precursor proteins to β-pleated sheet configurations. *J Neuroimmunol* 1988;20:95-110

118 Gajdusek DC. Subacute spongiform encephalopathies: transmissible cerebral amyloidoses caused by unconventional viruses. In: Fields BN, Knipe DM, Chanock RM et al, eds. *Virology*, 2nd Ed. New York: Raven Press, 1990;2289-324

119 Gajdusek DC, Gibbs CJ Jr. Brain amyloidoses–precursor proteins and the amyloids of transmissible and nontransmissible dementias: scrapie-kuru-CJD viruses as infectious polypeptides or amyloid-enhancing vector. In: Goldstein A, ed. *Biomedical Advances in Aging*. New York: Plenum Publishing Corporation, 1990;3-24

120 Glenner GG, Eanes ED, Bladen HA, Linke RP, Termine JD. Beta-pleated sheet fibrils – a comparison of native amyloid with synthetic protein fibrils. *J Histochem Cytochem* 1974;22:1141-58

121 Glenner GG. Amyloid deposits and amyloidosis. *N Engl J Med* 1980;302:1283-92

122 Gabizon R, Rosenmann H, Meiner Z et al. Mutation and polymorphism of the prion protein gene in Libyan Jews with Creutzfeldt-Jakob disease. *Am J Hum Genet* 1993;33:828-35

123 Kitamoto T, Tateishi J. Human prion diseases with variant prion protein. *Phil Trans R Soc Lond B* 1994;343: 391-8

124 Cuillé J, Chelle PL. Experimental transmission of trembling to the goat. *CR Seances Acad Sci* 1939;208: 1058-60

125 Dickinson AG, Stamp JT. Experimental scrapie in Cheviot and Suffolk sheep. *J Comp Pathol* 1969; 79:23-26

126 Hadlow WJ, Kennedy RC, Race RE. Natural infection of Suffolk sheep with scrapie virus. *J Infect Dis* 1982;146:657-64

127 Hadlow WJ, Kennedy RC, Race RE, Eklund CM. Virologic and neurohistologic findings in dairy goats affected with natural scrapie. *Vet Pathol* 1980;17:187-99

128 Gordon WS. Review of work on scrapie at Compton, England, 1952–1964. In: *Report of Scrapie Seminar, ARS 91-53*. Washington, DC: US Department of Agriculture, 1966;53-67

129 Pattison IH, Millson GC. Scrapie produced experimentally in goats with special reference to the clinical syndrome. *J Comp Pathol* 1961;71:101-8

130 Dickinson AG, Fraser H. An assessment of the genetics of scrapie in sheep and mice. In: Prusiner SB, Hadlow WJ, eds. *Slow Transmissible Diseases of the Nervous System,* Vol. 1. New York: Academic Press, 1979; 367-86

131 Bruce ME, Dickinson AG. Biological evidence that the scrapie agent has an independent genome. *J Gen Virol* 1987;68:79-89

132 Kimberlin RH, Cole S, Walker CA. Temporary and permanent modifications to a single strain of mouse scrapie on transmission to rats and hamsters. *J Gen Virol* 1987;68:1875-81

133 Dickinson AG, Outram GW. Genetic aspects of unconventional virus infections: the basis of the virino-hypothesis. In: Bock G, Marsh J, eds. *Novel Infectious Agents and the Central Nervous System.* Ciba Foundation Symposium 135. Chichester, UK: John Wiley and Sons, 1988;63-83

134 Dickinson AG, Meikle VMH, Fraser H. Identification of a gene which controls the incubation period of some strains of scrapie agent in mice. *J Comp Pathol* 1968;78:293-9

135 Fraser H, Dickinson AG. Scrapie in mice. Agent-strain differences in the distribution and intensity of grey matter vacuolation. *J Comp Pathol* 1973;83:29-40

136 Bruce ME, McBride PA, Farquhar CF. Precise targeting of the pathology of the sialoglycoprotein, PrP, and vacuolar degeneration in mouse scrapie. *Neurosci Lett* 1989;102:1-6

137 Hecker R, Taraboulos A, Scott M et al. Replication of distinct prion isolates is region specific in brains of transgenic mice and hamsters. *Genes Dev* 1992;6:1213-28

138 Dickinson AG, Bruce ME, Outram GW, Kimberlin RH. Scrapie strain differences: the implications of stability and mutation. In: Tateishi J, ed. *Proceedings of Workshop on Slow Transmissible Diseases.* Tokyo: Japanese Ministry of Health and Welfare, 1984;105-18

139 Kingsbury DT, Kasper KC, Stites DP et al. Genetic control of scrapie and Creutzfeldt-Jakob disease in mice. *J Immunol* 1983;131:491-6

140 Carp RI, Moretz RC, Natelli M, Dickinson AG. Genetic control of scrapie: incubation period and plaque formation in mice. *J Gen Virol* 1987;68:401-7

141 Carlson GA, Kingsbury DT, Goodman PA, et al. Linkage of prion protein and scrapie incubation time genes. *Cell* 1986;46:503-11

142 Hunter N, Hope J, McConnell I, Dickinson AG. Linkage of the scrapie-associated fibril protein (PrP) gene and *Sinc* using congenic mice and restriction fragment length polymorphism analysis. *J Gen Virol* 1987;68:2711-6

143 Race RE, Graham K, Ernst D, Caughey B, Chesebro B. Analysis of linkage between scrapie incubation period and the prion protein gene in mice. *J Gen Virol* 1990;71:493-7

144 Ziegler DR. In: O'Brien SJ, ed. *Genetic Maps – Locus Maps of Complex Genomes,* 6th Ed. Cold Spring Harbor, NY: Cold Spring Harbor Laboratory Press, 1993;4.42-5

145 Carlson GA, Westaway D, DeArmond SJ, Peterson-Torchia M, Prusiner SB. Primary structure of prion protein may modify scrapie isolate properties. *Proc Natl Acad Sci USA* 1989;86:7475-9

146 Fraser H. Neuropathology of scrapie: the precision of the lesions and their diversity. In: Prusiner SB, Hadlow WJ, eds. *Slow Transmissible Diseases of the Nervous System,* Vol. 1. New York: Academic Press, 1979;387-406

147 Taraboulos A, Jendroska K, Serban D et al. Regional mapping of prion proteins in brains. *Proc Natl Acad Sci USA* 1992;89:7620-4

148 DeArmond SJ, Yang SL, Lee A et al. Three scrapie prion isolates exhibit different accumulation patterns of the prion protein scrapie isoform. *Proc Natl Acad Sci USA* 1993;90:6449-53

149 DeArmond SJ, Mobley WC, DeMott DL et al. Changes in the localization of brain prion proteins during scrapie infection. *Neurology* 1987;37:1271-80

150 Casaccia-Bonnefil P, Kascsak RJ, Fersko R, Callahan S, Carp RI. Brain regional distribution of prion protein PrP 27–30 in mice stereotaxically microinjected with different strains of scrapie. *J Infect Dis* 1993;167:7-12

151 Bruce ME, McConnell I, Fraser H, Dickinson AG. The disease characteristics of different strains of scrapie in *Sinc* congenic mouse lines: implications for the nature of the agent and host control of pathogenesis. *J Gen Virol* 1991;72:595-603

152 Marsh RF, Bessen RA, Lehmann S, Hartsough GR. Epidemiological and experimental studies on a new incident of transmissible mink encephalopathy. *J Gen Virol* 1991;72:589-94

153 Bessen RA, Marsh RF. Biochemical and physical properties of the prion protein from two strains of the transmissible mink encephalopathy agent. *J Virol* 1992;66:2096-101

154 Bessen RA, Marsh RF. Identification of two biologically distinct strains of transmissible mink encephalopathy in hamsters. *J Gen Virol* 1992;73:329-34

155 Kirschbaum WR. Zwei eigenartige Erkrankungen des Zentralnervensystems nach Art der spastischen Pseudosklerose (Jakob). *Z Ges Neurol Psychiatr* 1924;92:175-220

156 Meggendorfer F. Klinische und genealogische Beobachtungen bei einem Fall von spastischer Pseudosklerose Jakobs. *Z Ges Neurol Psychiatr* 1930;128:337-41

157 Stender A. Weitere Beiträge zum "Kapitel Spastische Pseudosklerose Jakobs". *Zr Neurol Psychiatr* 1930;128: 528-43

158 Davison C, Rabiner AM. Spastic pseudosclerosis (disseminated encephalomyelopathy; corticopallidospinal degeneration). Familial and nonfamilial incidence (a clinico-pathologic study). *Arch Neurol Psychiatry* 1940; 44:578-98

159 Friede RL, DeJong RN. Neuronal enzymatic failure in Creutzfeldt-Jakob disease. A familial study. *Arch Neurol* 1964;10:181-95

160 Jacob H, Pyrkosch W, Strube H. Die erbliche Form der Creutzfeldt-Jakobschen Krankheit. *Arch Psychiatr Z Neurol* 1950;184:653-74

161 Masters CL, Gajdusek DC, Gibbs CJ Jr, Bernouilli C, Asher DM. Familial Creutzfeldt-Jakob disease and other familial dementias: an inquiry into possible models of virus-induced familial diseases. In: Prusiner SB, Hadlow WJ, eds. *Slow Transmissible Diseases of the Nervous System,* Vol. 1. New York: Academic Press, 1979;143-94

162 Masters CL, Gajdusek DC, Gibbs CJ Jr. The familial occurrence of Creutzfeldt-Jakob disease and Alzheimer's disease. *Brain* 1981;104:535-58

163 Rosenthal NP, Keesey J, Crandall B, Brown WJ. Familial neurological disease associated with spongiform encephalopathy. *Arch Neurol* 1976;33:252-59

164 Doh-ura K, Tateishi J, Sasaki H, Kitamoto T, Sakaki Y. Pro → Leu change at position 102 of prion protein is the most common but not the sole mutation related to Gerstmann-Sträussler syndrome. *Biochem Biophys Res Commun* 1989;163:974-9

165 Goldgaber D, Goldfarb LG, Brown P et al. Mutations in familial Creutzfeldt-Jakob disease and Gerstmann-Sträussler-Scheinker's syndrome. *Exp Neurol* 1989;106:204-6

166 Kretzschmar HA, Honold G, Seitelberger F et al. Prion protein mutation in family first reported by Gerstmann, Straussler, and Scheinker. *Lancet* 1991;337:1160

167 Kretzschmar HA, Kufer P, Riethmüller G et al. Prion protein mutation at codon 102 in an Italian family with Gerstmann-Sträussler-Scheinker syndrome. *Neurology* 1992;42:809-10

168 Collinge J, Brown J, Hardy J et al. Inherited prion disease with 144 base pair gene insertion. 2. Clinical and pathological features. *Brain* 1992;115:687-710

169 Owen F, Poulter M, Lofthouse R et al. Insertion in prion protein gene in familial Creutzfeldt-Jakob disease. *Lancet* 1989;51-2

170 Poulter M, Baker HF, Frith CD et al. Inherited prion disease with 144 base pair gene insertion. 1. Genealogical and molecular studies. *Brain* 1992;115:675-85

171 Owen F, Poulter M, Shah T et al. An in-frame insertion in the prion protein gene in familial Creutzfeldt-Jakob disease. *Mol Brain Res* 1990;7:273-6

172 Collinge J, Harding AE, Owen F et al. Diagnosis of Gerstmann-Sträussler syndrome in familial dementia with prion protein gene analysis. *Lancet* 1989;ii:15-7

173 Collinge J, Owen F, Poulter H et al. Prion dementia without characteristic pathology. *Lancet* 1990;336:7-9

174 Goldfarb LG, Brown P, McCombie WR, et al. Transmissible familial Creutzfeldt-Jakob disease associated with five, seven, and eight extra octapeptide coding repeats in the *PRNP* gene. *Proc Natl Acad Sci USA* 1991;88:10926-30

175 Owen F, Poulter M, Collinge J et al. A dementing illness associated with a novel insertion in the prion protein gene. *Mol Brain Res* 1992;13:155-7

176 Brown P. Infectious cerebral amyloidosis: clinical spectrum, risks and remedies. In: Brown F, ed. *Developments in Biological Standardization.* Basel, Switzerland: Karger, 1993;91-101

177 Laplanche JL, Chatelain J, Launay JM, Gazengel C, Vidaud M. Deletion in prion protein gene in a Moroccan family. *Nucleic Acids Res* 1990;18:6745

178 Vnencak-Jones CL, Phillips JA. Identification of heterogeneous *PrP* gene deletions in controls by detection of allele-specific heteroduplexes (DASH). *Am J Hum Genet* 1992;50:871-2

179 Palmer MS, Mahal SP, Campbell TA et al. Deletions in the prion protein gene are not associated with CJD. *Hum Mol Genet* 1993;2:541-4

180 Kahana E, Milton A, Braham J, Sofer D. Creutzfeldt-Jakob disease: focus among Libyan Jews in Israel. *Science* 1974;183:90-1

181 Alter M, Kahana E. Creutzfeldt-Jakob disease among Libyan Jews in Israel. *Science* 1976;192:428

182 Herzberg L, Herzberg BN, Gibbs CJ Jr et al. Creutzfeldt-Jakob disease: hypothesis for high incidence in Libyan Jews in Israel. *Science* 1974;186:848

183 Neugut RH, Neugut AI, Kahana E, Stein Z, Alter M. Creutzfeldt-Jakob disease: familial clustering among Libyan-born Israelis. *Neurology* 1979;29:225-31

184 Zilber N, Kahana E, Abraham MPH. The Libyan Creutzfeldt-Jakob disease focus in Israel: an epidemiologic evaluation. *Neurology* 1991;41:1385-9

185 Kahana E, Zilber N, Abraham M. Do Creutzfeldt-Jakob disease patients of Jewish Libyan origin have unique clinical features? *Neurology* 1991;41:1390-2

186 Goldfarb LG, Mitrova E, Brown P, Toh BH, Gajdusek DC. Mutation in codon 200 of scrapie amyloid protein gene in two clusters of Creutzfeldt-Jakob disease in Slovakia. *Lancet* 1990;336:514-5

187 Gabizon R, Meiner Z, Cass C et al. Prion protein gene mutation in Libyan Jews with Creutzfeldt-Jakob disease. *Neurology* 1991;41:160

188 Hsiao K, Meiner Z, Kahana E et al. Mutation of the prion protein in Libyan Jews with Creutzfeldt-Jakob disease. *N Engl J Med* 1991;324:1091-7

189 Goldfarb LG, Brown P, Mitrova E et al. Creutzfeldt-Jacob disease associated with the *PRNP* codon 200[Lys] mutation: an analysis of 45 families. *Eur J Epidemiol* 1991;7:477-86

190 Bertoni JM, Brown P, Goldfarb L, Gajdusek D, Omaha NE. Familial Creutzfeldt-Jakob disease with the *PRNP* codon 200[Lys] mutation and supranuclear palsy but without myoclonus or periodic EEG complexes. *Neurology* 1992;42(suppl 3):350 (abstr)

191 Chapman J, Ben-Israel J, Goldhammer Y, Korczyn AD. The risk of developing Creutzfeldt-Jakob disease in subjects with the *PRNP* gene codon 200 point mutation. *Neurology* 1994;44:1683-6

192 Spudich S, Mastrianni JA, Wrensch M et al. Complete penetrance of Creutzfeldt-Jakob disease in Libyan Jews carrying the E200K mutation in the prion protein gene. *Mol Med* 1995;1:607-13

193 Goldfarb LG, Haltia M, Brown P et al. New mutation in scrapie amyloid precursor gene (at codon 178) in Finnish Creutzfeldt-Jakob kindred. *Lancet* 1991;337:425

194 Goldfarb LG, Brown P, Haltia M et al. Creutzfeldt-Jakob disease cosegregates with the codon 178[Asn] *PRNP* mutation in families of European origin. *Ann Neurol* 1992;31:274-81

195 Brown P, Goldfarb LG, Kovanen J et al. Phenotypic characteristics of familial Creutzfeldt-Jakob disease associated with the codon 178[Asn] *PRNP* mutation. *Ann Neurol* 1992;31:282-5

196 Haltia M, Kovanen J, Goldfarb LG, Brown P, Gajdusek DC. Familial Creutzfeldt-Jakob disease in Finland: Epidemiological, clinical, pathological and molecular genetic studies. *Eur J Epidemiol* 1991;7:494-500

197 Fink JK, Warren JT Jr, Drury I, Murman D, Peacock BA. Allele-specific sequencing confirms novel prion gene polymorphism in Creutzfeldt-Jakob disease. *Neurology* 1991;41:1647-50

198 Lugaresi E, Medori R, Montagna P et al. Fatal familial insomnia and dysautonomia with selective degeneration of thalamic nuclei. *N Engl J Med* 1986;315:997-1003

199 Medori R, Montagna P, Tritschler HJ et al. Fatal familial insomnia: a second kindred with mutation of prion protein gene at codon 178. *Neurology* 1992;42:669-70

200 Goldfarb LG, Petersen RB, Tabaton M et al. Fatal familial insomnia and familial Creutzfeldt-Jakob disease: disease phenotype determined by a DNA polymorphism. *Science* 1992;258:806-8

201 Johnson RT. Prion disease. *N Engl J Med* 1992;326:486-7

202 Petersen RB, Tabaton M, Berg L et al. Analysis of the prion protein gene in thalamic dementia. *Neurology* 1992;42:1859-63

203 Wells GAH, Scott AC, Johnson CT et al. A novel progressive spongiform encephalopathy in cattle. *Vet Rec* 1987;121:419-20

204 Hope J, Reekie LJD, Hunter N et al. Fibrils from brains of cows with new cattle disease contain scrapie-associated protein. *Nature* 1988;336:390-2

205 Prusiner SB, Fuzi M, Scott M et al. Immunologic and molecular biological studies of prion proteins in bovine spongiform encephalopathy. *J Infect Dis* 1993;167:602-13

206 Fraser H, McConnell I, Wells GAH, Dawson M. Transmission of bovine spongiform encephalopathy to mice. *Vet Rec* 1988;123:472

207 Dawson M, Wells GAH, Parker BNJ. Preliminary evidence of the experimental transmissibility of bovine spongiform encephalopathy to cattle. *Vet Rec* 1990;126:112-3

208 Bruce M, Chree A, McConnell I, Foster J, Fraser H. Transmissions of BSE, scrapie and related diseases to mice. In: ed. *IXth International Congress of Virology.* Glasgow, Scotland, Aug 8–13, 1993:93

209 Dawson M, Wells GAH, Parker BNJ, Scott AC. Primary parenteral transmission of bovine spongiform encephalopathy to the pig. *Vet Rec* 1990; Sept 29: 338

210 Bruce M, Chree A, McConnell I et al. Transmission of bovine spongiform encephalopathy and scrapie to mice: strain variation and the species barrier. *Phil Trans R Soc Lond B* 1994;343:405-11

211 Baker HF, Ridley RM, Wells GAH. Experimental transmission of BSE and scrapie to the common marmoset. *Vet Rec* 1993;132:403-6

212 Collinge J, Palmer MS, Sidle KC et al. Unaltered susceptibility to BSE in transgenic mice expressing human prion protein. *Nature* 1995;378:779-83

213 Malmgren R, Kurland L, Mokri B, Kurtzke J. The epidemiology of Creutzfeldt-Jakob disease. In: Prusiner SB, Hadlow WJ, eds. *Slow Transmissible Diseases of the Nervous System,* Vol. 1. New York: Academic Press, 1979;93-112

214 Brown P, Cathala F, Raubertas RF, Gajdusek DC, Castaigne P. The epidemiology of Creutzfeldt-Jakob disease: conclusion of 15-year investigation in France and review of the world literature. *Neurology* 1987;37: 895-904

215 Harries-Jones R, Knight R, Will RG et al. Creutzfeldt-Jakob disease in England and Wales, 1980–1984: a case–control study of potential risk factors. *J Neurol Neurosurg Psychiatry* 1988;51:1113-9

216 Cousens SN, Harries-Jones R, Knight R et al. Geographical distribution of cases of Creutzfeldt-Jakob disease in England and Wales 1970–84. *J Neurol Neurosurg Psychiatry* 1990;53:459-65

217 Goldfarb L, Korczyn A, Brown P, Chapman J, Gajdusek DC. Mutation in codon 200 of scrapie amyloid precursor gene linked to Creutzfeldt-Jakob disease in Sephardic Jews of Libyan and non-Libyan origin. *Lancet* 1990; 336:637-8

218 Lacroute F. Non-Mendelian mutation allowing ureidosuccinic acid uptake in yeast. *J Bacteriol* 1971;106: 519-22

219 Wickner RB. [URE3] as an altered URE2 protein: evidence for a prion analog in *Saccharomyces cerevisiae*. *Science* 1994;264:566-9

220 Coschigano PW, Magasanik B. The URE2 gene product of *Saccharomyces cerevisiae* plays an important role in the cellular response to the nitrogen source and has homology to glutathione *S*-transferases. *Mol Cell Biol* 1991;11:822-32

221 Cox BS, Tuite MF, McLaughlin CS. The psi factor of yeast: a problem in inheritance. *Yeast* 1988;4:159-78

222 Chernoff YO, Lindquist SL, Ono B, Inge-Vechtomov SG, Liebman SW. Role of the chaperone protein Hsp104 in propagation of the yeast prion-like factor [psi⁺]. *Science* 1995;268:880-4

223 Deleu C, Clavé C, Bégueret J. A single amino acid difference is sufficient to elicit vegetative incompatibility in the fungus *Podospora anserina*. *Genetics* 1993;135:45-52

224 Schellenberg GD, Bird TD, Wijsman EM et al. Genetic linkage evidence for a familial Alzheimer's disease locus on chromosome 14. *Science* 1992;258:668-71

225 Van Broeckhoven C, Backhovens H, Cruts M et al. Mapping of a gene predisposing to early-onset Alzheimer's disease to chromosome 14q24.3. *Nature Genet* 1992;2:335-9

226 Van Broeckhoven C, Haan J, Bakker E et al. Amyloid β protein precursor gene and hereditary cerebral hemorrhage with amyloidosis (Dutch). *Science* 1990;248:1120-2

227 Mullan M, Houlden H, Windelspecht M et al. A locus for familial early-onset Alzheimer's disease on the long arm of chromosome 14, proximal to the α_1-antichymotrypsin gene. *Nature Genet* 1992;2:340-2

228 St George-Hyslop P, Haines J, Rogaev E et al. Genetic evidence for a novel familial Alzheimer's disease locus on chromosome 14. *Nature Genet* 1992;2:330-4

229 Levy E, Carman MD, Fernandez-Madrid IJ et al. Mutation of the Alzheimer's disease amyloid gene in hereditary cerebral hemorrhage, Dutch type. *Science* 1990;248:1124-6

230 Goate A, Chartier-Harlin MC, Mullan M et al. Segregation of a missense mutation in the amyloid precursor protein gene with familial Alzheimer's disease. *Nature* 1991;349:704-6

231 Corder EH, Saunders AM, Strittmatter WJ et al. Gene dose of apolipoprotein E type 4 allele and the risk of Alzheimer's disease in late onset families. *Science* 1993;261:921-3

232 Saunders AM, Schmader K, Breitner JCS et al. Apolipoprotein E ε4 allele distributions in late-onset Alzheimer's disease and in other amyloid-forming diseases. *Lancet* 1993;342:710-1

233 Saunders AM, Strittmatter WJ, Schmechel D et al. Association of apolipoprotein E allele ε4 with late-onset familial and sporadic Alzheimer's disease. *Neurology* 1993;43:1467-72

234 Rosen DR, Siddique T, Patterson D et al. Mutations in Cu/Zn superoxide dismutase gene are associated with familial amyotrophic lateral sclerosis. *Nature* 1993;362:59-62

235 Rogaev EI, Sherrington R, Rogaeva EA et al. Familial Alzheimer's disease in kindreds with missense mutations in a gene on chromosome 1 related to the Alzheimer's disease type 3 gene. *Nature* 1995;376:775-8

236 Sherrington R, Rogaev EI, Liang Y et al. Cloning of a gene bearing missense mutations in early-onset familial Alzheimer's disease. *Nature* 1995;375:754-60

237 Levy-Lahad E, Wasco W, Poorkaj P et al. Candidate gene for the chromosome 1 familial Alzheimer's disease locus. *Science* 1995;269:973-7

238 Levy-Lahad E, Wijsman EM, Nemens E et al. A familial Alzheimer's disease locus on chromosome 1. *Science* 1995;269:970-3

239 Carlson GA, Ebeling C, Torchia M, Westaway D, Prusiner SB. Delimiting the location of the scrapie prion incubation time gene on chromosome 2 of the mouse. *Genetics* 1993;133:979-88

240 Puckett C, Concannon P, Casey C, Hood L. Genomic structure of the human prion protein gene. *Am J Hum Genet* 1991;49:320-9

241 Palmer MS, Dryden AJ, Hughes JT, Collinge J. Homozygous prion protein genotype predisposes to sporadic Creutzfeldt-Jakob disease. *Nature* 1991;352:340-2

242 Fink JK, Peacock ML, Warren JT, Roses AD, Prusiner SB. Detecting prion protein gene-mutations by denaturing gradient gel-electrophoresis. *Hum Mutat* 1994;4:42-50

243 Crow TJ, Collinge J, Ridley RM et al. Mutations in the prion gene in human transmissible dementia. *Seminar on Molecular Approaches to Research in Spongiform Encephalopathies in Man, Medical Research Council, London, Dec 14, 1990 (Abstr)* 1990

244 Goldfarb LG, Brown P, Goldgaber D et al. Creutzfeldt-Jakob disease and kuru patients lack a mutation consistently found in the Gerstmann-Sträussler-Scheinker syndrome. *Exp Neurol* 1990;108:247-50

245 Goldfarb L, Brown P, Goldgaber D et al. Identical mutation in unrelated patients with Creutzfeldt-Jakob disease. *Lancet* 1990;336:174-5

246 Hsiao KK, Doh-ura K, Kitamoto T, Tateishi J, Prusiner SB. A prion protein amino acid substitution in ataxic Gerstmann-Sträussler syndrome. *Ann Neurol* 1989;26:137

247 Hsiao KK, Cass C, Schellenberg GD et al. A prion protein variant in a family with the telencephalic form of Gerstmann-Sträussler-Scheinker syndrome. *Neurology* 1991;41:681-4

248 Hsiao K, Prusiner SB. Inherited human prion diseases. *Neurology* 1990;40:1820-7

249 Tateishi J, Kitamoto T, Doh-ura K et al. Immunochemical, molecular genetic, and transmission studies on a case of Gerstmann-Sträussler-Scheinker syndrome. *Neurology* 1990;40:1578-81

250 Ripoll L, Laplanche JL, Salzmann M et al. A new point mutation in the prion protein gene at codon 210 in Creutzfeldt-Jakob disease. *Neurology* 1993;43:1934-8

251 Mastrianni JA, Iannicola C, Myers R, Prusiner SB. Identification of a new mutation of the prion protein gene at codon 208 in a patient with Creutzfeldt-Jakob disease (abstr). *Neurology* 1995;45 (suppl):201

252 Hsiao K, Dlouhy S, Farlow MR et al. Mutant prion proteins in Gerstmann-Sträussler-Scheinker disease with neurofibrillary tangles. *Nat Genet* 1992;1:68-71

253 Dlouhy SR, Hsiao K, Farlow MR et al. Linkage of the Indiana kindred of Gerstmann-Sträussler-Scheinker disease to the prion protein gene. *Nat Genet* 1992;1:64-7

254 Kitamoto T, Iizuka R, Tateishi J. An amber mutation of prion protein in Gerstmann-Sträussler syndrome with mutant *PrP* plaques. *Biochem Biophys Res Commun* 1993;192:525-31

255 Kitamoto T, Ohta M, Doh-ura K et al. Novel missense variants of prion protein in Creutzfeldt-Jakob disease or Gerstmann-Sträussler syndrome. *Biochem Biophys Res Commun* 1993;191:709-14

Transmissible Subacute Spongiform Encephalopathies:
Prion Diseases
L Court, B Dodet, eds
© 1996, Elsevier, Paris

Prnp gene dosage, allelic specificity and gene regulation in the transmissible spongiform encephalopathies

Jean Manson

Institute for Animal Health, BBSRC and MRC Neuropathogenesis Unit, Ogston Building, West Mains Road, Edinburgh EH9 3JF, Scotland

Summary – We have used gene targeting techniques to replace the endogenous murine prion protein gene (Prnp) with a series of mutated Prnp genes. By insertion of a neomycin gene into the coding region of Prnp, we have produced mice with one and two inactive Prnp genes (Prnp null). Infection of these mice with a number of strains of scrapie has shown that gene dosage affects the incubation period of disease, but does not affect the final pathology.

The Prnp null mice have been shown to have defects in synaptic transmission and circadian rest activity and sleep, indicating possible normal functions for the prion protein (PrP).

In order to investigate the development of spontaneous disease and disease specificity, in which specific Prnp alleles are involved and the relationship between Sinc and PrP, we have produced mice with point mutations in the endogenous Prnp gene. Mice in which the amino acid 101 of PrP has been altered from a proline to leucine residue have not developed any spontaneous neurodegenerative disease up to 500 days.

The transmissible spongiform encephalopathies such as scrapie, bovine spongiform encephalopathy and Creutzfeldt-Jakob disease are associated with alterations in the neural membrane protein PrP. The PrP is a protease-sensitive cell surface glycoprotein anchored in the membrane by a glycoinositol phospholipid [1], but during the course of scrapie infection, the PrP aggregates and accumulates in and around the cells of the brain as protease-resistant deposits. The distribution of PrP in infected brain is dependent on both the host genotype and the strain of scrapie [2].

Prion protein gene (Prnp) dosage and scrapie disease

Transgenic models have been produced to investigate the mechanisms of these diseases. Introduction of hamster Prnp genes into mice has shown that increasing the copy number of the Prnp gene reduces the incubation period of the disease, and the species type of PrP expressed alters the susceptibitity of the mice to specific isolates of scrapie [3]. Transgenic mice with high copy numbers of the $Prnp^b$ allele of the murine Prnp gene were also shown to have shorter incubation periods when injected with the Chandler scrapie isolate than their non-transgenic littermates [4]. Transgenic mice with high copy numbers of the murine Prnp gene containing a codon 101 proline to leucine mutation spontaneously develop neurodegeneration,

Key words: prion protein / null mice / gene dosage / allelic specificity / transmissible spongiform encephalopathies / normal function

spongiform changes in the brain and astrogliosis [5]. However, overexpression of the wild-type *Prnp* gene has also been shown to lead to a lethal neurological disease involving spongiform changes in the brain and muscle degeneration [6]. While these experiments have demonstrated that increasing the copy number of the *Prnp* gene leads to shortening the incubation periods of the disease, they have also demonstrated that random integration of multiple copies of the *Prnp* gene results in clinical artefacts which may not accurately reflect the disease process.

In view of these problems, we have replaced the murine *Prnp* gene with a series of mutated *Prnp* genes using gene targeting techniques. Endogenous murine genes can be modified by gene targeting in embryonic stem (ES) cells derived from early mouse embryos. Specific gene alterations can be made to ES cells in culture and, because these cells retain their totipotency, the mutated cells can be reintroduced into a blastocyst and contribute to the developing embryo. These transgenic models allow the effect of PrP copy number and the different *Prnp* alleles on disease susceptibility to be compared directly since the *Prnp* gene is in the correct gene location under the control of endogenous regulatory sequences and the sites of integration and expression levels between the different lines of mice are constant.

Production of PrP-deficient mice

In order to investigate both the normal function of PrP and its role in transmissible spongiform encephalopathies (TSE), we produced by gene targeting an inbred 129/Ola mouse line with an inactive *Prnp* gene. In mice heterozygous for this mutation *(Prnp+/−)*, PrP mRNA is reduced by approximately 50% throughout the brain compared with the wild-type mice *(Prnp+/+)*. The steady-state level of PrP is also significantly reduced in the heterozygotes compared to wild-type mice. PrP mRNA and protein are not detected in brains of mice homozygous for the mutation *(Prnp−/−)* [7].

Scrapie disease in PrP deficient mice

Onset of disease

Mice with no functional copies of the *Prnp* gene have been shown to survive scrapie infection when injected with ME7, 301V or 301C strains (table I). No disease symptoms were observed in these mice over 500 days after inoculation, whereas terminal stages of the the disease are detected in wild-type mice around 147, 154, and 227 days, respectively. These results have clearly demonstrated that *Prnp* gene expression is required for the development of scrapie disease. In mice with one functional copy of the *Prnp* gene, the time of onset of disease is delayed, but the amount by which the onset is delayed in the heterozygotes appears to differ depending on either the strain of scrapie or the effective dose of inoculum (table I, fig 1).

PrP accumulation in Prnp-*deficient mice*

In *Prnp+/−* mice infected with the ME7 strain of scrapie, PrP deposition starts in the same brain area as in the wild-type mice and can be detected as early as 50 days. The pattern of PrP deposition in the brain of heterozygotes follows an identical course, but builds up more slowly than in the wild-type mice. By the terminal stages of disease, the amount detected in the brain by immunohistochemical techniques is identical in the *Prnp+/−* and *Prnp+/+* mice [8]. These experiments indicate that protease resistant PrP (PrPSc) accumulation is not limited by the levels of protease sensisitive PrP (PrPC) and raise questions as to whether there is a causal relationship between PrPSc accumulation in the brain and the development of clinical disease.

Table I. Incubation period of *Prnp*[+/+], *Prnp*[+/-] and *Prnp*[-/-] mice infected with scrapie.

Strain	Prnp[+/+]		Prnp[+/-]		Prnp[-/-]	
	Days	Mice (n)	Days	Mice (n)	Days	Mice (n)
ME7	147 ± 2	30	293 ± 4	12	> 500	30
301C	154 ± 1	14	230 ± 2	17	> 500	21
301V	227 ± 3	17	320 ± 3	17	> 500	14

Groups of mice with two *(Prnp*[+/+]*)*, one *(Prnp*[+/-]*)* or no wild-type *Prnp* genes *(Prnp*[-/-]*)* were infected with ME7, 301V or 301C strains of scrapie. The time point at terminal stages of disease was calculated and used to give an estimate of the incubation period.

Severity of disease

The timing of disease onset and progression of disease symptoms is delayed in the PrP deficient mice [8, 9]. There appears however to be no difference in the severity of the clinical signs and the pathology in the brain at terminal stages of disease. In *Prnp*[+/-] mice injected with ME7, 301V and 301C strains of scrapie, the pattern and degree of vacuolation were similar to those detected in the wild-type mice, and disease symptoms, although delayed, were no less severe than those seen in the wild-type mice (fig 1). *Prnp* gene dosage therefore appears to affect the timing of disease, but does not affect the final pathology.

Allelic variants of *Prnp*

The *Sinc* gene has been shown to be the major gene controlling the survival time of mice exposed to scrapie [10, 11] and animals homozygous for the alleles of *Sinc*[s7] and *Sinc*[p7] have allelic forms of the *Prnp* gene. The *Prnp* allele with amino acids Leu[108] and Thr[189] is associated with short incubation times when infected with the Chandler isolate of scrapie *(Prnp*[a]*)*, whereas the allele with Phe[108] and Val[189] *(Prnp*[b]*)* is associated with long incubation periods with the same isolate of scrapie [12]. The *Sinc* gene has been shown to be closely linked to the *Prnp* gene by restriction fragment length polymorphisms analysis [13] and *Sinc* congenic mice encode different PrP proteins [14], suggesting that *Prnp* may indeed be the product of the *Sinc* locus.

In order to establish whether the *Sinc* gene is the *Prnp* gene, we have developed a double replacement gene targeting strategy to introduce point mutations into the *Prnp* gene. Using this technique, amino acids 108 and 189 in the murine embryonic stem cells from a *Sinc*[s7] mouse have been altered to Phe[108] and Val[189] which are associated with a *Sinc*[p7] mouse [15]. The effect of these mutations on the incubation period of a number of strains of scrapie will be used to establish whether the two genes are identical.

Allelic forms of *Prnp* have also been linked to incubation periods of scrapie in sheep [16], and are associated with the incidence of Gerstmann-Sträussler-Scheinker (GSS) and Creutzfeldt-Jakob disease in humans [17]. A mutation at codon 102 resulting in the replacement of a proline residue for a leucine has been shown to be linked to GSS disease in man [18]. Standard transgenic approaches have been used to produce mice expressing high levels of a chimaeric hamster-murine *Prnp* gene with the mutation in the equivalent amino acid (101) of the murine gene. These transgenic mice have been shown to spontaneously develop neurodegeneration,

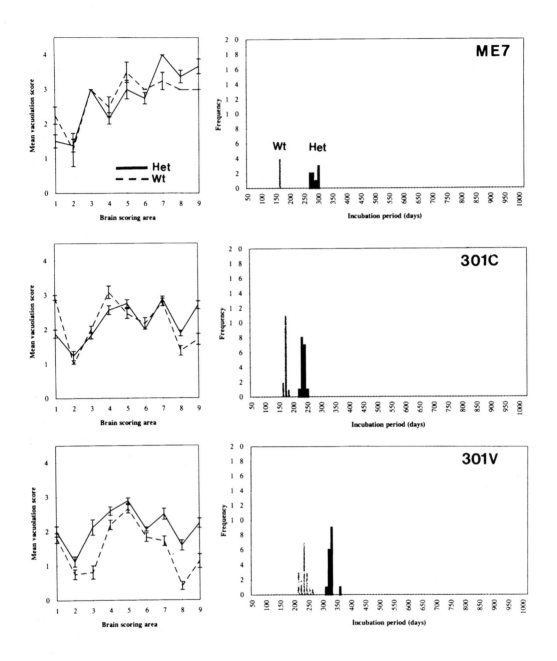

Fig 1. Lesion profiles and incubation periods for 129/Ola *Prnp*+/+ (Wt) and *Prnp*+/− (Het) mice infected with ME7, 301V and 301C strains of scrapie. The amount of vacuolation in nine areas of the brain at terminal stages of disease were used to calculate the lesion profile in the 129/Ola *Prnp*+/+ mice. A lesion profile was produced for the terminal stage of disease in the 129/Ola *Prnp*+/− mice and compared to the terminal 129/Ola *Prnp*+/+ profile for each strain of scrapie. The brain areas assessed were: medulla (1), cerebellum (2), superior colliculus (3), hypothalamus (4), thalamus (5), hippocampus (6), septum (7), thalamic cortex (8) and forebrain cortex (9). The time point at terminal stages of scrapie for a group of animals of each genotype was used to give an estimate of incubation periods of ME7, 301C and 301V strains in *Prnp*+/+ (Wt) and *Prnp*+/− mice.

spongiform changes in the brain and astrogliosis [5]. Whether this disease can be attributed specifically to the PrP mutation or to overexpression of the *Prnp* gene is difficult to establish since mice with low copy numbers of the mutant gene do not develop disease. We have used a double replacement gene targeting strategy to introduce a Pro-Leu mutation into amino acid 101 of the mouse PrP gene and have produced a colony of mice both heterozygous and homozygous for the mutant gene [15]. These mice are being assessed for any development of neurodegenerative disease. No symptoms have been detected in mice up to 500 days old. The brains of homozygous mutant mice at 50, 100 and 200 days show no abnormal pathology. Experiments are also being carried out to assess whether this mutation has led to any alteration in disease susceptibility or replication of infectivity in the transgenic mice.

Normal function of PrP

The normal function of PrP is not known. The *Prnp* gene is expressed from mid-gestation in the mouse embryo in the developing central and peripheral nervous tissue [19]. It continues to be expressed throughout development and mRNA can be detected in a number of non neuronal tissues, both within the embryo and in the extra-embryonic tissue. The *Prnp* gene is expressed at high levels in the adult brain and recently has been detected in astrocytes and oligodendrocytes [20]. Lower levels of PrP mRNA can be detected in other tissues such as the heart, lung and spleen [21]. These studies have suggested a role for PrP in promoting neuronal differentiation and in maintaining neuronal function in the differentiated neurones. However, the expression of PrP is not limited to neuronal cells since it can be detected in astrocytes and oligodentrocytes and in specific non-neuronal cells of the embryo, suggesting a more widespread role for PrP, perhaps as part of a cell signalling mechanism or as a cell adhesion molecule.

It was hoped that the generation of *Prnp* null mice would address the question of the normal function of PrP. The initial analysis of the *Prnp* null mice [7, 22] however, showed that mice without PrP appeared to develop and reproduce normally. It has now been shown that there is little or no phenotypic effect in many null mutant mice produced by gene targeting. This lack of phenotype is thought to be due to the organism compensating for the loss of the gene by alteration in expression of other genes or use of alternative developmental pathways. If genes are compensating for the loss of PrP in these mice there is at present no candidate gene, although alterations in gene expression patterns in the *Prnp* null mice may allow identification of such genes.

Long-term potentiation

More recently however, a number of phenotypic differences have been detected between *Prnp* null and wild-type mice. *Prnp* null mice have been shown to display weakened long-term potentiation (LTP) in the hippocampal CA1 subfield [23]. We have also shown abnormalities in the LTP in the inbred *Prnp* null mice. These mice consistently showed an absence of LTP in the CA1 region of the hippocampus and in its place a short-term potentiation which decays to control levels within one hour [24]. Since the two lines of mice have been produced by different targeting strategies, these results indicate that the impairment of synaptic plasticity is a result of the loss of PrP, and not an artefact of gene targeting and subsequent mouse production. Reintroduction of the high copy numbers of the human *Prnp* gene into *Prnp* null mice has been shown to restore the LTP response to that seen in the wild-type controls [25]. The effect on LTP can therefore be attributed to an absence of PrP, but it remains difficult to ascribe the differences between knockout and wild-type mice to a specific function of PrP or to compensatory effects of the organism in the absence of PrP.

Alterations in circadian rest activity and sleep in Prnp *null mice*

We have shown that mice devoid of PrP have alterations in both circadian activity rhythms and sleep [26]. *Prnp* null mice exhibit a longer circadian period than the wild-type mice. During prolonged recording in constant darkness, the circadian period of the null mice remains stable in contrast to the wild-type mice, in which the circadian period becomes more variable and progressively shorter. The *Prnp* null mice also show a more fragmented sleep than the wild-type mice and a much larger response to sleep deprivation. The increase in electroencephalogram slow wave activity during recovery sleep is almost twice as large in *Prnp* null as in wild-type mice. These results indicate that PrP may be involved in sleep regulation, and in this respect, some of the pathological symptoms in prion diseases such as fatal familial insomnia may be due to a loss of function of PrP rather than to an abnormal accumulation of PrP^{Sc}.

Prnp gene regulation and therapy

The expression level of the *Prnp* gene has been shown to alter the incubation period of scrapie in mice. Transgenic mice containing multiple copies of the *Prnp* gene have reduced incubation periods and mice with only one copy of the gene have extended incubation periods when compared to wild-type mice. We have therefore investigated the mechanisms by which the *Prnp* gene is normally regulated to establish whether abnormal regulation of the gene may be involved in disease susceptibility.

We isolated the promoter region of the mouse *Prnp* gene and sequenced a 5' region of the *Prnp* gene, which includes 1.2 kb 5' to exon 1, exons 1 and 2, and intron 1. Sequencing of this region from several strains of mice identified a polymorphism present in the 5' promoter region linked to *Sinc*, the gene controlling scrapie incubation periods in mice. We used this gene fragment and its deletions to examine promoter-mediated expression of a chloramphenicol acetyl transferase reporter gene in murine neuroblastoma cells. There are two major areas of promoter activity in this region of the *Prnp* gene, immediately upstream from exons 1 and exon 2. The 5' region of intron 1 also contains elements that are capable of suppressing promoter function [27].

Since the level of expression of the *Prnp* gene clearly alters the development of disease, identification of the sequences important in the regulation of *Prnp* gene expression will allow us to investigate whether cases of TSE not associated with mutations of the coding region of PrP are associated with mutations in regulatory sequences. An understanding of the regulation of *Prnp* gene expression will also allow us to assess the potential for therapeutic intervention in these diseases through modulation of the *Prnp* gene expression.

Acknowledgments

The work described in this chapter is also the work of the following: H Baybutt, L Aitchison, P McBride, I McConnell, J Hope (Institute for Animal Health, Neuropathogenesis Unit, Edinburgh); A Clarke (Dept of Pathology, Edinburgh University); A Johnston, N MacLeod (Dept of Physiology, Edinburgh University); D Melton, R Moore, N Redhead (Institute for Cell and Molecular Biology, Edinburgh University); and I Tobler (Institute for Pharmacology, University of Zurich).

References

1 Stahl N, Borchelt DR, Hsiao K, Prusiner SB. Scrapie prion protein contains a phosphoinositol glycolipid. *Cell* 1987;51:229-40

2   Bruce M, McBride PA, Farquhar CF. Precise targeting of the pathology of the sialoglycoprotein, PrP, and vacuolar degeneration in mouse scrapie. *Neurosci Lett* 1989;102:1-6

3   Scott M, Foster D, Mirenda C et al. Transgenic mice expressing hamster prion protein produce species-specific scrapie infectivity and amyloid plaques. *Cell* 1989;59:847-57

4   Westaway D, Mirenda CA, Foster D et al. Paradoxical shortening of scrapie incubation times by expression of prion protein transgenes derived from long incubation period mice. *Neuron* 1991;7:59-68

5   Hsiao KK, Scott M, Foster D, Groth DF, DeArmond SJ, Prusiner SB. Spontaneous neurodegeneration in transgenic mice with mutant prion protein. *Science* 1990;250:1587-90

6   Westaway D, DeArmond SJ, Cayetano-Canlas J et al. Degeneration of skeletal muscle, peripheral nerves and central nervous system in transgenic mice overexpressing wild type prion proteins. *Cell* 1994;76:117-29

7   Manson JC, Clarke AR, Hooper M, Aitchison L, McConnell I, Hope J. 129/Ola mice carrying a mutation in *PrP* that abolishes mRNA production are developmentally normal. *Mol Neurobiol* 1994;8:1-5

8   Manson JC, Clarke AR, McBride PA, McConnell I, Hope J. *PrP* gene dosage determines the timing but not the final intensity or distribution of lesions in scrapie pathology. *Neurodegeneration* 1994;3:311-40

9   Bueler H, Aguzzi A, Sailer A et al. Mice devoid of PrP are resistant to scrapie. *Cell* 1993;73:1339-47

10 Dickinson AG, Meikle MV, Fraser H. Identification of a gene which controls incubation period of some strains of scrapie in mice. *J Comp Pathol* 1968;78:293-9

11 Carlson GA, Kingsbury DT, Goodman PA et al. Linkage of prion protein and scrapie incubation time genes. *Cell* 1986;46:503-11

12 Westaway D, Goodman PA, Mirenda C, McKinley MP, Carlson GA, Prusiner SB. Distinct prion proteins in short and long incubation period mice. *Cell* 1987;51:651-62

13 Hunter N, Hope J, McConnell I, Dickson AG. Linkage of the scrapie associated fibril protein *(PrP)* gene and *Sinc* using congenic mice and restriction fragment length polymorphisms analysis. *J Gen Virol* 1987;68:2711-2

14 Hunter N, Dann JC, Bennett AD, Somerville RA, McConnell I, Hope J. Are *Sinc* and the *PrP* gene congruent? Evidence from *PrP* gene analysis in *Sinc* congenic mice. *J Gen Virol* 1992;73:2751-5

15 Moore R, Redhead N, Selfridge J, Hope J, Manson J, Melton D. Double replacement gene targeting for the production of a series of mouse strains with different prion protein alterations. *Nature Biotechnol* 1995;13: 999-1004

16 Goldmann W, Hunter N, Benson G, Foster J, Hope J. Scrapie-associated fibril proteins are encoded by lines of sheep selected for different alleles of the *Sip* gene. *J Gen Virol* 1991;72:2411-7

17 Collinge J, Palmer M. Human prion diseases. *Bailliere's Clin Neurol* 1994;3:241-57

18 Hsiao K, Baker H, Crow T et al. Linkage of a prion protein missense variant to Gerstmann-Sträussler syndrome. *Nature* 1989;338:342-5

19 Manson J, West JD, Thomson V, McBride P, Kaufman MH, Hope J. The prion protein gene: a role in mouse embryogenesis? *Development* 1992;115:117-22

20 Moser M, Colello RJ, Pott U, Oesch B. Developmental expression of the prion protein gene in glial cells. *Neuron* 1995;14:509-17

21 Oesch B, Westaway D, Walchi M et al. A cellular gene encodes scrapie PrP 27–30 protein. *Cell* 1985;40:735-46

22 Bueler H, Fischer M, Lang Y et al. The neuronal cell surface protein PrP is not essential for normal development and behaviour in mice. *Nature* 1992;356:577-82

23 Collinge J, Whittington M, Sidle K et al. Prion protein is necessary for normal synaptic function. *Nature* 1994;370:295-7

24 Manson J, Hope J, Clarke A, Johnston A, Black C, MacLeod N. *PrP* gene dosage and long term potentiation. *Neurodegeneration* 1995;4:113-5

25 Whittington M, Sidle K, Gowland I et al. Rescue of neurophysiological phenotype seen in *PrP* null mice by transgene encoding human prion protein. *Nature Genet* 1995;9:197-201

26 Tobler I, Gaus SE, Deboer T et al. Mice devoid of prion protein display altered circadian rhythms and sleep patterns. *Nature* 1996;380:639-42

27 Baybutt H, Manson J. Characterisation of two promoters for prion protein (PrP) gene expression in neuronal cells. *Gene* 1996; in press

Transmissible Subacute Spongiform Encephalopathies:
Prion Diseases
L Court, B Dodet, eds
© 1996, Elsevier, Paris

The role of protease-sensitive prion protein (PrPc) in prion diseases

Charles Weissmann[1], Alex Raeber[1], Marck Fischer[1], Andreas Sailler[1], Thomas Rülicke[2]

[1]*Institut für Molekularbiologie, Abteilung I, Universität Zürich, 8093 Zürich 1;* [2]*Biologisches Zentral-labor, Universitätsspital, Zürich, 8091 Zürich 2, Switzerland*

Prion, the transmissible agent that causes spongiform encephalopathies such as scrapie is believed to be a modified form of protease-sensitive prion protein (PrPc) and devoid of nucleic acid. After inoculation with mouse scrapie prions, homozygous $Prnp^{0/0}$ ('PrP knockout') mice remained free of scrapie symptoms for at least 2 years, while all wild-type controls died within 6 months. There was no propagation of prions in the $Prnp^{0/0}$ animals. Surprisingly, hetero-zygous $Prnp^{0/+}$ mice, which express PrPc at about half the normal level, also showed enhanc-ed resistance to scrapie disease, despite high levels of infectious agent and PrPSc in the brain early on. Introduction of murine PrP transgenes rendered $Prnp^{0/0}$ mice highly susceptible to mouse but not to hamster prions, while the introduction of Syrian hamster PrP transgenes rendered them susceptible to hamster prions, but to a much lesser extent to mouse prions. These experiments pave the way for the application of reverse genetics, making it possible to investigate which domains of the PrP protein are required to mediate susceptibility to scrapie in $Prnp^{0/0}$ mice. We prepared animals transgenic for genes encoding PrP with amino terminal deletions of various lengths and found that PrP lacking 48 amino acids is still biologically active.

To determine whether PrPc expression was the only prerequisite for propagation of prions, PrP-deficient mice were reconstituted with transgenes encoding PrP driven by heterologous promoters. Using an IRF-1 promoter/Eμ enhancer to drive PrP expression, we generated mice with high levels of PrP in B cells, T cells and dendritic cells, and low levels in the brain. Six months after inoculation with scrapie prions, these mice had accumulated roughly 10^7 LD_{50} (median lethal dose) units of scrapie infectivity in the spleen, but showed no infectivity in the brain. B cells, T cells, and non-B, non-T cells were purified using appropriate antibodies bound to magnetic beads. Infectivity was found in B and T cells, but no infectivity was found in circulating leucocytes. Scrapie-inoculated mice transgenic for a *Prnp* gene under the control of *lck* promoter overexpressed PrP on the surface of T cells, but surprisingly did not propagate prions in either thymus or spleen. The two apparently contradictory findings regarding T cells tentatively suggest that T cells can carry but not propagate infectivity. Moreover, it is clear that overexpression of PrP is not sufficient to allow prion propagation.

STRAIN BIOLOGY

Transmissible Subacute Spongiform Encephalopathies:
Prion Diseases
L Court, B Dodet, eds
© 1996, Elsevier, Paris

Dissecting prion strains using transgenic mice

George A Carlson

McLaughlin Research Institute, 1520 23rd Street South, Great Falls, MT 59405, USA

Summary – Although post-translational conversion of host protease-sensitive prion protein (PrP^C) to disease-specific prion protein (PrP^{Sc}) may be sufficient to account for prion replication, the existence of stable prion isolates or strains demands that the agent encode information independent of the host. Manipulation of the mouse prion protein gene *(Prnp)* copy number and the ratio of numbers of *a* to *b* alleles through genetic crosses involving transgenic and *Prnp*-ablated mice has provided new clues to the nature of prion-specified information. Prion strains had been classified by incubation time profiles defined, in part, by whether the *a* or *b* allele of *Prnp* was dominant in prolonging incubation time. It is now clear that incubation time properties of prions reflect PrP^C-PrP^{Sc} interactions dependent on the PrP sequences of both the prion donor and the host. Application of the chemical laws of mass action to interpret complex strain incubation time profiles may place constraints on models for prion replication.

Introduction

Experimental goat scrapie provided the first evidence for prion strains [1]. Distinct clinical manifestations, labeled nervous or drowsy, were produced by two isolates and bred true through subsequent intracerebral inoculations. Subsequently, both goat and sheep scrapies were transmitted to mice and hamsters, leading to isolation of a variety of scrapie strains [2]. Scrapie isolates that exhibited distinct and stable properties on repeated passage in a single inbred mouse strain provided compelling evidence for agent-specified information; based on the assumption that the transmissible agent was a virus, this information was assumed to be encoded by a polynucleotide [3]. However, prions appear to be devoid of functional nucleic acid and their predominant component is PrP^{Sc}, a disease-specific, post-translational derivative of the normal prion protein isoform, PrP^C [4]. β-Pleated sheet secondary structure predominates in PrP^{Sc}, with little, if any, found in PrP^C; the two isoforms may differ only in their conformations [5, 6]. PrP^{Sc} is necessary, and probably sufficient, for transmission of scrapie. No scrapie-specific nucleic acid has ever been detected in purified prion preparations [7]. Prion replication is thought to involve a process in which the PrP^{Sc} isoform serves as a template for conversion of PrP^C, possibly involving the unfolding of α-helices and refolding into β-sheet structures. Although undoubtedly different in detail, a prion-like conversion of one protein isoform by an altered form of the same protein has been postulated as a mechanism for some non Mendelian traits in yeast [8, 9].

Prion protein genotype and scrapie incubation time

The isolation and characterization of most mouse scrapie isolates predate the discovery of PrP [10] and its chromosomal *Prnp* gene [11, 12]. However, the major defining property of prion

---

Key words: incubation time / scrapie strains / genetics / prion protein

strains was incubation time profiles in mice as a function of *Prnp* genotype. The predominant influence of this single gene, previously known as *Sinc*, on mouse scrapie incubation time was described more than 30 years ago [14, 15], but the gene chromosomal location was unknown until the demonstration of linkage with *Prnp* [16]. Evidence indicates that *Prnp* itself, rather than a linked locus, controls scrapie incubation period. The *a* and *b* alleles of *Prnp* differ at codons 108 (Leu/Phe) and 189 (Thr/Val) [17]. The term allotype distinguishes between the PrP-A and PrP-B proteins encoded by the *a* and *b* alleles of *Prnp*, and reserves the term isoform to distinguish between PrP^C and PrP^Sc. It is important to stress that the effect of each *Prnp* allele on prion incubation time is dependent on the scrapie strain that is inoculated. Prolongation of incubation times for the Rocky Mountain Lab (RML) isolate by the *b* allele appears to be dominant because *a/a* homozygotes become ill ≤150 days after inoculation, while *a/b* and *b/b* mice show no signs of disease before 200 days. In contrast, the effect of the *a* allele is overdominant in prolonging the 22A incubation period; *Prnp^b* homozygous mice have shorter incubation times (approximately 200 days) than *a* homozygous animals (~400 days). F1 hybrids have longer 22A incubation times (~500 days) than either parent. Such isolate-specific patterns of allelic interaction with the prion incubation time gene were taken as evidence that the scrapie agent had a genome independent of the host [3, 18]. Sheep also show similar scrapie strain-dependent differences in the effects of particular *Prnp* gene alleles on incubation time [19–21].

Influence of PrP sequence matching and mismatching between host and donor

Passage history can have a profound influence on prion strain characteristics. As outlined above, 22A prions have a relatively short 200-day incubation time in mice homozygous for the *b* allele of *Prnp*, while *Prnp^a* mice become ill roughly 400 days after inoculation. This incubation time profile is stable only if 22A is maintained in *Prnp^b* homozygotes. Incubation time rapidly shortens with subsequent passages through mice expressing PrP-A. If scrapie were due to a virus, this shortening of the incubation period might reflect host selection for more rapidly replicating mutants present in the 22A isolate. The isolate that results from passage in *Prnp^a* homozygotes produces long incubation times in *Prnp^b* mice [3]. However, preferential interaction between homologous PrP allotypes provides a more viable interpretation.

The first demonstration of the effects of PrP homology between donor and host was the shorter incubation times produced by RML PrP^Sc-B prions (passed through mice homozygous for the *b* allele of *Prnp*) in *Prnp^b* homozygous mice than by RML prions composed of PrP^Sc-A [22]. PrP sequence identity between the host and the prion donor reduced the incubation time by over 100 days. It is important to note, however, that RML prion incubation periods in *Prnp^b* mice never became as short as in *a/a* homozygotes, regardless of passage history. The species barrier to scrapie transmission is a more dramatic demonstration of the effect of PrP matching between donor and host [23]. Only a small percentage of mice inoculated with Syrian hamster (SHa) prions become ill only after 500 days or more. Mice expressing *Prnp* transgenes from SHa do not exhibit the species barrier to hamster scrapie and incubation times can be as short as 50 days in high copy number mice [24, 25]. SHa*Prnp* transgenic (Tg) mice inoculated with SHa prions produce only SHa prions, while mouse prions elicit production of mouse but not SHa scrapie agent. Clearly, interaction between PrP^C and PrP^Sc is an essential feature of prion replication.

Prion strain properties reflect chemical interactions between PrP^C and PrP^{Sc}

Freedom from nature's constraint of three possible combinations for two alleles was provided by the availability of Tg and gene-ablated mice, and spurred reevaluation of old concepts surrounding prion strains. For more than 30 years, the term 'dominant' was used to describe the effects of the *b* allele of *Prnp* in prolonging scrapie incubation time [15, 16]. There also was compelling genetic evidence that *Prnp*, rather than a distinct incubation time gene, controlled the scrapie incubation period [17, 26, 27]. For these reasons, it seemed likely that expression of *Prnp^b* transgenes would prolong the RML prion incubation time of *Prnp^a* homozygous mice. Surprisingly, Tg*(Prnp^b)* mice had shorter incubation times than their non Tg littermates, posing an apparent paradox [28]. One interpretation of this result was the possibility that a distinct incubation time locus had been missed in earlier studies and was not included within the transgene construct. To test the distinct incubation time gene hypothesis, the influence of an 'authentic' *Prnp^b*-bearing chromosome on RML isolate incubation time was tested. Tg*(Prnp^b)15* mice have three copies of the *Prnp^b*-containing cosmid insert derived from I/LnJ mice. Tg*15* mice were crossed with B6.I-1 (B6.I-*Prnp^b*/Co) congenic mice and inoculated with RML isolate. B6.I mice have long RML incubation times and are homozygous for *I/LnJ*-derived alleles of *Prnp* and other tightly linked genes [27]; the genetic background is derived from the short incubation time B6 strain. All offspring of this cross are *Prnp* heterozygous, with half expected to be transgene-negative and half transgene-positive. *Prnp^a* homozygous mice with the transgene array had incubation times of 115 ± 3 days compared to 144 ± 5 days for transgene-negative *Prnp^a* mice. In comparison, transgene-positive *Prnp^a/Prnp^b* heterozygous mice had incubation times of 166 ± 2 days, not as long as 'authentic' heterozygous mice, but significantly longer than transgene-positive *Prnp^a* mice.

These and similar studies can be interpreted in terms of PrP^C concentration and PrP^C-PrP^{Sc} interactions. Incubation time is inversely proportional to the amount of PrP^C expressed [24, 25, 28, 29]. Mice carrying one normal *Prnp* allele and one ablated allele have very long incubation times. The apparent dominance of the *b* allele in prolonging RML incubation time in *Prnp* heterozygous mice is likely to be due to the reduced amount of PrP^C produced from a single copy of *Prnp^a*. Based on these and similar results, we suggest that RML prions convert PrP^C-A into PrP^{Sc}-A more rapidly than they convert PrP^C-B to PrP^{Sc}-B. *Prnp^b* transgene expression in *Prnp^a* homozygous mice would not be expected to affect the supply of PrP^C-A, and additional PrP^C, even the less efficiently converted PrP^C-B allotype, could only shorten incubation time. Thus PrP-A and PrP-B appear to function independently in supporting RML replication and their effects on incubation time are additive. To summarize, the supply of PrP^C-A is the primary determinant of RML incubation time [29, 30].

The 22A scrapie isolate produces disease more rapidly in *Prnp^b* mice than in *Prnp^a* mice, and *Prnp* heterozygous mice have longer incubation times than either parent. In one experiment [29], incubation times in B6 were 405 ± 2 days, in B6.I, 194 ± 10 days, and in their F1 hybrid, 508 ± 14 days. Incubation times in Tg*15* mice revealed that increasing the amount of PrP-B shortened incubation time. Tg*15* mice with three copies of *Prnp^b* in their transgene array (plus two endogenous *a* alleles) had much shorter incubation times than 'authentic' *Prnp^a/Prnp^b* heterozygotes (395 ± 12 days versus 508 ± 14); incubation times were further shortened by transgene homozygosity (286 ± 15 days). The incubation times of Tg*15* transgene homozygous mice carrying six copies of *Prnp^b* in addition to two copies of *Prnp^a* remained longer than those of B6.I mice with two copies of *Prnp^b*, however. This suggests that *Prnp^a* expression somehow intereferes with productive interaction between PrP^{Sc}-B in the 22A inoculum and PrP^{C}-B in the host.

Even though results using the 22A isolate are dramatically different from those obtained with the RML isolate, these too can be explained by allotype-specficity of PrP^{C}-PrP^{Sc} interactions. Dickinson and Outram [31] proposed a replication site hypothesis to account for overdominance, postulating that dimers of the *Sinc* gene product feature in the replication of the scrapie agent. Such a model assumes that PrP^{C}-B dimers are more readily converted to PrP^{Sc} than are PrP^{C}-A dimers and that PrP^{C}-A:PrP^{C}-B heterodimers would be very resistant to conversion to PrP^{Sc}. Increasing the ratio of PrP-B to PrP-A would shorten incubation times by favoring the formation of PrP^{C}-B homodimers. An alternative model invokes two aspects of PrP^{C}-PrP^{Sc} interaction, binding affinity and the rate at which benign PrP^{C} is converted to the malignant isoform, without the need to postulate PrP^{C} dimer formation. A higher affinity of 22A PrP^{Sc} for PrP^{C}-A than for PrP^{C}-B could account for the exceptionally long incubation time of *Prnp* heterozygous mice if PrP^{C}-A is inefficiently converted to PrP^{Sc}. According to this scenario, the reduction in the supply of 22A prions available for interaction with the PrP^{C}-B product of the single *Prnp^{b}* allele would slow disease progression. A similar mechanism could account for the relative rarity of individuals heterozygous for the Met/Val polymorphism at codon 129 of the human *PRNP* gene in spontaneous Creutzfeldt-Jakob disease [32].

Results using the 87V prion isolate support the possibility of non productive PrP^{C}-PrP^{Sc} interactions and favor the possibility that binding affinities and rate constants may be sufficient to explain incubation time profiles of prion strains. Only *Prnp^{b}* mice became uniformly infected by 87V prions with almost all *Prnp^{a}*, *Prnp^{a}*-*Prnp^{b}* and Tg*15* mice failing to develop any signs of illness. Only a single *Prnp^{a}* mouse out of 31 inoculated developed scrapie. In contrast to the results with other scrapie isolates, expression of the three copy *Prnp^{b}* transgene array did not alter scrapie susceptibility. The postulate of a high binding affinity of PrP^{C}-A for 87V PrP^{Sc}-B without conversion to PrP^{Sc}-A is sufficient to explain these results. These arguments on the postulated interactions between PrP^{C} and PrP^{Sc} that are involved in replication of the RML, 22A and 87V prion strains are summarized in table I.

Although the amino acid sequence of host PrP^{C} places major constraints on disease manifestation and incubation time, additional levels of complexity, either higher-order structure or cofactors, are necessary for prion diversity. For example, both PrP^{Sc}-A and PrP^{Sc}-B prions can 'encode' the RML prion-specified property of shorter incubation times in *Prnp^{a}* mice than in *Prnp^{b}* animals. In other words, even though interactions between homologous allotypes favor shorter incubation times [22], the incubation time profile and pathological changes induced by RML prions are relatively stable regardless of passage history.

Differences among prion isolates within a single mouse strain

Although PrP^{C}-PrP^{Sc} interactions can account for different incubation time profiles of prion strains, the nature of the strain-specific differences among strains is unknown. There is no doubt that 22A and RML produce dramatically different incubation time profiles in panels of mice expressing various amounts of PrP-A and PrP-B, even when each strain is maintained in *Prnp^{b}* mice. Striking differences in the types and distribution of pathological lesions within the same inbred mouse strain can also be produced by different isolates; only one example is the florid deposition of amyloid plaques produced by 87V in I/LnJ or B6.I-1 mice compared to the near total absence of plaques in these mouse strains inoculated with most other prion isolates [3, 29]. The arguments outlined above, that interactions between PrP^{C} and PrP^{Sc} governed by

Table I. Postulated interactions between PrP^C and PrP^{Sc} for three prion strains.

Prion strain	$PrP^C \rightarrow PrP^{Sc}$ conversion rate	Allelic interaction	Long incubation time dominance
RML	PrP-A > PrP-B	Additive	Decreased supply of PrP^C-A
22A	PrP-B > PrP-A	Interference	Decreased supply of PrP^C-B and active effect of PrP^C-A
87V	PrP-B >> PrP-A	Interference	Active effect of PrP^C-A, binding without conversion, decreased supply of prions

the laws of mass action can account for the incubation time profiles of prion strains without invoking an additional macromolecule, might also suggest that PrP^{Sc} can adopt a variety conformations that confer strain properties [30]. Different sized proteinase K-resistant PrP^{Sc} fragments result from infection with the hyper or drowsy hamster prion strains derived from transmissible mink encephalopathy [33], compatible with distinct conformations distinguishing PrP^{Sc} in the two strains. It remains unclear whether the strain-associated property of PrP^{Sc} sensitivity to proteinases is responsible for major differences in incubation periods or pathology. A key question, relating to both reproduction of strain-specific properties and prion replication itself, is whether simple physical interactions between PrP^C and PrP^{Sc} are sufficient or whether cofactors or metabolic activity are required. Recent studies suggest that in vitro admixture of radiolabeled PrP^C and excess PrP^{Sc} results in some proteinase K-resistant radiolabeled product [34]. Production of different sized proteinase-resistant fragments from labeled SHaPrP^C by the drowsy or hyper prion strains in the cell-free system also has been reported [34]. Unfortunately, excess PrP^{Sc} in the reaction mixture precludes determination of whether propagation of strain-specific proteinase resistance properties in the cell-free system reflects prion replication. Cell-free acquisition of PrP^{Sc} properties by PrP^C appears to require denaturation of both isoforms at the start of the reaction. Recent studies have shown that synthetic PrP peptides, which themselves are non infectious, can convert PrP^C to a PrP^{Sc}-like form in vitro [35]. This provides the opportunity to test whether in vitro altered PrP can transmit the disease. However, it is possible that the aggregation or 'seeded crystallization' observed in vitro may reflect late events in the disease process that are peripheral to prion replication.

Evidence from Tg mice suggests the involvement of an additional host component in prion replication [36]. In contrast to the abolition of the species barrier to hamster prions by expression of SHaPrP transgenes, human prions replicate poorly in Tg mice expressing human PrP. Chimeric *Prnp* transgenes, consisting of a human core flanked by mouse sequences, readily support human prion replication, suggesting the involvement of an additional component (protein X) in prion replication. Interestingly, the involvement of chaperone proteins has been demonstrated in the propagation of the yeast factor *[PSI+]* [9]. The participation of a chaperone or the existence of a PrP intermediate between PrP^C and PrP^{Sc} would not alter the basic conclusion that strain properties reflect interactions between PrP isoforms in the inoculum with PrP isoforms in the host. Definitive determination of the biochemical basis for scrapie strain behavior will probably require understanding the nature of the interaction between the cellular and scrapie isoforms of PrP, the physical differences between the two isoforms, and the participation of chaperones or other molecules in protein folding and unfolding.

Acknowledgments

This review is based on work with SB Prusiner, SJ DeArmond, D Westaway, SL Yang, M Scott, M Torchia, D Groth, G Telling, and many other members of the University of California, San Francisco Scrapie Group.

References

1 Pattison IH, Millson GC. Scrapie produced experimentally in goats with special reference to the clinical syndrome. *J Comp Path* 1961;71:101-8

2 Zlotnik I. Slow, latent, and temperate virus infections: 'Observations on the experimental transmission of scrapie of various origins to laboratory animals'. Washington, DC: NINDB Monograph No. 2, 1965

3 Bruce ME, Dickinson AG. Biological evidence that the scrapie agent has an independent genome. *J Gen Virol* 1987;68:79-89

4 Prusiner SB. Molecular biology of prion diseases. *Science* 1991;252:1515-22

5 Pan K-M, Baldwin M, Nguyen J et al. Conversion of alpha-helices into beta-sheets features in the formation of the scrapie prion proteins. *Proc Natl Acad Sci USA* 1993;90:10962-6

6 Safar J, Roller PP, Gajdusek DC, Gibbs CJ Jr. Conformation transitions, dissociation, and unfolding of scrapie amyloid (prion) protein. *J Biol Chem* 1993;269:20276-84

7 Kellings K, Meyer N, Mirenda C, Prusiner SB, Riesner D. Further analysis of nucleic acids in purified scrapie prion preparations by improved return refocussing gel electrophoresis (RRGE). *J Gen Virol* 1992;73:1025-9

8 Wickner RB. [URE3] as an altered URE2 protein: evidence for a prion analog in *Saccharomyces cerevisiae*. *Science* 1994;264:566-9

9 Chernoff YO, Lindquist SL, Ono B-I, Inge-Vechtomov SG, Liebman SW. Role of the chaperone protein Hsp104 in propagation of the yeast prion-like factor [psi⁺]. *Science* 1995;268:880-4

10 Bolton DC, McKinley MP, Prusiner SB. Identification of a protein that purifies with the scrapie prion. *Science* 1982;218:1309-11

11 Oesch B, Westaway D, Wächli M et al. A cellular gene encodes scrapie PrP 27-30 protein. *Cell* 1985;40: 735-46

12 Sparkes RS, Simon M, Cohn VH et al. Assignment of the human and mouse prion protein genes to homologous chromosomes. *Proc Natl Acad Sci USA* 1986;83:7358-62

13 Dickinson AG, Fraser H. Scrapie pathogenesis in inbred mice: an assessment of host control and response involving many strains of agent. In: Katz M, Meuler V, eds. *Slow Virus Infections of the CNS*. New York: Springer-Verlag, 1977;3-14

14 Dickinson AG, MacKay JMK. Genetical control of the incubation period in mice of the neurological disease, scrapie. *Heredity* 1964;19:279-88

15 Dickinson AG, Meikle VMH, Fraser HG. Identification of a gene which controls the incubation period of some strains of scrapie agent in mice. *J Comp Pathol* 1968;78:293-9

16 Carlson GA, Kingsbury DT, Goodman PA et al. Linkage of prion protein and scrapie incubation time genes. *Cell* 1986;46:503-11

17 Westaway D, Goodman PA, Mirenda CA, McKinley MP, Carlson GA, Prusiner SB. Distinct prion proteins in short and long scrapie incubation period mice. *Cell* 1987;51:651-62

18 Dickinson AG, Meikle VMH. Host-genotype and agent effects in scrapie incubation: change in allelic interaction with different strains of agent. *Mol Gen Genet* 1971;112:73-9

19 Goldman W, Hunter N, Foster JD, Salbaum JM, Beyreuther K, Hope J. Two alleles of a neural protein gene linked to scrapie in sheep. *Proc Natl Acad Sci USA* 1990;87:2476-80

20 Westaway D, Zuliani V, Cooper CM et al. Homozygosity for prion protein alleles encoding glutamine-171 renders sheep susceptible to natural scrapie. *Genes Dev* 1994;8:959-69

21 Westaway D, Carlson GA, Prusiner SB. On safari with PrP: prion diseases of animals. *Trends Microbiol* 1995; 3:141-7

22 Carlson GA, Westway D, DeArmond SJ, Peterson-Torchia M, Prusiner SB. Primary structure of prion protein may modify scrapie isolate properties. *Proc Natl Acad Sci USA* 1989;86:7475-9

23 Pattison IH. Experiments with scrapie with special reference to the nature of the agent and the pathology of the disease. Washington DC: NINDB Monograph No 2. 1965;249-57

24 Scott M, Foster D, Mirenda C et al. Transgenic mice expressing hamster prion protein produce species-specific scrapie infectivity and amyloid plaques. *Cell* 1989;59:847-57

25 Prusiner SB, Scott M, Foster D et al. Transgenetic studies implicate interactions between homologous PrP isoforms in scrapie prion replication. *Cell* 1990;63:673-86

26 Carlson GA, Goodman PA, Lovett M et al. Genetics and polymorphism of the mouse prion gene complex: control of scrapie incubation time. *Mol Cell Biol* 1988;8:5528-40

27 Carlson GA, Ebeling C, Torchia M, Westaway D, Prusiner SB. Delimiting the location of the scrapie prion incubation time gene on chromosome 2 of the mouse. *Genetics* 1993;133:979-88

28 Westaway D, Mirenda CA, Foster D et al. Paradoxical shortening of scrapie incubation times by expression of prion protein transgenes derived from long incubation time mice. *Neuron* 1991;7:59-68

29 Carlson GA, Ebeling C, Yang SL et al. Prion isolate specified allotypic interactions between the cellular and scrapie proin proteins in congenic and transgenic mice. *Proc Natl Acad Sci USA* 1994;91:5690-4

30 Carlson GA, DeArmond SJ, Torchia M, Westaway D, Prusiner SB. Genetics of prion diseases and prion diversity in mice. *Philos Trans R Soc Lond B* 1994;343:363-9

31 Dickinson AG, Outram GW. The scrapie replication-site hypothesis and its implications for pathogenesis. In: Prusiner SB, Hadlow WJ, eds. *Slow Transmissible Diseases of the Nervous System*, vol 2. New York: Academic Press, 1979; 13-31

32 Palmer MS, Dryden AJ, Hughes JT, Collinge J. Homozygous prion protein genotype predisposes to sporadic Creutzfeldt-Jakob disease. *Nature* 1991;352:340-2

33 Bessen RA, Marsh RF. Biochemical and physical properties of the prion protein from two strains of the transmissible mink encephalopathy agent. *J Virol* 1992;66:2096-101

34 Bessen RA, Kocisko DA, Raymond GJ, Nandan S, Lansbury PT, Caughey B. Non-genetic propagation of strain-specific properties of scrapie prion protein. *Nature* 1995;375:698-700

35 Kaneko K, Peretz D, Pan KM et al. Prion protein (PrP) synthetic peptides induce cellular PrP to acquire properties of the scrapie isoform. *Proc Natl Acad Sci USA* 1995;92:11160-4

36 Telling GC, Scott M, Hsiao KK et al. Transmission of Creutzfeldt-Jakob disease from humans to transgenic mice expressing chimeric human-mouse prion protein. *Proc Natl Acad Sci USA* 1995;92:9936-40

Transmissible Subacute Spongiform Encephalopathies:
Prion Diseases
L Court, B Dodet, eds
© 1996, Elsevier, Paris

Transmission and strain typing studies of scrapie and bovine spongiform encephalopathy

Moira Bruce, Aileen Chree, Irene McConnell, Karen Brown, Hugh Fraser

Institute for Animal Health, BBSRC and MRC, Neuropathogenesis Unit, Ogston Building, West Mains Road, Edinburgh EH9 3JF, Scotland, UK

Bovine spongiform encephalopathy (BSE) has been transmitted to mice directly from cattle and indirectly from experimentally-infected sheep, goats and pigs. Closely similar patterns of incubation periods and neuropathology were seen in panels of mouse strains infected from each of these species. Transmission to mice has also confirmed that domestic cats and two species of antelopes have been accidentally infected with the BSE agent. However, transmission from sheep with natural scrapie has given variable results, with no individual sources resembling BSE. Therefore, these methods can be used to investigate any suspicion that BSE has spread to other species. For example, transmission has been set up from two cases of Creutzfeldt-Jakob disease (CJD) in dairy farmers who had BSE in their herds, alongside transmission from CJD cases with no occupational link with BSE. No results are available yet from these experiments.

A striking feature of both primary transmission of BSE to mice and some natural scrapie transmissions is the apparent equivalence of intracerebral and peripheral routes of infection. The incubation periods seen following intraperitoneal challenge are as short as or even shorter than incubation periods following intracerebral injection. In contrast, the intraperitoneal incubation period with mouse-passaged scrapie or BSE is longer than the intracerebral incubation period, usually by about 50%. These observations suggest that a major component of the species barrier is the inability of the agent associated with foreign tissue components to establish infection directly in the brain. This implies an obligatory processing or replication phase in peripheral tissues before infection can invade the central nervous system. This interpretation is supported by a recent study in which severe combined immunodeficient (SCID) mice were found to be relatively resistant to BSE from cow brain, even when challenged by intracerebral injection.

References

Bruce ME. Scrapie strain variation and mutation. *Br Med Bull* 1993;49:822-38
Bruce ME, Chree A, McConnell I, Foster J, Pearson G, Fraser H. Transmission of bovine spongiform encephalopathy and scrapie to mice: strain variation and the species barrier. *Philos Trans R Soc Lond Series B* 1994;343:405-11

YEAST MODELS

Transmissible Subacute Spongiform Encephalopathies:
Prion Diseases
L Court, B Dodet, eds
© 1996, Elsevier, Paris

[*URE3*] as a model for prions

Elisabeth Guillemet, Eric Fernandez-Bellot, François Lacroute,
Christophe Cullin*

Laboratoire Propre Associé à l'Université Pierre-et-Marie-Curie, Centre de Génétique Moléculaire du CNRS, 91190 Gif-sur-Yvette, France

Summary – During the 1960s, when F Lacroute's laboratory was involved in the study of the pyrimidine pathway, several mutants were isolated that allowed the uptake of ureidosuccinate. In particular, two kinds of mutants, named *urei2* and [*UREI3*], were studied. *urei2* mutants present all the characteristics of Mendelian mutations, while [*UREI3*] exhibit some radically different properties. The behavior of [*UREI3*] can be summarized as follows: i) it is dominant; ii) it displays non Mendelian segregation; iii) it can be acquired in a wild-type cell by cytoduction with a [*UREI3*] cell; iv) it can exist only in a *urei2* wild-type background; and v) it can be reversibly cured in the presence of guanidium chloride. All these properties were previously reported, but no convincing arguments could explain them. During the last 20 years, the *urei2* gene has been cloned and characterized. Recently, Wickner repeated the experiments using both genetics and molecular biology tools. In light of the results, he postulated that [*UREI3*] could be a yeast prion. [*UREI3*] would be a modified form of Ure2p. In this form, the protein would not be active, thereby explaining the phenotype which is strictly equivalent to *urei2⁻*. The [*UREI3*] modified protein would also have the ability to convert newly synthesized Ure2p into [*UREI3*]. This property fits well with the observed behavior of [*UREI3*] previously summarized. Here we present in more detail some of the experiments which argue for such a hypothesis and new data concerning the protein Ure2p.

Introduction

The yeast *Saccharomyces cerevisiae* is a eukaryotic cell which can grow either as a haploid or a diploid cell. This aspect of its life cycle is a wonderful gift for the geneticist, and during the past 20 years, it has been combined with the techniques of molecular biology, producing a vast and powerful palette of tools for the study of yeast genetics. In particular, it is now possible to study any given gene by deletion or replacement. In this way, several alleles of a target gene created in vitro can be reintroduced into the yeast genome and studied in vivo. If the mutated gene confers some phenotype, the relationship between this target gene and the others can be studied not only by the classic genetic approach, but also by using the multicopy plasmid library to look for suppressors. If the mutated gene does not confer any suitable properties to the yeast cell, it is still possible to look for an interacting gene by searching for co-lethal mutants [1] or by the two-hybrid system [2]. As the yeast genome is now completely sequenced, the identification of a gene requires the knowledge of only a small part of its

---

*Correspondence and reprints

Key words: yeast / ure2 / prion

sequence. Starting from a gene, the analysis of the encoded protein is greatly facilitated. A very large collection of plasmids permits the overexpression of a target gene and an easy purification by adding different tags [3]. Finally, the yeast *S cerevisiae* has proven to be a model cell for the study of all kinds of functions once considered strictly limited to 'higher' eukaryotes. In this paper, we will present the remarkable properties of [*URE3*] (the brackets indicate phenotype), a yeast prion (although yeast is not likely to have spongiform encephalopathy!).

Discovery and general properties of [*URE3*]

During the 1960s, Lacroute and his collaborators were involved in the study of the yeast pyrimidine biosynthetic pathway. Several mutants affecting the production of uridine-monophosphate (UMP) were isolated since they were auxotrophic for uracil [4]. Here, we focus our attention on the *ura2* mutants. The *ura2* gene codes for two distinct enyzmatic activities that permit the production of ureidosuccinate (USA). When this mutant is incubated in rich medium lacking uracil but supplemented with USA, the yeast cells do not grow because USA is not imported into the cell in these growth conditions. However, when a poor source of nitrogen is used instead of NH_3 (which is found in rich medium), the same mutant is able to grow [5]. These unexpected results demonstrate the ability of *S cerevisiae* to import USA by a mechanism regulated by the nitrogen source. A comparison of the chemical structure of USA and allantoate (which is a poor nitrogen source) shows that the molecules share a common structure (fig 1). This homology explains the import of USA by the allantoate permease and thus the relationship between the complementation of a *ura2*⁻ mutant by USA and nitrogen catabolic repression (NCR).

Yeast NCR

S cerevisiae is able to discriminate between rich and poor nitrogen sources. In the presence of both, the rich one is preferentially used, and the enzymes required for the import and degradation of the poor nitrogen source are repressed. In the presence of a poor nitrogen source, allantoate permease will be expressed at a sufficient level for importing allantoate. This mechanism is well correlated with the ability for a *ura2*⁻ strain to grow in the presence of USA. When a rich nitrogen source such as NH_3 is used, the permease is repressed, USA cannot be taken up, and the mutant cell will not grow. When the nitrogen source used is a poor one, the permease is derepressed and the USA can be taken up by yeast cells, thus permitting

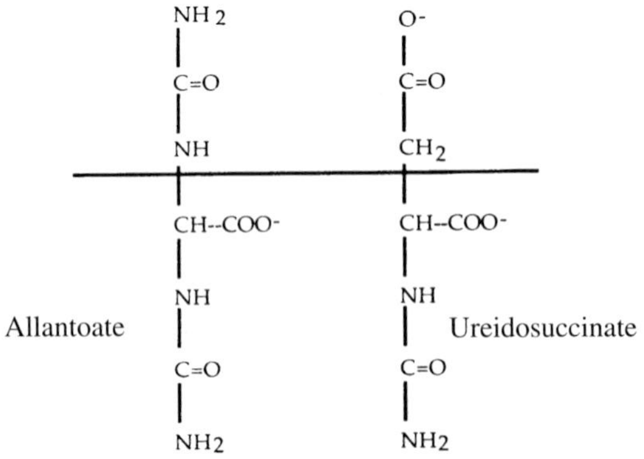

Fig 1. Chemical structure of ureidosuccinate and allantoate. Both share a common structure previously recognized by the allantoin pump.

the production of UMP and finally the growth of the *ura2* mutant. The NCR requires Gln3p as a transcriptional activator [6]. This protein is under the control of Ure2p [7]. In the presence of a rich nitrogen source, the high level of glutamine permits inhibition of Gln3p by Ure2p. This inactivation is not observed when a poor source of nitrogen is used. Gln3p is then able to activate the genes that contain the specific upstream activating sequence (UAS) (fig 2).

Selection of ure mutants

Lacroute and other members of his group looked for ura2 mutants able to grow in the presence of USA and NH₃. Such mutants were expected to be defective in the control of NCR.They named these mutants *ureX* since they were able to import USA. The *ure2* mutants appeared to be classic mutants. The phenotype segregates 2:2 as expected for a Mendelian element. Our understanding of the NCR highlights the role of *ure2*. A yeast cell mutated at this locus may be unable to produce a fully functional Ure2p. In this case, Gln3p is always active without any requirement for the nitrogen source.

Genetic properties of the [URE3] strain

More curiously, [URE3] mutants did not behave as normal Mendelian mutants [8]. When such mutants were crossed with a wild-type cell, the corresponding diploid appeared to be [URE3] (that is, the uptake of USA was still efficient in the presence of NH₃). The yeast diploid can live for a very long time in the diploid state, but under nitrogen starvation, a meiotic step is observed, leading to the formation of an ascus containing four spores. These four spores (the products of meiosis) can then be separated and analyzed. When a diploid [URE3] resulting

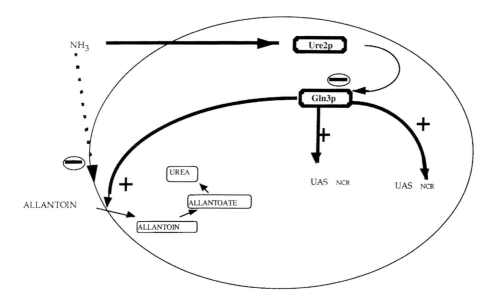

Fig 2. Schematic representation of nitrogen NCR in *S cerevisiae*. In the presence of a rich nitrogen source such NH₃ or glutamine, the protein Ure2p prevents the activation of target genes by the transcriptional activator Gln3p. This inhibition is lost when the cells are grown in the presence of a poor nitrogen source such as allantoin.

from a cross between a [*URE3*] cell and a wild-type strain was allowed to sporulate, very often all four spores appeared to be [*URE3*]. This behavior can be observed when the phenotype is related to cytoplasmic elements such as mitochondria. To test the possibility that [*URE3*] might be under the control of a cytoplasmic element, Aigle and Lacroute [9] attempted to transfer the cytoplasm of a [*URE3*] cell into a wild-type cell. The cytoduction technique (which permits the mixing of cytoplasms without nuclear fusion) confirmed that [*URE3*] was not related to a chromosomal character since a wild-type cell could become [*URE3*] without any modification of its nuclear genotype. Aigle and Lacroute [9] then showed that [*URE3*] was not related to mitochondria or [*PSI*] (which now appears to be another yeast prion [10]). More recently, Wickner reported that [*URE3*] was unrelated to the other well-known cytoplasmic elements of *S cerevisiae* [11]. Interestingly, [*URE3*] cannot be maintained in a strain carrying the *ure2-1* allele. This result was confirmed by Wickner in a strain bearing a complete deletion of the *ure2* gene. All these properties are summarized in table I.

At that time, genetic techniques were not developed enough that one could learn more about [*URE3*]. In 1991, the *ure2* gene was cloned and the corresponding sequence was described by Coschigano and Magasanick [7].

[*URE3*] and the prions

More recently, Wickner [11] repeated all the experiments carried out by Aigle and Lacroute. He used all the latest molecular biology tools for this purpose and unambiguously demonstrated the same properties for [*URE3*]. He also showed that the frequency at which [*URE3*] appeared depended on the level of expression of *ure2* since the frequency of [*URE3*] increases substantially in a strain transformed by a multicopy plasmid bearing *ure2*. Based on these genetic observations, Wickner argued that [*URE3*] was an altered form of Ure2p. According to this prion model, the preexisting Ure2p could be converted into the [*URE3*] state [11]. This modification would abolish the normal function of Ure2p and thus lead to a phenotype identical to the one observed in the *ure2⁻* mutant strain. In the [*URE3*] state, the protein would acquire a new feature that is a characteristic of prions: it could convert Ure2p into the altered form (fig 3). This property explains perfectly the results of cytoduction experiments.

In an original study on *ure2* recently published by Masison and Wickner [12], it was shown that Ure2p behaves differently in a wild-type and in a [*URE3*] yeast strain. In a crude extract, the sensitivity of Ure2p to proteinase K (PK) was shown to be higher in a wild-type cell than in a [*URE3*] one. Western blot experiments demonstrated that the species which cross-react with antibodies raised against Ure2p remained present in PK-treated extracts of [*URE3*] cells, but were absent from similarly treated wild-type extracts [12].

This difference could be due to a modification of the structure of Ure2p in [*URE3*] cells. As the protein presents the same electrophoretic mobility in both wild-type and [*URE3*] strains, a modification of the primary structure of Ure2p remains very unlikely.

Table I. Genetic properties of [*URE3*].

Dominant
Invasive (transmission by cytoduction)
Incompatible with the *ure2-1* allele and with a deletion of *ure2*
Reversible curing by growth in the presence of 5 mM guanidine

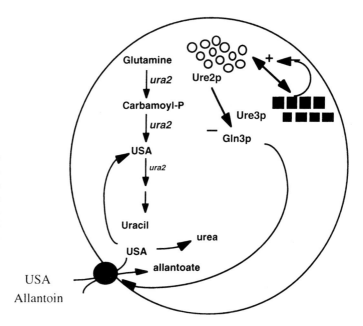

Fig 3. The prion hypothesis for [*URE3*]. The phenotype corresponds to a loss of function of Ure2p but the genetic experiments argue for a model in which no mutation is required. The protein would exist in a functional form, but also in a non functional form. This altered form would have acquired the remarkable property of being able to convert the wild-type protein into the abnormal form similar to the mechanism described for prions.

Ure2p contains two domains

In the same study, Masison and Wickner [12] created several deletions of *ure2* that demonstrated certain structure–function relationships in the protein. They determined the ability of each modified gene to complement a Δ*ure2* strain (a strain carrying a complete deletion of a *ure2* gene). The results clearly showed that the large C-terminal part of the protein was sufficient to retain the function of the *ure2* gene in the NCR. Moreover, this domain appears to be not only sufficient, but also strictly required for the function of *ure2*. Even small deletions in the C-terminal part completely abolished *ure2* activity. The same alleles were then tested for their ability to increase the frequency of appearance of [*URE3*] in a wild-type strain. In this case, the non functional alleles of *ure2* that retain the N-terminal part of *ure2* were still able to increase the frequency of [*URE3*]. When the N-terminal part of the molecule was removed, this induction was completely lost. Surprisingly, the frequency of induction was actually higher for the altered allele than that measured when the fully functional gene was used. This result has led to the idea that the catalytic domain could be considered an inhibitor of the [*URE3*] inducing domain for the transformation of a wild-type cell into a [*URE3*] cell. This domain (called the prion domain) corresponds to the N-terminal sequence of Ure2p and has no homology with the other proteins present in the data banks. The sole remarkable feature is the high percentage of clustered glutamine residues (fig 4).

If we consider the proposed prion mechanism and these results, Ure2p is assumed to interact with itself. Our group has recently tested this possibility by using the two-hybrid system.

The two-hybrid system

The two-hybrid technique initially developed by Fields and Song [2] is based on the modular nature of the GAL4 (galactosidase 4) *trans*-activator. This protein contains two domains: a

MMNNNGNQVSNLSNALRQVNIGNRNSNTTTDQSNINFEFSTGVNNNNNNNSSSNNNNVQNNNSGRNGSQN

NDNENNIKNTLEQHRQQQQAFSDMSHVEYSRITKFFQEQPLEGYTLFSHRSAPNGFKVAIVLSELGFHYNTIFLDFNLGEHRA

PEFVSVNPNARVPALIDHGMDNLSIWESGAILLHLVNKYYKETGNPLLWSDDLADQSQINAWLFFQTSGHAPMIGQALHFRYFH

SQKIASAVERYTDEVRRVYGVVEMALAERREALVMELDTENAAAYSAGTTPMSQSRFFDYPVWLVGDKLTIADLAFVPWNNVV

DRIGINIKIEFPEVYKWTKHMMRRPAVIKALRGEZ

Fig 4. Primary structure of Ure2p. The prion domain is highlighted in bold type.

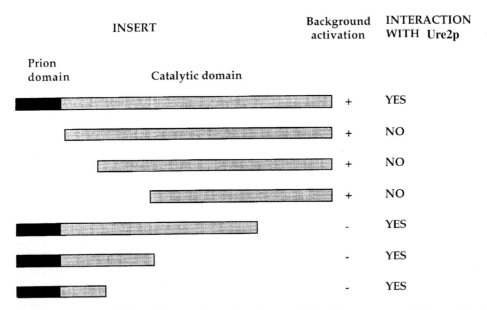

Fig 5. Each insert was tested for its ability to activate alone the transcription of a reporter gene (background activation). In a separate experiment, we looked for this activation due to the interaction with Ure2p.

DNA-binding domain and a transcription domain. Each domain can be expressed as an independent protein, but in this case, they cannot reassociate and the target gene recognized by the binding domain will not be transcribed. In order to reconstitute a fully functional activator, the experimenter creates a vector in which the gene product of interest is fused to the GAL4-binding domain. This gene will be considered as the bait. In a different vector, a second hybrid protein is constructed, containing the activating domain and the product of a cloned gene (which will be called the prey). If both hybrids are expressed in the same cell, and if the bait interacts with the prey, then a functional activator will be reassembled as a dimer. Consequently, a reporter gene such as *lacZ* containing the specific GAL4 UAS will be transcribed, thus announcing the interaction between the bait and the prey.

In this system, when *ure2* is used as a bait without a prey, significant β-galactosidase activity is detected which indicates that *ure2* alone is sufficient to act as an activating domain. When this test was carried out with a strain bearing *ure2* as both bait and prey, the level of β-galactosidase activity increased sixfold (data not shown). This increase probably results from the interaction of Ure2p with itself, but the background expression due to the Ure2p bait alone makes the interpretation of the results somewhat complicated. Since we wanted to use this construction to screen a library of yeast prey fused with the activating domain, we needed to reduce the background activity level. We tested several partial deletions of *ure2* as baits. Our results (fig 5) show that the background activation requires the functional catalytic domain of *ure2*. For deletions that remove part of it, this activity was not observed, but a clear β-galactosidase activity was detected when these baits were tested with *ure2* as prey. This test confirms the possibility that *ure2* interacts with itself. Together, these experiments give us interesting tools to specify the interaction of Ure2p with itself. We now want to extend our results by screening a two-hybrid library.

Conclusion

The behavior of *[URE3]* is consistent with the prion paradigm and can be considered an attractive model for the replication of such molecules. Yeast cells offer some remarkable tools that will permit the selection of genes implicated in the control of this mechanism. Together with biochemical approaches, we hope to get a much clearer picture of this system. Whether it will be closely related to the conversion of the prion protein still remains to be established. Yeast as an expressing system for mammalian cDNA provides a capital tool for this purpose.

Acknowledgments

We are very grateful to R Karess for looking over the English.

References

1 Basson ME, Moore RL, O'Rear J, Rine J. Identifying mutations in duplicated functions in *Saccharomyces cerevisiae*: recessive mutations in HMG-CoA reductase genes. *Genetics* 1987;117:645-55

2 Fields S, Song O. A novel genetic system to detect protein–protein interactions. *Nature* 1989;340:245-6

3 Rose AB, Broach JR. Propagation and expression of cloned genes in yeast: 2-microns circle-based vectors. *Meth Enzymol* 1990;234-79

4 Lacroute F. Regulation of pyrimidine biosynthesis in *Saccharomyces cerevisiae*. *J Bacteriol* 1968;95:824-32

5 Drillien R, Lacroute F. Ureidosuccinic acid uptake in yeast and some aspects of its regulation. *J Bacteriol* 1972;109:203-8

6 Courchesne WE, Magasanik B. Regulation of nitrogen assimilation in *Saccharomyces cerevisiae*: roles of the URE2 and GLN3 genes. *J Bacteriol* 1988;170:708-13

7 Coschigano PW, Magasanik B. The *URE2* gene product of *Saccharomyces cerevisiae* plays an important role in the cellular response to the nitrogen source and has homology to glutathione *S*-transferases. *Mol Cell Biol* 1991;11:822-32

8 Lacroute F. Non-Mendelian mutation allowing ureidosuccinic acid uptake in yeast. *J Bacteriol* 1971;106:519-22

9 Aigle M, Lacroute F. Genetical aspects of [URE3], a non-mitochondrial, cytoplasmically inherited mutation in yeast. *Mol Gen Genet* 1975;136:327-35

10 Ter-Avanesyan AM, Dagkesamanskaya AR, Kushnirov VV, Smirnov VN. The *SUP35* omnipotent suppressor gene is involved in the maintenance of the non-Mendelian determinant [*psi+*] in the yeast *Saccharomyces cerevisiae*. *Genetics* 1994;137:671-6

11 Wickner RB. [URE3] as an altered URE2 protein: evidence for a prion analog in *Saccharomyces cerevisiae* [see comments]. *Science* 1994;264:566-9

12 Masison D, Wickner RB. Prion-inducing domain of yeast Ure2p and protease resistance of Ure2p in prion-containing cells. *Science* 1995;270:93-5

Transmissible Subacute Spongiform Encephalopathies:
Prion Diseases
L Court, B Dodet, eds
© 1996, Elsevier, Paris

The yeast prion protein Sup35p: a paradigm for studying the maintenance and infectivity of mammalian prions?

Mick F Tuite[1], Vitaly V Kushnirov[2], Ian Stansfield[1], Simon Eaglestone[1], Michael D Ter-Avanesyan[2]

[1]*Research School of Biosciences, University of Kent, Canterbury, Kent CT2 7NJ, UK;* [2]*Institute of Experimental Cardiology, 3rd Cherepkovskaya Str 15A, 121552 Moscow, Russia*

Summary – Two extrachromosomal genetic elements in the yeast *Saccharomyces cerevisiae* (namely [*PSI*] and [*URE3*]) represent the first examples of a prion-like phenomenon in this lower eukaryote. The [*PSI*] determinant is the product of the *SUP35* gene (Sup35p), a protein that plays a critical role in translation termination as an essential subunit (eRF-3) of the release factor. Here we review the genetic behavior of the [*PSI*] determinant and suggest that further study of this phenomenon in yeast will shed light on the mechanism of establishment, maintenance and infectivity of the mammalian scrapie prion protein, PrP^Sc.

Introduction

The transmissible spongiform encephalopathies of mammalian species [1, 2; recent reviews] may not represent the only examples of protein-based 'infectivity'; two examples of extrachromosomally inherited determinants of the yeast *Saccharomyces cerevisiae*, [*URE3*] and [*PSI*], have also been ascribed to an underlying prion-like 'infectious' protein [3–5]. We review here the properties of one of these determinants *([PSI])* and argue that: the protein underlying this determinant, namely Sup35p, shares common structural features with mammalian prion proteins (PrP); that the mechanism(s) by which the yeast PrP autocatalytically alters its conformation and hence function might be analogous to the mechanism proposed for mammalian PrP; and that the N-termini of the two yeast PrP may represent a domain critical for the switch from the PrP to the disease-specific PrP^Sc form.

Unusual genetic behavior of the yeast non-Mendelian [*PSI*] determinant

The yeast non Mendelian genetic determinant [*PSI*] is defined by the structural status of the protein encoded by the nuclear *SUP35* gene, ie, Sup35p [3–5]. The phenotype associated with the [*PSI*] determinant is very different from that for the [*URE3*] determinant (which defines an ability to utilize ureidosuccinate as a nitrogen source); in a [*PSI*+] strain there is an increase, in comparison with a [*psi*⁻] strain, in the efficiency with which weak nonsense suppressor tRNAs are able to translate their cognate termination codons in vivo [6, 7]. The [*PSI*] determinant is propagated in a similar way to the [*URE3*] determinant; it segregates to approaching

Key words: *yeast* **(Saccharomyces cerevisiae)** */ Sup35p / prion /* **[PSI]** *determinant*

100% of the meiotic spores from a [*PSI⁺*] × [*psi⁻*] genetic cross, is transferred by cytoduction and may be reversibly cured by growth in the presence of a number of non mutagenic agents including the protein denaturing agent guanidine hydrochloride [7]. Thus, the [*PSI*] and [*URE3*] determinants share a number of common genetic properties, via their mode of inheritance by meiotic transmission and mitotic transmission.

A possible relationship between the [*PSI*] determinant and the product of the *SUP35* gene was first suggested by the observation that nuclear allosuppressor mutants (*sal3*), having the same phenotype as [*PSI⁺*] strains, map to the *SUP35* gene [8], although mutations within several other chromosomal loci, namely *SAL1, SAL2, SAL4(SUP45)* and *SAL5* give identical phenotypes to *sal3* mutants [9]. Evidence for a direct link between [*PSI*] and Sup35p came from the finding that overexpression of the *SUP35* gene in a [*psi⁻*] strain resulted in an elevated frequency of conversion of the [*psi⁻*] strain to [*PSI⁺*] [10]. Subsequently, it was discovered that a point mutation in the *SUP35* gene caused loss of the [*PSI⁺*] phenotype [11]. The *SUP35* gene was originally identified by its omnipotent suppressor alleles, ie, translational suppression of all three classes of nonsense mutation [12, 13]. The way in which mutations in the *SUP35* gene are able to influence translational suppression has recently become apparent with the discovery that Sup35p is eRF-3, an integral component of the yeast translation termination release factor [14].

Wickner [3] was the first to suggest that Sup35p may be acting like a prion in defining the [*PSI*] determinant. Perhaps the most significant observation in relation to its putative prion-like mode of inheritance is that [*PSI⁺*] cells may be induced de novo by overexpression of the *SUP35* gene in a [*psi⁻*] strain [10]. Similarly, overexpression of the *URE2* gene results in a significant elevation in the spontaneous frequency with which [*URE3*] cells arise [3]. Overexpression of normal mammalian PrP^C proteins may trigger formation of the PrP^Sc (scrapie) state [15]. The parallel between the yeast and higher eukaryotic proteins in this respect is therefore striking.

Domain structure of Sup35p and its relationship to the [*PSI*] determinant

Three domains can be easily distinguished in the 685 amino acid Sup35p protein by their amino acid sequence; they may also represent distinct functional units (fig 1) [16]. The C-terminal C domain (aa 254–685) is essential for viability and shows significant amino acid identity to the translation elongation factor EF-1α. In contrast, the N-terminal part of Sup35p is unique in sequence, non essential for viability and may be further divided into two domains, the extreme N-terminal N domain (1–114) and the middle M domain (115–253), both of which are distinct in their amino acid composition and predicted secondary structure [16].

Overexpression of the N domain alone or the intact (N + M + C) Sup35p protein results in the suppression of nonsense mutations in vivo [16], indicating a defect in translation termination. On the other hand, deletion of the N domain or both the N + M domains gives rise to a dominant antisuppressor phenotype, which is indicative of an increase in the efficiency of translation termination leading to a reduction in the efficiency of tRNA-mediated nonsense suppression. Several of the *SUP35* deletion mutants alleles studied by Ter-Avanesyan et al [16] have one other relevant phenotype, namely they cause [*PSI⁺*] strains to become [*psi⁻*]. Engineering the [*psi⁻*] strains carrying various *sup35* deletion alleles to express either full-length Sup35p or the N domain of *Sup35*p does not restore the 'allosuppressor', ie, [*PSI⁺*] phenotype, yet these strains can still transmit the [*PSI⁺*] determinant to [*psi⁻*] strains by mitotic or meiotic transmission [17]. Thus, Sup35p is essential for the manifestation of the [*PSI⁺*] phe-

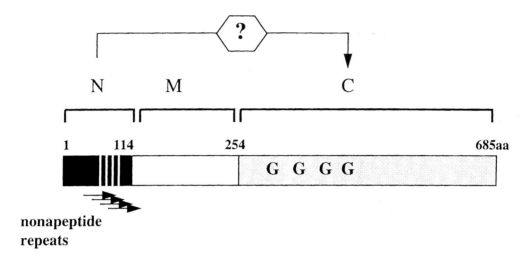

Fig 1. The domain organization of Sup35p, the product of the *SUP35* gene which defines the [*PSI*] determinant. The location of the nonapeptide repeat sequences at the N-terminus (arrows) and the guanosine triphosphate-binding sites (G) in the C-terminus are indicated as are the location of the N, M and C domains as defined by Ter-Avanesyan et al [16]. aa: amino acids.

notype, while the N domain of Sup35p is necessary and sufficient for the maintenance of the [*PSI*⁺] state and for its mitotic and meiotic transmission.

A dominant nuclear mutation which prevents the maintenance of the [*PSI*⁺] state (Psi No More, *PNM2*) is an allele of the *SUP35* gene [11]. Sequence analysis of the *PNM2* allele identified a single base-pair change in the N domain of Sup35p in the second of the five tandem peptide repeats present in this region (fig 1, 2) giving rise to a Gly^{58} to Asp^{58} amino acid substitution [11]. This further demonstrates the importance of the N domain of Sup35p in the maintenance of the [*PSI*⁺] determinant. No mutation equivalent to *PNM2* has been described for the [*URE3*] determinant. The identity of the protein encoded by a second *PNM* gene *(PNM1)* [18] remains to be defined.

Yeast prion-like proteins show structural similarity to mammalian PrP

The most striking amino acid similarity between the yeast and mammalian PrP is found at the N-termini of these proteins. Yeast Sup35p and mammalian and avian PrP contain short tandemly arranged nona- or octapeptide repeats (fig 2). In the distantly related yeast species *Pichia pinus*, the Sup35p protein also has a related peptide repeat at its N-terminus [19]. The amino acid sequences of these repeats are not totally conserved between species, yet they do show clear similarities which suggest that they may define important structural features shared by PrP and the *S cerevisiae* and *P pinus* Sup35p.

A comparison of the predicted structural features in the N-terminal sequences of Sup35p of *S cerevisiae* and *P pinus* indicate that they have a similar amino acid composition, being rich in Q, N, P, G, Y and S residues, and their predicted secondary structure consists almost exclusively of β-turns (data not shown). These features are also typical for the PrP repeat region from all species for which sequence information is available. The potential importance of the nonapeptide repeats in Sup35p is underscored by the finding that the *PNM2* mutation in the second of these repeats has important dominant effects on [*PSI*⁺] maintenance and inheritance [11].

Sup35p repeat peptides; *S cerevisiae*

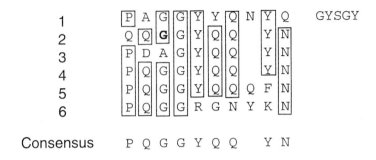

1	P	A	G	G	Y	Y	Q	N	Y	Q	GYSGY
2	Q	Q	**G**	G	Y	Q	Q		Y	N	
3	P	D	A	G	Y	Q	Q		Y	N	
4	P	Q	G	G	Y	Q	Q		Y	N	
5	P	Q	G	G	Y	Q	Q	Q	F	N	
6	P	Q	G	G	R	G	N	Y	K	N	

Consensus P Q G G Y Q Q   Y N

Sup35p repeat peptides; *Pichia pinus*

N Q G G Y $[X]_4$

N R G G Y $[X]_2$

N R G G Y $[X]_3$

N R G G Y $[X]_{10}$

N Q G G Y $[X]_{10}$

Mammalian/avian PrP octapeptide repeats

human $[P\ H\ G\ G\ G\ W\ G\ Q]_5$

rat $[P\ H\ G\ G\ G\ W\ G\ Q]_5$

mouse $[P\ H\ G\ G\ G\ W\ G\ Q]_5$

bovine $[P\ H\ G\ G\ G\ W\ G\ Q]_5$

chicken $[P\ H\ N\ P\ G\ Y]_6$

Fig 2. The peptide repeats present in the N-terminal N domain of Sup35p from two yeast species, *S cerevisiae* and *P Pinus*. The mammalian and avian PrP peptide repeats are also shown for comparison. The location of the *PNM2* mutation is indicated in bold in the second repeat of the *S cerevisiae* sequences [11].

The role of Sup35p domains in the maintenance and expression of [*PSI*]

The N-terminal N domain of Sup35p appears to define a complete functional unit in the context of [*PSI⁺*] maintenance and inheritance. Deletion of the N domain results in the elimination of the [*PSI⁺*] determinant, while separate expression of the N and C domains in the same cell will result in maintenance of the [*PSI⁺*] determinant [17]. [*PSI⁺*] presumably represents an alternative conformational state of the N domain of Sup35p, which in turn is stably maintained by a self-perpetuating mechanism analogous to that proposed for propagation of mammalian PrP [3, 20, 21].

The C-terminal domain of Sup35p is important for manifestation of the [*PSI*⁺] phenotype and may be considered the functional domain with respect to the cellular role of Sup35p since deletion of this domain is lethal [16]. The recent discovery that Sup35p interacts with a second protein, Sup45p, to mediate translation termination at all three stop codons in *S cerevisiae* [14] infers that Sup35p functions in translation termination. When the C-terminal domain of Sup35p is expressed in the absence of the N-terminal domain, an antisuppressor phenotype is observed. This would suggest that the N-terminal domain may act to regulate the C-terminal domain of Sup35p in *cis* during translation termination. If the N domain is absent, highly efficient termination occurs, yet when it is attached to the C-terminal domain, depending on which conformational state the N domain takes up, termination can be maintained at the correct physiological level (ie, the [*psi*⁻] state) or have a much reduced efficiency (ie, the [*PSI*⁺] state). In other words, the N domain of Sup35p, in a [*PSI*⁺] strain, acts as a repressor in *cis* of the function of the C-terminal domain (fig 2). How the efficiency of termination is regulated in this way or indeed whether or not it has any physiological relevance, is open to question.

This model also explains an apparently contradictory observation, namely that [*PSI*⁺] is genetically dominant to [*psi*⁻], but when cell-free lysates from [*PSI*⁺] and [*psi*⁻] strains are mixed, the [*psi*⁻] 'phenotype' is observed in vitro [22]. This can be interpreted as the [*psi*⁻] form of Sup35p being functionally dominant in vitro since it allows for the formation of active release factor complexes and this outcompetes exogenously supplied nonsense suppressor tRNAs. Under these in vitro conditions presumably the autocatalytic conversion of Sup35p from the [*psi*⁻] form to the [*PSI*⁺] form does not occur and this might suggest that the prion form of Sup35p requires on-going translation of the *SUP35* mRNA, ie, the prion-like switch in conformation occurs cotranslationally. A precedent for this kind of mechanism for proteins other than PrP does exist; the mammalian p53 tumor suppressor protein can exist in a wild-type and a mutant conformation and cotranslation of a mutant p53 protein with a wild-type one drives the wild-type protein into the mutant conformation [23].

Can studying Sup35p and [PSI] tell us anything new about mammalian prions?

There is a wealth of genetic data which supports the hypothesis that Sup35p (and Ure2p) are yeast PrP, yet the case needs to be strengthened by supporting physicobiochemical data. There are two diagnostic physical properties of higher eukaryotic prions: resistance of a C-terminal core of the PrP (PrP 27–30) to degradation by proteinase K, and the aggregation of the prion form of the protein. Recent studies have shown that the Ure2p protein is indeed more proteinase K resistant in [*URE3*] strains than in wild-type strains [24], while Paushkin et al [25] have very recently shown that not only is Sup35p more resistant to proteinase K digestion in [*PSI*⁺] strains than [*psi*⁻] strains, but also that most of the Sup35p protein mainly exists in a sedimentable form in a [*PSI*⁺] strain, while in a [*psi*⁻] strain it is in a predominantly soluble cytoplasmic form.

The identification of another protein required for the establishment and maintenance of a yeast PrP was first described for Sup35p with the molecular chaperone (heat-shock protein) Hsp104 playing a role in maintaining the [*PSI*⁺] state [26]. This raises the question of whether this or another molecular chaperone is required for the PrP^C to PrP^Sc switch in mammalian cells. Telling et al [27] have now shown that there is at least one other protein (designated protein 'X') which is required for the formation of homologous prions in transgenic mice, although the identity of this protein remains to be described. The exploitation of yeast genetics should allow the identification of other genes whose products are required for the establishment and maintenance of the yeast prions, ie, *PNM* genes.

The genetic data described above infer a central role for the N domain of Sup35p in establishing and maintaining the prion form of Sup35p. The importance of the octapeptide repeats in the higher eukaryotic prions is less convincing. For example, certain amino acid substitutions in the C-terminal domain of human PrP increase the probability of onset of Creutzfeldt-Jakob disease (CJD) [21]. Nevertheless, the importance of the octapeptide repeats in the N-terminal domain for the conversion of PrPC to PrPSc is emphasized by the observations that PrP alleles with additional repeats are generally the cause of onset of CJD [28].

The existence of PrP in yeast and mammalian cells with potentially 'switchable' N-terminal domains suggests that the inheritance of distinct functional states of a protein may occur more widely in nature than for just PrP and related proteins. However, many such cases may go unnoticed because the protein-driven 'conformational switch' does not generate such an easily scorable phenotypic change as is provided by the prion-based encephalopathy diseases and the [*PSI*] and [*URE3*] determinants in yeast. In addition, it remains to be seen whether the conformational switches are simply spontaneous 'misfolding' events or whether the establishment of the prion form of these proteins is used as a means of regulating their activity. The discovery of prion-like proteins in yeast now provides a more amenable genetic system in which to further probe some of these issues.

Acknowledgments

This work was supported by an International Fellowship from The Wellcome Trust (VVK), a Fogarty International Centre grant (No SRG 1 RO3 TW00129-01 to MDT-A), an INTAS grant (to MDT-A and MFT) and a project grant from the Wellcome Trust (to MFT).

References

1 Prusiner SB. Biology and genetics of prion diseases. *Annu Rev Microbiol* 1994;48:655-86
2 Weissmann C. Molecular biology of prion diseases. *Trends Cell Biol* 1994;4:10-4
3 Wickner RB. [URE3] as an altered ure2 protein: evidence for a prion analogue in *Saccharomyces cerevisiae*. *Science* 1994;264: 566-9
4 Tuite MF. Psi no more for yeast prions. *Nature* 1994;370:327-8
5 Cox BS. Prion-like factors in yeast. *Current Biol* 1994;4:744-8
6 Cox BS. *Psi*, a cytoplasmic suppressor of super-suppressor in yeast. *Heredity* 1965;20:505-21
7 Cox BS, Tuite MF, McLaughlin CS. The *psi* factor of yeast: a problem of inheritance. *Yeast* 1988;4:159-78
8 Crouzet M, Tuite MF. Genetic control of translational fidelity in yeast: molecular cloning and analysis of the allosuppressor gene *SAL3*. *Mol Gen Genet* 1987;210:581-3
9 Cox BS. Allosuppressors in yeast. *Genet Res* 1977;30:187-205
10 Chernoff YO, Derkach IL, Inge-Vechtomov SG. Multicopy *SUP35* gene induces the *de novo* appearance of *psi*-like factors in the yeast *Saccharomyces cerevisiae*. *Curr Genet* 1993;24:268-70
11 Doel SM, McCready SJ, Nierras CR, Cox BS. The dominant *PNM2* mutation which eliminates the psi factor of *Saccharomyces cerevisiae* is the result of a missense mutation in the *SUP35* gene. *Genetics* 1994;137:659-70
12 Inge-Vechtomov SG, Adrianova VM. Recessive super-suppressors in yeast. *Genetika* 1970;6:103-15
13 Hawthorne DC, Leupold U. Suppressor mutations in yeast. *Curr Top Microbiol Immunol* 1974;64:1-47
14 Stansfield I, Jones KM, Kushnirov VV et al. The products of the *SUP45* (eRF1) and the *SUP35* (eRF3) genes interact to mediate translation termination in *Saccharomyces cerevisiae*. *EMBO J* 1995;14:4365-73
15 Westaway D, DeArmond SJ, Cayetano-Canlas J et al. Degeneration of skeletal muscle, peripheral nerves and the central nervous system in transgenic mice over-expressing wild-type PrP. *Cell* 1994;76:117-29
16 Ter-Avanesyan MD, Kushnirov VV, Dagkesamanskaya AR et al. Deletion analysis of the *SUP35* gene of the yeast *Saccharomyces cerevisiae* reveals two non-overlapping functional regions in the encoded protein. *Mol Microbiol* 1993;7:683-92
17 Ter-Avanesyan MD, Dagkesamanskaya AR, Kushnirov VV, Smirnov VN. The *SUP35* omnipotent suppressor gene is involved in the maintenance of the non-Mendelian determinant [*PSI*+] in the yeast *Saccharomyces cerevisiae*. *Genetics* 1994;137:671-6

18 Young CSH, Cox BS. Extrachromosomal elements in a super-suppression system of yeast. 1. A nuclear gene controlling inheritance of the extrachromosomal elements. *Heredity* 1975;26:413-22

19 Kushnirov VV, Ter-Avanesyan MD, Didichenko SA et al. Divergence and conservation of *SUP2* (*SUP35*) gene of yeasts *Pichia pinus* and *Saccharomyces cerevisiae*. *Yeast* 1990;6:461-72

20 Weissmann C. A unified theory of prion propagation. *Nature* 1991;352:679-83

21 Prusiner SB. Molecular biology of prion diseases. *Science* 1991;252:1515-22

22 Tuite MF, Cox BS, McLaughlin CS. A ribosome-associated inhibitor of *in vitro* nonsense suppression in yeast. *FEBS Lett* 1987;225:205-8

23 Milner J, Metcalf EA (1991) Cotranslation of activated mutant p53 with wild-type drives the wild-type p53 into the mutant conformation. *Cell* 1991;55:765-74

24 Masison DC, Wickner RB. Prion-inducing domain of yeast Ure2p and protease resistance of Ure2p in prion-containing cells. *Science* 1995;270:93-5

25 Paushkin SV, Kushnirov VV, Smirnov VN, Ter-Avanesyan MD. Propagation of the yeast prion-like [*PSI+*] determinant is mediated by oligomerisation of the *SUP35*-encoded polypeptide chain release factor. *EMBO J* 1996;15:3127-32

26 Chernoff YO, Lindquist SL, Ono B-I, Inge-Vechtomov SG, Liebman SW. Role of the chaperone protein Hsp104 in propagation of the yeast prion-like factor [PSI+]. *Science* 1995;268:880-4

27 Telling G, Scott M, Mastrianni J et al. Prion propagation in mice expressing human and chimaeric PrP transgenes implicates the interaction of cellular PrP with another protein. *Cell* 1995;83:79-90

28 Baldwin MA, Cohen FE, Prusiner SB. Prion protein isoforms, a convergence of biological and structural investigations. *J Biol Chem* 1995;270:19197-200

PROTEINS AND PEPTIDES

Transmissible Subacute Spongiform Encephalopathies:
Prion Diseases
L Court, B Dodet, eds
© 1996, Elsevier, Paris

A model of 'conformational education' by molecular chaperones as an hypothesis for prion propagation

Jean-Pierre Liautard

Département Biologie Santé, Inserm U431, Case Courrier 100, Université Montpellier II, place Eugène-Bataillon, 34095 Montpellier, France

Summary – It is generally thought that a protein molecule adopts its three-dimensional structure spontaneously under physiological conditions. However, experimental results on chaperone-assisted folding, crystallographic studies of CD2 and structure analysis of prion protein have revealed that a protein can adopt more than one conformation. This result can be explained in two ways: either protein folding is a thermodynamically irreversible process or there are thermodynamic differences between the systems analysed.

As far as the first hypothesis is concerned, a model for protein folding that uses the thermodynamics of irreversible processes, proposed several years ago, was applied to the problem. It is shown that in certain circumstances a protein can interact with another molecule during folding. This molecule exhibits the characteristics of a chaperone. Thus, a model of protein folding assisted by a chaperone has been developed. According to this model, the chaperone lengthens the life of secondary structures, thus increasing the probability of including this structure into the folded protein. It is shown that chaperones may induce a change in conformation by their preference for a certain type of local secondary structure. The structural modifications resulting from this model consist primarily of changes in secondary structure. As a consequence, a protein may adopt different conformations during folding under the control of chaperone proteins. Because this phenomenon takes place during folding it resembles education (contrary to conversion that results from a change from one three-dimensional structure to another). This development leads to the proposition that structural information is associated with a chaperone. Experimental results in agreement with the model are discussed.

As a consequence of this model, some proteins will behave as autochaperones, meaning that they could be involved in their own folding. Experimental evidence for the existence of such proteins has been published. These proteins could propagate their own conformational information. The behaviour of autochaperones has been analysed and offers a possible explanation for the aetiology of prion diseases. The chaperone model involves a folding step escorted by a chaperone. The word 'education' should be used for such an action instead of 'conversion'. Conversion supposes that the protein is still folded and that the tertiary structure changes under the influence of another protein. Education implies help during folding.

Introduction

It is thought that prion infection is due solely to a protein, that is, the protein-only hypothesis [1]. This very unusual feature is in contradiction to the two major paradigms of present biology. The molecular biology paradigm states that the only informational molecules are nucleic acids. Infection implies multiplication and thus it is clearly biological information that is injected into the cell. The infectious agent reproduces its own characteristics. Therefore, if infectious prions are devoid of nucleic acids the protein must be able to transport some kind of information. It has been suggested that such information may result from the structure of

---

Key words: prions / chaperone / model / thermodynamic / education

the protein [2, 3]. However, this hypothesis is based on the experimental finding that the prion protein can exist in at least two conformations. This finding is in contradiction to the paradigm of structural biology stated by Anfinsen: "a protein folds to the most stable conformation". Evidently there is only one most stable conformation. It is therefore necessary to reconcile the prion finding with the physicochemical knowledge concerning protein folding.

Starting with the protein-only hypothesis and the now clearly established finding that prion protein exists in two different conformations, protease-sensitive prion protein (PrP^C) and disease-specific prion protein (PrP^{Sc}), I propose a model of chaperone-assisted protein folding that exhibits most of the properties of prion propagation.

What is the physicochemical meaning of the existence of a protein with two conformations?

Experimental results have clearly established that prion protein exists in two different conformations [4, 5]. This raises a major question. How can the same protein sequence fold into different conformations? This question mainly concerns the field of thermodynamics. The Anfinsen paradigm states that a protein folds into a unique structure. The explanation of this phenomenon has been associated with the reversible character of protein folding. Indeed, if each folding step is a reversible process, the protein will always find the minimum energy level. If a misfolded structure is attained, a reversal is possible and the misfolding will be corrected. As a consequence, if two conformations exist for a single sequence this means that the two conformations correspond to two different thermodynamic systems, that is, with two different sets of thermodynamic parameters. Furthermore, this hypothesis precludes the possibility that a protein exists under a kinetically stable structure. Indeed, by returning to the same thermodynamic system the protein should always find the same structure. Thus, hypothesis 1 of figure 1 implies a change in the thermodynamic parameters. This hypothesis is not tenable because the two different structures have been found in vitro under comparable conditions. It is therefore necessary to introduce an irreversible step into the system in order to understand the existence of the two structures. One possibility is to introduce an irreversible step between the two conformations. This means, for prion, that conversion from PrP^C to PrP^{Sc} is an irreversible step. This corresponds to hypothesis 2 of figure 1. An irreversible step has been introduced at this stage by Huang et al [6]. Such an irreversible step was introduced by these authors without any thermodynamic analysis, and is simply given to render the model of conversion possible. Without this step it is clear that increasing the amount of PrP^{Sc} will also increase the amount of PrP^C and thus result in a complete degradation of PrP^{Sc}. This is clearly not the case. The amyloid formation model proposed by Lansbury and co-workers [7, 8] should also involve a thermodynamically irreversible step. However, the polymerization theory merely involved an energetic barrier that had to be bypassed to obtain the amyloid. This is achieved by nucleation in a saturated solution. The time lapse to polymerization is driven by the formation energy of the nucleus. This result implies, however, that at least one step in protein folding is irreversible in accordance with hypothesis 3 in figure 1. Thus, a simple analysis of the two-structure phenomenon suggests that protein folding involves an irreversible process.

Protein folding is an irreversible process

It is generally assumed that protein folding is a reversible process; however, many recent findings challenge this dogma. As explained previously, the existence of two conformations in

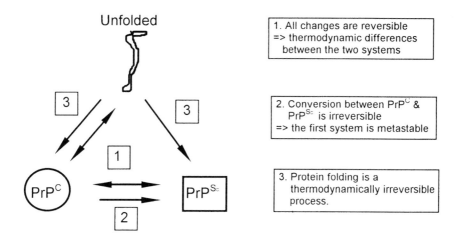

Fig 1. Different possibilities that lead to two conformations from a single sequence.

a protein implies that structure is reached under irreversible conditions, otherwise an equilibrium would be established between the two forms. Recently, complete structure determination by nuclear magnetic resonance of recombinant CD2 has revealed that this protein can exist under two different structures that can be purified [9]. The authors also showed that denaturation of one form leads to the formation of the other structure suggesting that one structure is more stable than the other. A consequence of this is that one structure has kinetic stability [9]. If the protein exhibits kinetic stability, this means that the structure was attained by a kinetic rather than a reversible thermodynamic process. The existence of kinetic stability for proteins has been known for many years. Indeed, subtilisin and protein alpha-lytic present structures after processing of the propeptide that can no longer be reached without this propeptide [10, 11]. Under these conditions the propeptide is thought to function as an internal chaperone [12]. The propeptide assists the protein during folding and drives the protein to a structure in which the propeptide is no longer necessary for the stability of the structure achieved.

Another example of a protein driven to a specific structure is provided by chaperone-assisted folding. It has been clearly established that without a chaperone some proteins cannot fold properly and therefore precipitate [13, 14]. This means that a kinetically favourable folding pathway is impaired by the chaperone. Driving the protein to the biologically native conformation is an irreversible process as this requires hydrolysis of adenosine triphosphate (ATP) (see [15] for a discussion of the irreversibility of chaperone folding). We have previously obtained experimental results suggesting that protein folding is an irreversible process because of the formation of two different structures of tropomyosin [16].

Furthermore, thermodynamic analysis suggests that protein folding is generally an irreversible process. Indeed, the thermodynamic reversibility criterion is the conservation of entropy in the system during the evolution. What happens when a protein folds spontaneously? According to Prigogine [17], the change in entropy dS of the system will be:

$$dS/dt = v(A/T) \quad (1)$$

where v is the speed of folding and A the chemical affinity necessary to bring together the amino acids. Clearly v should be positive (otherwise folding does not occur) and the affinity should be positive to promote this folding. Thus, $dS/dt > 0$, folding is an irreversible process.

Consequences of the involvement of an irreversible step in protein folding

The introduction of a thermodynamically irreversible step has one major consequence: folding is a kinetically driven process. Indeed, the speed of folding is, as a preliminary approximation, proportional to the entropy increase (see equation 1). As a consequence, the structure attained can be different from the less energetic structure.

I have previously proposed a model of protein folding that involves an irreversible step. The model is based on the following scheme:

$$U < = = = = > SS \text{-----} > MG < = = = = > N \quad (2)$$

where U is the unfolded form, SS is the secondary structure formation, MG represents the molten globule and N the native protein.

This means that unfolded protein very rapidly acquires secondary structures that are in equilibrium with unfolded protein. Thus, more than one secondary structure can be formed (see fig 2). Because these secondary structures are in equilibrium, the quantities of each different possible secondary structure depend on their respective energies. This proportion can be calculated according to Boltzmann's law. The secondary structures interact with each other, and these interactions take place in a thermodynamically irreversible manner (see equation 2). This implies that different secondary structures can interact giving rise to different conformations, as schematized in figure 2. Thus, the introduction of an irreversible step during folding results in the possibility of more than one final conformation for one polypeptide.

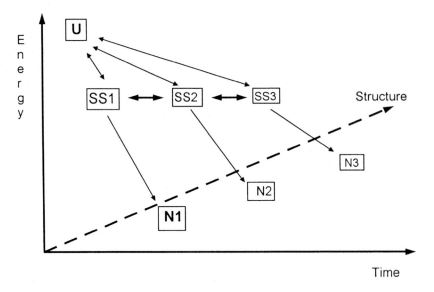

Fig 2. One sequence can fold into different conformations. The folding is represented as a chemical reaction with three coordinates. The first is energy, and this decreases during the reaction. The second coordinate is time. The third coordinate represents the chemical reaction; ie, formation of the structure. The unfolded (U) molecule rapidly acquires secondary structures (here SS1, SS2 and SS3). These secondary structures are in rapid equilibrium with each other and each can interact with another part of the molecule that will stabilize the secondary structure in a 'native' conformation (N1 or N2 or N3). Thus, the main difference between the structures of the resulting proteins will be the incorporation of a different secondary structure.

In some cases, however, the secondary structures formed cannot interact rapidly enough and two proteins exhibiting incomplete folding will interact together resulting in precipitation. It has been shown that molecular chaperone blocks the protein precipitation during folding [14]. This phenomenon can be explained using the protein-folding model described [18].

Molecular chaperone-assisted protein folding

The model of protein folding presented above (called the thermo-kinetic model) assumes that proteins fold alone, without interaction with other proteins. On the other hand, it is evident that under certain conditions (mentioned above and detailed below), interaction between the proteins could compete with the folding. This phenomenon can be analysed using the thermo-kinetic model of protein folding [18]. The thermodynamics of irreversible processes indicate that the speed of folding determines the final interactions between the amino acids according to the following equation [18]:

$$S_{ai \to aj} = \sum_n \left[\prod_m (P^{ss}_{ai+m} \cdot P^{ss}_{aj+m}) (A_{ai+n \to aj+n}) \right] \quad [3]$$

where P^{ss}_{ai} is the probability (according to Boltzmann's law) that amino acid a_i is in the secondary structure ss, thus allowing for interaction; n is the number of amino acids around a_i that influence the speed of association and m should be taken to be between 0 and n. $A_{i \to j}$ represents the chemical affinity between amino acids a_i and a_j.

Consider the case of an interaction between two molecules: an unfolded molecule I, and a folded molecule J: the molecular chaperone. At the beginning of the folding, part of molecule I is in equilibrium between many possible secondary structures (fig 2). Because the quantity of the reactive secondary structure is too low ($P^{ss} \ll 1$) or alternatively because the affinity between two parts of the molecule is too low ($A_{ik \to il} \approx 0$), the folding is slow or an incorrect fold can take place. Suppose that another molecule (ie, a chaperone) is able to interact efficiently with one of the secondary structures (ie, either α-helix or β-sheet), this means that $A_{i \to j} \gg A_{ik \to il}$, ie, affinity with molecular chaperone ($A_{i \to j}$) is greater than intramolecular affinity ($A_{ik \to il}$). Consider the speed of interaction $V_{ai \to aj}$ between amino acid a_i from molecule I and a_j from molecule J (the chaperone) which is still governed by equation 3. As molecule J (the chaperone) is folded, $P^{ss}_{aj+m} = 1$ for all m but, as pointed out earlier, molecule I cannot fold without a molecular chaperone and its P^{ss}_{aj+m} is very low for all m. Hence, spontaneous correct folding does not occur, and the folding speed is governed by molecule J. Thus, the conformation of a_i and its neighbours should be that of the corresponding residues a_j and its neighbours, otherwise the affinity $A_{i \to j}$ is very low and interaction cannot occur. The chaperone lengthens the life of secondary structures and increases the probability of including this structure in the folded protein. As a consequence, molecule J determines the structure of molecule I. On the other hand, if no molecular chaperone is present, the folding could be driven in another direction which could result in misfolding or aggregation. The molecular chaperone is therefore present to avoid misfolding of the protein. An interesting consequence of this model, however, is the possibility that different chaperones stabilize different secondary structures and thus change the structure of the protein during folding. The possibility that a peptide interacts with different chaperones in different conformations (ie, either α-helix or β-sheet) has been demonstrated experimentally [19, 20]. Landry and Gierasch and co-workers [19, 20] found that a polypeptide that is unstructured when free in aqueous solution adopts an α-helical conformation on binding to GroEL and an extended structure on interaction with heat-shock protein 70 (Hsp70) [20]. The authors suggest that chaperonin binds amphipathic α-helices, the hydrophobic surfaces of which would presumably be accessible in the non native state but inaccessible in the native state. This suggests that more favourable interaction occurs resulting from better distribution of the affinity (A_i in the thermodynamic model) of the different amino acids. Clearly, the peptide presents potentiality (ie, P^{ss}) for α-helical or β-sheet conformations, but these secondary structures are stabilised only when they interact with chaperonin or Hsp70, respectively. These findings for peptide models are also probably true for proteins. For instance, it has been shown that OmpA associated with SecB chaperone exhibits secondary structure. Similarly, the conformation of GroEL-bound rhodanese and dihydrofolate reductase resembles that of the molten globule, characterised by containing all or part of the secondary structure [21].

The part of the protein stabilised in the defined secondary structure is then released from the chaperone in the correct conformation (thus, $P^{ss}_{ai+1} = 1$) that increases the possibility of internal interaction (see equation 3). The role of ATP – which further implies an irreversible step – in this reaction is probably to change the conformation of the chaperone to allow it to detach.

Because the model presented involves a thermodynamically irreversible step the protein could be kinetically trapped during folding in an energetic well. This would result in a new conformation and this change could be induced by chaperone as it favours adoption of a certain type of local secondary structure. Thus, a protein in some cases could change conformation under the control of chaperone protein. If the new conformation (misfolded) is not sufficiently soluble, it may precipitate and will be eliminated or alternatively lead to pathological structures (eg, amyloid). In contrast, if the new conformation is sufficiently soluble the protein can survive in the cell. Ultimately the two conformations could exert different biological functions. This result suggests that structural information can be transmitted by proteins. Thus, certain types of biological information can be transmitted, under specific circumstances, without the intervention of nucleic acid. Therefore, the structure of a protein is informative in itself.

The special case of autochaperone

A chaperone can catalyze its own folding (I call these molecules 'autochaperones'). Such chaperone folding, dependent on itself, has been found experimentally [22]. I have analysed the consequences of accidental misfolding of a chaperone on the folding of the forthcoming proteins [2]. Under these conditions the misfolded chaperone should invade the cell.

The thermodynamics of irreversible processes indicate that the speed of folding determines the final interactions between the amino acids (see eq 3). Consider the case of an interaction between two molecules (unfolded molecule I, and folded molecule J: the molecular chaperone) of the same kind of protein. The speed of interaction $V_{ai \to aj}$ between two amino acids, one (a_i) from molecule I and the other (a_j) from molecule J is governed by equation 3.

As molecule J, the molecular chaperone, is folded, $P^{ss}_{aj+m} = 1$ for all m, but as molecule I cannot fold without a molecular chaperone, its P^{ss}_{aj+m} is very low for all m. Hence, spontaneous correct folding does not occur, and the folding speed is governed by the molecule J. Thus, the conformation of a_i and its neighbours should therefore be that of the corresponding residues a_j and its neighbours, otherwise the affinity $A_{ai \to aj}$ will be very low and interaction cannot occur. Therefore, if molecule J is misfolded, molecule I will also be misfolded so that interaction between amino acids occurs, and the new misfolded molecule should have the same structure as its molecular chaperone. A consequence of this model is that it is the secondary structure that changes during misfolding. Indeed, the competition that takes place between the secondary structure and the chaperone stabilises the secondary structure.

Using the deductions presented here, a model can be developed to simulate autochaperone misfolding invasion. Consider the number M_t of misfolded molecular chaperones at time t when a new molecule is synthesised. This number will depend on the number of misfolded chaperones at time $t-1$ (called M_{t-1}),. Furthermore, there is a probability (very low but not zero) that spontaneous misfolding will occur (if folding of the chaperone is not aided) noted P_i (it is easy to demonstrate that P_i is directly proportional to P^{ss} in eq 3). Otherwise, the folding will be performed on a molecular chaperone and the probability of misfolding is $[M_{t-1}/(F_{t-1} + M_{t-1})]$, where F_{t-1} is the number of correctly folded molecules ($F_0 = 100\%$ at time t_0). The elimination of the misfolded chaperone, as physiological degradation of the protein, may also be calculated. The simplest relationship for this degradation is kM_{t-1} (k is a constant). This analysis is summarised in figure 3 and in the following equation:

$$M_t < \text{---} M_{t-1} + P_i + [M_{t-1}/(F_{t-1} + M_{t-1})] - (kM_{t-1}) \qquad (4)$$

Characteristics of the autochaperone model

A 'primary' characteristic of the chaperone model is the invasion of a misfolded chaperone. Computer simulation were performed by iteration according to equation 4. Ten different simulations were performed in order to show the probabilistic process equivalent to a sporadic appearance (fig 4). The number of iterations was proportional to time. Figure 4A depicts this sporadic appearance of misfolded chaperones after lag periods of various durations.

A 'second' characteristic of the model is the influence of the mutation on the onset of the invasion. Indeed, according to the model, point mutations change the value of P^{ss} and thus can favour misfolding by increasing P_i. The calculations simulate this phenomenon (fig 4B). Increasing the probability P_i fourfold results in reducing the invasion lag period by about one-third.

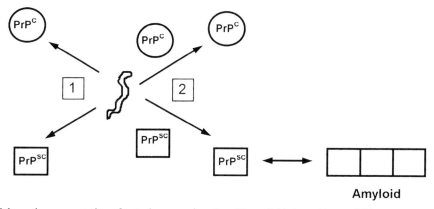

Fig 3. Schematic representation of autochaperone invasion. The unfolded peptide can fold spontaneously into the conformation PrP^C or PrP^{Sc}. **1**: spontaneous folding: the probability of folding into PrP^{Sc} conformation is given by P_i in equation 4. **2**: Chaperone-assisted folding: the folding can be assisted by a chaperone as either PrP^C or PrP^{Sc}. The chaperone directs the unfolded peptide into its own conformation. PrP^{Sc} is not highly soluble and precipitates when folded into amyloid structures.

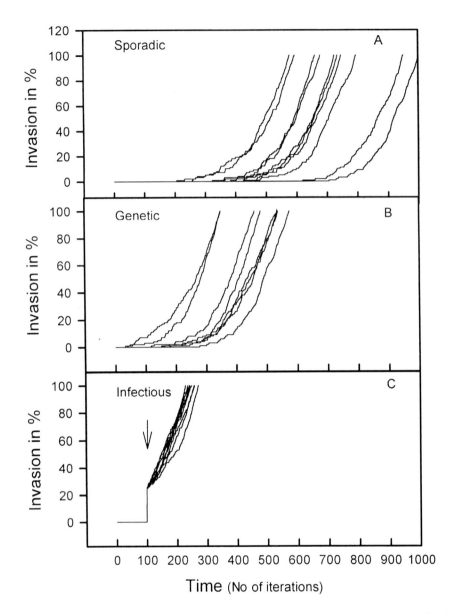

Fig 4. Computer simulation of autochaperone invasion. The simulation was performed using equation 4. In **A**, the parameters correspond to sporadic cases. In **B**, the P_i parameters were increased in order to take into account the increase in β-sheet resulting from mutations. This corresponds to the genetic disease. The main characteristic is reduced invasion time. **C** represents invasion by an added autochaperone. This mimics the infectious process.

A 'third' consequence of the model is invasion by a misfolded chaperone added from outside. External addition of *M* (the misfolded autochaperone) is equivalent to an infection. The results of this simulation, presented in figure 4C, model such an infection. After injection of *M* (arrow) the invasion appears very rapidly, with a short lag period. Furthermore, a relationship is seen to exist between the quantity of misfolded chaperone added and the time of infection (fig 5).

Misfolded molecular chaperones behave as prions

The characteristics of invasion by a misfolded autochaperone resemble those exhibited by prion disease. A direct comparison shows that the model could explain most of the features of this disease.

Disease features resulting from the chaperone model are: 1) localization is mainly in non dividing cells (brain) or in organs that cannot eliminate the rogue chaperone (k must be small). 2) Sporadic appearance, which is a consequence of Boltzmann's law governing the thermodynamic probability of a conformation in the protein folding (parameter P_i, a function of P^{ss} in eq 3). Thus, it is a random process. 3) Genetic transmission of susceptibility resulting from the increased probability of misfolding associated with a different amino acid. Point mutation of the chaperone may be at the origin of the heritability by increasing the probability of spontaneous misfolding (parameter P_i). Different mutations will give different speeds of invasion (governed by P_i). 4) The possibility of infection exists if the rogue chaperones manage to enter

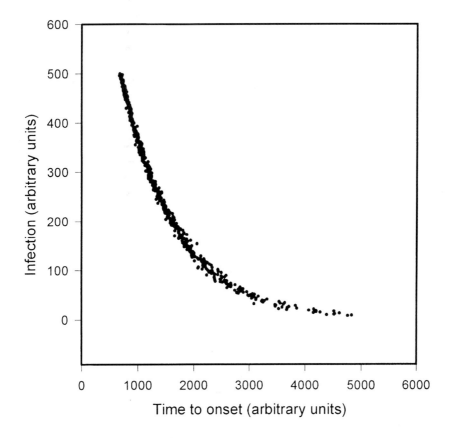

Fig 5. Relationship between the quantity of infectious units and the incubation time. The simulation was performed using equation 4. Different quantities of the misfolded chaperone were added at time zero and invasion time was calculated.

the cells or the organ considered (the seeding effect that increases M_t). 5) Furthermore, the relationship between the quantity of misfolded chaperone added and the time before onset resembles that found by experimental infection of prion diseases (fig 5). 6) Any increases in synthesis will decrease the time lag to sporadic appearance of the disease (the number of 'iterations' increasing per time unit).

Are prions misfolded molecular chaperones

To validate the model, it will be necessary to prove that prion proteins are really chaperones. The proof that prions are indeed molecular chaperones should be experimental. No such evidence has yet been found. However, Caughey and collaborators [23] succeeded in changing the structure of PrPC to PrPSc in vitro. This result was interpreted by the author as a 'crystallization-like mechanism'. However, this interpretation suffers a drawback. In the model chosen, nucleus formation is the rate-limiting step. Addition of a tiny quantity of nucleus should overcome the time necessary to produce a spontaneous formation of PrPSc. This should be achieved by adding a very small amount of PrPSc to the PrPC solution. The author found, on the contrary, that a huge amount of PrPSc was necessary to transform PrPC into PrPSc. This result did not fit very well the crystallisation model of Lansbury and co-workers [7, 8]. In contrast, if PrPSc is considered as a chaperone the 50-fold excess is not exceptional. Furthermore, in order to obtain a conversion the author was required to dissolve the PrPC in 2.8 M guanidium-chloride. At this concentration the protein is at least partially denatured. Thus, folding could be chaperone-dependent.

On the other hand, sequence and structural homology between prion molecules and genuine molecular chaperones have been found [24]. Sequences were examined for convergence by studying the periodicity in the sequences by Fourier transform analysis. The results show that prion, Hsp70, Hsc70 and SSA1 all have strikingly similar glycine repeat sequences [24]. These are all heat-shock proteins and the best known molecular chaperones are also heat-shock proteins. Furthermore, it has been shown that prion and Hsp70 also share secondary structure distribution [24]. Another observation is also in good agreement with the model, that is, that numerous mutations are associated with prion disease. Mutations M109F, A117V, D178N, F198S and E200K fall in the predicted α-helices and each mutation decreases the α-helix formation [24]. These mutations increase the stability of the local β-sheet and thus (as deduced from equation 3) the quantity of β-sheet available for the long-range interaction.

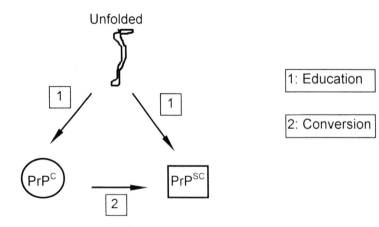

Unfolded

1: Education

2: Conversion

PrPC

PrPSC

1

1

2

Fig 6. A schematic representation of education versus conversion.

Education versus conversion

An interesting consequence of the model is the finding that structural changes take mainly secondary structures into account. In all cases it is the change in secondary structure that leads to the conformational modification. This is possible because the protein is denatured and the secondary structure is stabilised by the long-range interactions. The direct changes of an α-helix to a β-sheet as proposed by the conversion models [6, 8] require excessive amounts of energy. The only possible way is to denature the protein then renature it in another conformation. If another protein is involved in this process, that protein is a chaperone.

Here, I should underline a conceptual difference between the conversion model and the chaperone model (fig 6). The chaperone model involves a folding step escorted by a chaperone. The word 'education' should be used for such an action instead of 'conversion' Conversion supposes that the protein is still folded and that the tertiary structure changes under the influence of another protein.

References

1 Prusiner SB. Molecular biology of prion diseases. *Science* 1991;252:1515-22
2 Liautard JP. Are prions misfolded molecular chaperones? *FEBS Lett* 1991;294:155-7
3 Cohen FE, Pan KM, Huang Z, Baldwin M, Fletterick RJ, Prusiner SB. Structural clues to prion replication. *Science* 1994;264:530-1
4 Pan KM, Baldwin M, Nguyen J et al. Conversion of α-helices into β-sheets features in the formation of the scrapie prion proteins. *Proc Natl Acad Sci USA* 1993;90:10962-6
5 Safar J, Roller PP, Gajdusek C, Gibbs CJ. Conformational transition, dissociation and unfolding of scrapie amyloid (prion) protein. *J Biol Chem* 1993;268:20276-84
6 Huang Z, Prusiner SB, Cohen FE. Structures of prions proteins and conformational models for prion diseases. In: Prusiner S, ed. *Prions, Prions, Prions.* Berlin: Springer, 1996;49-67
7 Come JH, Fraser PE, Lansbury PT. A kinetic model for amyloid formation in the prion diseases: importance of seeding. *Proc Natl Acad Sci USA* 1993;90:5959-63
8 Jarrett JT, Lansbury PT. Seeding one-dimensional crystallization of amyloid: a pathogenic mechanism in Alzheimer's disease and scrapie. *Cell* 1993;73:1055-8
9 Murray AJ, Lewis SJ, Barclay AN, Brady RL. One sequence, two folds: a metastable structure of CD2. *Proc Natl Acad Sci USA* 1995;92:7337-41
10 Baker D, Sohl JL, Agard DA. A protein-folding reaction under kinetic control. *Nature* 1992;356:263-5
11 Shinde U, Li Y, Chatterjee S, Inouye M. Folding pathway mediated by an intramolecular chaperone. *Proc Natl Acad Sci USA* 1993;90:6924-8
12 Shinde U, Inouye M. Intramolecular chaperones and protein folding. *Trends Biochem Sci* 1993;18:442-6
13 Hendrick JP, Hartl FU. Molecular chaperone functions of heat-shock proteins. *Annu Rev Biochem* 1993;62:349-84
14 Hartl FH, Hlodan R, Langer T. Molecular chaperones in protein folding: the art of avoiding sticky situations. *Trends Biochem Sci* 1994;19:20-5
15 Gulukota K, Wolynes PG. Statistical mechanics of kinetic proofreading in protein folding in vivo. *Proc Natl Acad Sci USA* 1994;91:9292-6
16 Ferraz C, Heitz F, Sri Widada J, Caron E, Cave A, Liautard JP. Conformational stability of human skeletal tropomyosins modified by site-directed mutagenesis. *Protein Eng* 1991;4:561-8
17 Prigogine I. *Introduction à la thermodynamique des processus irréversibles.* Paris: Dunod, 1968
18 Liautard JP. A thermo-kinetic model for protein folding. *CR Acad Sci* 1990;311:385-9
19 Landry SJ, Gierasch LM. The chaperonin GroEL binds a polypeptide in an alpha-helical conformation. *Biochemistry* 1991;307:359-62
20 Landry SJ, Jordan R, McMachen R, Gierash L. Different conformation for the same polypeptide bound the chaperones DnaK and GroEL. *Nature* 1992;355:455-7
21 Martin J, Langer T, Boteva R, Schramel A, Horwich A, Hartl FU. Chaperonin-mediated protein folding at the surface of GroEL through a 'molten globule'-like intermediate. *Nature* 1991;352:36-42
22 Cheng MY, Hartl FU, Horwich AL. The mitochondrial chaperonin hsp60 is required for its own assembly. *Nature* 1990;348:455-8

23 Bessen R, Kocisko DA, Raymond GJ, Nandan S, Lansbury PT, Caughey B. Non-genetic propagation of strain-specific properties of scrapie prion protein. *Nature* 1995;375:698-700
24 Liautard JP. Prions and molecular chaperones. *Arch Virol* 1993;7(suppl):227-43

Transmissible Subacute Spongiform Encephalopathies:
Prion Diseases
L Court, B Dodet, eds
© 1996, Elsevier, Paris

Formation of PrP-res in vitro

Suzette A Priola, Bruce Chesebro, Byron Caughey

Laboratory of Persistent Viral Diseases, National Institute of Allergy and Infectious Diseases, Rocky Mountain Laboratories, 903 South 4th Street, Hamilton, MT 59840, USA

Summary – Infection with agents of the transmissible spongiform encephalopathies (TSE) is characterized by the conversion of normal, protease-sensitive prion protein (PrP^C) into a protease-resistant form, PrP-res. The mechanism by which this conversion occurs is unknown, although conformational changes in PrP rather than covalent modifications appear to be involved. PrP-res itself has been hypothesized to be the infectious agent in the TSE, although this hypothesis has not been proven. Recent data suggests that the TSE agent which causes bovine spongiform encephalopathy in cattle may have crossed the species barrier into humans, underscoring the importance of determining how the formation of PrP-res is controlled. Using cells persistently infected with mouse scrapie, we have mapped the region of PrP^C important in the formation of PrP-res and have shown that generation of PrP-res can be controlled by a single amino acid residue. We have developed a cell-free system to model the conversion of PrP^C to PrP-res which demonstrates both species and strain specificity at the protein level. These data provide evidence that scrapie species barriers and scrapie strains can be dictated by specific PrP protein interactions.

Species-specific formation of PrP-res in vitro

The transmissible spongiform encephalopathies (TSE) are a group of rare, neurodegenerative diseases which include scrapie in sheep, bovine spongiform encephalopathy (BSE) in cattle, and kuru, Creutzfeldt-Jakob disease (CJD) and Gerstmann-Sträussler-Scheinker disease (GSS) in humans. Studies on the molecular basis of pathogenesis of the TSE have focused on the conversion of normal, protease-sensitive prion protein (PrP^C) into a protease-resistant form (PrP-res) usually associated with disease. Experiments in transgenic (Tg) mice have demonstrated that high levels of expression of hamster PrP^C in mice normally resistant to infection with hamster scrapie render the mice susceptible to hamster scrapie agent [1]. This suggests that interactions between homologous PrP molecules could control the scrapie species barrier [2]. Subsequent studies in Tg mice, as well as in vitro experiments in cells persistently infected with mouse scrapie, mapped the region of the PrP important in the conversion process to the middle third of the *Prnp* gene [3, 4].

Using a tissue culture system, we studied the molecular basis of the barrier to infection that exists between the TSE agents of different species at the level of PrP-res formation. Mouse and hamster PrP^C or chimeric mouse/hamster PrP^C molecules were expressed in mouse neuroblastoma cells persistently infected with mouse scrapie. These cells replicate mouse scrapie infectivity, express mouse PrP^C and accumulate mouse PrP-res [5, 6]. Formation of PrP-res from the recombinant PrP^C molecules was assayed using the 3F4 monoclonal antibody (mAb)

Key words: scrapie / prion protein

which reacts with hamster PrP, but not with the mouse PrP normally synthesized in mouse neuroblastoma cells. Using this system, we were able to show that PrP-res formation could be controlled by a region of PrP^C encompassing approximately 75 amino acid residues [7]. The importance of this region of the molecule is underscored by the fact that several mutations within and around this region have been associated with disease in human TSE (fig 1A). We mapped a single amino acid residue which controls PrP-res formation to position 139 of the Syrian hamster *Prnp* gene. PrP^C molecules expressing either leucine or isoleucine at this position are converted to PrP-res, while PrP^C molecules expressing methionine are not (S Priola, unpublished data).

Heterologous PrP molecules interfere with accumulation of PrP-res

Following the introduction of the TSE agent from one host to another, disease incubation times are initially prolonged [8, 9]. Studies in Tg mice expressing high levels of hamster PrP^C have suggested that non-homologous interactions between PrP molecules can delay disease incubation times [2]. We have expressed non homologous PrP^C molecules in cells persistently infected with mouse scrapie and demonstrated that PrP-res accumulation was significantly inhibited [10]. We mapped this interference in PrP-res formation to a single amino acid residue located at position 111 in mouse PrP (position 112 in human PrP; fig 1A). These data directly demonstrated, for the first time, that heterologous PrP molecules could significantly interfere with PrP-res formation [10]. Furthermore, when mouse neuroblastoma cells normally susceptible to mouse scrapie expressed heterologous hamster PrP^C, they became resistant to mouse scrapie infection (S Priola, unpublished data). Therefore, expression of a non homologous PrP molecule appeared to prevent in vitro infection by a TSE agent.

Cell-free conversion of PrP^C to PrP-res

We have been able to precisely define PrP amino acid residues which play a significant role in the conversion of PrP^C to PrP-res [7, 10]. However, our tissue culture system limits our studies to the species barrier which exists between two defined strains of mouse and hamster scrapie. We recently developed a cell-free assay [11, 12] which allows us to study the ability of PrP-res from any species to convert PrP^C from any other species.

Mouse or hamster PrP-res derived from scrapie-infected brain was incubated with immuno-precipitated, metabolically labeled mouse or hamster PrP^C or mouse/hamster PrP^C chimeras expressed in uninfected mouse neuroblastoma cells [11, 12]. The mixture was then treated with proteinase K (PK) and analyzed for PK-resistant radiolabeled products. Hamster PrP-res was able to convert hamster, but not mouse, PrP^C to protease-resistant forms. In contrast, mouse PrP-res was able to convert both mouse and hamster PrP^C to protease-resistant products, although the sizes of these products differed (fig 1B) [11]. These results demonstrated a similar specificity to that observed in vivo where mouse Chandler scrapie agent could infect both mice and hamsters [8, 9], but the 263K strain of hamster scrapie agent could only infect hamsters [13]. Furthermore, only those constructs which contained a region of hamster PrP from residues 112-188 were converted to PK-resistant products (fig 1B). These results are in agreement with our studies using these same chimeric constructs in mouse scrapie-infected mouse neuroblastoma cells [7] which mapped the species specificity of PrP-res formation to an amino acid at position 138 within this same region (fig 1A, boxed residue). Thus, similar PrP–PrP interactions appeared to be involved in PrP-res generation in cultured mouse neuroblastoma cells and in the cell-free conversion system.

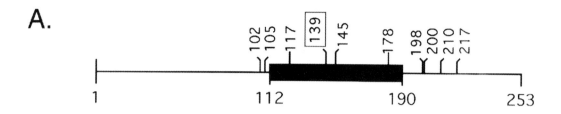

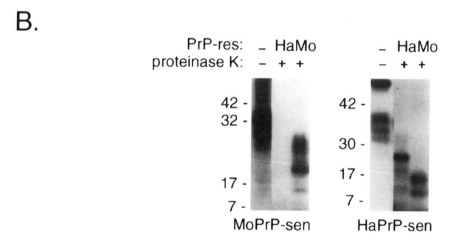

Fig 1. Species specificity in the in vitro formation of PrP-res. A: Single amino acid mutations associated with familial human transmissible spongiform encephalopathies. A map of the 253-amino acid human *PRNP* gene is shown. The numbers represent amino acid residues. The shaded area indicates the region of protease-sensitive prion protein (PrP^C) primary sequence which dictates the conversion to protease-resistant forms [3, 10, 12]. Boxed residue 139 is the amino acid residue which has been shown to control conversion of mouse PrP^C to mouse PrP-res in tissue culture [10]. Several mutations associated with familial human TSE are clustered within and around this central region of PrP [31]. Mutations associated with Gerstmann-Sträussler-Scheinker disease: Pro→Leu 102; Pro→Leu 105; Tyr→Amber stop codon 145; Phe→Ser 198; Gln→Arg 217; mutations associated with Creutzfeldt-Jakob disease: Asp→Asn 178; Glu→Lys 200; Val→Ile 210; mutations associated with fatal familial insomnia; Ala→Val 117; Asp→Asn 178. B) The cell-free conversion of PrP^C to PrP-res is species specific. As described previously [11, 12], radiolabeled mouse PrP^C (MoPrP^C) or Syrian hamster PrP^C (SHaPrP^C) was incubated in the absence (–) or presence of Mo or SHa PrP-res. The samples were then treated with PK as indicated and PK-resistant bands were detected using sodium dodecyl sulfate-polyacrylamide gel electrophoresis (SDS-PAGE) fluorography. Conversion to PK-resistant forms is indicated by a decrease in size of the radiolabeled product, while a lack of conversion gives no detectable PK-resistant bands. Molecular mass markers (in kDa) are indicated on the left side of each panel.

Strain-specific formation of PrP-res in the cell-free system

To date, there are no molecular data which explain the variable biological properties of different scrapie strains. One possibility is that PrP-res itself might be able to pass on its own strain-specific characteristics to PrP^C via direct PrP–PrP interactions. Stable variations in PrP-res structure rather than specific mutations in an agent-specific nucleic acid could therefore provide a molecular basis for scrapie strains.

Strain-specific differences in PrP-res have been described for two strains of hamster scrapie originally derived from mink infected with transmissible mink encephalopathy [14]. In both strains, PrP-res is derived from hamster PrP^C, but PrP-res derived from hamster brains infected with the 'drowsy' strain is approximately 1 kDa smaller than PrP-res derived from hamster brains infected with the 'hyper' strain [15]. PrP-res from these two sources was used to test whether the strain-specific size difference could be detected in PrP-res generated in the cell-free system. The results showed that the PK-resistant products, like the PrP-res species which induced their formation, differed by 1 kDa in size [16]. Since the PrP^C precursor was identical in both cases, the different sources of PrP-res used in these experiments must influence the final size of the protease-resistant product. These results provide a possible molecular basis for this scrapie strain-specific characteristic, consistent with the idea that self-propagation of PrP-res may proceed via distinctive three-dimensional structures.

Modeling the mechanism of conversion in the cell-free system

The mechanism by which PrP^C is converted to PrP-res is unclear. The heterodimer theory proposes that a monomer of PrP-res interacts with a monomer of PrP^C and induces the conversion of the PrP^C monomer to PrP-res. The two PrP-res molecules dissociate and the process repeats [17–19]. The difficulty with the heterodimer model is that monomeric PrP-res has never been detected. The seeded polymerization model of conversion suggests that an ordered nucleus of aggregated PrP-res acts as a seed for the polymerization of PrP^C to PrP-res [20, 21]. Converting activity in the heterodimer model would be associated with monomeric PrP-res, while converting activity in the seeded polymerization model would be associated with larger, multimeric forms of PrP-res.

Sedimentation and ultrafiltration analysis of the converting activity in hamster PrP-res preparations was assayed in both denaturing and non denaturing conditions. In all cases tested, all of the converting activity was associated with particles several times larger in size than monomeric PrP [22]. The data clearly demonstrated that monomeric PrP-res was not required for conversion to occur, although it does not rule out the possibility that under other conditions monomeric PrP-res may be able to convert PrP^C to PrP-res. The tight association of converting activity with multimeric forms of PrP-res is inconsistent with the heterodimer model and consistent with the nucleated polymerization model of PrP-res formation.

Potential intermediates in the conversion of PrP^C to PrP-res

Our data from the cell-free conversion assay suggest that multimeric forms of PrP are important in the conversion process [22]. We recently defined a dimer of hamster PrP^C in mouse neuroblastoma cells which has properties intermediate between those of PrP^C and PrP-res. This dimeric PrP is PK-sensitive but it forms large aggregates. This molecule is responsible for the 24-kDa PK-resistant band detected in the cell-free conversion assay [11]. A dimeric PrP-res molecule is expressed in vivo only in scrapie-infected hamster brains [23]. The hamster PrP dimer therefore has the properties expected of an intermediate in the conversion process, although it may not be absolutely required for conversion to occur [11, 23].

Future studies

The ultimate question in the TSE diseases remains the exact nature of the infectious agent and whether or not it is composed solely of PrP-res. This hypothesis can be definitively proven by the de novo generation of PrP-res from PrP^C derived from an uninfected source under conditions that do not permit the replication of a more conventional infectious agent and demonstrating that this PrP-res causes disease. As yet, it has not been possible to address this issue using either of the model systems described. In order to convincingly demonstrate new infectivity, conditions have to be developed which allow either the de novo generation of PrP-res from PrP^C in the absence of any initial PrP-res or which allow the conversion to proceed under conditions in which the newly converted PrP-res is in vast excess to the initial PrP-res.

The specificity of the cell-free conversion system and its close correlation with in vivo scrapie species barriers and scrapie strains suggests that this system may be useful as a rapid test for the susceptibility of any species to the TSE agents of another species. One experiment, given the possibility that the BSE agent has crossed the species barrier into humans in England [24], is to test whether human PrP^C can be converted to protease-resistant forms by BSE PrP-res. It would also be interesting to determine which amino acid residue may control the ability of bovine PrP-res to convert PrP^C from other species. However, given the likelihood that other factors may influence TSE species barrier effects in vivo, caution should be used in extrapolating data from the in vitro to the in vivo situation. Identifying effective inhibitors of the conversion process is of the highest priority and the cell-free conversion system could provide a rapid assay system for testing compounds which have been shown to inhibit PrP-res formation in tissue culture [25, 26] and delay disease in vivo [27–30], as well as analyzing other possible inhibitors.

References

1 Scott M, Foster D, Mirenda C et al. Transgenic mice expressing hamster prion protein produce species-specific scrapie infectivity and amyloid plaques. *Cell* 1989;59:847-57
2 Prusiner SB, Scott M, Foster D et al. Transgenetic studies implicate interactions between homologous PrP isoforms in scrapie prion replication. *Cell* 1990;63:673-86
3 Scott MRD, Butler DA, Bredesen DE, Walchli M, Hsiao KK, Prusiner SB. Prion protein gene expression in cultured cells. *Protein Eng* 1988;2:69-76
4 Scott M, Groth D, Foster D et al. Propagation of prions with artificial properties in transgenic mice expressing chimeric PrP genes. *Cell* 1993;73:979-88
5 Race RE, Fadness LH, Chesebro B. Characterization of scrapie infection in mouse neuroblastoma cells. *J Gen Virol* 1987;68:1391-9
6 Race RE, Caughey B, Graham K, Ernst D, Chesebro B. Analyses of frequency of infection, specific infectivity, and prion protein biosynthesis in scrapie-infected neuroblastoma cell clones. *J Virol* 1988;62:2845-9
7 Priola SA, Caughey B, Race RE, Chesebro B. Heterologous PrP molecules interfere with accumulation of protease-resistant PrP in scrapie-infected murine neuroblastoma cells. *J Virol* 1994;68:4873-8
8 Kimberlin RH, Walker CA. Evidence that the transmission of one source of scrapie agent to hamsters involves separation of agent strains from a mixture. *J Gen Virol* 1978;39:487-96
9 Kimberlin RH, Cole S, Walker CA. Temporary and permanent modifications to a single strain of mouse scrapie on transmission to rats and hamsters. *J Gen Virol* 1987;68:1875-81
10 Priola SA, Chesebro B. A single hamster amino acid blocks conversion to protease-resistant PrP in scrapie-infected mouse neuroblastoma cells. *J Virol* 1995;69:7754-8
11 Kocisko DA, Come JH, Priola SA et al. Cell-free formation of protease-resistant prion protein. *Nature* 1994;370: 471-4
12 Kocisko DA, Priola SA, Raymond GJ, Chesebro B, Lansbury PT, Jr, Caughey B. Species specificity in the cell-free conversion of prion protein to protease-resistant forms: a model for the scrapie species barrier. *Proc Natl Acad Sci USA* 1995;92:3923-7

13 Kimberlin RH, Walker CA, Fraser H. The genomic identity of different strains of mouse scrapie is expressed in hamsters and preserved on reisolation in mice. *J Gen Virol* 1989;70:2017-25

14 Bessen RA, Marsh RF. Biochemical and physical properties of the prion protein from two strains of transmissible mink encephalopathy agent. *J Virol* 1992;66:2096-101

15 Bessen RA, Marsh RF. Distinct PrP properties suggest the molecular basis of strain variation in transmissible mink encephalopathy. *J Virol* 1994;68:7859-68

16 Bessen RA, Kocisko DA, Raymond GJ, Nandan S, Lansbury PT, Jr, Caughey B. Nongenetic propagation of strain-specific phenotypes of scrapie prion protein. *Nature* 1995;375:698-700

17 Griffith JS. Self-replication and scrapie. *Nature* 1967;215:1043-4

18 Bolton DC, Bendheim PE. A modified host protein model of scrapie. In: Bock G, Marsh J, eds. *Novel Infectious Agents and the Central Nervous System.* Chichester: John Wiley & Sons, 1988;164-81

19 Prusiner SB. Molecular biology of prion diseases. *Science* 1991;252:1515-22

20 Brown P, Goldfarb LG, Gajdusek DC. The new biology of spongiform encephalopathy: infectious amyloidoses with a genetic twist. *Lancet* 1991;337:1019-22

21 Jarrett JT, Lansbury PT, Jr. Seeding "one-dimensional crystallization" of amyloid: a pathogenic mechanism in Alzheimer's disease and scrapie? *Cell* 1993;73:1055-8

22 Caughey B, Kocisko DA, Raymond GJ, Lansbury PT. Aggregates of scrapie associated prion protein induce the cell-free conversion of protease-sensitive prion protein to the protease-resistant state. *Chem Biol* 1995;2:807-17

23 Priola SA, Caughey B, Wehrly K, Chesebro B. A 60-kDa prion protein (PrP) with properties of both the normal and scrapie-associated forms of PrP. *J Biol Chem* 1995;270:3299-305

24 Will RG, Ironside JW, Zeidler M et al. A new variant of Creutzfeldt-Jakob disease in the UK. *Lancet* 1996;347:921-5

25 Caughey B, Race RE. Potent inhibition of scrapie-associated PrP accumulation by Congo red. *J Neurochem* 1992;59:768-71

26 Caughey B, Raymond GJ. Sulfated polyanion inhibition of scrapie-associated PrP accumulation in cultured cells. *J Virol* 1993;67:643-50

27 Ehlers B, Diringer H. Dextran sulphate 500 delays and prevents mouse scrapie by impairment of agent replication in spleen. *J Gen Virol* 1984;65:1325-30

28 Kimberlin RH, Walker CA. Suppression of scrapie infection in mice by heteropolyanion 23, dextran sulfate, and some other polyanions. *Antimicrob Agents Chemother* 1986;30:409-13

29 Ladogana A, Casaccia P, Ingrosso L et al. Sulphate polyanions prolong the incubation period of scrapie-infected hamsters. *J Gen Virol* 1992;73:661-5

30 Caughey B, Ernst D, Race RE. Congo red inhibition of scrapie agent replication. *J Virol* 1993;67:6270-2

Transmissible Subacute Spongiform Encephalopathies:
Prion Diseases
L Court, B Dodet, eds
© 1996, Elsevier, Paris

Conformational transitions of solubilized prion protein PrP 27–30

Detlev Riesner[1], Klaus Kellings[1], Karin Post[1], Martin Pitschke[1], Holger Wille[2], Hana Serban[2], Darlene Groth[2], Michael A Baldwin[2], Stanley B Prusiner[2,3]

[1]*Institut für Physikalische Biologie und Biologisch-Medizinisches, Forschungszentrum, Heinrich-Heine-Universität Düsseldorf, 40225 Düsseldorf, Germany;* [2]*Departments of Neurology and* [3]*Biochemistry and Biophysics, University of California, San Francisco, CA 94143-0518, USA*

Summary – Structural studies of the disease-specific prion protein (PrP^{Sc}) have been slowed due to its lack of solubility under conditions where infectivity is retained. We examined a variety of detergents for their ability to disperse the protease-resistant, purified prion protein (PrP 27–30), the N-terminally truncated form of PrP^{Sc}. Addition of ionic detergents such as sodium dodecyl sulfate or sarkosyl followed by sonication and ultracentrifugation yielded three fractions: an insoluble fraction of infectious aggregates, a soluble but non-infectious fraction of PrP oligomers, and a lipid-rich, infectious fraction at the meniscus. PrP oligomers appear under the electron microscope as 10-nm diameter spherical particles. Since these PrP particles are soluble, they can be studied by quantitative biophysical measurements. Of particular interest are the conformational transitions which were predicted on the basis of the prion-hypothesis. Biochemical characteristics of PrP^{Sc} like proteinase K-resistance, β-sheet secondary structure and polymeric aggregation could be induced by different chemical treatments, but do not have a strict relationship. Infectivity, however, could not yet be regenerated.

Introduction

Prions are proteinaceous particles that are largely, if not entirely, composed of an abnormal isoform of the prion protein (PrP) designated disease-specific prion protein (PrP^{Sc}). A protease-resistant polypeptide, PrP 27–30, can be derived from PrP^{Sc} by limited proteolysis with retention of infectivity. Both PrP^{Sc} and the cellular isoform, protease-sensitive prion protein (PrP^{C}), are encoded by a chromosomal gene; PrP^{Sc} is produced from the cellular isoform by a posttranslational process [for review, see 1].

Several possible modes of action have been considered that might explain the transformation of the cellular isoform PrP^{C} into PrP^{Sc}. Among these are a second component, a covalent post-translational modification or a conformational change. Despite intensive search, no second component such as a scrapie-specific nucleic acid has been found; quantitative analyses of residual nucleic acids have eliminated oligonucleotides larger than about 80 nucleotides [2, 3]. Proteolysis and peptide mapping failed to identify any covalent post-translational modification

---

Key words: prion protein / conformational transition / infectivity

that could account for the conversion [4]. However, conformational differences were identified by infrared (IR) spectroscopy and circular dichroism (CD). PrP^C is predominantly an α-helical protein, whereas PrP^{Sc} exhibits substantial β-sheets and reduced amounts of α-helix [5, 6]. Treatment of PrP 27–30 or PrP^{Sc} with solvents such as hexafluoroisopropanol (HFIP) which increase the α-helix content of PrP^{Sc} resulted in substantial reduction in scrapie infectivity [7, 8].

Conformational studies of the PrP isoforms are complicated by solubility differences in these proteins. For example, PrP^C is soluble in non-denaturing detergents, whereas PrP^{Sc} is not. To obtain high-resolution structural data, conditions for solubilization of PrP^{Sc} under which scrapie prion infectivity is retained need to be identified. Early attempts to solubilize scrapie infectivity were unsuccessful [9] and the insolubility of prions slowed the development of effective purification protocols [10, 11]. Once the insolubility of scrapie infectivity was known, it was used to enrich fractions for infectivity which led to the discovery of PrP 27–30 [12]. Some studies were misoriented by the assumption that scrapie is caused by a virus and that the smallest scrapie prion particles were likely to be quite large despite contradictory results with ionizing radiation [13]. The insolubility of PrP^{Sc} has also prevented determination of its molecular homogeneity or heterogeneity. Since as many as 10^5 PrP molecules correlate with one infectious unit, even fewer PrP molecules could represent the essential portion of the infectious agent, but are hidden among many more PrP 27–30 molecules. A homogeneous population of 10^5 PrP^{Sc} molecules is the other alternative.

Solubilization of PrP^{Sc} and PrP 27–30 was effective when mixed with phospholipids to form liposomes or detergent-lipid-protein complexes (DLPC) [14]. The use of phospholipids to solubilize scrapie prions has not been useful in structural studies of PrP because such high concentrations of lipid are required. In some studies, scrapie infectivity was reported to have sedimentation coefficients as low as 2–3S [15, 16]. Recent claims that PrP 27–30 can be recovered after sodium dodecyl sulfate–polyacrylamide gel electrophoresis (SDS–PAGE) with retention of prion infectivity [17] could not be confirmed [18].

To identify conditions for solubilization of PrP 27–30, we examined a variety of ionic and non-ionic detergents under conditions that were likely to retain scrapie prion infectivity. Sonication of the prion rods in the presence of 0.2–0.3% SDS produced a soluble fraction that did not sediment during centrifugation at 100,000 g for 1 h. Unlike PrP 27–30 amyloid polymers, this soluble fraction had a high α-helical content and relatively little β-sheet, contained ~10-nm diameter spherical particles and had an observed sedimentation coefficient of about 6S. The spheres were composed of four to six PrP 27–30 molecules; they contained little, if any, scrapie infectivity. A lipid-rich, infectious fraction located at the meniscus after ultracentrifugation remained insoluble. Although our studies did not yield infectious preparations of soluble prions, they provided different fractions of PrP 27–30 which could be characterized by infectivity, protein conformation and conformational transitions.

Materials and methods

Chemicals and enzymes

HFIP and acetonitrile were of spectroscopic grade from Aldrich (Milwaukee, WI, USA). Lipids were from Avanti Polar Lipids (Alabaster, AL, USA). Dodecylmaltoside (dodecyl-β-D-maltopyranoside), lauryldimethylamineoxide (LDAO) and thesit were kindly provided by W Welte (Freiburg, Germany); the other detergents were from Calbiochem (San Diego, CA, USA). Benzonase^{TM} and proteinase K were from Merck (Darmstadt, Germany).

Preparation of PrP 27–30

PrP 27–30 was purified from the brains of scrapie-infected Syrian hamsters (Lak:LVG from Charles River Laboratories) as described previously [19]. The final purification step was either sucrose gradient centrifugation or ultrafiltration in which less than 300-kDa proteins were removed and PrP was obtained in the residue [20]. Prions from both sources gave similar results.

Bioassays

Bioassays of scrapie prions were performed in Syrian hamsters using an incubation time interval procedure [19]. The same correlation between the inoculated dose and the incubation time was found for homogenates, purified prion rods from sucrose gradient centrifugation and DLPC, indicating that a variety of chemical treatments and fractionation procedures did not alter the relationship between the inoculated dose and the incubation time.

Buffers

Sonication buffer A (20 mM MOPS, 20 mM Tris, pH 7.2); sonication buffer B (10 mM sodium phosphate, pH 7.0); TBST (10 mM Tris-HCl, pH 8.0, 0.15 M NaCl, 0.01% Tween-20).

Sonication and centrifugation

Sonication was carried out with a cup horn sonicator (Branson) at 40 W for 5 min at temperatures not exceeding 17 °C; details are described elsewhere [21]. After sonication, the sample was centrifuged in a 1.5-mL Eppendorf spin tube (Beckman Instruments, Palo Alto, CA, USA) in a tabletop TL-100 ultra-centrifuge (Beckman Instruments). Two protocols were applied. The 'soluble fraction' protocol: sonication was carried out in 440 μL of buffer followed by centrifugation in thick-wall polycarbonate tubes in a TLS-55 swinging-bucket rotor at 40,000 rpm for 115 min at 4 °C (100,000 g) and a 220-μL supernatant fraction was collected from just below the meniscus. The 'total supernatant' protocol: sonication was carried out in 250-μL buffer and either the whole sample was loaded onto the sucrose gradient or 220–230 μL of the supernatant were separated from the pellet by careful pipetting after centrifugation using a fixed-angle rotor TLA-100.3 at 40,000 rpm (60,000 g) or at 50,000 rpm (100,000 g) for 1 h at 4 °C.

Sucrose gradient centrifugation

Sucrose gradient centrifugation (5–20% sucrose in 10 mM sodium phosphate buffer, pH 7) was performed in a SW-60 rotor (4-mL polyallomer tubes) at 60,000 rpm for 5.5 h. It was estimated that the largest particles in the supernatant from the solubilization centrifugation (~25S) would just reach the bottom of the gradient under these conditions. The details are given elsewhere [21].

PrP analysis by PAGE, Western blots and quantitation

PrP samples were analyzed by SDS–PAGE (12% acrylamide) from which the proteins were transferred onto an Immobilon P membrane (Millipore), incubated with primary anti-PrP antibody then with the alkaline phosphatase-conjugated secondary antibody. In most cases, PrP was determined from dot blots visualized with the chemiluminescent system (ECL; Amersham) and scanned with a densitometer (for details see [21]).

Digestion by proteinase K

Samples were incubated with either 50 μg of proteinase K/mL at 37 °C for 1 h or 10 μg of protease K/mL at 37 °C for 1 to 30 min.

Circular dichroism spectroscopy

CD spectra were recorded with a Jasco model 720 spectropolarimeter using 1- or 5-mm path length cylindrical cells at room temperature. All measurements were carried out in buffer B (10 mM sodium phosphate, pH 7.0) containing 0.2% SDS, which allowed readings to be made at 190 nm. Signal averaging allowed satisfactory CD spectra to be obtained with PrP 27–30 concentrations as low as ~10 μg/mL. Quantitative secondary structure assignments were carried out with a computer program kindly provided by W Curtis Johnson, University of Oregon [22]. Further details are described elsewhere [21].

Electron microscopy

The samples were adsorbed for about 20 s onto glow-discharged, carbon-coated grids. Excess material and sucrose were washed away with three drops each of 0.1 M and 0.01 M ammonium acetate. The grids were then stained with 2% uranyl acetate and viewed under a JEOL 100CX II electron microscope at 80 kV. The diameter of the particles was measured parallel to one side of each print, independently of the orientation of the particles.

Results

Sonication of PrP 27–30 with detergents

Prion rods (~300 μg of PrP 27–30/mL) were sonicated under controlled conditions in buffer A with detergents that were representative of ionic, non ionic, zwitterionic and glycosidic classes. After sonication, the samples were centrifuged in a fixed-angle rotor at 60,000 or 100,000 g at 4 °C for 1 h and the supernatant was sampled according to the total supernatant protocol. The effectiveness of the detergent was evaluated on the basis of the fraction of the PrP 27–30 and the scrapie infectivity remaining in the supernatant, as determined by Western blots with anti-PrP antibodies and bioassays in Syrian hamsters, respectively (table I). The most effective detergents were SDS and sarkosyl, both of which resulted in 20–30% of PrP

Table I. Efficiency of solubilization of PrP 27–30 by sonication in buffer A containing various detergents[a].

Detergent in sonication buffer	Supernatant		
	PrP content (%)	Infectivity (%)	log_{10} ID_{50}/mL
None	1–2	2	6.4
SDS (0.05–10%)	20–30[b]	8–50[b]	7.4–8.4[b]
Sarkosyl (2%)	20	50	9.1
Zwittergent 3-12 (1–3%)	10	2	7.2–7.6
LDAO (1%)	5	< 0.1	4.9
Thesit (3%)	10	7	7.4
Dodecylmaltoside (3%)	10	6	7.6
DLPC	1–2	4	7.1

[a]The PrP 27–30 was determined with serial dilutions by Western blots and the infectivity by bioassays in Syrian hamsters. After solubilization, centrifugation according to the total supernatant protocol, the PrP 27–30 content and the infectivity of supernatant and pellet were added; the percentage of the soluble fraction is listed. [b]The values are not related to the SDS concentrations in a monotonic way. They refer to SDS concentration > 0.2%. ID_{50}: 50% infectious dose

27–30 and a similar fraction of the infectivity in the supernatant fraction. SDS concentrations from 0.05 up to 10% were investigated, and the effectiveness of solubilization saturated around 0.3%, in good agreement with the critical micelle concentration (~0.24%). The finding that SDS and sarkosyl did not inactivate scrapie infectivity at low concentrations is in agreement with earlier findings [10]. For all subsequent experiments, sonication was carried out with 0.2 or 0.3% SDS.

The effectiveness of solubilization was not modified when centrifugation was changed from 60,000 to 100,000 g with otherwise identical parameters. Three different sonication time protocols (5 min, three periods of 5 min with 5-min breaks, and 30 min continuously) yielded very similar results. The initial temperature was set either at 7 °C (with an increase to a maximum of 17 °C during a 5-min sonication period) or at 30 °C, increasing to 37 °C; the efficiency did not change. In an attempt to minimize denaturation, we applied a 5-min sonication, starting at 7 °C with an increase to no more than 17 °C.

Chaotropic agents known to denature PrP 27–30 and destroy infectivity were used at low concentrations, eg, 1 M urea or 0.5 M guanidine hydrochloride (Gdn HCl) [18]. The effect of HFIP on prion infectivity was evaluated at a concentration of 10% (v/v). Sucrose was also added up to 25% (w/v), either alone or in combination with the other reagents. After sonication in 0.3% SDS, the fraction of PrP 27–30 remaining in the supernatant was considerably more sensitive to digestion with proteinase K than that in the prion rods found in the pellet. Bioassays of these digested supernatant fractions showed reductions in prion titers by factors of 5, 16, 3, 50 and 5 in five independent experiments. While such changes are not significant, there was always a small reduction in titer. In contrast, HFIP diminished infectivity drastically. Other reagents did not appear to be promising. Consequently, further studies were not performed.

Sucrose gradient centrifugation analysis

Sucrose gradient centrifugation was performed to determine whether PrP 27–30 from the supernatant fraction after sonication and infectivity cosedimented. To minimize contamination of the supernatant fraction by any particles floating in the meniscus or particles loosely adherent to the pellet, centrifugation of the SDS-treated sonicated prion rods was performed in a swinging-bucket rotor, and only 50% of the supernatant was collected with a pipette from the upper portion of the liquid column (soluble fraction protocol). The gradients were preformed with either no detergent or 0.2% SDS to test for reaggregation in the absence of SDS. Gradients were fractionated from either the top or the bottom to assess possible artifacts in the profile arising from the fractionation protocol. Each fraction from the sucrose gradient was analyzed for PrP 27–30 content and scrapie infectivity. Profiles in the presence and absence of 0.2% SDS are similar, each profile represents average values from four or five independent experiments (fig 1A).

The majority of PrP 27–30 was found in fractions 2 to 5 and sedimented as a symmetrical peak of ~6S, but only ~0.02% of the infectivity was recovered in these fractions. It is noteworthy that the supernatant collected by the soluble fraction protocol contained only 6% of the total infectivity, of which 50% was loaded onto the gradient. Because the specific infectivity in fractions 2 to 5 was nearly 3 orders of magnitude lower than that of the prion rods, we conclude that soluble PrP 27–30 prepared under these conditions is devoid of prion infectivity.

The low recovery of infectivity in the supernatant loaded onto the gradients in figure 1 compared with the nearly 50% infectivity in the supernatant in the initial SDS-containing soni-

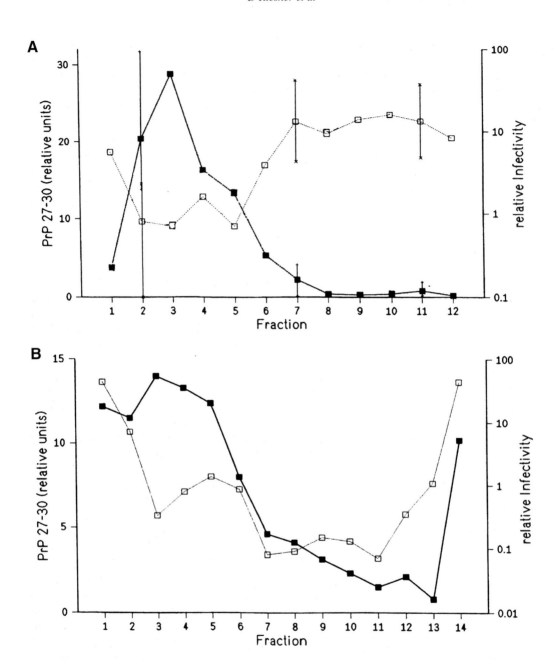

Fig 1. Analysis by 5–20% sucrose gradient centrifugation of the soluble fraction (**A**) or of the total sample (**B**) after sonication in 0.2% SDS. The sedimentation profiles show PrP 27–30 (■) and infectivity (□). The numbers in A are the means of four independent experiments. Relative units are depicted (in %) for PrP 27–30 and infectivity; the sum of all fractions was set at 100%. Error bars are defined by the highest and the lowest values respectively; they are shown for three fractions only, but are characteristic of all fractions. The larger error bars in fraction 2 are due to the flank position in the PrP peak or due to the very low and therefore inaccurate infectivity data, respectively. Modified from Riesner et al [21].

cation experiments (table I) appeared contradictory. Therefore, prion rods were sonicated in 0.2% SDS and loaded directly onto the sucrose gradient without prefractionation by differential centrifugation. Again, most of the PrP 27–30 was found in fractions 2 to 5 (fig 1B), but some PrP 27–30 was also found in the meniscus fraction. Most notably, $\sim 10^7$ 50% infectious doses (ID_{50}) of prion infectivity per mL were found in the meniscus and the adjacent fraction as well as at the bottom of the gradient, in contrast to fractions 3 to 5 which contained most of the PrP 27–30 and whose prion titers were lower by nearly a factor of 1,000.

Prions in the meniscus fractions would seem to be of either small size or low density. Since the supernatant fractions prepared with the soluble fraction protocol would have eliminated the prions of low density but not those of small size, we conclude that the absence of infectivity in the meniscus fraction in figure 1A indicates a low-density population of prions. Most likely, the low density of the prions would be due to lipids and possibly detergents bound to PrP 27–30. The origin of the lipids has to be the prion rods.

Ultrastructural analysis

Fractions 2 to 5 of the sucrose gradients (fig 1) contained numerous spherical particles (fig 2A and 3E). Since these fractions contained only PrP 27–30 as judged by SDS–PAGE, and because no particles other than spheres were found by electron microscopy (EM), we concluded that the spheres were composed of PrP 27–30. Most of them ranged from 8 to 12 nm in diameter; a substantial portion was ~ 10 nm in diameter (fig 2B). We were unable to identify any consistent substructure using negative staining.

Since particles in fractions 2 to 5 were of spherical shape, we estimated the molecular mass from the average sedimentation coefficient ($\sim 6S$) determined from the position of the peak fraction. According to the method of Durchschlag [23], the partial specific volume (v) for a sedimenting particle was calculated to be 0.71 cm^3 g^{-1}, based on the following composition of PrP 27–30: protein, 54% ($v = 0.71$ cm^3 g^{-1}); carbohydrate, 18% ($v = 0.65$ cm^3 g^{-1}); glycolipid anchor, 7% ($v = 0.59$ cm^3 g^{-1}) and 21% SDS, ie, 0.4 g of SDS per gram protein ($v = 0.82$ cm^3 g^{-1}) according to Tanford [24] and assuming hydration as 0.5 g of water per gram of PrP 27–30 (the values for SDS binding and hydration being rough estimates only). Applying the standard method for evaluation of a spherical molecule of 6S [25], a molecular mass of 117 kDa was obtained, which better corresponded to pentameric PrP 27–30. Allowing for errors and uncertainties in this calculation, we concluded that each of these spheres contained between four and six PrP 27–30 molecules. On the other hand, the observed sedimentation values for PrP 27–30 and the spheres suggested that PrP 27–30 is unlikely to form particles other than the spheres visualized under EM. Samples from selected fractions of the sucrose gradient centrifugation before and after sonication with 0.2% SDS (fig 1) were examined by EM. We confirmed that samples obtained before sonication consisted of prion rods which were 100–300 nm in length (fig 3A). Rods were still present after sonication in SDS, but were smaller (50–150 nm); in addition, PrP 27–30 was dispersed into small spherical particles and elongated aggregates (fig 3B).

The meniscus fraction of the sucrose gradient contained a heterogeneous array of particles (fig 3C) and aggregates of small rods (fig 3D). The finding of rod fragments in the meniscus fractions implies that the density of these fragments is below 1.018 g/cm^3 (5% sucrose) and that the rod fragments are associated with substantial amounts of lipid. Fraction 3 which had the highest concentration of PrP 27–30, contained only small, ~ 10-nm diameter spherical particles (fig 3E). These spheres were indistinguishable from those shown in figure 2A. The pellet fraction contained rod fragments that withstood the sonication procedure (data not shown).

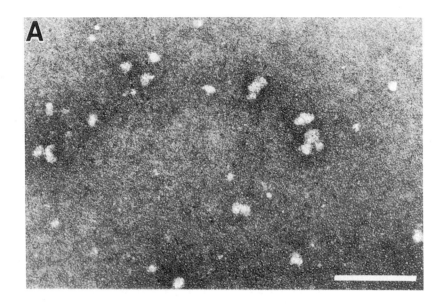

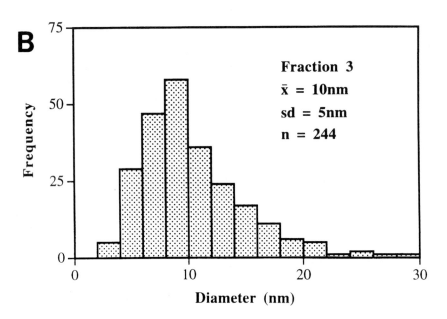

Fig 2. A. Electron micrograph of the peak fraction of sucrose gradient centrifugation as shown in figure 1A (bar = 100 nm); **B.** Size distribution of the spherical particles in **A**. Modified from Riesner et al [21]. sd: standard deviation.

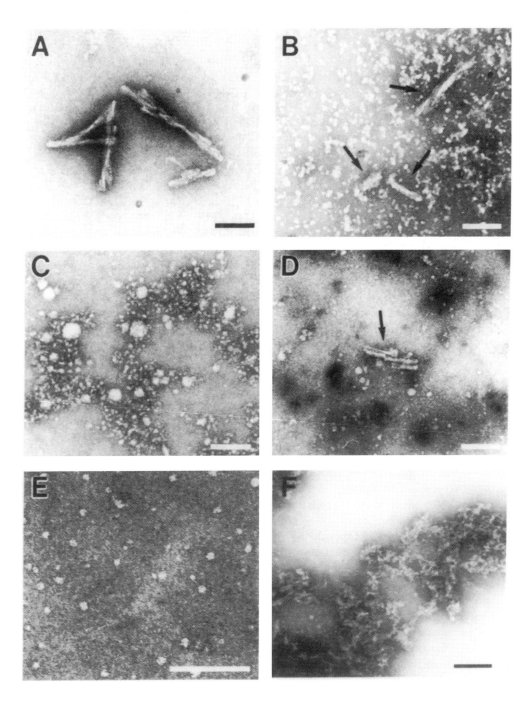

Fig 3. Electron micrographs of different fractions of PrP 27–30. Prion rods before sonication (**A**) and after sonication (**B**) in 0.2% SDS; (**C**) and (**D**) meniscus fractions from the sucrose gradient illustrated in figure 1B; (**E**) fraction 3 from the same gradient; (**F**) after addition of 25% acetonitrile to the supernatant prepared according to the total supernatant protocol. The arrows point to residual rod fragments; the bar corresponds to 100 nm. Figure modified from Riesner et al [21].

Conformational transitions

Conformational analyses were performed on the supernatant and the resuspended pellet fractions. After the supernatant sample was removed from the 100,000 g centrifugation for the sucrose gradient (soluble fraction protocol, fig 1A), the remaining supernatant was collected and examined by CD spectroscopy. The pellet was resuspended in an equal volume of the same buffer. The supernatant generated a spectrum characteristic of a protein of high α-helical content, whereas the spectrum of the pellet was characteristic of structures containing high β-sheet content. These spectra were similar to those obtained from fractions prepared using the total supernatant protocol in which the supernatant (fig 4A) showed minima at 208 and 222 nm, characteristic of α-helical structures, whereas the pellet fraction gave a maximum at 195 nm and a minimum at 218 nm, representative of β-sheets. The spectrum of the supernatant showed no changes over a 24-h period suggesting that PrP 27–30 remained soluble over this time interval. In contrast, the ellipticity of the resuspended pellet decreased notably over 1 h in accordance with the insolubility of the rods. When the prion rods were sonicated for up to 1 h without SDS, the supernatant fraction gave a lower ellipticity spectrum with a characteristic β-sheet signal suggesting that the conformation of PrP 27–30 in 'broken rods' was unaltered by sonication alone. However, the amplitudes in the CD spectra cannot be correlated with molar ellipticities.

Addition of acetonitrile is known to induce β-sheets in peptides. Increasing the acetonitrile concentration from 0 to 50% in increments of 10% induced a progressive shift from α-helix to β-sheet (fig 4B). The proportions of α-helical and β-sheet structures determined from the CD spectra were calculated (table II). The α-helical content of the supernatant fractions ranged from 40 to 60%; this range of values might be due to variations in the SDS sonication procedure as well as the protocol used to separate the supernatant from the pellet after centrifugation in a fixed-angle rotor. The addition of 10 or 20% acetonitrile led to a small reduction in the α-helical content and to a concomitant increase in β-sheets; 30% acetonitrile produced approximately equal amounts of α-helices and β-sheets. The spectra of the resuspended pellet resembled that of the supernatant to which 30% acetonitrile had been added (table II). A concentration of 40–50% acetonitrile seemed to denature PrP 27–30. The addition of 30– 50% acetonitrile to the supernatant precipitated PrP 27–30 after exposure overnight as assessed by 100,000 g centrifugation. The increase in the β-sheet content correlated with the formation of insoluble aggregates in other experiments (data not shown).

Since infectious prion rods containing PrP 27–30 are rich in β-sheets, we investigated whether the increase in β-sheets due to acetonitrile was accompanied by an increase in infectivity. The addition of 10–20% acetonitrile had no effect on prion infectivity, while 30% acetonitrile reduced infectivity by ~75%. Thus, the conformational change induced by acetonitrile is not equivalent to that occurring during the formation of PrP^{Sc}. The bioassays were started before the separation of the SDS-sonicated rods in infectious lipid-rich fraction and non infectious spherical particle content was known. With that knowledge, the infectivity tests would have been performed on isolated spherical particles, after reaggregation into β-sheets by 30% acetonitrile. Thus, we cannot exclude completely at present that part of the reaggregated PrP 27–30 regained infectivity. The addition of 40–50% acetonitrile diminished prion titers by a factor of 10^3 to 10^4. Acetonitrile at concentrations above 30% would act as a denaturant for PrP.

EM was also used to monitor the effect of acetonitrile on the supernatant following SDS sonication and centrifugation. No rods were found in the presence of 25% acetonitrile; only amorphous aggregates were observed (fig 3F).

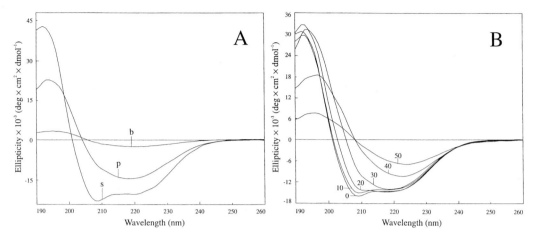

Fig 4. Circular dichroism spectra. **A**. Supernatant (s) and pellet (p) prepared according to the total supernatant proto-col in 10 mM phosphate buffer, pH 7.0, containing 0.2% SDS; 'broken rods' (b) were obtained by sonication for 1 h in the absence of SDS and without centrifugation. **B**. The supernatant after addition of 0 to 50% (in increments of 10%) acetonitrile. Ellipticities were calibrated according to an amino-acid analysis of the sample and were cor-rected for dilution with acetonitrile. Modified from Riesner et al [21].

Characteristic fractions containing PrP 27–30 were tested for proteinase K resistance. The majority of the PrP 27–30 in the meniscus fraction was digested within the first minute, whereas the remaining PrP 27–30 was stable for 30 min (fig 5A). This behavior may reflect the array of structures found by EM (fig 4C and D) with aggregates of rods being responsible for the residual protease K resistance. In contrast, the soluble PrP 27–30 from the peak frac-tions was digested nearly completely within the first minute (fig 5B). The pellet obtained after SDS sonication of prion rods followed by centrifugation in a fixed-angle rotor exhibited substantial protease K resistance (fig 5C). When the peak fraction was treated with 30% aceto-nitrile and precipitable PrP 27–30 recovered by a 100,000 *g* centrifugation, measurable-but-modest proteinase K resistance was regained (fig 5D).

Table II. The effect of acetonitrile (AcN) on the infectivity of the supernatant after solubilization centrifugation and on the secondary structure as determined by CD spectroscopy.

Fraction	AcN	$\log_{10} ID_{50}/mL$	Infectivity[a]	Secondary structure (%)			
	(%)		(%)	α-helix	β-sheet	Turns	Other
Pellet	0	9.6	83	29	31	10	30
Supernatant	0	8.9	17	61	5	16	18
	10	9.1	26	57	9	15	19
	20	8.9	17	53	18	9	20
	30	8.4	5	37	33	5	25
	40	5.9	0.02	26	36	6	32
	50	5.0	0.002	22	29	13	36

[a]The total infectivity of pellet and supernatant (100%) after solubilization centrifugation was 9.7 $\log_{10}$ ID$_{50}$/mL.

Fig 5. Western blots showing variable proteinase K resistance of PrP 27–30 fractions from solubilization experiments. **A.** From the meniscus of the gradient in figure 1B. **B.** From fraction 3 of the gradient in figure 1A. **C.** Pellet from soluble fraction protocol. **D.** Fraction 3 (cf, fig 5B) with addition of 30% acetonitrile. The incubation times were 0, 1, 5 and 30 min at 37 °C at a final concentration of 10 μg of proteinase K per mL. Modified from Riesner et al [21].

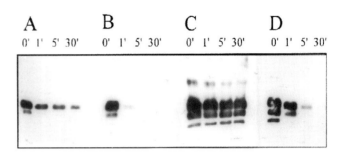

Discussion

Considerable evidence supports the hypothesis that a conformational change in PrP underlies both the propagation of prions and the pathogenesis of prion diseases [5, 24]. Although no experimental studies have demonstrated the refolding of PrP^C into PrP^Sc in cell-free systems, the studies reported here address this issue from the perspective of modifying the structure of PrP 27–30 under relatively gentle conditions. The conformational changes of PrP 27–30 were characterized by chemical parameters like solubility, protease K resistance and secondary structure, and by the biological property of infectivity.

Disruption of prion rods

After sonication of prion rods in the presence of 0.2% SDS, ~25% of the PrP 27–30 was found in the supernatant after centrifugation at 100,000 g for 1 h. The pellet contained the majority of PrP 27–30, consisting of prion rods which withstood the SDS-sonication procedure, but the rods were smaller as measured by EM (data not shown). The CD spectrum of this pellet fraction showed a more pronounced β-sheet signal than that of prion rods subjected to sonication in the absence of SDS, a finding most probably the consequence of the larger particle size when sonication is done in the absence of SDS.

The data in table I suggested that a major part of scrapie infectivity as well as PrP 27–30 content was transferred into the supernatant containing PrP 27–30 in a soluble state, but sucrose gradient centrifugations (fig 1) demonstrated that the supernatant was heterogeneous. Soluble PrP 27–30 formed a fairly homogeneous population of spherical particles sedimenting at 6S and lacking infectivity, whereas lipid-rich infectious prions were identified in the meniscus fraction of sucrose gradients when the total supernatant procedure was used to generate the sample loaded onto the gradient (fig 1B). It should be emphasized that lipids or other compounds which are responsible for the low density of the prions were not added during sonication, but are constituents of the prion rods. A heterogeneous array of particles (including spheres of different sizes, aggregates of spheres and even small prion rod-like structures) were found by EM of lipid-rich fractions (fig 3). A portion of the PrP 27–30 (fig 5) as well as the prion infectivity was also destroyed by proteinase K digestion, whereas residual PrP 27–30 and infectivity resisted limited digestion. It is possible that the rod-shaped structures in the lipid-rich fraction represent a proteinase K-resistant subpopulation of PrP 27–30 molecules that are associated with prion infectivity. An explanation of our results could be that lipids may stabilize the conformation of PrP 27–30 and thus preserve prion infectivity. One would expect

310

that PrP 27–30 in the lipid-rich fraction assumes the β-sheet structure, but this could not yet be measured. The stabilizing effect of lipids on the preservation of infectivity might be similar to that reported previously for DLPC [14]. In DLPC, the β-sheet content of PrP 27–30 was similar to that measured for the prions rods by Fourier transform infrared spectroscopy [7].

Formation of spherical particles

The PrP 27–30 molecules that form the spherical oligomers resemble PrP^C in several aspects: like PrP^C, PrP 27–30 in spheres is soluble, sensitive to proteinase K, α-helical and non-infectious; both molecules have a low β-sheet content [5, 7]. The transformation of prion rods into spherical particles has to be differentiated from the denaturation of PrP 27–30 which has been studied several times before, most recently by Safar et al [6] and Oesch et al [26]. Why such spheres form is unclear, but elucidating the mechanism of their formation may provide insight into the molecular forces that participate in the polymerization of PrP 27–30 into prion rods. The spheres of PrP 27–30 reported here resemble morphological spheres generated by sonication of prion rods in the absence of SDS; a 'non-sedimenting' fraction contained spherical particles 8–20 nm in diameter as judged by EM, but the yield was so low that the particles could not be further characterized [27].

   Some investigators attempted to determine the size of infectious prions using sucrose gradient centrifugation after extraction of Creutzfeldt-Jakob disease (CJD)-infected mouse brain preparations with sarkosyl and sonication [28]. A non-infectious peak of ~7S containing PrP was found which was interpreted as being composed of 'monomeric or dimeric' PrP. From these findings, the authors concluded that PrP can be separated from infectivity and, therefore, cannot be the major or essential component of the infectious prion. Indeed, the 7S PrP peak in these studies is likely to be composed of the same spherical oligomers of PrP 27–30 reported here. In other studies, spherical particles were described with an average diameter of 12 nm in brain homogenates from scrapie-infected Syrian hamsters and not from uninfected controls [29]. Similar spheres have also been found in homogenates prepared from the brains of patients who died of CJD [30]. On the basis of these ultrastructural findings, the investigators suggest that those 'virus-like' particles are the causative agents of scrapie and CJD. It seems likely that these 12-nm spheres are identical to the spherical particles shown in figure 2A, but the particles described herein are oligomers of PrP 27–30 and definitely not 'virus-like'.

Conformational transitions of PrP 27–30

A portion of PrP 27–30 from prion rods could be converted into PrP spheres with high α-helical content and converted back into β-sheets by addition of 25–30% acetonitrile. The sedimentation behavior demonstrated that the β-sheet conformation is connected with aggregation and thus insolubility, and only the α-helical state of PrP spheres is soluble. Our experiments showed however that β-sheet formation is faster than the larger forming aggregates, since β-sheet spectra could be measured within minutes, whereas aggregates formed overnight. In addition, proteinase K resistance is connected with aggregation, whereas soluble PrP spheres lose proteinase K resistance (fig 5). The aggregates reformed by 25% acetonitrile are morphologically different from prion rods and do not posses scrapie infectivity as measured by bioassay in hamsters.

   Our findings may be relevant in regard to a recent study in which PrP^C was reported to acquire proteinase K resistance in the presence of a 50-fold excess of PrP 27–30 [31]. The protocol involves exposure of PrP^C to 3 M Gdn HCl prior to mixing with PrP^Sc which was either exposed to 3 M Gdn HCl or not. Whether the conformation of PrP 27–30 in 0.2% SDS after sonication is equivalent to that of PrPᶜ or PrPˢᶜ in 3 M Gdn HCl remains to be established but seems to be unlikely if the analysis by Safar et al [6] is taken into account. These authors have characterized by spectroscopic methods the 'molten globules' structure, which was induced either by 2 M Gdn HCl or at a pH < 2. Our results are similar to their report in that a structure of higher α-helix content could be induced. The marked difference is, however, in the infectivity, since they described their intermediates as infectious, whereas we separated infectivity from the soluble α-helical state and could not restore infectivity. Since our studies show that reaggregation can be accompanied by the acquisition of proteinase K resistance without restoration of prion infectivity, it is possible that some of the results of the mixing studies noted above could be explained by protein aggregation in which PrPᶜ is not converted into PrPˢᶜ. It should be noted that subsequent investigation failed to confirm the renaturation from 3 M Gdn HCl [32]. The relatively uniform size of the spheres suggests that the PrP 27–30 molecules comprising them possess a unique metastable conformation which produces this distinct quaternary structure. Whether conditions that renature the PrP 27–30 in the spheres and restore scrapie infectivity can be identified remains unknown, but is certainly worth further investigation.

Acknowledgments

This work was supported by research grants from the National Institutes of Health (NIH), USA (NS22786), the Ministerium für Wissenschaft und Forschung (MWF) of NRW, Germany and the Bundesministerium für Forschung und Technologie (BMFT), Germany.

References

1 Prusiner SB. Molecular biology of prion diseases. *Science* 1991;252:1515-22
2 Kellings K., Meyer N, Mirenda C, Prusiner SB, Riesner D. Further analysis of nucleic acids in purified scrapie prion preparations by improved return refocussing gel electrophoresis (RRGE). *J Gen Virol* 1992;73:1025-9
3 Riesner D. The search for a nucleic acid component to scrapie infectivity. *Semin Virol* 1991;2:215-26
4 Stahl N, Baldwin MA, Teplow DB et al. Structural analysis of the scrapie prion protein using mass spectrometry and amino acid sequencing. *Biochemistry* 1993;32:1991-2002
5 Pan KM, Baldwin MA, Nguyen J, Gasset M. Conversion of α-helices into β-sheets features in the formation of the scrapie prion proteins. *Proc Natl Acad Sci USA* 1993;90:10962-6
6 Safar J, Roller PP, Gajdusek DC, Gibbs CJ Jr. Conformational transitions, dissociation, and unfolding of scrapie amyloid (prion) protein. *J Biol Chem* 1993;268:20276-84
7 Gasset M, Baldwin MA, Fletterick RJ, Prusiner SB. Perturbation of the secondary structure of the scrapie prion protein under conditions associated with changes in infectivity. *Proc Natl Acad Sci USA* 1993;90:1-5
8 Safar J, Roller PP, Gajdusek DC, Gibbs CJ Jr. Thermal-stability and conformational transitions of scrapie amyloid (prion) protein correlate with infectivity. *Protein Sci* 1993;2:2206-16
9 Millson GC, Manning EJ. The effect of selected detergents on scrapie infectivity. In: Prusiner SB, Hadlow WJ, eds. *Slow Transmissible Diseases of the Nervous System* Vol 2. New York: Academic Press 1979;409-2
10 Prusiner SB, Groth DF, Cochran SP, Masiarz FR, McKinley MP, Martinez HM. Molecular properties, partial purification, and assay by incubation period measurements of the hamster scrapie agent. *Biochemistry* 1980; 19:4883-91
11 Siakotos AN, Gajdusek DC, Gibbs Jr CJ, Traub RD, Bucana C. Partial purification of the scrapie agent from mouse brain by pressure disruption and zonal centrifugation in sucrose–sodium chloride gradients. *Virology* 1976;70:230
12 Prusiner SB, Bolton DC, Groth DF, Bowman KA, Cochran SP, McKinley MP. Further purification and characterization of scrapie prions. *Biochemistry* 1982;21:6942-50
13 Alper T, Haig DA, Clarke MC. The exceptionally small size of the scrapie agent. *Biochem Biophys Res Commun* 1966;22:278
14 Gabizon R, McKinley MP, Prusiner SB. Purified prion proteins and scrapie infectivity copartition into liposomes. *Proc Natl Acad Sci USA* 1987;84:4017-21

15 Malone TG, Marsh RF, Hanson RP, Semancik JS. Membrane-free scrapie activity. *J Virol* 25, 1978;933-5

16 Prusiner SB, Hadlow WJ, Eklund CM, Race RE, Cochran SP. Sedimentation characteristics of the scrapie agent from murine spleen and brain. *Biochemistry* 1978;17:4987-92

17 Brown P, Liberski PP, Wolff A, Gajdusek DC. Conservation of infectivity in purified fibrillary extracts of scrapie-infected hamster brain after sequential enzymatic digestion or polyacrylamide gel electrophoresis. *Proc Natl Acad Sci USA* 1990;87:7240-4

18 Prusiner SB, Groth D, Serban A, Stahl N, Gabizon R. Attempts to restore scrapie prion infectivity after exposure to protein denaturants. *Proc Natl Acad Sci USA* 1993;90:2793-7

19 Prusiner SB, Cochran SP, Groth DF, Downey DE, Bowman KA, Martinez HM. Measurement of the scrapie agent using an incubation time interval assay. *Ann Neurol* 1982;11:353-8

20 Hecker R, Taraboulos A, Scott M et al. Replication of distinct prion isolates is region specific in brains of transgenic mice and hamsters. *Genes Dev* 1992;6:1213-28

21 Riesner D, Kellings K, Post K et al. Disruption of prion rods generates 10-nm spherical particles having high α-helical content and lacking scrapie infectivity. *J Virol* 1996;70:1714-22

22 Johnson WC Jr. Protein secondary structure and circular dichroism: a practical guide. *Proteins* 1990;7:205-14

23 Durchschlag H. Specific volumes of biological macromolecules and some other molecules of biological interest. In: Hinz HJ, ed. *Thermodynamic Data for Biochemistry and Biotechnology* Berlin: Springer-Verlag, 1986: 45-128

24 Tanford C. *The Hydrophobic Effect: Formation of Micelles and Biological Membranes* (2nd edn). New York: John Wiley & Sons Inc, 1980

25 Cantor CR, Schimmel PR. *Biophysical Chemistry*. San Francisco: WH Freeman and Company, 1980

26 Oesch B, Jensen M, Nilsson P, Fogh J. Properties of the scrapie prion protein: quantitative analysis of protease resistance. *Biochemistry* 1994;33:5926-31

27 McKinley MP, Braunfeld MB, Bellinger CG, Prusiner SB. Molecular characteristics of prion rods purified from scrapie-infected hamster brains. *J Infect Dis* 1986;154:1212-22

28 Sklaviadis TK, Manuelidis L, Manuelidis EE. Physical properties of the Creutzfeldt-Jakob disease agent. *J Virol* 1989;63:1212-22

29 Ozel M, Diringer H. Small virus-like structure in fractions from scrapie hamster brain. *Lancet* 1994;343:894-5

30 Ozel M, Xi YG, Baldauf E, Diringer H, Pocchiari M. Small virus-like structure in brains from cases of sporadic and familial Creutzfeldt-Jakob disease. *Lancet* 1994;344:923-4

31 Kocisko DA, Come JH, Priola SA et al. Cell-free formation of protease-resistant prion protein. *Nature* 1994;370:471-4

32 Kaneko K, Paretz D, Pan KM et al. Prion protein (PrP) synthetic peptides induce cellular PrP to acquire properties of the scrapie isoform. *Proc Natl Acad Sci USA* 1995;92:1160-4

Transmissible Subacute Spongiform Encephalopathies:
Prion Diseases
L Court, B Dodet, eds
© 1996, Elsevier, Paris

Conformational change in the prion protein can be modeled by peptides

Michael A Baldwin[1], Fred E Cohen[2,3], Stanley B Prusiner[1,2]*

Departments of [1]Neurology, [2]Biochemistry and Biophysics, [3]Molecular and Cellular Pharmacology, University of California, San Francisco, CA 94143-0518, USA

Summary – The normal cellular protease-sensitive prion protein (PrP^C) is predominantly α-helical, whereas the disease-specific isoform (PrP^{Sc}) is rich in β-sheet. Denaturation reduces the β-sheet and destroys prion infectivity. The absence of any detectable chemical differences between PrP^C and PrP^{Sc} suggests that the pathogenicity derives solely from the α-helix→β-sheet conversion. Synthetic peptides for four regions of the amino acid sequence predicted to be structured have proven to be useful in modeling some aspects of this conformational change. Three of the four peptides of 13–17 amino acids for each of these regions formed amyloid fibrils. Longer peptides of up to 56 amino acids containing the first two regions were employed to optimize conditions for the α-helix→β-sheet conversion, as monitored by Fourier transform infrared strectroscopy (FTIR), circular dichroism and proton nuclear magnetic resonance (NMR) in solution. Dispersal in sodium dodecyl sulfate prevented aggregation and encouraged α-helix formation, allowing proton NMR measurements that confirmed the extent of the α-helices. Site–specific isotopically labeled peptides were prepared for some FTIR studies and for ^{13}C cross–polarization magic angle spinning NMR in the solid state, which revealed the locations of the β-strands. Rotational resonance in double–labeled peptides allowed the measurement of internuclear distances. The results of these studies proved helpful in guiding attempts at an in vitro conversion of PrP^C to PrP^{Sc}. Like PrP^{Sc}, the 56 amino acid peptide for residues 90-145 proved to be capable of causing a change in the physical properties of PrP^C and of a recombinant protein corresponding to residues 90–231, such that these entities became insoluble, protease resistant and rich in β-sheet, all of which are characteristics of PrP^{Sc}.

Introduction

The disease-specific scrapie prion protein (PrP^{Sc}) was first isolated and sequenced as the protease-resistant core designated PrP 27–30 [1–3]; later, the full-length PrP^{Sc} molecule was purified and sequenced [4–6]. The construction of oligonucleotide probes based on the N-terminal sequence of PrP 27–30 resulted in the molecular cloning of cDNAs encoding PrP [7, 8] which were in turn used to retrieve clones carrying the chromosomal *Prnp* gene, expressing a protein of 254 amino acids [9, 10]. Edman sequencing revealed post-translational trimming and loss of a 22-amino acid signal peptide, whereas PrP 27–30 proved to be N-terminally truncated by proteolysis, predominantly between residues 89 and 90 [3]. Because of its accumulation in infected brains, insolubility and protease resistance, PrP^{Sc} was purified more easily than the cellular isoform (PrP^C) which is sensitive to proteases and soluble in non dena-

---

*Correspondence and reprints

Key words: prion / prion protein / peptide

turing detergents [11]. Once sufficient PrP^C was purified, it was possible to demonstrate that it had the same N-terminal sequence as PrP^{Sc}, even though the two isoforms had different physical properties [6].

Covalent post-translational modifications were identified for both PrP^C and PrP^{Sc}, including N-linked complex oligosaccharides at the two consensus sites [12, 13], a single disulfide bond between the only two cysteines in the mature protein [6], and the loss of a C-terminal signal peptide with the attachment of a glycosyl-phosphatidylinositol anchor containing a hetero-geneous sialylated glycan [14]. Peptide mapping of PrP^{Sc} confirmed the anticipated amino acid sequence and revealed no further post-translational modifications [15]. More recently, analysis by mass spectrometry of the intact proteins suggests that the covalent structures of PrP^C and PrP^{Sc} are probably identical [16]. Nevertheless, the two isoforms possess very different physical properties; circular dichroism (CD) and Fourier transform infrared (FTIR) spectroscopic studies showed that this reflects dramatic differences in the secondary structure. PrP^C is rich in α-helices and devoid of β-sheets [17], whereas PrP^{Sc} has a high β-sheet content [17–21].

As yet there are no high-resolution structural data for either isoform, partly due to the diffi-culty of isolating sufficient amounts of purified protein for analysis and also because of the physical properties of PrP^{Sc}, which has a propensity to aggregate. Furthermore, there is no identified sequence homology with any other family of proteins. Despite this, a structure was predicted for PrP^C based on the spectroscopic observation that it is predominantly α-helical. The significant sequence differences between the prion proteins from different species, parti-cularly comparing mammalian and avian sequences [22], allowed sequence alignment and the prediction of four α-helical regions [23, 24]. Assuming PrP^C to be a four-helix bundle protein, there were 300,000 possible arrangements of the predicted helices [25]. Fortunately, this number could be reduced to 200 by constraints imposed by the presence of a disulfide bond between helices 3 and 4, the length of the loops and the requirement for the glycans to be at the surface. Realistic surface-to-volume ratios reduced this to four families, one of which was favored due to a close association between the locations of the known pathogenic mutations and their predicted destabilizing effects on the interhelix packing; almost all of the mutations lie within or close to the predicted helices [25]. Thus, there is a detailed model for the struc-ture of PrP^C, leading to predictions of the structure of PrP^{Sc} in which helices 1 and 2 are converted into four β-strands [26]. This model is supported by the following observations which suggest that the critical region involved in the conversion of PrP^C to PrP^{Sc} is most likely the N-terminus of PrP 27–30:

– It contains the most highly conserved region of PrP, even showing high homology between mammalian and avian forms [23].

– It is bounded by the N-terminus of infectious PrP 27–30 and an amber mutation in a patient who died from a prion disease [27].

– It is closely related to truncated species identified in PrP plaques, such as residues 58–150 [28].

– It contains the region in which sequence variations have the greatest influence on the barriers to transmission of disease between species, ie, residues 90–130 [29].

Peptides as models of the putative helices

The four putative helices were synthesized as individual peptides of 13 to 17 amino acids [24]. Some of these were found to have quite limited solubility in water, from which they tended to precipitate as amyloid fibrils that could be observed by electron microscopy and stained with Congo red dye to give green-gold birefringence under polarized light. Their secondary structures were investigated by FTIR in the amide I region (1,700–1,600 cm^{-1}) [24]. The FTIR technique readily distinguishes between β-sheet structures which absorb most strongly at low frequencies of ~1,620–1,630 cm^{-1} and α-helices or coils which absorb at ~1,645–1,660 cm^{-1} [30]. After deposition from the helix-promoting solvent hexafluoroisopropanol (HFIP), the peptides gave FTIR spectra in which the amide I bands indicated high α-helix content. The addition of water induced a rapid change in three of the four samples, H1, H3 and H4, with dramatic increases in the fraction of β-sheet. Only H2 remained largely unaltered [24]. Later experiments using CD spectroscopy confirmed these assignments, except that H2 was shown to be predominantly unstructured rather than α-helical [31].

The observation that certain regions of PrP prepared as small isolated peptides could exist in different conformational states was highly intriguing, particularly as the conversion of PrP^C into PrP^{Sc} involves a similar conformational change. These studies were extended to longer peptides that included flanking regions or that incorporated two of the predicted helices, such as Syrian hamster peptide SHa90-145, which includes both H1 and H2. In some cases the addition of the flanking regions had a dramatic effect on the behavior of the peptides; for example, an N-terminal five-residue extension of H1 from 109–122 to 104–122 (104H1) increased the solubility and completely removed the tendency for the peptide to be structured in aqueous solution [32]. The longer H1/H2 peptides denoted SHa109–141 and SHa90–145 were also unstructured in water, but the addition of 140 mM sodium chloride promoted β-sheet and fibril formation [31]. In contrast, high sodium dodecyl sulfate (SDS) concentrations induced α-helix formation in 104H1 (104–122), SHa109-141 and SHa90-145, as seen by CD spectroscopy, but not in H2 (129–141) alone. Nuclear magnetic resonance (NMR) showed that, in the presence of SDS, the H2 region of the longer peptides was indeed stabilized in its predicted α-helical form, apparently by the presence of the H1 helix [31]. This observation provided support for the prediction that the first two helices lie parallel to each other and are linked by hydrophobic interactions.

The application of NMR to the conformational analysis of peptides and proteins in solution is well established, but there is much less experience of NMR of peptides or proteins in the solid state [33]. However, PrP^{Sc} in its native state is largely insoluble and thus peptide models in solution are unrepresentative of this PrP isoform. Chemical shifts of ^{13}C-labeled peptides in the solid state were used to characterize the secondary structure at individual residues in H1. Lyophilization from acetonitrile gave a β-sheet structure, whereas lyophilization from HFIP showed that, in short peptides, this same section of the PrP sequence between residues 112 and 120 could also exist in an α-helical conformation. Exposure to humid air caused a shift to the β-sheet structure that was complete within 2 h, confirming that the α-helical form is only metastable in the solid state [34]. These experimental findings establish that the same PrP peptides can adopt either structure, thereby extending the earlier spectroscopic observations using FTIR and CD. Rotational resonance between pairs of ^{13}C nuclei in double-labeled peptides allowed the measurement of internuclear distances, giving preliminary indications of a turn between two strands in this region [34].

The structure of prion rods composed largely of PrP 27–30 was compared with that of linear peptide aggregates by X-ray fiber diffraction. The intersheet spacing of peptide H1 (5.91–7.99 Å) and of a shorter peptide corresponding to the core of H1 (5.13 Å) was significantly less than that of either PrP 27–30 (8.82 Å) or the longer peptide SHa90-145 (9.15 Å). The similarity between the spacings of the longer peptide and PrP 27–30 is consistent with the hypothesis that the H1/H2 region represents the β-sheet domain of PrP^{Sc} [35].

Peptides can induce conformational change

The results of transgenetic studies argue that conversion of PrP^C into PrP^{Sc} requires an interaction between PrP^C and PrP^{Sc} molecules that preexist within an infected cell or are introduced exogenously through inoculation of prions. Pathogenic mutations in the *Prnp* gene produce prion diseases, presumably by destabilizing PrP^C which results in its spontaneous conversion into PrP^{Sc} [36]. In the absence of such mutations, the spontaneous conversion of PrP^C into PrP^{Sc} is probably a rare event which could account for the low frequency with which sporadic Creutzfeldt-Jakob disease is encountered. It seems likely that during this interaction, the conformation of the PrP^C is changed to that of PrP^{Sc}, with the concomitant generation of infectivity. In general, PrP^{Sc} retains its full length, but in some instances the molecule undergoes limited proteolysis which results in its polymerization into amyloid [37]. Ordered polymers of PrP form amyloid plaques that are diagnostic of prion diseases when present, but are an inconsistent feature [38]. Considerable data argue that the intracellular accumulation of PrP^{Sc} is responsible for central nervous system dysfunction and the attendant neuropathologic changes [39, 40]. Underlying the formation of PrP^{Sc} is a profound change in the secondary structure which involves a decrease in α-helix and an increase in β-sheet. Attempts to convert PrP^C into PrP^{Sc} with an accompanying increase in scrapie infectivity have not been successful to date.

Some features of the conversion of PrP^C into PrP^{Sc} have been modeled with PrP peptides [32]. The H1 peptide was shown to convert other PrP peptides from α-helix or coil conformations to β-sheet structure. These include the H2 peptide, the only region of the putative PrP secondary structure that did not spontaneously adopt β-sheet as an isolated peptide in aqueous buffers. Longer, more hydrophilic peptides containing the H1 sequence were also converted by interaction with H1. The secondary structure of the more soluble peptide 104H1 (104–122) in aqueous solution was found to be primarily random coil. A small fraction of this extended peptide converted to β-sheet spontaneously in buffers containing acetonitrile; however, addition of ~1% H1 induced an $\alpha \rightarrow \beta$ shift in a substantially larger fraction of 104H1 molecules; eg, in 10% acetonitrile the addition of 1% H1 was sufficient to convert 25% of the 104H1 into β-sheet [32]. This conversion can be viewed as catalytic in that a small amount of one peptide was capable of inducing the conformational change in a large amount of another.

These interactions are most efficient between peptides of homologous sequences. H1 peptides for both mouse (Mo) and Syrian hamster (SHa), differing by only two amino acids, form β-sheets whereas SHa 104H1 peptide is random coil. The SHa 104H1 is converted rapidly by incubation with SHa H1, whereas Mo H1 induces a slow conversion that is incomplete after 24 h. Whether or not this slowed conversion of SHa 104H1 by Mo H1 is a reflection of the prion 'species barrier' remains to be established. In support of such a contention are studies with transgenic (Tg) mice expressing chimeric *Prnp* genes. While the region of PrP containing H1 and H2 is clearly an important factor in crossing the species barrier, it is not the only factor. It is noteworthy that even the most efficient conformational conversions of these peptides did not give rise to infectivity as monitored by intracerebral inoculation.

Once transgenetic studies had demonstrated that PrP^C must bind to PrP^{Sc} during the formation of nascent PrP^{Sc}, attempts to demonstrate the conversion of PrP^C into PrP^{Sc} in vitro were made [41]. In those studies, neither binding of PrP^C to PrP^{Sc} nor conversion of PrP^C into PrP^{Sc} could be demonstrated using equimolar amounts of the two PrP isoforms in the absence of any denaturants. More recently, when [^{35}S]PrP^C immunoprecipitated from mouse neuroblastoma (N2a) cells was mixed with a 50-fold excess of PrP^{Sc}, the radiolabeled PrP^C became insoluble and exhibited resistance to limited proteolysis. This was achieved after partial denaturation of the PrP^C and PrP^{Sc} in 3 M guanidine hydrochloride [42]. While the authors interpreted these results as conversion of PrP^C into a PrP^{Sc}-like molecule, the data are equally compatible with the binding of PrP^C to PrP^{Sc} in the absence of any conversion. The binding of PrP^C to PrP^{Sc} was species-specific [43] and the length of the protease-resistant core of [^{35}S]PrP^C was strain-dependent [44]. Attempts to separate the bound [^{35}S]PrP^C from the PrP^{Sc} using α-PrP monoclonal antibody, synthetic PrP peptides, detergents or phospholipids were unsuccessful [45]; thus, no conclusion can be drawn as to whether PrP^C is merely bound to PrP^{Sc} or its structure was actually altered. Furthermore, the large excess of PrP^{Sc} required for these studies precluded any assessment of scrapie prion infectivity by bioassay.

An alternative approach for converting PrP^C into PrP^{Sc} in vitro was developed using synthetic PrP peptides [45]. The object of these studies was to devise a system in which no PrP^{Sc} was added in order to be able to measure whether or not scrapie infectivity had been generated de novo. When [^{35}S]PrP^C was mixed with a synthetic PrP peptide corresponding to PrP residues SHa90–145, complexes containing PrP^C and the peptide were formed. These PrP^C–peptide complexes exhibited many properties that are characteristic of PrP^{Sc}, including insolubility in non denaturing detergents, resistance to proteolytic digestion and high β-sheet content. Smaller PrP peptides in the random coil conformation formed complexes with PrP^C at a much lower efficiency. As with the PrP^C–PrP^{Sc} complexes which required an excess of PrP^{Sc}, it was necessary to use a large excess of peptide. However, the pelleted material was found to contain approximately equimolar quantities of peptide and protein (Kaneko et al, unpublished data). As was observed for the PrP peptide–peptide and PrP^C–PrP^{Sc} complexes, the interactions were most efficient when the sequences were homologous [45].

Despite the physical changes in the protein, no prion infectivity has been induced to date by interaction with peptides. This may reflect reversibility of the complex formation and loss of protease resistance which suggests that a complete conversion of PrP^C molecules to PrP^{Sc} has not taken place. It is likely that there is an equilibrium between the PrP^C structure and an intermediate that has partial PrP^{Sc} character, but is not stable in the absence of excess peptide. Experiments with Tg mice having chimeric *Prnp* genes strongly implicate the requirement for another cellular factor to induce complete and irreversible conversion [46]. It remains to be determined whether such a factor can be isolated and added to the PrP^C–peptide mixtures to assist conversion in vitro.

Cytotoxic activity

It has been noted that there are many similarities between prion diseases and Alzheimer's disease [47]. Cytotoxicity of cultured neurons is well established for several amyloid-forming Aβ peptides. Prior incubation of the peptides to encourage β-sheet formation enhances this effect, and it has been suggested that the generation of free radicals and oxidation may play a role [48]. As yet it is not clear whether these findings have any direct relevance to the development of Alzheimer's disease. Similar neurotoxic effects have been described for the peptide corresponding to PrP residues 106–126, whereas other peptides which together accounted for almost all of the mature PrP sequence did not possess this property [49]. Again, although this observation is intriguing, its relevance to the induction of prion diseases is unclear.

Recombinant PrP

A major reason for the use of chemically synthesized peptides as models for the behavior of PrP has been the difficulty of isolating sufficient pure PrP from natural sources to carry out a comprehensive analysis. This was compounded by difficulties with the overexpression of the recombinant protein. Despite investigations on expression in *Escherichia coli*, yeast, baculovirus and Chinese hamster ovary cells, only recently has it proved possible to express and purify substantial amounts of PrP, albeit a truncated molecule corresponding to PrP residues 90–231 (termed rPrP) which is analogous to the amino acid backbone of PrP 27–30 [50]. The same conformational pluralism observed with the synthetic PrP peptide SHa90–145 has been observed with rPrP. Studies are in progress to determine if a β-sheet form of rPrP will induce PrPSc formation in inoculated rodents.

Conclusion

Despite the isolation of both the cellular and pathogenic isoforms of PrP more than a decade ago, these proteins have been particularly intractable to high-resolution structural studies. This is due to a combination of unfavorable physical properties such as the insolubility of PrPSc, the heterogeneity of the post-translational modifications and the difficulty of purifying significant quantities of either isoform, particularly PrPC. Consequently, the structural clues provided by synthetic PrP peptides corresponding to regions of putative secondary structure have been valuable.

The initial analysis of the peptides was generally confined to those regions predicted to exhibit secondary structure [24]. Such peptides displayed conformational transitions in accord with the ambiguous predictions generated from amino acid sequence analyses. To extend those studies, larger domains of PrP embodying multiple regions of secondary structure were synthesized [31]. The peptide SHa90–145 which embraces the two most N-terminal putative helical regions has been quite useful in both structural studies as well as investigations of PrPC–peptide complex formation [45]. That the SHa90–145 peptide can induce protease resistance in PrPC suggests that this, or a related peptide, might be able to initiate the formation of PrPSc. While some features of the transition of PrPC to PrPSc have been simulated with peptides, no prion infectivity has been created to date.

Studies of prions continue to provide a fascinating challenge to conventional paradigms in biology and disease. Because the peptides are synthesized chemically and thus are devoid of PrPSc, any infectivity measured by bioassay represents de novo generation of prions. Such a system should allow unambiguous definition of the constituents of the prion particle.

References

1 Prusiner SB, Bolton DC, Groth DF et al. Further purification and characterization of scrapie prions. *Biochemistry* 1982;21:6942-50
2 Bolton DC, McKinley MP, Prusiner SB. Identification of a protein that purifies with the scrapie prion. *Science* 1982;218:1309-11
3 Prusiner SB, Groth DF, Bolton DC, Kent SB, Hood LE. Purification and structural studies of a major scrapie prion protein. *Cell* 1984;38:127-34
4 Hope J, Morton LJD, Farquhar CF et al. The major polypeptide of scrapie-associated fibrils (SAF) has the same size, charge distribution and N-terminal protein sequence as predicted for the normal brain protein (PrP). *EMBO J* 1986;5:2591-7
5 Bolton DC, Bendheim PE, Marmorstein AD, Potempska A. Isolation and structural studies of the intact scrapie agent protein. *Arch Biochem Biophys* 1987;258:579-90

6   Turk E, Teplow DB, Hood LE, Prusiner SB. Purification and properties of the cellular and scrapie hamster prion proteins. *Eur J Biochem* 1988;176:21-30

7   Oesch B, Westaway D, Wälchli M et al. A cellular gene encodes scrapie PrP 27-30 protein. *Cell* 1985;40:735-46

8   Chesebro B, Race R, Wehrly K et al. Identification of scrapie prion protein-specific mRNA in scrapie-infected and uninfected brain. *Nature* 1985;315:331-3

9   Basler K, Oesch B, Scott M et al. Scrapie and cellular PrP isoforms are encoded by the same chromosomal gene. *Cell* 1986;46:417-28

10 Westaway D, Cooper C, Turner S et al. Structure and polymorphism of the mouse prion protein gene. *Proc Natl Acad Sci USA* 1994;91:6418-22

11 Meyer RK, McKinley MP, Bowman KA et al. Separation and properties of cellular and scrapie prion proteins. *Proc Natl Acad Sci USA* 1986;83:2310-4

12 Endo T, Groth D, Prusiner SB, Kobata A. Diversity of oligosaccharide structures linked to asparagines of the scrapie prion protein. *Biochemistry* 1989;28:8380-8

13 Haraguchi T, Fisher S, Olofsson S et al. Asparagine-linked glycosylation of the scrapie and cellular prion proteins. *Arch Biochem Biophys* 1989;274:1-13

14 Stahl N, Borchelt DR, Hsiao K, Prusiner SB. Scrapie prion protein contains a phosphatidylinositol glycolipid. *Cell* 1987;51:229-40

15 Stahl N, Baldwin MA, Teplow DB et al. Structural analysis of the scrapie prion protein using mass spectrometry and amino acid sequencing. *Biochemistry* 1993;32:1991-2002

16 Baldwin MA, Wang R, Pan KM et al. Matrix-assisted laser deborption/imization mass spectrometry of membrane proteins: the scrapie prion protein. In: Angeletti RH, ed. *Techniques in Protein Chemistry IV.* San Diego, CA: Academic Press, 1993;41-5

17 Pan KM, Baldwin M, Nguyen J et al. Conversion of α-helices into β-sheets features in the formation of the scrapie prion proteins. *Proc Natl Acad Sci USA* 1993;90:10962-6

18 Caughey BW, Dong A, Bhat KS et al. Secondary structure analysis of the scrapie-associated protein PrP 27–30 in water by infrared spectroscopy. *Biochemistry* 1991;30:7672-80

19 Gasset M, Baldwin MA, Fletterick RJ, Prusiner SB. Perturbation of the secondary structure of the scrapie prion protein under conditions associated with changes in infectivity. *Proc Natl Acad Sci USA* 1993;90:1-5

20 Safar J, Roller PP, Gajdusek DC, Gibbs CJ, Jr. Conformational transitions, dissociation, and unfolding of scrapie amyloid (prion) protein. *J Biol Chem* 1993;268:20276-84

21 Safar J, Roller PP, Gajdusek DC, Gibbs CJ. Thermal-stability and conformational transitions of scrapie amyloid (prion) protein correlate with infectivity. *Protein Sci* 1993;2:2206-16

22 Harris DA, Falls DL, Johnson FA, Fischbach GD. A prion-like protein from chicken brain copurifies with an acetylcholine receptor-inducing activity. *Proc Natl Acad Sci USA* 1991;88:7664-8

23 Gabriel JM, Oesch B, Kretzschmar H, Scott M, Prusiner SB. Molecular cloning of a candidate chicken prion protein. *Proc Natl Acad Sci USA* 1992;89:9097-101

24 Gasset M, Baldwin MA, Lloyd D et al. Predicted α-helical regions of the prion protein when synthesized as peptides form amyloid. *Proc Natl Acad Sci USA* 1992;89:10940-4

25 Huang Z, Gabriel JM, Baldwin MA et al. Proposed three-dimensional structure for the cellular prion protein. *Proc Natl Acad Sci USA* 1994;91:7139-43

26 Huang Z, Prusiner SB, Cohen FE. Scrapie prions: a three-dimensional model of an infectious fragment. *Folding Design* 1996;1:13-9

27 Kitamoto T, Iizuka R, Tateishi J. An amber mutation of prion protein in Gerstmann-Sträussler syndrome with mutant PrP plaques. *Biochem Biophys Res Commun* 1993;192:525-31

28 Tagliavini F, Prelli F, Ghisto J et al. Amyloid protein of Gerstmann-Sträussler-Scheinker disease (Indiana kindred) is an 11-kD fragment of prion protein within N-terminal glycine at codon 58. *EMBO J* 1991;10:513-9

29 Schätzl HM, Da Costa M, Taylor L, Cohen FE, Prusiner SB. Prion protein gene variation among primates. *J Mol Biol* 1995;245:362-74

30 Byler DM, Susi H. Examination of the secondary structure of proteins by deconvolved FTIR spectra. *Biopolymers* 1986;25:469-87

31 Zhang H, Kaneko K, Nguyen JT et al. Conformational transitions in peptides containing two putative α-helices of the prion protein. *J Mol Biol* 1995;250:514-26

32 Nguyen J, Baldwin MA, Cohen FE, Prusiner SB. Prion protein peptides induce α-helix to β-sheet conformational transitions. *Biochemistry* 1995;34:4186-92

33 Spencer RGS, Halverson KJ, Auger M et al. An unusual peptide conformation may precipitate amyloid formation in Alzheimer's disease: application of solid-state NMR to the determination of protein secondary structure. *Biochemistry* 1991;30:10382-7

34 Heller J, Larsen R, Ernst M et al. Application of rotational resonance to inhomogeneously broadened systems. *Chem Phys Lett* 1996;251:223-9

35 Nguyen JT, Inouye H, Baldwin MA et al. X-ray diffraction of scrapie prion rods and PrP peptides. *J Mol Biol* 1995;252:412-22

36 Cohen FE, Pan KM, Huang Z et al. Structural clues to prion replication. *Science* 1994;264:530-1

37 Prusiner SB, McKinley MP, Bowman KA et al. Scrapie prions aggregate to form amyloid-like birefringent rods. *Cell* 1983;35:349-58

38 Roberts GW, Lofthouse R, Allsop D et al. CNS amyloid proteins in neurodegenerative diseases. *Neurology* 1988;38:1534-40

39 Prusiner SB, Scott M, Foster D et al. Transgenetic studies implicate interactions between homologous PrP isoforms in scrapie prion replication. *Cell* 1990;63:673-86

40 Brandner S, Isenmann S, Raeber A et al. Normal host prion protein necessary for scrapie-induced neurotoxicity. *Nature* 1996;379:339-43

41 Raeber AJ, Borchelt DR, Scott M, Prusiner SB. Attempts to convert the cellular prion protein into the scrapie isoform in cell-free systems. *J Virol* 1992;66:6155-63

42 Kocisko DA, Come JH, Priola SA et al. Cell-free formation of protease-resistant prion protein. *Nature* 1994;370: 471-4

43 Kocisko DA, Priola SA, Raymond GJ et al. Species specificity in the cell-free conversion of prion protein to protease-resistant forms: a model for the scrapie species barrier. *Proc Natl Acad Sci USA* 1995;92:3923-7

44 Bessen RA, Kocisko DA, Raymond GJ et al. Non-genetic propagation of strain-specific properties of scrapie prion protein. *Nature* 1995;375:698-700

45 Kaneko K, Peretz D, Pan KM et al. Prion protein (PrP) synthetic peptides induce cellular PrP to acquire properties of the scrapie isoform. *Proc Natl Acad Sci USA* 1995;92:11160-4

46 Telling GC, Scott M, Mastrianni J et al. Prion propagation in mice expressing human and chimeric PrP transgenes implicates the interaction of cellular PrP with another protein. *Cell* 1995;83:79-90

47 Price DL, Borchelt DR, Sisodia SS. Alzheimer's disease and the prion disorders: amyloid β-protein and prion protein amyloidoses. *Proc Natl Acad Sci USA* 1993;90:6381-4

48 Hensley K, Carney JM, Mattson MP et al. A model for beta-amyloid aggregation and neurotoxicity based on free radical generation by the peptide: relevance to Alzheimer's disease. *Proc Natl Acad Sci USA* 1994;91: 3270-4

49 Forloni G, Angeretti N, Chiesa R et al. Neurotoxicity of a prion protein fragment. *Nature* 1993;362:543-6

50 Mehlhorn I, Groth D, Stöckel J et al. High-level expression and characterization of a purified 142-residue polypeptide of the prion protein. *Biochemistry* 1996;35:5528-37

Transmissible Subacute Spongiform Encephalopathies:
Prion Diseases
L Court, B Dodet, eds
© 1996, Elsevier, Paris

Chemicophysical properties and biological activities of synthetic prion protein (PrP) peptides and their relationship to human cerebral PrP amyloidoses

Fabrizio Tagliavini[1], Frances Prelli[2], Mario Salmona[3], Bernardino Ghetti[4], Giorgio Giaccone[1], Orso Bugiani[1], Blas Frangione[2], Gianluigi Forloni[3]

[1]*Istituto Nazionale Neurologico Carlo Besta, via Celoria 11, 20133 Milan, Italy;* [2]*Department of Pathology, New York University Medical Center, 530 First Avenue, New York, NY 10016, USA;* [3]*Istituto di Ricerche Farmacologiche Mario Negri, via Eritrea 62, 20157 Milan, Italy;* [4]*Department of Pathology and Laboratory Medicine, Indiana University School of Medicine, Indianapolis, IN 46202-5251, USA*

Summary – To investigate the role of prion protein (PrP) in the pathogenesis of prion diseases, we analyzed the effects of the exposure of primary cultures of neurons and astrocytes to synthetic peptides homologous to consecutive segments of the amyloid protein purified from brain tissue of patients with Gerstmann-Sträussler-Scheinker disease (ie, the octapeptide repeat region, residues 89–106, 106–126 and 127–147). At variance with other PrP fragments, the peptide PrP 106–126 caused neuronal death by apoptosis and proliferation, and hypertrophy of astrocytes with increased expression of glial fibrillary acidic protein. Structural analyses showed that PrP 106–126 is able to adopt different conformations in distinct environments; however, it has a high propensity to form stable β-sheet structures and to polymerize into amyloid-like fibrils that are partially resistant to proteinase K digestion. These data suggest that the region corresponding to residues 106–126 of human PrP may feature in the conformational transition from normal to abnormal PrP, and that cerebral accumulation of peptides including this sequence may by responsible for the neuropathological changes that occur in prion diseases.

Introduction

Prion diseases are neurodegenerative conditions characterized by the accumulation of abnormal isoforms of the prion protein (PrP) and the deposition of PrP amyloid in the brain [1, 2]. Distinct from normal PrP, the disease-specific PrP molecules are resistant to protease digestion [1, 2]; they are thought to be derived from protease-sensitive precursors by a post-translational process that involves a conformational change with a shift from α-helix to β-sheet secondary structure [3–5].

Amyloid formation occurs to the highest degree in Gerstmann-Sträussler-Scheinker (GSS) disease and in PrP-cerebral amyloid angiopathy (PrP-CAA). GSS disease is an autosomal

---

Key words: prion protein / prion protein peptides / prion protein amyloid / amyloidogenesis / neurotoxicity

dominant disorder that shows a wide spectrum of clinical presentations (eg, ataxia, spastic paraparesis, parkinsonism and dementia) and is associated with variant *PRNP* genotypes resulting from the combination of a mutation at codons 102 (P > L), 105 (P > L), 117 (A > V), 198 (F > S) and 217 (Q > R) with common polymorphisms at codon 129 (M/V) and codon 219 (E/K) [6]. PrP-CAA is a cerebrovascular amyloidosis associated with a mutation at *PRNP* codon 145 (Y > stop) [7].

The clinical variability of GSS disease is related to the distribution and extent of amyloid deposition as well as the occurrence of associated lesions [6]. Uni- and multicentric amyloid deposits may be found throughout the brain. In most cases, they are especially abundant in the cerebellar cortex; however, in a few instances (eg, some patients with the codon-117 mutation) the cerebellum is not involved. In addition to amyloid deposits, many patients with P > L substitution at codon 102 show severe spongiform changes in the cerebral cortex, while patients with codon 145, codon 198 or codon 217 mutations display neurofibrillary tangles composed of paired helical filaments in the neocortex, archicortex and subcortical nuclei [6–10].

Amyloid deposition is consistently accompanied by hypertrophy and proliferation of astrocytes and microglial cells, neuritic abnormalities and neuronal loss leading to variable degrees of atrophy of the affected regions [6]. The close topographical relationship between amyloid deposits and tissue changes suggests that PrP amyloid plays a role in the pathogenesis of nerve cell degeneration and glial cell reaction.

On this basis, we carried out studies to define the composition of amyloid fibrils in GSS disease families with different *PRNP* mutations, to identify the PrP sequences that are central to amyloid formation, and to evaluate the biological effects of PrP peptides on nerve and glial cells in vitro.

The amyloid protein in GSS disease

The biochemical composition of PrP amyloid was first determined in brain tissue samples obtained from patients of the Indiana kindred of GSS disease [11, 12] carrying an F > S substitution at PrP residue 198 [13, 14]. The amyloid preparations contained two major peptides of ~11 and ~7 kDa spanning residues 58–150 and 81–150 of PrP, respectively [11, 12].

The finding that the amyloid protein was an N- and C-terminal truncated fragment of PrP was verified by immunolabeling brain sections with antisera raised against synthetic peptides homologous to PrP residues 23–40, 90–102, 127–147 and 220–231. The amyloid cores were immunoreactive with the antisera to the mid-region of the molecule while only the periphery of the cores was immunolabeled by antibodies to N- or C-terminal domains [15].

In GSS-198, the amyloid protein does not include the region containing the amino acid substitution. To establish whether amyloid peptides originate from mutant or both mutant and wild-type PrP, we analyzed patients heterozygous for M/V at codon 129 and used V^{129} as a marker of the mutant allele, since in this family V^{129} is in phase with mutant S198. Amino-acid sequencing and mass spectrometry of peptides generated by digestion of the amyloid protein with endoproteinase Lys-C showed that the samples contained only peptides with V^{129}, suggesting that only mutant PrP was involved in amyloid formation [12].

Thereafter, we characterized the amyloid protein in GSS disease kindreds with other *PRNP* mutations (ie, A > V at codon 117 and Q > R at codon 217) and found that the smallest amyloid subunit was a ~7 kDa N- and C-terminal truncated fragment of PrP, whose size and sequence were similar to those of GSS-198. In all instances the amyloid protein was derived from the mutant allele [12]; in patients with GSS-117 it contained the variant V[117] [16].

Assembly and conformation of PrP peptides in vitro

To determine which residues are important for polymerization of PrP peptides into amyloid fibrils and which conditions promote peptide assembly, we investigated the fibrillogenicity of synthetic peptides homologous to consecutive segments of the amyloid protein purified from GSS disease-affected brains (ie, the octapeptide repeat region, residues 89–106, 106–126 and 127–147 of PrP). We found that peptides within the PrP sequence 106–147 readily assembled into fibrils, while peptides corresponding to the N-terminal segment of the amyloid protein did not [17]. In particular, the peptide PrP 106–126 was highly fibrillogenic and formed dense meshworks of straight filaments ultrastructurally similar to those observed in GSS disease-affected brains. The fibrillary assemblies generated by PrP 106–126 were partially resistant to proteinase K digestion and exhibited tinctorial and optical properties of in situ amyloid; that is, birefringence under polarized light after Congo red staining and fluorescence after thioflavine S treatment. In addition, they showed an X-ray diffraction pattern consistent with that of native amyloid fibrils, with reflections corresponding to H-bonds between neighboring polypeptide chains in a cross β-configuration [17, 18].

Circular dichroism spectroscopy revealed that peptide PrP 106–126 is able to adopt different conformations in relation to the microenvironment [19]. It showed primarily a β-sheet secondary structure in phosphate buffer, pH 5.0, a combination of β-sheet and random coil in phosphate buffer, pH 7.0, a random coil conformation in deionized water and an α-helical structure in the presence of micelles formed by a 5% sodium dodecyl sulfate solution. The addition of α-helix stabilizing solvents (eg, trifluoroethanol or hexafluoropropanol) to a solution of PrP 106–126 in deionized water induced a conformational shift from random coil to α-helix, but did not modify the β-sheet structure of the peptide previously suspended in phosphate buffer, pH 5.0 [19]. These data suggest that the PrP region including residues 106–126 may feature in the conformational transition from normal to abnormal PrP.

Biological effects of PrP peptides in vitro

To test the hypothesis that the accumulation of PrP amyloid protein causes nerve cell degeneration and glial cell reaction in GSS disease and PrP-CAA, we studied the effects of the exposure of primary cultures of neurons and astrocytes to the synthetic peptides used for fibrillogenesis studies.

The prolonged exposure of rat hippocampal neurons to micromolar concentrations of PrP 106–126 resulted in marked neuronal loss [20]. Conversely, the other PrP peptides and a scrambled sequence of PrP 106–126 did not have such an effect. The neurotoxicity of PrP 106–126 was dose-dependent; the toxic response was first detected at a concentration of 20 μM. It was statistically significant at 40 μM and resulted in virtually complete neuronal loss at 60 or 80 μM. Fluorescence microscopy following treatment with DNA-binding fluorochromes (eg, Hoechst 33258) as well as electron microscopy showed that neurons chronically exposed to PrP 106–126 presented a typical apoptotic morphology, with condensation of the chromatin and fragmentation of the nucleus. Agarose-gel electrophoresis of DNA extracted from cultured cells after 7 days of treatment with the peptide showed an apoptotic pattern of DNA fragmentation, resulting from cleavage of nuclear DNA in internucleosomal regions [20].

The prolonged treatment of rat astroglial cultures with PrP 106–126 induced a remarkable increase in the size and density of astroglial processes [21]. Conversely, no effects were observed following treatment of cultures with the other PrP peptides. The hypertrophy of astrocytes was associated with a prominent increase in glial fibrillary acidic protein (GFAP) transcripts as revealed by Northern blot analysis. Densitometric quantification of GFAP mRNA normalized for the level of β-actin message showed that this increment was dependent on the peptide concentration, being significant at 10 μM and resulting in a three- and five-fold increase above control values at 25 and 50 μM, respectively. The rise of GFAP transcripts was accompanied by a substantial increase in GFAP, as determined by Western blot analysis [21]. The hypertrophy of astroglial cells was associated with a 1.5-fold increase in cell number at a peptide concentration of 50 μM. The proliferation rate of astrocytes was much higher when cultures were kept in serum-free medium for the duration of the experiment [22]. The proliferative effect of PrP 106–126 was abolished by cotreatment of astroglial cultures with nicardipine; that is, a blocker of L-type voltage-sensitive calcium channels. Microfluorimetric analysis of intracellular calcium levels in single astrocytes showed that PrP 106–126 induced a rapid increase in cytosolic calcium concentrations, while the scrambled peptide did not. This effect was absent when calcium was removed from the medium and was prevented by preincubation of cultures with nicardipine. These data suggest that PrP 106-126 stimulates astroglial proliferation via an increase in intracellular calcium concentration, through the activation of L-type voltage-sensitive calcium channels [22].

In summary, our studies showed that the amyloid protein in GSS disease is an N- and C-terminal truncated fragment of PrP originating from mutant molecules. This fragment contains a sequence (ie, residues 106–126) that can adopt different conformations in distinct environments, although it has a high propensity to form stable β-sheet structures. When synthesized as a peptide, the sequence PrP 106–126 is fibrillogenic and partially resistant to protease digestion; it is toxic to neurons while having a growth-promoting activity on astroglial cells. This sequence is an integral part of PrP peptides that accumulate in the central nervous system of patients with prion diseases, and might be a major contributor to the molecular characteristics and the pathogenic properties of disease-specific PrP isoforms and PrP amyloid.

Acknowledgments

This work was supported by the Italian Ministry of Health, Department of Social Services, by the US National Institutes of Health (Grant NS29822) and by Telethon-Italy (Grant E.250).

References

1 Prusiner SB. Molecular biology of prion diseases. *Science* 1991;252:1515-22
2 Prusiner SB. Genetic and infectious prion diseases. *Arch Neurol* 1993;50:1129-53
3 Caughey BW, Dong A, Bhat KS, Ernst D, Hayes SF, Caughey WS. Secondary structure analysis of the scrapie-associated protein PrP 27-30 in water by infrared spectroscopy. *Biochemistry* 1991;30:7672-80
4 Pan KM, Baldwin M, Nguyen J et al. Conversion of α-helices into β-sheets features in the formation of the scrapie prion protein. *Proc Natl Acad Sci USA* 1993;90:10962-6
5 Safar J, Roller PP, Gajdusek DC, Gibbs CJ. Conformational transitions, dissociation, and unfolding of scrapie amyloid (prion) protein. *J Biol Chem* 1993;268:20276-84
6 Ghetti B, Piccardo P, Frangione B et al. Prion protein amyloidosis. *Brain Pathol* 1996;6:127-47
7 Ghetti B, Piccardo P, Spillantini MG et al. Vascular variant of prion protein cerebral amyloidosis with τ-positive neurofibrillary tangles: the phenotype of the stop codon 145 mutation in *PRNP*. *Proc Natl Acad Sci USA* 1996;93:744-8

8 Ghetti B, Tagliavini F, Giaccone G et al. Familial Gerstmann-Sträussler-Scheinker disease with neurofibrillary tangles. *Mol Neurobiol* 1994;8:41-8

9 Giaccone G, Tagliavini F, Verga L et al. Neurofibrillary tangles of the Indiana kindred of Gerstmann-Sträussler-Scheinker disease share antigenic determinants with those of Alzheimer's disease. *Brain Res* 1990;530:325-9

10 Tagliavini F, Giaccone G, Prelli F et al. A68 is a component of paired helical filaments of Gerstmann-Sträussler-Scheinker disease, Indiana kindred. *Brain Res* 1993;616:325-8

11 Tagliavini F, Prelli F, Ghiso J et al. Amyloid protein of Gerstmann-Sträussler-Scheinker disease (Indiana kindred) is an 11-kD fragment of prion protein with an N-terminal glycine at codon 58. *EMBO J* 1991;10: 513-9

12 Tagliavini F, Prelli F, Porro M et al. Amyloid fibrils in Gerstmann-Sträussler-Scheinker disease (Indiana and Swedish kindreds) express only PrP peptides encoded by the mutant allele. *Cell* 1991;79:695-703

13 Dlouhy SR, Hsiao K, Farlow MR et al. Linkage of the Indiana kindred of Gerstmann-Sträussler-Scheinker disease to the prion protein gene. *Nature Genet* 1992;1:64-7

14 Hsiao K, Dlouhy SR, Farlow MR et al. Mutant prion proteins in Gerstmann-Sträussler-Scheinker disease with neurofibrillary tangles. *Nature Genet* 1992;1:68-71

15 Giaccone G, Verga L, Bugiani O et al. Prion protein preamyloid and amyloid deposits in Gerstmann-Sträussler-Scheinker disease, Indiana kindred. *Proc Natl Acad Sci USA* 1992;89:9349-53

16 Tagliavini F, Prelli F, Porro M et al. Only mutant PrP participates in amyloid formation in Gerstmann-Sträussler-Scheinker disease with Ala > Val substitution at codon 117. *J Neuropathol Exp Neurol* 1991;54:416

17 Tagliavini F, Prelli F, Verga L et al. Synthetic peptides homologous to prion protein residues 106–147 form amyloid-like fibrils in vitro. *Proc Natl Acad Sci USA* 1993;90:9678-82

18 Selvaggini C, De Gioia L, Cantù L et al. Molecular characteristics of a protease-resistant, amyloidogenic and neurotoxic peptide homologous to residues 106–126 of the prion protein. *Biochem Biophys Res Commun* 1993;194:1380-6

19 De Gioia L, Selvaggini C, Ghibaudi E et al. Conformational polymorphism of the amyloidogenic and neurotoxic peptide homologous to residues 106–126 of the prion protein. *J Biol Chem* 1994;269:7859-62

20 Forloni G, Angeretti N, Chiesa R et al. Neurotoxicity of a prion protein fragment. *Nature* 1993;362:543-5

21 Forloni G, Del Bo R, Angeretti N et al. A neurotoxic prion protein fragment induces rat astroglial proliferation and hypertrophy. *Eur J Neurosci* 1994;6:1415-22

22 Florio T, Grimaldi M, Scorziello A et al. The prion protein fragment 106–126 increases intracellular calcium levels through a dihydropyridine-sensitive mechanism and induces cortical type I astrocyte proliferation in vitro. *Soc Neurosci Abstr* 1995;21:494

IN VITRO EXPRESSION

Transmissible Subacute Spongiform Encephalopathies:
Prion Diseases
L Court, B Dodet, eds
© 1996, Elsevier, Paris

Molecular chaperones and RNA aptamers as interactors for prion proteins

Stefan Weiss, Frank Edenhofer, Roman Rieger, Daniela Proske, Michael Famulok, Evelyn Fisch, Ernst-Ludwig Winnacker*

Laboratorium für Molekulare Biologie-Genzentrum-Institut für Biochemie der LMU München, Feodor-Lynen-strasse 25, D-81377 Munich, Germany

Summary – Prion proteins (PrP) play a crucial role in the pathogenesis of transmissible spongiform encephalopathies including scrapie, bovine spongiform encephalopathy and Creutzfeldt-Jakob disease. RNA aptamers and proteins specifically interacting with PrP from the Syrian golden hamster were identified using in vitro selection and the yeast two-hybrid system. The human molecular chaperone Hsp60 was identified as an interactor for the protease-sensitive prion protein (PrPC) in the *Saccharomyces cerevisiae* two-hybrid screen. Pull-down assays demonstrated that Hsp60 as well as its prokaryotic homologue GroEL also interact in vitro with recombinant PrPC and rPrP 27–30, the recombinant proteinase K-sensitive counterpart to the protease-resistant prion protein, PrP 27–30. Recombinant PrP peptides allowed the identification of the PrP-binding site for Hsp60 and GroEL. To date, no antibodies have been isolated which are able to distinguish between PrPC and disease-specific prion protein (PrPSc). In vitro selection (SELEX) employing recombinant PrPC identified RNA aptamers which bind to PrPC but not to rPrP 27–30. The aptamers may harbor guanosine quartets stacked upon each other which results in an extremely stable structure. Prion-specific aptamers could function as a diagnostic tool for the detection of prion diseases.

Molecular chaperones interact with prion proteins

Prions are proteinaceous infectious particles which are thought to replicate in the absence of nucleic acids [1–3]. Neither the 'protein only' or 'heterodimer' hypothesis [3] nor the 'nucleation-dependent polymerization model' [4] allows the involvement of nucleic acids in the propagation of prions. A central event in prion propagation is the conversion of the protease-sensitive prion protein (PrPC) into the disease-specific prion protein (PrPSc) concomitant with significant structural alterations [5]. α-Helical structures in PrPC are thought to convert into β-sheets in PrPSc and PrP 27–30, respectively. Molecular chaperones have been considered to be involved in prion replication [6]. Experiments involving transgenic mice suggested that chaperones might participate in the life cycle of prions [7]. We applied the yeast two-hybrid system [8] to screen for proteins specifically interacting with prion proteins. Glutathione S-transferase (GST)::PrPC fused to *Lex*A was used as the 'bait' (fig 1A) to screen a human HeLa cDNA library fused to the acidic activation domain B42. If any of the encoded proteins inter-

---

*Correspondence and reprints

***Key words:** aptamers / chaperones / prions*

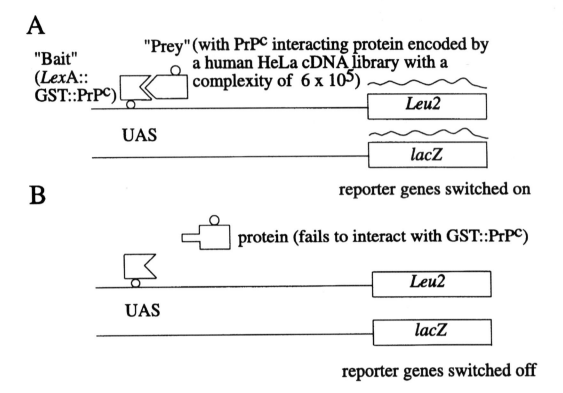

Fig 1. Schematic presentation of the *Saccharomyces cerevisiae* two-hybrid system [8]. **A.** The interaction between the 'prey' and the 'bait' activates the reporter system leading to complementation of leucine auxotrophy and blue-colored colonies in the presence of X-Gal. **B.** No interaction between 'prey' and 'bait'. The reporter system remains inactive.

acts with GST::PrP^C (fig 1A), the reporter system containing a *LacZ* and a *Leu2* expression cassette is switched on, leading to the complementation of leucine auxotrophy and to blue-colored colonies in the presence of X-Gal. When no interaction takes place, the reporter system remains inactive (fig 1B).

Yeast strain EGY 48 was transformed with pSH-GST::PrP^C [9], resulting in the expression of the LexA::GST::PrP^C bait. Yeast cells expressing LexA::GST::PrP^C were cotransformed with 'prey' and reporter plasmids. Yeast transformants were screened for both phenotypes and 55 positive clones sequenced; 17% encoded for the human chaperone Hsp60 (fig 2A). LexA::GST (fig 2A) and other control baits such as LexA::Bicoid did not interact with the chaperone [9]. To confirm this interaction in vitro we employed recombinant molecular chaperones as well as the recombinant prion proteins PrP^C [10] and rPrP 27–30 [11]. The latter represents the recombinant proteinase K-sensitive counterpart to PrP 27–30, the main component in scrapie prion preparations [12–14]. Hsp60 and its bacterial homolog GroEL [15] interact with PrP^C (fig 2B) and rPrP 27–30 (fig 2B) both fused to GST [9]. GST as well as the molecular chaperone Hsp70 interacted neither with Hsp60 nor with GroEL ([9], fig 2B),

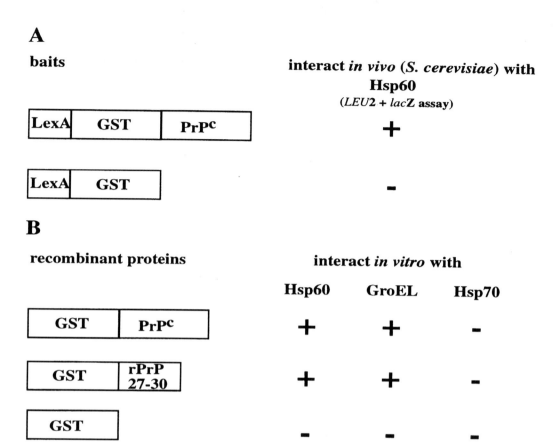

Fig 2. Interaction of prion proteins with molecular chaperones. **A.** Interactions in the *S cerevisiae* two-hybrid system. **B.** In vitro interactions employing recombinant proteins.

demonstrating the specificity of the PrP^C-Hsp60/GroEL interaction. To further map the interaction site between Hsp60 and the prion protein, we used recombinant prion peptides ([10], fig 3) fused to GST. Only petide P$_{180-210}$ interacts with Hsp60 and GroEL, suggesting that the proposed α-helical regions H3 and H4 [16] might be involved in prion propagation ([9]; fig 3). The role of molecular chaperones in the life cycle of prions, however, remains speculative. The generation of infectious prion proteins from uninfectious recombinant PrP molecules in the presence of molecular chaperones would be the ultimate proof that chaperones participate in prion replication. The subcellular site(s) where the interaction of prion proteins with molecular chaperones occur is also unknown. PrP^C is localized on the surface of neuronal cells via its glycophosphoinositol (GPI) anchor. Hsp60 has also been identified in membrane fractions of mammalian cells [17] as well as in compartments of the secretory pathway such as the endoplasmic reticulum, the Golgi apparatus, condensing vacuoles and secretory granules [18]. PrP^C also uses the secretory pathway on its way to the cell surface [19].

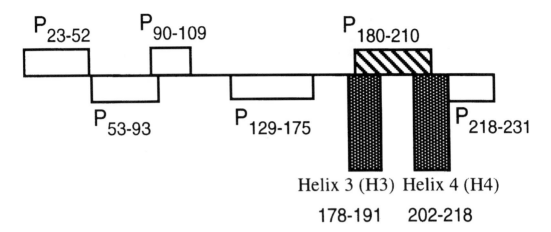

Fig 3. Interaction of prion peptides with molecular chaperones. Only peptide $P_{180-210}$ (hatched box) interacts with the molecular chaperones Hsp60 and GroEL. This peptide contains, at least in part, the two α-helices H3 and H4 (dotted boxes, predicted for prion proteins [16]).

RNA aptamers interact with PrP^C but not with rPrP 27-30

In vitro selection or systematic evolution of ligands by exponential enrichment (SELEX) is a combinatorial method to identify RNA molecules that bind with high affinity to defined molecular targets (for review, see [20, 21]). By means of this technique, nucleic acids have been isolated for a variety of protein targets including the human immunodeficiency virus-1 (HIV-1) proteins reverse transcriptase [22], integrase [23] and rev [24] as well as human α-thrombin [25]. In vitro selection schemes have been developed which apply novel fusion-protein technologies with different tags, such as GST.

In vitro selection requires the generation of a complex DNA pool (fig 4). In vitro transcription generates an RNA pool with a complexity of approximately 10^{15} different molecules. Preselection removes RNA molecules interacting with the tag itself, the carrier (which is sepharose or agarose) and antibodies or molecules conjugated to the carrier (fig 4). Free RNA molecules are subsequently incubated with the immobilized target protein (fig 4). RNA molecules interacting specifically with the target protein are termed RNA aptamers (adopted from the Latin aptus = to fit). In order to amplify the aptamers, the RNA is eluted, extracted and reverse transcribed into cDNA. The cDNA is polymerase chain reaction amplified and subsequently subjected to another in vitro transcription reaction (fig 4). The enriched RNA pool is used as input for further cycles of selection and amplification. After several cycles, RNA aptamers specifically binding to the target protein are enriched and the sequence of the selected RNA aptamers can be determined. Using GST fusion proteins, DNA and RNA aptamers have been selected directed against a *Pax*-paired domain [26] and the *Drosophila* sex-lethal protein [27], respectively.

No antibodies have ever been detected that can distinguish between PrP^C and PrP^{Sc} or the N-terminally truncated version PrP 27–30 [28]. In order to search for RNA aptamers able to bind to the recombinant prion protein PrP^C [10] but not rPrP 27–30 [11], we used an in vitro selection scheme as described in figure 4 (Weiss et al, unpublished results). After several

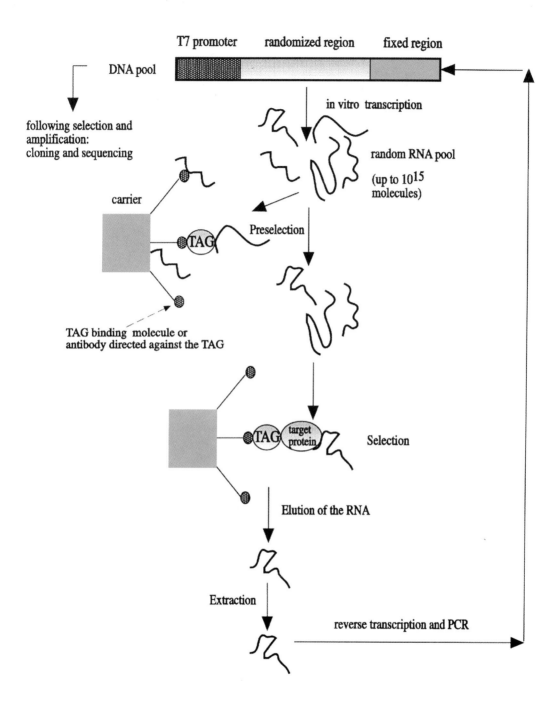

Fig 4. Schematic illustration of an in vitro selection scheme using a recombinant tagged protein and a randomized RNA pool. Only a few RNA molecules from a complex pool of 1×10^{15} molecules interact with the target protein. After elution, extraction, reverse transcription, polymerase chain reaction (PCR) amplification and in vitro transcription, the RNA is again subjected to a new cycle of selection. After several cycles of selection and amplification, RNA aptamers can be obtained that specifically interact with the target protein.

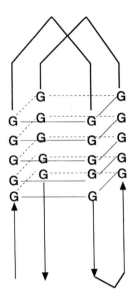

Fig 5. Nucleic acid containing a guanosine (G)-quartet structure. Several G quartets are stacked upon each other to form chelation cages. The nucleic acid forms three loops in an antiparallel manner. A prion aptamer has been selected that bind specifically to PrP^C but not to rPrP 27–30 and could represent a similar structure.

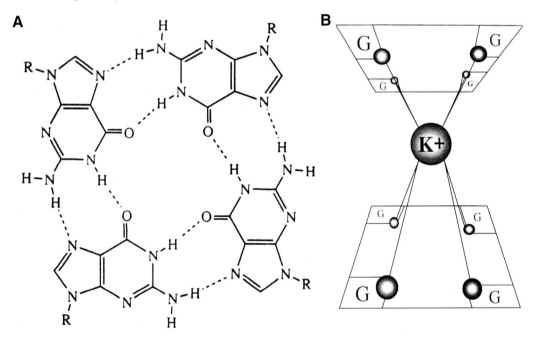

Fig 6. A. Hydrogen bonding in a guanosine (G)-tetrad. Hydrogen bonds are between nitrogen or oxygen atoms. Alternatively, G-tetrads are isostructural adenosine tetrads stabilized by only four hydrogen bonds (N6 to N1). **B.** Schematic presentation of a G-tetrad. One potassium ion complexes four oxygens of the upper and four oxygens of the lower G-tetrad. Alternatively, the structure can be stabilized in the presence of Li^+, Na^+ or Cs^+. NH_4^+ ions can also stabilize the structure. In this case, two protons of the NH_4^+ ion can be hydrogen-bonded to two oxygens in the upper and to two oxygens in the lower plane of the cage.

rounds of selection and amplification, the RNA pool was enriched and the selected RNA aptamers were able to interact specifically with PrP^C but not with rPrP 27–30 (Weiss et al, unpublished results). Sequence analysis demonstrates that the selected prion aptamers may contain several guanosine (G) quartets similar to those shown in figure 5 (Weiss et al, unpublished results).

G-tetrads are held together by Hoogsteen base pairing through hydrogen bonds between nitrogen or oxygen atoms (fig 6A; [29]). Two G-tetrads stacked upon each other form an eight-coordinate chelation cage with an alkali-cation located within the axial channel, complexing four oxygens of the upper and four oxygens of the lower G quartet (fig 6B). The stabilizing monovalent cations represent K^+, Na^+, Li^+ or Cs^+. NH_4^+ has also been shown to stabilize G tetrads [30]. G quartets have been demonstrated in telomeric DNA sequences found in species such as *Tetrahymena* [31] (for review, see [32]). G quartets in quadruple helical RNA structures have been suggested to be involved in the dimerization of HIV-1 genomic RNA [33, 34]. Guanine-rich sequences which are able to form G-tetrads were found in immunoglobulin switch regions, in gene promoters and in chromosomal telomers which are thought to bring the four homologous chromatids together during meiosis and to protect the DNA against degradation [29].

To date, no diagnostic assay for the detection of transmissible spongiform encephalopathies is available, relaying on antibodies distinguishing between PrP^C and PrP^{Sc}. Investigations are under way to assay the RNA aptamers described in samples of prion-free and prion-containing tissue. Because of their unusually stable structure, RNA aptamers can be various types of tissues. Furthermore, the high specificity and affinity of the aptamers recommends their use in a wide range of diagnostic applications.

Acknowledgments

We are indebted to M Groschup, Tübingen, Germany, for providing the PrPc antibody. This work was supported by grant WI 319/10-2 from the Deutsche Forschungsgemeinschaft to ELW and the Bundesministerium für Bildung, Forschung, Wissenschaft und Technologie grant 01-KI-9459/9 to S Weiss, F Edenhofer and EL Winnacker.

References

1 Alper T, Haig DA, Clarke MC. The exceptionally small size of the scrapie agent. *Biochem Biophys Res Commun* 1966;22:278-84

2 Alper T, Cramp WA, Haig DA, Clarke MC. Does the agent of scrapie replicate without nucleic acid? *Nature* 1967;214:764-6

3 Prusiner SB. Novel proteinaceous infectious particles cause scrapie. *Science* 1982;216:136-44

4 Lansbury PT Jr, Caughey B. The chemistry of scrapie infection: implications of the 'ice 9' metaphor. *Chem Biol* 1995;2:1-5

5 Pan KM, Baldwin M, Nguyen J et al. Conversion of α-helices into β-sheets features in the formation of the scrapie prion proteins. *Proc Natl Acad Sci USA* 1993;90:10962-6

6 Liautard JP. Prions and molecular chaperones. *Arch Virol* 1993;7:227-43

7 Telling GC, Scott M, Mastrianni J et al. Prion propagation in mice expressing human and chimeric PrP transgenes implicates the interaction of cellular PrP with another protein. *Cell* 1995;83:79-90

8 Gyuris J, Golemis E, Chertkov H, Brent R. Cdi1, a human G1 and S phase protein phosphatase that associates with Cdk2. *Cell* 1993;75:791-803

9 Edenhofer F, Rieger R, Famulok M, Wendler W, Weiss S, Winnacker EL. Prion protein PrP^C interacts with molecular chaperones of the Hsp60 family. *J Virol* 1996;70:4724-8

10 Weiss S, Famulok M, Edenhofer F et al. Overexpression of active Syrian golden hamster prion protein PrP^C as a glutathione S-transferase fusion in heterologous systems. *J Virol* 1995;69:4776-83

11 Weiss S, Rieger R, Edenhofer F, Fisch E, Winnacker EL. Recombinant prion protein rPrP 27–30 from Syrian golden hamster reveals proteinase K sensitivity. *Biochem Biophys Res Commun* 1996;219:173-9

12 Prusiner SB, Groth DF, Bolton DC, Kent SB, Hood LE. Purification and structural studies of a major scrapie prion protein. *Cell* 1984;38:127-34

13 Prusiner SB, McKinley MP, Bowman KA et al. Scrapie prions aggregate to form amyloid-like birefringent rods. *Cell* 1983;35:349-58

14 Prusiner SB, McKinley MP, Groth DF et al. Scrapie agent contains a hydrophobic protein. *Proc Natl Acad Sci USA* 1981;78:6675-9

15 Braig K, Otwinowski Z, Hedge R et al. The crystal structure of the bacterial chaperonin GroEL at 2.8 Å. *Nature* 1994;371:578-86

16 Huang Z, Gabriel JM, Baldwin MA, Fletterick RJ, Prusiner SB, Cohen FE. Proposed three-dimensional structure for the cellular prion protein. *Proc Natl Acad Sci USA* 1994;91:7139-43

17 Ross WR, Bertrand WS, Morrison AR. Identification of a processed protein related to the human chaperonins (hsp 60) protein in mammalian kidney. *Biochem Biophys Res Commun* 1992;185:683-7

18 Velez-Granell CS, Arias AE, Torres-Ruiz JA, Bendayan M. Molecular chaperones in pancreatic tissue: the presence of cpn10, cpn60 and hsp70 in distinct compartments along the secretory pathway of the acinar cells. *J Cell Sci* 1994;107:539-49

19 Borchelt DR, Taraboulos A, Prusiner SB. Evidence for synthesis of scrapie prion proteins in the endocytic pathway. *J Biol Chem* 1992;267:16188-99

20 Tuerk C, Gold L. Systematic evolution of ligands by exponential enrichment: RNA ligands to bacteriophage T4 DNA polymerase. *Science* 1990;249:505-10

21 Famulok M, Szostak JW. In vitro selection of specific ligand-binding nucleic acids. *Angew Chem Int Ed Engl* 1992;31:979-88

22 Tuerk C, MacDougal S, Gold L. RNA pseudoknots that inhibit human immunodeficiency virus type 1 reverse transcriptase. *Proc Natl Acad Sci* 1992;89:6988-92

23 Allen P, Worland S, Gold L. Isolation of high-affinity RNA ligands to HIV-1 integrase from a random pool. *Virol* 1995;209:327-36

24 Symensma TL, Giver L, Zapp M, Takle GB, Ellington A. RNA aptamers selected to bind human immunodeficiency virus type 1 Rev in vitro are rev responsive in vivo. *J Virol* 1996;70:179-87

25 Kubik MF, Stephens AW, Schneider D, Marlar RA, Tasset D. High-affinity RNA ligands to human α-thrombin. *Nucleic Acids Res* 1994;22:2619-26

26 Epstein L, Cai J, Glaser T, Jepeal L, Maas R. Identification of a *Pax* paired domain recognition sequence and evidence for DNA-dependent conformational changes. *J Biol Chem* 1994;269:8355-61

27 Sakashita E, Sakamoto H. Characterization of RNA binding specificity of the *Drosophila* sex-lethal protein by in vitro ligand selection. *Nucleic Acids Res* 1994;22:4082-6

28 Groschup MH, Langeveld J, Pfaff E. The major species specific epitope in prion proteins of ruminants. *Arch Virol* 1994;136:423-31

29 Sen D, Gilbert W. Formation of parallel four-stranded complexes by guanine rich motifs in DNA and its implications for meiosis. *Nature* 1988;334:364-6

30 Howard FB, Miles HT. A stereospecific complex of poly(I) with ammonium ion. *Biopolymers* 1982;21:147-57

31 Sundquist WI, Klug A. Telomeric DNA dimerizes by formation of guanine tetrads between hairpin loops. *Nature* 1989;342:825-9

32 Williamson JR. G-quartets in biology: reprise. *Proc Natl Acad Sci USA* 1993;90:3124

33 Sundquist WI, Heaphy S. Evidence for interstrand quadruplex formation in the dimerization of human immunodeficiency virus 1 genomic RNA. *Proc Natl Acad Sci USA* 1993;90:3393-7

34 Weiss S, Häusl G, Famulok M, König B. The multimerization state of retroviral RNA is modulated by ammonium ions and affects HIV-1 full-length cDNA synthesis in vitro. *Nucleic Acids Res* 1993;21:4879-85

Transmissible Subacute Spongiform Encephalopathies:
Prion Diseases
L Court, B Dodet, eds
© 1996, Elsevier, Paris

A cell culture model of familial prion diseases

David A Harris, Sylvain Lehmann

Department of Cell Biology and Physiology, Washington University School of Medicine, St Louis, MO 63110, USA

Summary – Familial forms of prion diseases are linked to insertional and point mutations in the gene encoding the prion protein (PrP). To develop a cell culture model of these disorders, we constructed stably transfected lines of Chinese hamster ovary cells that express murine homologues of each of six human PrPs carrying a disease-specific mutation. We find that these mutant PrPs are aberrantly attached to the plasma membrane, as revealed by retention on the cell surface following cleavage of their glycolipid anchors. In addition, the mutant proteins display the biochemical hallmarks of PrP^{Sc}, the pathogenic isoform of PrP. Our results provide the first evidence that a PrP^{Sc}-like molecule can be generated de novo in cultured cells, and they establish our system as a useful model to analyze the cellular and molecular mechanisms underlying inherited prion disorders.

Introduction

Prion diseases are highly unusual because they can arise not only by infection by also by inheritance [1]. Gerstmann-Sträussler-Scheinker (GSS) disease, fatal familial insomnia (FFI) and about 10% of the cases of Creutzfeldt-Jakob disease (CJD) are inherited in an autosomal dominant manner with nearly complete penetrance. These cases all result from one of several insertional and point mutations in the human gene that encodes the prion protein (PrP) [2].

Although several lines of cultured cells have been successfully infected with scrapie prions [3, 4], there has been no cell culture of genetic forms of prion diseases. To develop such a model, we have constructed stably transfected lines of Chinese hamster ovary (CHO) cells that express mouse PrP (MoPrP) molecules carrying mutations homologous to those associated with inherited prion diseases of humans. Table I lists the mutants we have examined to date, including one insertion mutation and five point mutations. As a negative control, we have also analyzed M128V, the mouse homologue of the nonpathogenic met/val polymorphism at codon-129 of human *PRNP*. We have carried out a detailed investigation of the biosynthesis, membrane topology and biochemical properties of these mutant PrPs.

Mutant PrPs are not released from the cell surface by phospholipase

We find that mutant PrPs are glycosylated and transported to the cell surface after synthesis with kinetics that are not markedly different from those of the wild-type protein [12]. Wild-type PrPs are attached to the plasma membrane exclusively by a glycosyl phosphatidylinositol (GPI) anchor at their C-terminus, and can be released by treatment with the bacterial enzyme phosphatidylinositol-specific phospholipase C (PIPLC) which removes the diacylglycerol

---

Key words: prion / mutant / cell culture

Table I. Mutant MoPrPs expressed in CHO cells, their human homologues, and associated phenotypes.

MoPrP	Human PrP	Phenotype	Reference
PG11	6 octapeptide insertion	CJD	Poulter et al [5]
P101L	P102L	GSS disease	Hsiao et al [6]
M128V	M129V	Normal	Owen et al [7]
D177N/Met^{128}	D178N/Met^{129}	FFI	Goldfarb et al [8]
D177N/Val^{128}	D178N/Val^{129}	CJD	Goldfarb et al [8]
F197S/Val^{128}	F198S/Val^{129}	GSS disease	Dlouhy et al [9]
E199K	E200K	CJD	Hsiao et al [10], Goldfarb et al [11]

portion of the anchor. Surprisingly, however, each of six MoPrPs carrying a disease-associated mutation is retained on the cell surface after treatment with PIPLC [12, 13] (fig 1). This phenomenon is specific for disease-associated mutations, since M128V MoPrP is released as efficiently as wild-type MoPrP.

We have investigated this effect further using PG11 MoPrP [12]. Failure of the phospholipase to release PG11 MoPrP is not due to the fact that the protein lacks a GPI anchor, since it metabolically incorporates the radioactive anchor precursors [³H]ethanolamine, [³H]palmitate and [³H]stearate. Moreover, inefficient release does not result from a chemical modification, such as acylation of the inositol ring, that renders the anchor resistant to enzymatic cleavage, since [³H]fatty acid label incorporated into the protein is removed by treatment with PIPLC.

Abnormal membrane association of mutant PrPs

These results indicate that the PrP molecules carrying pathogenic mutations have a second mechanism of membrane attachment in addition to their GPI anchor. The nature of this secondary attachment remains to be determined, although several possible models can be envisioned [12] (fig 2). First, the polypeptide chain of the mutant PrP might be integrated into the lipid bilayer (fig 2B). Consistent with this possibility, we find that PG11 PrP displays operational properties of an integral membrane protein even after cleavage of its GPI anchor, including inextractability in alkaline carbonate buffer, and partitioning into the detergent phase after lysis in Triton X-114 [12]. In addition, transmembrane forms of PrP have been observed when PrP mRNA is translated in vitro [15].

A second hypothesis is that the mutant proteins bind tightly to other membrane-associated molecules (fig 2C). This interaction, which would have to be alkaline stable, could involve other surface proteins, or carbohydrate-containing molecules such as glycosaminoglycans [16].

Third, it is possible that the mutant proteins aggregate on the cell surface, rendering the GPI anchors of a small percentage of the molecules inaccessible to PIPLC, or exposing hydrophobic surfaces that interact with the lipid bilayer.

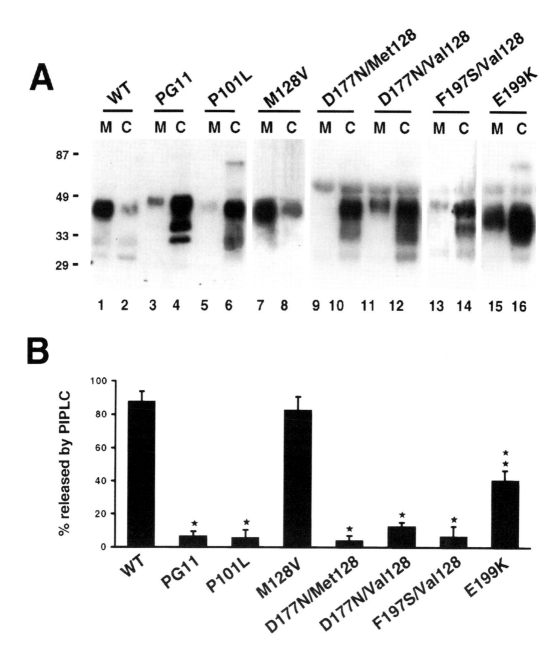

Fig 1. MoPrPs carrying disease-related mutations are not released from the cell surface by PIPLC. **A.** CHO cells expressing each protein were biotinylated with the membrane-impermeant reagent sulfobiotin-X-NHS, and were then incubated with PIPLC prior to lysis. MoPrP in the PIPLC incubation media (M lanes) and cell lysates (C lanes) was immunoprecipitated, and visualized by developing blots with horseradish peroxydase-streptavidin and enhanced chemiluminescence. **B.** The amount of PrP released by PIPLC was plotted as a percentage of the total amount of PrP (medium + cell lysate). Each bar represents the mean ± SD. Values that are significantly different from wild-type (WT) MoPrP by t-test are indicated by single ($P < 0.01$) and double ($P < 0.001$) asterisks (reprinted from Harris [14]).

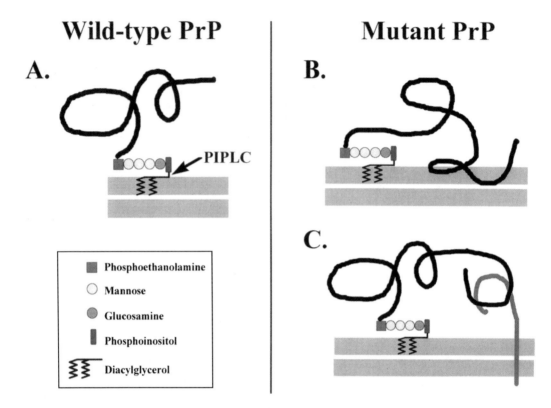

Fig 2. Models for the membrane topology of wild-type and mutant PrPs. **A.** Wild-type PrP is anchored to the membrane exclusively by its GPI anchor, the core structure of which is illustrated, along with the site cleaved by PIPLC. The polypeptide chain of mutant PrP may be integrated into the lipid bilayer (**B**), or bind tightly to another membrane-associated molecule **C.** In both cases, the mutant protein would be retained on the cell surface after cleavage of the GPI anchor by PIPLC (reprinted from Harris [14]).

Although most of our studies have focused on mutant PrPs, we have recently found that disease-specific PrP isoform (PrPSc) produced in scrapie-infected neuroblastoma cells also remains associated with the cell membrane after treatment with PIPLC [13], a feature consistent with a large body of literature suggesting that prion infectivity is tightly membrane-bound [1]. Our results suggest a profound similarity between mutant and infectious PrPs in their mode of membrane attachment, and they argue that alterations in the membrane topology of PrP may be a general feature of prion diseases.

Mutant PrPs display PrPSc-like properties

PrPSc can be distinguished biochemically from the normal form of the prion protein (PrPC) by its insolubility in non-denaturing detergents, and by its relative resistance to digestion by proteinase K which produces a core fragment of 27–30 kDa [1]. We have found that mutant PrPs carrying pathogenic mutations display both of these properties [13, 17].

Detergent insolubility was assessed by centrifuging lysates of metabolically labeled CHO cells at 265,000 × g for 40 min, a protocol that sediments PrPSc but not PrPC. Under these conditions, significantly more of each of the PrPs carrying a disease-associated mutation sedimented than did wild-type MoPrP (fig 3A). Again, M128V MoPrP behaved like the wild-type protein.

To assess the protease resistance of the MoPrPs, lysates of metabolically labeled cells were treated for 10 min with 3.3 μg/mL of proteinase K. We observed that all the MoPrPs carrying pathogenic mutations were cleaved by the protease to yield fragments that migrated between 27 and 30 kDa, the same size as the protease-resistant core of PrPSc (fig 3B). In contrast, wild-type MoPrP, as well as M128V MoPrP, were completely degraded under these conditions. In other experiments, we found that proteinase K treatment of PG11 MoPrP removes the octapeptide repeat region of the molecule, arguing that the protease cleavage site is similar for mutant and infectious forms of PrPSc [17]. Although some forms of PrPSc from scrapie-infected brain and neuroblastoma cells can withstand digestion conditions that are harsher than those employed here, we note that the protease resistance of PrPSc can vary considerably, depending on the scrapie strain from which it is derived, and on the cell type where it is synthesized [18, 19].

Mutant PrPs display 'strain-specific' features

We noted several biochemical features that distinguished the mutant PrPs from each other. First, the protease-resistant fragments of the D177N/Met[128] and E199K proteins consistently migrated 1–2 kDa more slowly than the fragments of the other mutant proteins (fig 3B), suggesting that the cleavage site may not be identical for all of the mutants. Second, the glycosylation patterns of the mutants were different from each other, although all of the disease-related mutants displayed a higher proportion of lower molecular mass glycoforms than wild-type PrP (fig 1A). Third, E199K MoPrP was more PIPLC-releasable, and less detergent-insoluble than the other mutants (figs 1B and 3A). These observations raise the possibility that mutant PrPs possess 'strain-specific' molecular properties that account for the different clinical and neuropathological phenotypes with which they are associated. PrPs from the brains of patients with inherited prion diseases also show mutation-specific variations in proteinase K cleavage site and glycoform distribution [20].

Kinetics of PrPSc generation

To analyze the kinetics with which the mutant PrPs were converted into PrPSc, we measured the acquisition of detergent insolubility and protease resistance by PG11 MoPrP after pulse-labeling [17]. We found that the amount of detergent-insoluble protein increased from ~15 to ~35% of the initial label during the first hour of chase (fig 4A). The amount of protease-resistant PG11 protein also increased during the chase period (fig 4B), although the rise occurred more gradually, peaking at 6 h. We also observed that the metabolic decay of detergent-insoluble and protease-resistant PG11 was quite slow, with substantial amounts of both forms remaining by 16 h of chase (fig 4A and B). We conclude from these results that PG11 MoPrP acquires PrPSc-like properties in a post-translational process that confers enhanced metabolic stability on the protein. These kinetic properties are reminiscent of those that have been described for PrPSc in scrapie-infected neuroblastoma cells, which is synthesized over 1–3 h, and is stable for > 24 h after pulse-labeling [21, 22].

We are currently extending these kinetic studies in an attempt to define individual steps in the generation of PrPSc. We have found that PG11 MoPrP becomes hydrophobic, as measured by Triton X-114 phase partitioning of the PIPLC-treated protein, within 30 min after pulse-

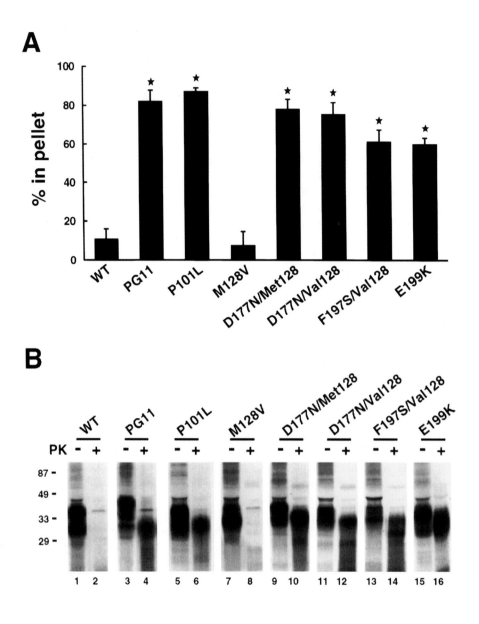

Fig 3. MoPrP carrying disease-related mutations are detergent insoluble and protease resistant when expressed in cultured CHO cells. **A.** Lysates of metabolically labeled CHO cells expressing each MoPrP were centrifuged first at 16,000 × g for 5 min, and then at 265,000 × g for 40 min. MoPrP in the supernatants and pellets from the second centrifugation was quantified by immunoprecipitation, and the percentage of PrP in the pellet was calculated. Each bar represents the mean ± SD of values from three experiments. Values that are significantly different from wt MoPrP by *t*-test (*P* < 0.001) are indicated by an asterisk. **B.** Proteins in lysates of metabolically labeled cells were either digested at 37 °C for 10 min with 3.3 μg/mL of proteinase K (+ lanes), or were untreated (– lanes), prior to recovery of MoPrP by immunoprecipitation and analysis by sodium dodecyl sulfate-polyacrilamide gel electrophoresis. Five times as many cell equivalents were loaded in the + lanes as in the – lanes (reprinted from Harris [14]).

labeling (Lehmann and Harris, unpublished data). This observation, combined with those mentioned earlier, suggests a sequence of events in which the PG11 molecule rapidly becomes hydrophobic, then detergent-insoluble, and only later protease-resistant (fig 4C). This scheme has led us to hypothesize that mutant PrPs are synthesized initially in the PrP^C conformation, and then gradually acquire PrP^{Sc}-like properties in a step-wise manner with distinct biochemical intermediates. It will be interesting now to further define these intermediates, and identify the cellular compartments where they are generated.

Conclusion

Our results provide the first demonstration that a PrP^{Sc}-like molecule can be generated de novo in uninfected cultured cells. The biochemical similarities between mutant PrP^{Sc} generated in our system and authentic infectious PrP^{Sc} include detergent insolubility, protease resistance, slow metabolic generation and turnover, 'strain-like' structural variations and aberrant association with cellular membranes. We have detected these properties in six different MoPrP mutants whose human homologues are associated with each of the major phenotypes of inherited prion disease (table I), but not in wild-type MoPrP, or in a MoPrP mutant (M128V) whose

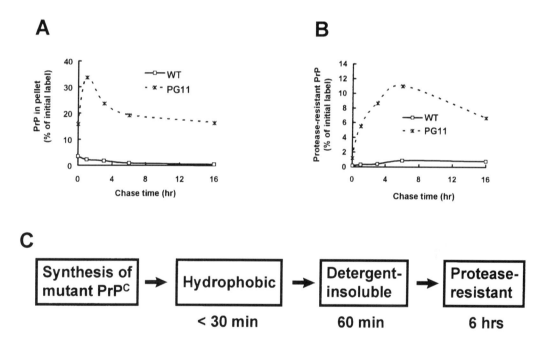

Fig 4. Kinetics of the acquisition of PrP^{Sc}-like properties by PG11 MoPrP. **A.** CHO cells expressing PG11 MoPrP were labeled for 30 min with [^{35}S]methionine and chased for the indicated times. At the end of each chase period, cell lysates were centrifuged as described in figure 3A, and MoPrP immunoprecipitated from supernatants and pellets of the 265,000 × g spin. The amount of PrP in the pellet at each time point was plotted as a percentage of the total amount of radioactive PrP present at the end of the labeling period. **B.** CHO cells were labeled for 45 min with [S^{35}]methionine, chased for the indicated times and then lysed. Proteins in cell lysates were digested at 37 °C for 10 min with proteinase K (3.3 μg/mL), and MoPrP recovered by immunoprecipitation. The amount of protease-resistant PrP (M$_r$ 27–30 kDa) at each time point was plotted as a percentage of the total amount of radioactive PrP present at the end of the labeling period. **C.** A scheme for generation of PrP^{Sc} from mutant PrP (A and B are reprinted from Lehmann and Harris [17]).

human homologue represents a nonpathogenic polymorphism. We are currently testing whether mutant PrPs synthesized in CHO cells are infectious. The close biochemical similarities between mutant and infectious PrP[Sc] that are reported here strongly support the hypothesis that all prion diseases involve common cellular and molecular mechanisms, which our system will now permit us to analyze.

Acknowledgments

This work was supported by grants to DAH from the National Institutes of Health and the Alzheimer's Association. SL is the recipient of a postdoctoral fellowship for physicians from the Howard Hughes Medical Institute, and of awards from INSERM (Institut National de la Santé et de la Recherche Médicale, Paris, France) and the Simone and Cino del Duca Foundation (Paris, France).

References

1 Prusiner SB. Prions. In: Fields BN, Knipe DM, Howley PM, eds. *Virology*, 3rd edn. Philadelphia, New York: Lipincott-Raven Press, 1996;2901-50
2 Prusiner SB, Hsiao KK. Human prion diseases. *Ann Neurol* 1994;35:385-95
3 Race RE, Caughey B, Graham K, Ernst D, Chesesbro, B. Analyses of frequency of infection, specific infectivity, and prion protein biosynthesis in scrapie-infected neuroblastoma cell clones. *J Virol* 1988;62:2845-9
4 Taraboulos A, Serban D, Prusiner SB. Scrapie prion proteins accumulate in the cytoplasm of persistently infected cultured cells. *J Cell Biol* 1990;110:2117-32
5 Poulter M, Baker HF, Frith CD et al. Inherited prion disease with 144 base pair gene insertion. 1. Genealogical and molecular studies. *Brain* 1992;115:675-85
6 Hsiao K, Baker HF, Crow TJ et al. Linkage of a prion protein missense variant to Gerstmann-Sträussler syndrome. *Nature* 1989;338:342-5
7 Owen F, Poulter M, Collinge J, Crow TJ. Codon 129 changes in the prion protein gene in Caucasians. *Am J Hum Genet* 1990;46:1215-6
8 Goldfarb LG, Petersen RB, Tabaton M et al. Fatal familial insomnia and familial Creutzfeldt-Jakob disease: disease phenotype determined by a DNA polymorphism. *Science* 1992;258:806-8
9 Dlouhy SR, Hsiao K, Farlow MR et al. Linkage of the Indiana kindred of Gerstmann-Sträussler-Scheinker disease to the prion protein gene. *Nature Genet* 1992;1:64-7
10 Hsiao K, Meiner Z, Kahana E et al. Mutation of the prion protein in Libyan Jews with Creutzfeldt-Jakob disease. *N Engl J Med* 1991;324:1091-7
11 Goldfarb LG, Brown P, Mitrova E et al. Creutzfeldt-Jakob disease associated with the *PRNP* codon 200[LYS] mutation: an analysis of 45 families. *Eur J Epidemiol* 1991;7:477-86
12 Lehmann S, Harris DA. A mutant prion protein displays an aberrant membrane association when expressed in cultured cells. *J Biol Chem* 1995;270:24589-97
13 Lehmann S, Harris DA. Mutant and infectious prion proteins display common biochemical properties in cultured cells. *J Biol Chem* 1996;271:1633-7
14 Harris DA. Cell biological insights into prion diseases. In: Kretzschmar H, Riesner D, eds. *Prion Diseases: Current State and Perspectives in Prion Research.* New York: Chapman and Hall, 1996 (in press)
15 Hay B, Barry RA, Lieberburg I, Prusiner SB, Lingappa VR. Biogenesis and transmembrane orientation of the cellular isoform of the scrapie prion protein. *Mol Cell Biol* 1987;7:914-20
16 Shyng SL, Lehmann S, Moulder KL, Harris DA. Sulfated glycans stimulate endocytosis of the cellular isoform of the prion protein, PrP[C], in cultured cells. *J Biol Chem* 1995;270:30221-9
17 Lehmann S, Harris DA. Two mutant prion proteins expressed in cultured cells acquire biochemical properties reminiscent of the scrapie isoform. *Proc Natl Acad Sci USA* 1996;93:5610-4
18 Bessen RA, Marsh RF. Biochemical and physical properties of the prion protein from two strains of the transmissible mink encephalopathy agent. *J Virol* 1992;66:2096-101
19 Meiner Z, Halimi M, Polakiewicz RD, Prusiner SB, Gabizon R. Presence of prion protein in peripheral tissues of Libyan Jews with Creutzfeldt-Jakob disease. *Neurology (NY)* 1992;42:1355-60
20 Monari L, Chen SG, Brown P et al. Fatal familial insomnia and familial Creutzfeldt-Jakob disease: different prion proteins determined by a DNA polymorphism. *Proc Natl Acad Sci USA* 1994;91:2839-42
21 Caughey B, Raymond GJ. The scrapie-associated form of PrP is made from a cell surface precursor that is both protease- and phospholipase-sensitive. *J Biol Chem* 1991;266:18217-23
22 Borchelt DR, Taraboulos A, Prusiner SB. Evidence for synthesis of scrapie prion proteins in the endocytic pathway. *J Biol Chem* 1992;267:16188-99

Transmissible Subacute Spongiform Encephalopathies:
Prion Diseases
L Court, B Dodet, eds
© 1996, Elsevier, Paris

A theoretical model of prion dynamics in spongiform encephalopathies

Joab Chapman, Amos D Korczyn

Department of Physiology and Pharmacology, Sackler Faculty of Medicine, Tel Aviv University, Ramat Aviv 69978, Israel

Summary – We propose a two-stage theory for the pathogenesis of prion diseases which suggests that the pathogenic form of the prion protein (PrP) is membrane bound, partially protease-resistant, and formed spontaneously in sporadic and genetic disease. Once a critical concentration of this form is reached, it may proliferate by interaction with the normal form of the protein, possibly in lysosomes. This preclinical stage is followed by a fulminant stage in which there is rapid degeneration of neuronal tissue. The available data on the age-dependent penetrance of *PRNP* mutations is analyzed, assuming that either the spontaneous transformation or a direct interaction of pathological and normal proteins is the step determining the age at onset of disease.

Introduction

Spongiform encephalopathy (SE) in sheep, a disease known as scrapie, has been recognized for over a century [1], and similar diseases such as kuru and Creutzfeldt-Jakob disease (CJD) were subsequently identified in humans [2]. Following the key discovery of their transmissibility by veterinary workers in scrapie [1] and by Gajdusek and his coworkers in kuru and CJD [2], it seemed that eventually a pathogenic virus would be identified. However, attempts at isolating a specific nucleic acid have been unsuccessful, and the puzzling observation that 'infectivity' depends on the protein components in the purified fractions led Prusiner to the logical but unorthodox conclusion that the disease may be transmitted by means of a protein alone [3]. In order to simplify and, no doubt, facilitate the propagation of this notion, Prusiner coined the term 'prions' to designate such proteinaceous infectious particles. Purification and eventual sequencing of the protein revealed that it is coded for by a host chromosomal gene [4]. The prion protein (PrP) has two main forms [5]: a normal component of many cell types which is termed PrP^C (protease-sensitive PrP) and the key component of the infectious particles which is termed PrP^{Sc} (disease-specific PrP). PrP^{Sc} differs from PrP^C by being detectable in brains affected by SE, its protease resistance, its presence in cytoplasmic vesicles but not on the cell surface [6], its having more β-pleated sheet structure [7], its propensity to form amyloid, being strongly bound to membranes and having a much longer half-life [5, 8]. These significant physical, chemical and biological differences occur despite the fact that the amino acid sequences of the two PrP isoforms seem to be identical and that no post-translational differences have been identified between them [9]. In this paper, we propose a model based on the present knowledge about prion proliferation. This model is then utilized

Key words: Creutzfeldt-Jakob disease / prion / genetics / pathogenesis

to explain features of prion diseases in humans. Basic experiments are proposed to confirm or reject the model, and epidemiological evidence from familial CJD is analyzed in an attempt to provide further information on the molecular mechanisms involved.

The model: molecular basis

In the interests of parsimony and in light of most experimental evidence, we shall make the basic assumption that all diseases associated with PrP are the result of the metabolism of this protein alone and do not involve an additional (nucleic acid) component. Under this assumption, theoretical considerations and experimental evidence indicate that the proliferation of PrP^{Sc} may result from an interaction between normal and abnormal forms of PrP. As has been previously proposed [5], this interaction may be simplified into a chemical reaction equation:

$$\text{Equation 1: } PrP^C + PrP^{Sc} \Rightarrow 2PrP^{Sc}$$

The exact stoichiometry of this equation is however open to conjecture. This simplified chemical reaction would fit many theories of prion multiplication, including a crystallization process proposed by Gajdusek et al [10]. A key step in the confirmation of this theory would be the in vitro replication of the pathological form of PrP by exposure of PrP^C to a minute amount of PrP^{Sc}. A recent experiment has demonstrated conversion of PrP^C to a protease-resistant form by exposing it to PrP^{Sc} under non physiological, partially denaturing conditions [11]. It is still not clear however whether the protein formed under these conditions is indeed pathogenic. The difficulty in replicating prion proliferation in vitro may explain why unorthodox proposals, such as differences in conformation between two identical proteins under identical conditions have been advanced, and 'foldases', chaperones and other macromolecules have been assumed to be involved in this reaction [5]. We believe a more conventional and straightforward model should be sought which would lay the basis for performing the crucial in vitro PrP conversion.

As has been outlined above, the two forms of PrP differ in a number of physical and chemical properties. It is of interest, therefore, to note that two forms of PrP are found in cell-free translation studies: a transmembrane (lipid phase) form which may span the bilayer twice (at the transmembrane and amphipathic helix domains) and a secretory (water soluble) form [5]. The fact that PrP has the potential to exist in both forms may underlie the pathogenesis of the disease and suggest that these two forms may exist in vivo, PrP^{Sc} being similar to the transmembrane form and PrP^C to the secretory form. A central issue in this theory is the site at which the PrP^{Sc}–PrP^C interaction represented in reaction 1 actually takes place. Most PrP^C is found on the cell surface, while most PrP^{Sc} is observed intracellularly in cytoplasmic vesicles [12]. A reaction between them could possibly occur at any cellular membrane, but this would then require PrP to be able to switch freely between its two states. We propose that the two most likely moments of prion proliferation take place when the protein is being synthesized or metabolized.

In cell-free expression systems, there is evidence that the conversion of PrP^C to PrP^{Sc} occurs in lysosomes after internalization from the plasma membrane as a step leading to protein degradation [13]. It is important to note that in these model systems the level of PrP^{Sc} proliferation is low, without overt pathology in the cells themselves. Thus, such a low level of pathology may be compatible with normal function of brain cells over long periods.

The second moment during which interaction between two molecules of PrP can occur is the process (the course) of the protein synthesis in the rough endoplasmic reticulum (RER) where it passes through the membrane following insertion of a signal peptide. Though there is currently little evidence to support such a possibility, we propose that if PrP^{Sc} is present in the RER membrane at this point it interacts with the newly synthesized polypeptide chain, causing it to adopt a transmembrane conformation. The accumulation of significant amounts of PrP^{Sc} in membrane structures may impede their recycling and result in the observed accumulation of vesicular structures in the cytoplasm in the very early stages of the development of prion diseases [14]. This type of prion proliferation would produce larger amounts of PrP^{Sc} than those generated in the lysosomal pathway. It also could adversely affect the synthesis of other proteins and thus would have a fulminant detrimental effect on brain cell function. It is possible that in cell culture studies in which PrP^{Sc} infection is not lethal, the concentration of intracellular PrP^{Sc} is insufficient for protein synthesis to be affected.

The model in human forms of prion diseases

In humans, at least three forms of acquired prion disease have been demonstrated. One is due to direct inoculation with disease-affected brain components, such as in kuru and in iatrogenic cases. These forms are conceptually similar to experimental scrapie. The second form is genetic, and related to the inheritance of mutations in the *PRNP* gene which codes for PrP. From clinical data in humans and from evidence that transgenic mice carrying a *Prnp* mutation develop SE, it seems likely that *PRNP* mutations are in fact a primary factor in inducing disease. The third, and most common, form of human prion disease is represented by sporadic cases of CJD. Although it has been proposed that these cases are externally acquired or are the result of somatic *PRNP* mutations, there is no direct or epidemiological evidence supporting a source of infection in sporadic CJD, nor have mutations ever been detected in any of these cases.

According to the model outlined in previous sections, all three forms of prion disease in humans would be initiated by minute quantities of PrP^{Sc} which would first initiate a low level of replication leading to the accumulation of a 'critical mass' of PrP^{Sc} in a cell (or a group of cells) sufficient for the production of this protein to start with interaction with newly synthesized PrP (equation 1). A key question is the source of PrP^{Sc} in each of the human forms of prion diseases. In iatrogenic transmission, the pathological protein is inoculated into the brain or peripheral blood. In sporadic and genetic cases of CJD, spontaneous conversion of PrP^{C} to PrP^{Sc} may occur. This may be summarized in the following equation:

$$\text{Equation 2: } \mathrm{PrP^{C}} \Rightarrow \mathrm{PrP^{Sc}}$$

Under normal conditions, the probability of this process occurring to a significant extent must be relatively low, resulting in only one sporadic case of CJD per 1,000,000 per year. In contrast, a carrier of any one of the known mutations in the *PRNP* gene has a cumulative probability approaching 1 of developing the disease over lifespan [15]. Thus, the mutations greatly increase the probability of the pathogenic process depicted in equation 2. A genetic factor that may influence reaction 1 is homozygosity of the allelic forms at *PRNP* codon 129 which is associated with higher levels of sporadic CJD [16].

We propose that initially there is accumulation by spontaneous transformation as depicted in equation 2 followed by proliferation of PrP^{Sc} as in equation 1. At a critical level of PrP^{Sc}, cellular functions such as protein synthesis are affected and widespread cell death occurs. The model also predicts that the age at onset of clinical symptoms would mainly be determined by the kinetics of equations 1 and 2. Since in experimental model the concentration of PrP^{Sc} does not change significantly following clinical manifestion of disease [17], the duration of disease may be determined predominantly by the toxicity of PrP^{Sc}.

The forms of prion disease in non-human species do not reveal novel modes of transmission. The much publicized outbreak of bovine spongiform encephalopathy is now thought to have been due to ingestion of PrP^{Sc}–containing feed. Scrapie in sheep is either due to direct contact with PrP^{Sc} or has a genetic basis, and though this is still open to debate, both mechanisms are consistent with the proposed model. The central difficulty presented by studies in animal models is the existence of strains of scrapie which retain distinct features, such as the incubation time, through a number of passages [18]. A possible explanation for this phenomenon is that several conformations of PrP^{Sc} exist and each of these interacts with newly synthesized PrP, causing it to adopt the same specific conformation. Each conformation of PrP^{Sc} interacts with PrP^{C} according to reaction 1, but at a different rate leading to different incubation times [19].

Genetic forms of human prion disease

Mutations in the *PRNP* gene could influence either the predisposition of mutant PrP^{C} (muPrP^{C}) to spontaneously adopt the muPrP^{Sc} state (by a reaction analogous to that depicted in equation 2) or the interaction of muPrP^{Sc} with either PrP^{C} or muPrP^{C} (by reactions similar to that depicted in equation 1). There is evidence, from cell-free expression studies [20] and from cell lines carrying mutations [21], that muPrP^{C} is less soluble and has increased protease resistance which are properties characteristic of PrP^{Sc}. These properties of muPrP^{C} may predispose it to convert to muPrP^{Sc} either spontaneously (equation 2) or by interaction with PrP^{Sc} or muPrP^{Sc} (equation 1). The reaction with the fastest kinetics predominates and high levels of its products will accumulate in the central nervous system (CNS). It seems logical to assume that the reaction catalyzed by PrP^{Sc} (equation 1) proceeds at a faster rate than the spontaneous reaction depicted in reaction 2. This is supported by the fact that inoculation of PrP^{Sc} into the brain in iatrogenic and experimental disease greatly hastens the accumulation of more PrP^{Sc}. It is interesting to note therefore that there is some evidence [22] that most PrP^{Sc} in the genetic cases is muPrP^{Sc}, indicating that the muPrP^{Sc}–muPrP^{C} interaction has the fastest kinetics. This is also compatible with data obtained from patients who are homozygous for the E200K mutation, but have a disease course which is not more fulminant than that found in heterozygotes [23].

Genetic prion diseases seem to fall roughly into two general groups. One group, typified by codon E200K mutation, is very similar to classical sporadic CJD with relatively late onset (mean at the sixth decade) and rapidly progressing disease leading to death within months. Brain pathology is characterized by widespread spongiform changes and is compatible with potential clinical involvement of the entire CNS. A second group, typified by Gerstmann-Sträussler-Scheinker disease (GSS) and fatal familial insomnia (FFI), has an earlier age at onset of disease and a slower progression, leading to death within several years of onset. Brain pathology is characterized by the relative rarity of spongiform changes, the presence of amyloid deposits (GSS) and a predilection for certain brain areas such as the cerebellum (GSS) or the thalamus (FFI).

Using the model outlined above, we propose that those mutations with earlier age at onset of disease result in a more rapid spontaneous conversion of muPrP^{C} to muPrP^{Sc}. Since diseases with a tendency towards early onset seem to have a longer clinical course, it is possible that mutations which facilitate accumulation of PrP^{Sc} attenuate its toxicity. Another possibility is that the production of muPrP^{Sc} in GSS and FFI is confined to specific brain areas, thus limiting damage to the CNS and prolonging survival. The molecular mechanisms underlying this area specificity are unknown, but seem to be determined by genetic variation such as the *PRNP* codon 129 polymorphism [24].

Another relevant issue is the age-dependent penetrance of the *PRNP* mutations. In addition to their obvious practical value for genetic counseling, these calculations may shed some light on the pathogenesis of the disease. According to the model proposed, the onset of clinical disease is dependent on the accumulation of a critical amount of PrP^{Sc}. The probability for this event to occur is linked to the concentration of PrP^{Sc} as proposed in equation 3:

$$\text{Equation 3: } q_x = k_1[\text{PrP}^{Sc}]^n$$

where q_x is the incidence of CJD and n the order of the reaction. The concentration of PrP^{Sc} can be presented for either equation 1 or 2: in equation 1, the rate of production of PrP^{Sc} is proportional to the concentration of PrP^{Sc} by the power of m (m being the order of the reaction):

$$\text{Equation 4: } \frac{d[\text{PrP}^{Sc}]}{dt} = k_2[\text{PrP}^{Sc}]^m$$

In equation 2, the concentration of PrP^{Sc} is dependent on the concentration of PrP^{C} which remains constant, even during clinical CJD. In this reaction, therefore:

$$\text{Equation 5: } \frac{d[\text{PrP}^{Sc}]}{dt} = k_3$$

We have previously described the age-dependent incidence of the *PRNP* E200K mutation [15] and demonstrated that the incidence of CJD clearly rises with age, excluding the possibility that the mean concentration of PrP^{Sc} is constant with age. A best fit regression analysis indicated that the best fit for q_x was with (age)^2. If equation 1 were the step determining the rate of PrP^{Sc} accumulation, this result would be compatible with $n = 1$ and $m = 1$ in equations 3 and 4, ie, equation 1 would have first order kinetics and the probability of developing CJD would be directly proportional to the concentration of PrP^{Sc}. If, on the other hand, equation 2 were the rate-limiting step in PrP^{Sc} accumulation, the incidence studies would be compatible with $n = 2$ in equation 3, ie, the probability of manifesting clinical CJD would be proportional to the square of the concentration of PrP^{Sc}.

All these calculations reflect the probability of a sufficient amount of protein accumulating in a limited volume of the brain which would imply that sporadic and genetic cases of CJD begin focally. Clinical and pathological data demonstrating substantial heterogeneity in CJD support this assumption [25].

We postulate that if similar analyses of other *PRNP* mutations are performed, they will reveal similar curves with k values inversely correlating to onset of disease. Since sporadic CJD is proposed to be another form of spontaneous transformation of PrP to PrP^{Sc}, its age-dependent incidence should conform to a similar pattern.

References

1 Parry HB. Scrapie: a transmissible hereditary disease of sheep. *Nature* 1960;185:441-3
2 Gajdusek DC. Unconventional viruses and the origin and disappearance of kuru. *Science* 1977;197:943-60
3 Prusiner SB. Prions and neurodegenerative diseases. *N Engl J Med* 1987;317:1571-81

4 Oesch B, Westaway D, Waichil M et al. A cellular gene encodes scrapie PrP 27-30 protein. *Cell* 1985;40: 735-46

5 Prusiner SB. Molecular biology of prion diseases. *Science* 1991;252:1515-22

6 Piccardo P, Safar J, Ceroni M, Gajdusek MD, Gibbs CJ. Immunohistochemical localization of prion protein in spongiform encephalopathies and normal brain tissue. *Neurology* 1990;40:518-22

7 Pan KM, Baldwin M, Nguyen J et al. Conversion of α-helices in β-sheets features in the formation of the scrapie prion proteins. *Proc Natl Acad Sci USA* 1993;90:10962-6

8 Meyer RK, McKinley MP, Bowman KA et al. Separation and properties of cellular and scrapie prion proteins. *Proc Natl Acad Sci USA* 1986;83:2310-4

9 Stahl N, Baldwin MA, Tepolow DB et al. Structural analysis of the scrapie prion protein using mass spectrometry and amino acid sequencing. *Biochemistry* 1993;32:1991-2002

10 Brown P, Goldfarb LG, Gajdusek DC. The new biology of spongiform encephalopathy: infectious amyloidoses with a genetic twist. *Lancet* 1991;337:1019-22

11 Kocisko DA, Come JH, Priola SA, et al. Cell-free formation of protease-resistant prion protein. *Nature* 1994;370:471-4

12 Safar J, Ceroni M, Piccardo P et al. Subcellular distribution and physicochemical properties of scrapie associated precursor protein and relationship with scrapie agent. *Neurology* 1990;40:503-8

13 Caughey B, Raymond GJ. The scrapie-associated form of PrP is made from a cell surface precursor that is both protease- and phospholipase-sensitive. *J Biol Chem* 1991;266:18217-23

14 Liberski PP, Yanagihara R, Gibbs CJ, Gajdusek DC. Appearance of tubulovesicular structures in experimental Creutzfeldt-Jakob disease and scrapie precedes the onset of clinical disease. *Acta Neuropathol* 1990;79:349-54

15 Chapman J, Ben-Israel J, Goldhammer Y, Korczyn AD. The risk of developing Creutzfeldt-Jakob disease in subjects with the *PRNP* gene codon 200 point mutation. *Neurology* 1994;44:1683-6

16 Palmer MS, Dryden AJ, Hughes JT, Collinge J. Homozygous prion protein genotypes predispose to sporadic Creutzfeldt-Jakob disease. *Nature* 1991;352:340-2

17 Jendroska K, Heinzel FP, Torchia M, et al. Proteinase-resistant prion protein accumulation in Syrian hamster brain correlates with regional pathology and scrapie infectivity. *Neurology* 1991;41:1482-90

18 Bruce ME, Dickinson AG, Fraser H. Cerebral amyloidosis in scrapie in the mouse: effect of agent strain and mouse genotype. *Neuropathol Appl Neurobiol* 1976;2:471-8

19 Bessen RA, Kocisko DA, Raymond GJ, Nandan S, Lansbury PT, Caughey B. Non-genetic propagation of strain-specific properties of scrapie prion protein. *Nature* 1995;375:698-700

20 Brown P, Chapman J, Cervenakova L, Goldfarb LG, Korczyn AD. Cell-free expression of the *PRNP* gene containing a large insert associated with Creutzfeldt-Jakob disease. *Neurology* 1996;46(suppl. 2):A154

21 Lehmann S, Harris DA. Two mutant prion proteins expressed in cultured cells acquire biochemical properties reminiscent of the scrapie isoform. *Proc Natl Acad Sci USA* 1996;93:5610-14

22 Gabizon R, Telling G, Meiner Z, Halimi M, Kahana I, Prusiner SB. Insoluble wild-type and protease-resistant mutant prion protein in brains of patients with inherited prion disease. *Nature Med* 1996;2:59-64

23 Hsiao K, Meiner Z, Kahana E et al. Mutation of the prion protein in Libyan Jews with Creutzfeldt-Jakob disease. *N Engl J Med* 1991;324:1091-7

24 Goldfarb LG, Peterson RB, Tabaton M et al. Fatal familial insomnia and familial Creutzfeldt-Jakob disease: disease phenotype determined by a DNA polymorphism. *Science* 1992;258:806-8

25 Chapman J, Brown P, Goldfarb LG, Arlazaroff A, Gajdusek DC, Korczyn AD. Clinical heterogeneity and unusual presentation of Creutzfeldt-Jakob disease in Jewish patients with *PRNP* codon 200 mutations. *J Neurol Neurosurg Psychiatry* 1993;56:1109-12

Transmissible Subacute Spongiform Encephalopathies:
Prion Diseases
L Court, B Dodet, eds
© 1996, Elsevier, Paris

Susceptibility to hamster scrapie agent in transgenic mice with neuron-specific expression of a hamster prion protein minigene

Bruce Chesebro, Suzette A Priola, Richard E Race

Laboratory of Persistent Viral Diseases, Rocky Mountain Laboratories, National Institute of Allergy and Infectious Diseases, NIH, 903 South 4th Street, Hamilton, 59840, USA

Summary – The effect of cell type-restricted Syrian hamster prion protein (SHaPrP) expression on susceptibility to the hamster scrapie agent was studied in transgenic mice expressing a 1-kilobase (kb) hamster cDNA clone containing the 0.76-kb SHa*Prnp* open reading frame under control of the neuron-specific enolase promoter. In these mice, SHaPrP was only detected in brain tissue, with the highest levels found in neurons of the cerebellum, hippocampus, thalamus and cerebral cortex. These transgenic mice were susceptible to infection by the 263K strain of hamster scrapie with an average incubation period of 93 days, compared to 72 days in normal hamsters. In contrast, non-transgenic mice were not susceptible to this agent. These results indicated that neuron-specific expression of a 1-kb SHa*Prnp* minigene including the SHa*Prnp* open reading frame was sufficient to mediate susceptibility to hamster scrapie, and that SHaPrP expression in non-neuronal brain cells was not necessary to overcome the transmissible spongiform encephalopathies species barrier.

Introduction

Transmissible spongiform encephalopathies (TSE) are degenerative brain diseases which occur naturally in primates and ruminants, and include scrapie of sheep, bovine spongiform encephalopathy, and several human diseases such as Creutzfeldt-Jakob disease, kuru, and Gerstmann-Sträussler-Scheinker disease. An important feature of TSE diseases is the accumulation in brain of a proteinase K-resistant protein, known as PrP-res or PrPSc (disease-specific PrP), which is associated with the pathogenic process [1–3]. PrP-res is post-translationally derived from a normal host proteinase K-sensitive PrP molecule (PrP-sen) [4–6] by an as yet undefined mechanism. PrP may play a critical role in interspecies transmission of TSE. Many species show resistance to disease induction by TSE agents derived from other species. This resistance or 'species barrier' is manifested either by total lack of disease induction or by a prolonged incubation period prior to onset of clinical disease. Genetic studies have indicated that the *Prnp* genotype strongly influences the host susceptibility to TSE agents [7–9]. Transgenic mice with a 40 kilobase (kb)–transgene expressing high levels of Syrian hamster PrP (SHaPrP)-sen are susceptible to disease when inoculated with the hamster scrapie agent, while normal mice are resistant [10, 11]. This observation suggests that transmission of scrapie may be dependent on interactions between the host PrP-sen and PrP-res derived from the donor of the inoculated agent. Interactions between PrP molecules from different species have

---

Key words: scrapie / transgenic mice / neuron-specific enolase / degenerative brain disease

been found to inhibit generation of PrP-res in both scrapie-infected cells [12] and cell-free systems [13, 14]. Such inhibitory interactions may provide a biochemical explanation for the species barrier.

TSE agents are known to replicate in both lymphoreticular organs and brain, but the precise cell types involved are not known. PrP-sen expression has recently been demonstrated in astrocytes and oligodendrocytes [15], and neurons [16–19]. However, it is unclear which of these three cell types are critical to TSE agent replication in brain. The present work shows that transgenic mice expressing the hamster *Prnp* gene under control of the neuron-specific enolase (NSE) promoter [20, 21] are highly susceptible to the hamster scrapie agent. In these mice, SHaPrP expression was found exclusively in neurons and not in glial cells or cells within the spleen or lymph nodes. Thus, hamster PrP expression in neurons was sufficient to abrogate the TSE species barrier. Furthermore, because the transgene used in these experiments contained only 1 kb of hamster DNA including the open reading frame of hamster *Prnp*, the susceptibility of these transgenic mice to hamster scrapie was mediated by the hamster *Prnp* gene itself rather than by the additional 39 kb of transgene DNA used in previous studies [10, 11, 22].

Results

Generation and characterization of transgenic (Tg) mice

The Tg*52NSE* line of transgenic mice was derived by inoculation of a construct containing the NSE promoter plus a 1-kb cDNA containing the SHa*Prnp* open reading frame as previously described [21]. To analyze SHaPrP mRNA expression in the brain of transgenic mice, Northern blotting was performed using a probe reactive with both mouse and hamster *Prnp* genes. Tg*52NSE* mice expressed a 1.6-kb PrP mRNA band corresponding to the predicted SHa*Prnp* transgene in addition to the 2.3-kb PrP mRNA band expressed in non transgenic mice (fig 1). The intensity of the 1.6-kb SHaPrP band was 4 to 5 times higher than that of the 2.3-kb endogenous mouse PrP band.

To study tissue specificity of SHaPrP protein expression, a variety of tissues from SHa*Prnp* transgenic mice and normal hamsters were analyzed by immunoblotting, using monoclonal antibody 3F4 which has strong reactivity for SHaPrP and no reactivity for mouse PrP [23]. In Tg*52NSE* mice, the SHaPrP protein was detected in brain, but not in six other tissues studied (fig 2). In contrast, Tg*10* mice contained a cosmid SHa*Prnp* transgene plus 40 kb of flanking DNA [10, 21], and like normal hamsters, expressed SHaPrP protein in a variety of other tissues in addition to brain. Thus, the transgene in Tg*52NSE* mice showed a restricted pattern of expression as expected from the use of the NSE promoter [20].

In situ hybridization was used to demonstrate the localization of SHaPrP mRNA in a variety of neuronal populations in Tg*52NSE* mouse brain (data not shown). The highest expression was seen in Purkinje cells of the cerebellum, neurons of the dentate gyrus, and pyramidal neurons of the hippocampus, whereas granular layer neurons of the cerebellum, cells in various layers of the cerebral cortex and cells with large nuclei in the dorsal portion of the thalamus also expressed moderate amounts of SHaPrP mRNA [21]. These results were consistent with neuron-specific expression of the SHa*Prnp* transgene in Tg*52NSE* mice.

Susceptibility of Tg52NSE and Tg10 mice to hamster scrapie agent

To determine the influence of SHaPrP expression on susceptibility to hamster scrapie agent, animals were inoculated intracerebrally with hamster scrapie strain 263K. All Tg*52NSE* mice

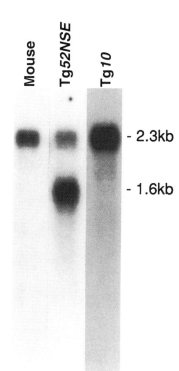

Fig 1. Expression of SHaPrP mRNA in transgenic mouse brain. Poly-adenylated RNA (2 μg) extracted from the brain of a normal non transgenic mouse, a Tg*52NSE* mouse, and a Tg*10* mouse [21] was separated on an agarose/formaldehyde gel, blotted and hybridized to a ^{32}P-labeled DNA probe derived from SHaPrP. Under the conditions used, this probe hybridizes to RNA derived from either mouse PrP (MoPrP) or SHaPrP. The 2.3-kb mRNA common to all three lanes represents MoPrP mRNA derived from the endogenous Mo*PrnP* gene. The SHaPrP mRNA derived from Tg*10* brain is also present at 2.3-kb. The unique 1.6-kb mRNA band in the Tg*52NSE* lane is derived from the NSE-SHaPrP construct and is the size expected for the fully spliced transcript [21].

died between 70 and 118 days postinoculation while non-Tg littermates were clinically normal 400 days postinoculation (fig 3). Affected mice exhibited a 1–3 day clinical course characterized by a 'stilted' gait, mild ataxia and inactivity. Tg*10* mice had a more variable and protracted clinical course lasting several weeks or even months and died between 160 and 405 days postinoculation (fig 3). The clinical symptoms in Tg*10* mice included whole body tremors, ataxia, and progression to paralysis and death. In hamsters, clinical symptoms included ataxia, tremor and somnolence. Clinically ill Tg*52NSE* mice and Tg*10* mice had histopathological findings typical of scrapie, with astrocytosis and spongiosis being the most prominent findings and easily detectable SHaPrP-res (data not shown). In summary, Tg*52NSE* and Tg*10* mice were highly susceptible to hamster 263 K scrapie agent, but both the tempo and symptoms of clinical disease were different in the two transgenic strains.

In order to determine whether the expression of SHaPrP would modify the pathogenesis of mouse scrapie, Tg*52NSE* and Tg*10* mice were also inoculated with the Chandler strain of mouse-adapted scrapie agent. Tg*52NSE* mice died 180 ± 3 days postinoculation ($n = 8$), while normal littermates died 160 ± 3 days postinoculation ($n = 11$). Tg*10* mice died 201 ± 4 days postinoculation ($n = 17$), while their normal littermates died 164 ± 4 days after inoculation ($n = 14$). Thus, expression of SHaPrP in both Tg*52NSE* and Tg*10* mice delayed the onset of clinical disease induced by the mouse scrapie agent, suggesting that expression of SHaPrP could partially interfere with the development of mouse scrapie.

Discussion

In previous experiments, Tg mice expressing SHaPrP were made using a 40-kb cosmid clone containing the SHa*Prnp* gene [10, 11], but because of the large amount of DNA in these transgenes, it was not possible to prove that SHaPrP expression was the only genetic factor involved in the induction of susceptibility to hamster scrapie in these mice. Furthermore, earlier attempts to produce Tg mice using only the SHa*Prnp* open reading frame were not successful [10]. In contrast, the present experiments succeeded in getting high levels of SHaPrP expression by using a transgene containing only 1 kb of hamster DNA including the 762-base pair open reading frame of SHa*PrnP* together with the NSE promoter. Thus, the high susceptibility of Tg*52NSE* mice to hamster scrapie demonstrates that this SHa*Prnp* minigene including the open reading frame itself is the critical element in inducing susceptibility to the hamster scrapie agent in vivo.

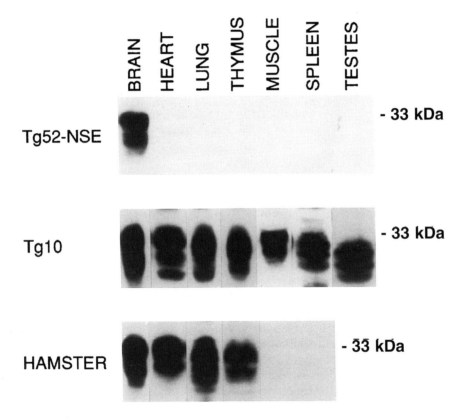

Fig 2. Western blot detection of SHaPrP-sen in tissues from a Tg*52NSE* mouse, a Tg*10* mouse and a normal hamster. Suspensions of various tissues were separated on 15% sodium dodecyl sulfate gels. Proteins were transferred to Immobilon nylon membranes and immunoblots performed using antibody 3F4 which recognizes hamster PrP but not mouse PrP. Blots were developed using the Enhanced Chemiluminescence reagent system (Amersham). Immuno-reactive bands shown represent SHaPrP-sen. SHaPrP-sen was detected in all of the indicated tissues from Tg*10* mice. In Tg*52NSE* mice, SHaPrP-sen was detected only in brain. Normal hamsters had SHaPrP-sen in brain, heart, lung and thymus but not in muscle or spleen. Hamster testes were not assayed.

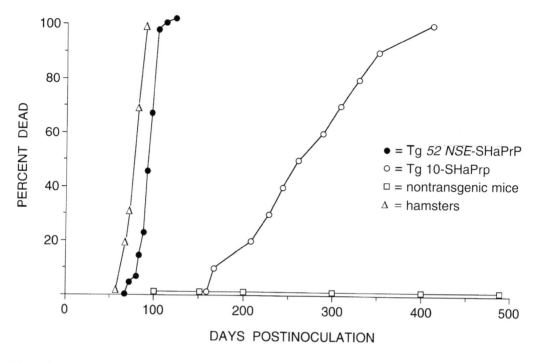

Fig 3. Kinetics of death of Tg52*NSE* mice (*n* = 60), Tg*10* mice (*n* = 65), non transgenic mice (*n* = 25) and hamsters (*n* = 10) after intracerebral inoculation with hamster scrapie strain 263K.

Several previous reports indicate that astrocytes and splenic follicular dendritic cells (FDC) may be the earliest sites of PrP-res accumulation following scrapie infection [15, 24, 25]. These sites might also be important in restriction of agent replication following interspecies transmission of TSE agents [25]. However, based on the present findings, SHaPrP expression in astrocytes or FDC was not required to mediate the susceptibility of mice to intracerebral inoculation with hamster scrapie. Nevertheless, SHaPrP expression in cells such as astrocytes, FDC or even other cell types might also be sufficient to overcome the scrapie species barrier, and FDC in spleen and lymph nodes might be particularly involved in interspecies transmission following intraperitoneal inoculation of the agent. Our current experiments are testing some of these possibilities.

In contrast to the present and previous [10, 11] data obtained with Tg mice, we have not been successful in infecting mouse neuroblastoma cells in vitro with hamster scrapie, even when clones expressing high levels of SHaPrP were utilized [12]. In mouse neuroblastoma cells infected with mouse scrapie, the expression of hamster PrP interfered with generation of mouse PrP-res [12], and by analogy, similar interactions may also be capable of blocking exogenous infection by the hamster scrapie agent in these cell lines. Although in the present experiments, inhibitory effects of SHaPrP expression on infection of Tg52*NSE* mice with mouse scrapie resulted in a significantly increased incubation period, all mice eventually developed clinical disease. In the converse experiment of infection of Tg52*NSE* by hamster scrapie agent, the incubation period was longer than in normal hamsters (fig 3), but again all mice were susceptible. At present, we have no adequate explanation for the differences in interspecies infection experiments between in vitro mouse neuroblastoma cell lines and transgenic mice in vivo. However, one important factor might be that the mice are 100,000-fold more sensitive to scrapie infection than are mouse neuroblastoma cells [26, 27].

Although experiments with PrP Tg mice have provided helpful insights into the importance of PrP in scrapie pathogenesis and interspecies transmission, it is possible that some phenomena observed in Tg mice are the result of abnormal overexpression of PrP. For example, some Tg mice overexpressing mouse or non mouse PrP have developed a spontaneous degenerative brain disease in the absence of scrapie infection [22, 28]. In other instances, transgenic mice expressing a different mouse *Prnp* gene usually associated with resistance to most mouse scrapie strains had increased sensitivity to mouse scrapie [29]. Both of these unexpected examples of neurodegenerative disease in *Prnp* transgenic mice are thought to involve overexpression of PrP, but the detailed pathogenic mechanisms involved in each are not known. Thus, it will be important in the future to confirm conclusions derived from transgenic mice by using *Prnp* null mice in which the mouse *Prnp* gene can be replaced in its normal context in the mouse genome by a single copy of a mutant or foreign *Prnp* gene.

References

1   Bolton DC, McKinley MP, Prusiner SB. Identification of a protein that purifies with the scrapie prion. *Science* 1982;218:1309-11
2   Diringer H, Gelderblom H, Hilmert H, Ozel M, Edelbluth C, Kimberlin RH. Scrapie infectivity, fibrils and low molecular weight protein. *Nature* 1983;306:476-8
3   Prusiner SB. Novel proteinaceous infectious particles cause scrapie. *Science* 1982;216:136-44
4   Caughey B, Raymond GJ. The scrapie-associated form of PrP is made from a cell surface precursor that is both protease- and phospholipase-sensitive. *J Biol Chem* 1991;266:18217-23
5   Stahl N, Baldwin MA, Teplow DB et al. Structural studies of the scrapie prion protein using mass spectrometry and amino acid sequencing. *Biochemistry* 1993;32:1991-2002
6   Borchelt DR, Scott M, Taraboulos A, Stahl N, Prusiner SB. Scrapie and cellular prion proteins differ in the kinetics of synthesis and topology in cultured cells. *J Cell Biol* 1990;110:743-52
7   Carlson GA, Kingsbury DT, Goodman PA et al. Linkage of prion protein and scrapie incubation time genes. *Cell* 1986;46:503-11
8   Hunter N, Hope J, McConnell I, Dickinson AG. Linkage of the scrapie-associated fibril protein (PrP) gene and *Sinc* using congenic mice and restriction fragment length polymorphism analysis. *J Gen Virol* 1987;68:2711-6
9   Race RE, Graham K, Ernst D, Caughey B, Chesebro B. Analysis of linkage between scrapie incubation period and the *prion protein* gene in mice. *J Gen Virol* 1990;71:493-7
10 Scott M, Foster D, Mirenda C et al. Transgenic mice expressing hamster prion protein produce species-specific scrapie infectivity and amyloid plaques. *Cell* 1989;59:847-57
11 Prusiner SB, Scott M, Foster D et al. Transgenetic studies implicate interactions between homologous PrP isoforms in scrapie prion replication. *Cell* 1990;63:673-86
12 Priola SA, Caughey B, Race RE, Chesebro B. Heterologous PrP molecules interfere with accumulation of protease-resistant PrP in scrapie-infected murine neuroblastoma cells. *J Virol* 1994;68:4873-8
13 Kocisko DA, Come JH, Priola SA et al. Cell-free formation of protease-resistant prion protein. *Nature* 1994; 370:471-4
14 Kocisko DA, Priola SA, Raymond GJ, Chesebro B, Lansbury PT, Jr, Caughey B. Species specificity in the cell-free conversion of prion protein to protease-resistant forms: a model for the scrapie species barrier. *Proc Natl Acad Sci USA* 1995;92:3923-7
15 Moser M, Colello RJ, Pott U, Oesch B. Developmental expression of the *prion protein* gene in glial cells. *Neuron* 1995;14:509-17
16 Kretzschmar HA, Prusiner SB, Stowring LE, DeArmond SJ. Scrapie prion proteins are synthesized in neurons. *Am J Pathol* 1986;122:1-5
17 Brown HR, Goller NL, Rudelli RD, Merz GS, Wisniewski HM, Robakis NK. The mRNA encoding the scrapie agent protein is present in a variety of non-neuronal cells. *Acta Neuropathol* 1990;80:1-6

18 Manson J, West JD, Thomson V, McBride P, Kaufman MH, Hope J. The *prion protein* gene: a role in mouse embryogenesis? *Development* 1992;115:117-22

19 Manson JC, Clarke AR, McBride PA, McConnell I, Hope J. *PrP* gene dosage determines the timing but not the final intensity or distribution of lesions in scrapie pathology. *Neurodegeneration* 1994;3:331-40

20 Forss-Petter S, Danielson PE, Catsicas S et al. Transgenic mice expressing beta-galactosidase in mature neurons under neuron-specific enolase promoter control. *Neuron* 1990;5:187-97

21 Race RE, Priola SA, Bessen RA et al. Neuron-specific expression of a hamster *prion protein* minigene in transgenic mice induces susceptibility to hamster scrapie agent. *Neuron* 1995;15:1183-91

22 Westaway D, DeArmond SJ, Cayetano-Canlas J et al. Degeneration of skeletal muscle, peripheral nerves, and the central nervous system in transgenic mice overexpressing wild-type prion proteins. *Cell* 1994;76:117-29

23 Kascsak RJ, Rubenstein R, Merz PA et al. Mouse polyclonal and monoclonal antibody to scrapie-associated fibril proteins. *J Virol* 1987;61:3688-93

24 Diedrich JF, Bendheim PE, Kim YS, Carp RI, Haase AT. Scrapie-associated prion protein accumulates in astrocytes during scrapie infection. *Proc Natl Acad Sci USA* 1991;88:375-9

25 Muramoto T, Kitamoto T, Hoque MZ, Tateishi J, Goto I. Species barrier prevents an abnormal isoform of prion protein from accumulating in follicular dendritic cells of mice with Creutzfeldt-Jakob disease. *J Virol* 1993;67:6808-10

26 Race RE, Caughey B, Graham K, Ernst D, Chesebro B. Analyses of frequency of infection, specific infectivity, and prion protein biosynthesis in scrapie-infected neuroblastoma cell clones. *J Virol* 1988;62:2845-9

27 Race RE, Ernst D. Detection of proteinase K-resistant prion protein and infectivity in mouse spleen by 2 weeks after scrapie agent inoculation. *J Gen Virol* 1992;73:3319-23

28 Hsiao KK, Scott M, Foster D, Groth DF, DeArmond SJ, Prusiner SB. Spontaneous neurodegeneration in transgenic mice with mutant prion protein. *Science* 1990;250:1587-90

29 Westaway D, Mirenda CA, Foster D et al. Paradoxical shortening of scrapie incubation times by expression of prion protein transgenes derived from long incubation period mice. *Neuron* 1991;7:59-68

NON PRION MODELS

Transmissible Subacute Spongiform Encephalopathies:
Prion Diseases
L Court, B Dodet, eds
© 1996, Elsevier, Paris

The protein, the prion and the infectious agent: a dialectic presentation of usable concepts

Nicole Grégoire[1], Dominique Dormont[2], Jean Nicoli[3]

[1]*Laboratoire de Biochimie et Circulation cérébrale, Sce du Pr Salamon, CHU Timone, 13385 Marseille cedex 5;* [2]*Laboratoire de Neurovirologie, CEA, DRM, BP 6, 92265 Fontenay-aux- Roses;* [3]*Laboratoire de Biologie Moléculaire, Institut de Médecine Tropicale, SSA, Jardin du Pharo, boulevard Charles Livon, 13007 Marseille, France*

Summary – Prion diseases are a tragedy and have to conform to the Aristotelian rule of the three unities: action, place and time. 'Action unity' seems to be satisfied by the unity of pathogenesis which is an amyloidogenesis. 'Place unity' is also respected since amyloidogenesis occurs in one endosomal compartment. 'Time unity' is less clearly determined, but the pathogenic process is a continuous process. It is a Shakespearian tragedy, however, and characters (one, two or several) must be defined. The following concepts, generally accepted, will be examined: transmissibility versus infectivity, virus versus protein amplification, heterodimerization versus nucleation, autologous versus heterologous chaperone molecules, signal transduction and cellular traffic.

Introduction

The session is devoted to non protein models of spongiform encephalopathies. The problem raised would thus seem to illustrate a Hegelian dialectic: the thesis of the all-protein irrefutably opposed to the antithesis of the all-infectious. This contrast may appear to be oversimplified. This work, therefore, is aimed at interpreting the concepts that can and cannot be used; to do so, it is beneficial to recall meaning, fields and implications.

Spongiform encephalopathies can be compared to a tragedy, with action, place and duration, but also characters. To begin with, they resemble a Greek tragedy: the action is a pathogenesis linked to a unique protein – the prion protein (PrP); the unique place is a post-Golgian endosomal compartment; the process, once started, is continuous. In this way, the three Aristotelian unities (action, place and time) are more or less satisfied (fig 1).

As for the characters, the tragedy is less simple: from Greek it becomes Shakespearian. The knot of the tragedy is neuronal death, probably by apoptosis. However, there is no a priori assurance that this terminal process reproduces all the mechanisms leading to it. The difficulty lies in that the only known measurements are directly linked to this terminal process (disease, mortality), or indirectly (appearance of protease-resistant PrP [PrP-res], glial reaction). According to the primary dialectic, in the all-protein model, the tragedy has two characters: the killer PrP protein and the cellular victim. In the all-infectious one, a third character is added: the instigator, like Iago in Othello, in virus form.

***Key words:** prion*

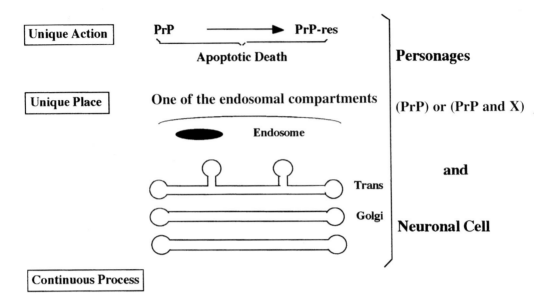

Fig 1. The three Aristotelian unities (action, place and time) of the tragedy of prion diseases.

In keeping with a dialectic presentation, we will compare usable concepts. We will accept with no further examination the unicity of sporadic forms (scrapie, Creutzfeldt-Jakob disease), the infectious forms (kuru, bovine spongiform encephalopathy [BSE]) and familial forms (familial Creutzfeldt-Jakob disease, Gerstmann-Sträussler-Scheinker disease, fatal familial insomnia [FFI]) (fig 2).

Transmissibility and infectivity (table I)

Ultraviolet- and ionizing radiation-resistance implies a small target. This excludes micro-organisms, and the viroids (small uncoated RNA) have been otherwise excluded. The virus cannot be formally excluded from this criterion. Lwoff's definition [1] should be recalled: a virus is distinguished from organisms because it has a single type of nucleic acid and no Lipmann's, systems. A virus is distinguished from genetic structures (such as transposons, plasmids) because it is infectious, that is to say its genome can be multiplied.

Transmissibility implies simply the passage from one individual to another. This means that FFI and other familial forms are transmissible [2]. Transmissibility does not inevitably imply

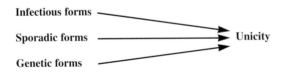

Fig 2. Unicity of the three epidemiologic forms of prion diseases.

Table I. Criteria distinguishing a virus from genetic elements and organisms.

	Virus	*Genetic elements*	*Organisms*
Nucleic acid (NA)	One (RNA or DNA)	One (DNA)	Two (DNA + RNA)
Lipmann's systems	No	No	No
Infectivity	Yes	No	Possible
Multiplication	Multiple copies of viral NA	Replication of NA	Binary division

	Transmissibility	versus	Multiplication	
	no multiplication (ie, fungal or botulic toxin)		multiplication of NA = Virus multiplication of protein (ie, tetanus toxin)	

'A virus is a virus because it is a virus'. From André Lwoff [1].

multiplication. In epidemic forms (scrapie, kuru, BSE) and in experimental diseases, multiplication is also clearly observed (with an increase in infectivity); it has never been demonstrated to be multiplication of a nucleic acid, however.

Viral genes versus cellular genes (table II)

The existence of strains implies a genetic determinism [3], but it does not make it possible to locate the gene(s) in a viral genome or at the cellular level. At least two strain characteristics have been related to a cellular gene: the incubation period is linked to a gene neighboring, connected or identical to the *Prnp* gene [4]. In addition, transgenic experiments revealed the existence of the species barrier linked to the *Prnp* gene [5, 6]. The location and profile of the brain lesion are unknown.

At this point we have an amplification which could be nucleic acid or proteic amplification.

Nucleic acid versus protein amplification (table III)

The amplification observed daily in the laboratory at the times of strain passages could be due to the amplification of an external genome.

The transgenic model may be of no great use here, since the number of gene copies, and thus the expression rate, is high, while it is known that the *Prnp* gene is unique and its transcription rate remains unchanged.

The yeast model of prion diseases [7] proves that a second amplification level is possible: the amplification of a protein conformational structure. Prion diseases are pathogenetically

Table II. Strain characteristics and their linkage to cellular or viral genes.

Evidence of strain's existence	*Cellular gene* (Prnp)	*Viral genes*
Incubation duration	Yes	?
Species barrier	Yes	?
Brain lesion localization	?	?

Table III. Nucleic acid amplification and protein amplification.

	Nucleic acid amplification (virus hypothesis)	Protein amplification (prion hypothesis)
Model	Virus, viroids	Yeast's prion
Mechanism	Multiple cycles of genome replication	Multiplication of a protein conformational structure
Target	1st genetic code	2nd genetic code

disorders of the 'second genetic code', the code which governs the folding of the proteic molecule. Anfisen's classic experiment on ribonuclease made it possible to identify this second code as the nucleic acid code since the primary sequence of the protein determined the conformation adopted. Molecular chaperone demonstration has established that a protein can adopt several distinct configurations, despite an identical primary sequence. The second genetic code does not depend here on the first genetic code. The example of the yeast genes [PSI] and [URE3], moreover, ensures the possibility of a protein conformational structure amplification.

Nucleation versus heterodimerization (fig 3)

Two concepts can explain the amplification of a protein conformational structure: 1) nucleation [8], which is a quasi-cristallization; here, the thermodynamic process induces conformational change in proteins which themselves aggregate in the network; and 2) dimerization [9], which implies the formation of protein dimers, the PrP-res form inducing a conformation identical to that of its normal homologue. ·

Both cases implicate an instructionist theory. Analogous theories had previously been elaborated to explain antibody diversity. They gain strength with the identification of molecular chaperones.

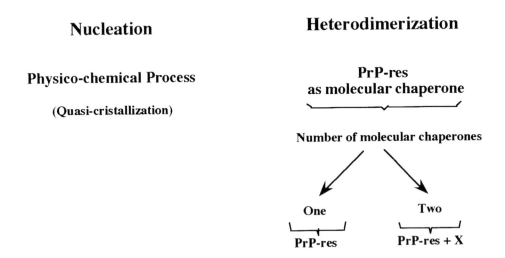

Fig 3. The two concepts of protein amplification: nucleation and dimerization, hypothesized in [8] and [9].

Two or three characters?

The hypothesis of an inducing process has been previously raised at a transcriptional level [10]. It is possible that such an inducer exists, but in the form of a molecular chaperone distinct from the PrP [11]. The PrP molecule would have two fields of interaction: a central domain of interaction with another PrP, and a C-terminal interaction domain with an X protein (second molecular chaperone).

Conclusion

We must return to the scene of the tragedy. It has been shown that the normal PrP is linked to a calcium-voltage-dependent canal [12], and that the PrP-res protein could be found at the cell surface [13]. Recent data seem to indicate that the conformational change occurs in one of the trans-Golgian endosomal compartments. Two hypotheses are thus possible: that of a cellular traffic disorder, and that of the existence of inducing transduction signals on the cel-lular surface. Growth hormone could constitute such a signal. Our experimental results do not permit confirmation of this hypothesis.

Returning to the title of this text, we now have a coherent story, the different conceptual levels of which are well understood. It remains to be said that this story is a scientific story and can therefore be falsified, according to Popper. The demonstration of the existence of a specific virus would not invalidate the established results of the unicity of pathogenesis and the role of the PrP in pathogenesis; however, it would surely invalidate the postulate of the unicity of the three different forms of scrapie diseases.

References

1 Lwoff A. The concept of virus (The third Majority Stephenson Memorial Lecture). *J Gen Microbiol* 1957;17: 239-53

2 Tateishi J, Kitamoto T. Inherited prion diseases and transmission to rodents. *Brain Pathol* 1995; 5:53-9

3 Bruce ME, McConnell I, Fraser H, Dickinson AG. The disease characteristics of different strains of scrapie in *Sinc* congenic mouse lines: implications for the nature of the agent and host control of pathogenesis. *J Gen Virol* 1991;72:595-603

4 Carlson GA, Hsiao KK, Oesh B, Westaway D, Prusiner SB. Genetics of prion infections. *Trends Genetics* 1991;7:61-5

5 Prusiner SB, Scott M, Foster D et al. Transgenic studies implicate interactions between homologous PrP isoforms in scrapie prion replication. *Cell* 1990;63:673-86

6 Scott M, Groth D, Foster D et al. Propagation of prions with artificial properties in transgenic mice expressing chimeric PrP genes. *Cell* 1993;73:979-88

7 Wickner RB. [URE 3] as an altered ure 2 protein: evidence for a prion analog in *Saccharomyces cerevisiae*. *Science* 1993;264:566-9

8 Come JM, Fraser PE, Langsbury PT Jr. A kinetic model for amyloid formation in the prion diseases: importance of seeding. *Proc Natl Acad Sci USA* 1993;90:5959-63

9 Cohen FE, Pan KM, Huang Z, Baldwin M, Fletterick RJ, Prusiner SB. Structural clues to prion replication. *Science* 1994;264:530-1

10 Nicoli J, Pisano R. In: Court LA, Dormont D, Brown P, Kinsbury DT, eds. *Unconventional Virus Diseases of the Central Nervous System.* Paris: CEA, Département de protection sanitaire, Service de Documentation, 1986;385-93

11 Telling GC, Scott M, Mastrianni J et al. Prion propagation in mice expressing human and chimeric *PrP* transgenes implicates the interaction of cellular PrP with another protein. *Cell* 1995;83:79-90

12 Whatley SA, Powell JF, Politopoulou G, Campell IC, Brammer MJ, Percy NS. Regulation of intracellular free calcium levels by the cellular prion protein *Neuroreport* 1995;6:2333-7

13 Lehman S, Harris DA. A mutant prion protein displays an aberrant membrane association when expressed in cultured cells. *J Biol Chem* 1995;270:24589-97

Transmissible Subacute Spongiform Encephalopathies:
Prion Diseases
L Court, B Dodet, eds
© 1996, Elsevier, Paris

Small virus-like particles in transmissible spongiform encephalopathies

Muhsin Özel, Elizabeth Baldauf, Michael Beekes, Heino Diringer*

Robert Koch Institut, Bundesinstitut für Infektionskrankheiten und nicht übertragbare Krankheiten, Nordufer 20, 13353 Berlin, Germany

Summary – Earlier studies have described small virus-like, disease-specific particles from brains of scrapie-infected hamsters [9] and from various cases of Creutzfeldt Jakob disease [10]. First infectivity studies with filtrates of such preparations of scrapie-associated fibrils and porcine circovirus as an internal standard to size are presented and discussed here.

Introduction

The pathogenesis of transmissible spongiform encephalopathies (TSE) can be described as a virus-induced amyloidosis [1]. This concept implies that the aggregation of a glycoprotein located at the surface of nerve cells is critically involved in the pathogenesis of these diseases, as it is transformed into an amyloid upon interaction with a hitherto undetected virus. The glycoprotein may function as a receptor for the virus.

Amyloid and infectivity are intimately associated with each other and copurify [2–4]. Separation of infectivity and amyloid can apparently be achieved to some extent (see Manuelidis, this volume). In contrast to the prion hypothesis, the virus concept strictly refutes that the amyloid protein itself is the infectious agent and responsible for transmissibility of the disease [5]. It predicts that transmissibility is due to a virus, ie, a foreign nucleic acid coding for a coat protein. It closely resembles a third concept, the virino hypothesis according to which the foreign nucleic acid uses the host-derived amyloid as a protective coat. The virus concept can link TSE of animal and man [6], and a recent report [7] on a new form of Creutzfeldt-Jakob disease (CJD) in Britain – data published 1 month after the congress – substantiates this idea [8]. It is therefore necessary to continue the search for an agent causing TSE rather than to concentrate solely on the conformation of the amyloid protein (considered by some to be a prion). Earlier studies have described virus-like, disease-specific particles from brains of scrapie-infected hamsters [9] and from various cases of CJD [10]. The first infectivity studies with filtrates of such preparations are presented here.

---

*Correspondence and reprints.

Key words: virus / amyloidosis / scrapie

Methods

Fractions of scrapie-associated fibrils (SAF) and filtered extracts of particles were obtained as previously reported [10]. Before filtration they were mixed with a purified preparation of porcine circo-virus (PCV) [11]. Samples were analyzed for infectivity before and after filtration.

The infectivity of PCV was assayed in susceptible cells [12]. After heating for 3 min at 70 °C to inactivate PCV and intracerebral inoculation into hamsters, scrapie infectivity was assayed by incubation period measurement.

Results

Figure 1a-c shows virus-like particles in association with SAF. In figure 1a two particles can be seen within the uncoiled SAF. Figure 2 represents an electron micrograph obtained from a

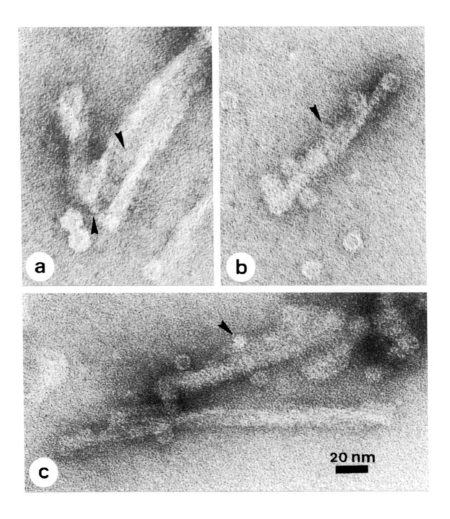

Fig 1. Electron micrographs of SAF isolated from scrapie-infected hamster brains and negatively stained with 1% uranyl acetate. **a–c**: virus-like particles (arrowheads) and ferritin associated with SAF. **a**: two virus-like particles within which the uncoiled SAF can be seen (magnification × 400,000).

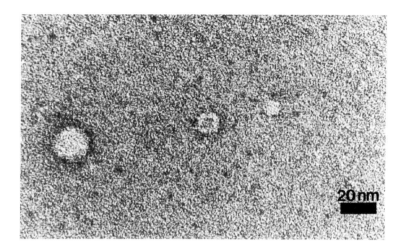

Fig 2. Electron micrograph of a 15-nm filtrate of the mixture of PCV and an SAF preparation. The material was concentrated 200-fold by centrifugation and negatively stained with 1% uranyl acetate. This figure demonstrates the extraordinarily small size of the particle as compared to ferritin and PCV, the smallest virus known (magnification × 500,000). From left to right: PCV, ferritin, virus-like particle.

15-nm filtrate concentrated 200-fold. It shows the particle together with ferritin and PCV, demonstrating the extraordinarily small size of the particle as compared to PCV, the smallest virus known.

Table I represents two filtration studies. Between 10 and 100% of the infectivity of the PCV could be recovered after filtration through 50- and 30-nm filters. A sharp cutoff occurs with 15-nm filters, allowing recovery of less than 1% of the PCV infectivity.

One to 90% of the scrapie infectivity could be recovered in the 50- and 30-nm filtrate. No infectivity was detected in the 15-nm filtrate, indicating that no infectivity or less than 0.01–0.1% of the infectivity had passed the filter.

Table I. Filtration of a mixture of scrapie agent and PCV.

		PCV		Scrapie agent*		
Experiment	Material	LD_{50}/mL	% recovery	Incubation period (dpi)	LD_{50} in 50 µL	% recovery
I	Unfiltered	6×10^4	$\equiv 100$	88 ± 2 (5/5)	10^3-10^4	$\equiv 100$
	50 nm	6×10^4	100	112 ± 13 (5/5)	10^0-10^1	1
(GH 370)	30 nm	3.3×10^4	50	112 ± 13 (5/5)	10^0-10^1	1
	15 nm	5.3×10^2	0.9	> 300 (0/5)	$< 10^0$	< 0.1
II	Unfiltered	2×10^6	$\equiv 100$	75 ± 5 (5/5)	10^4-10^5	$\equiv 100$
	50 nm	4×10^4	4	85 ± 5 (5/5)	10^3-10^4	90
(GH 376)	30 nm	1×10^5	10	85 ± 5 (5/5)	10^3-10^4	90
	15 nm	2.5×10^3	0.1	> 200 (0/5)	$< 10^0$	< 0.01

*Filtrates were used undiluted for infectivity assays, but were concentrated 200-fold for search of particles by electron microscopy. dpi: days post-infection; LD_{50}: 50% lethal dose.

Discussion

Although virus-like particles can be isolated from brain material of scrapie-infected hamsters or CJD patients, the pathogenetic function of these structures remains unclear. PCV was used as an internal standard to size scrapie infectivity by filtration [13]. The expected cutoff at 15 nm for PCV with a particle size of 17 nm proves that filtration worked correctly. Scrapie infectivity was also withheld by filters of this pore size, but passed 50- and 30-nm filters with the same efficiency. Earlier studies on size determination by filtration depending on sizing filters report values of between 25 and 100 nm [14–18]. Our results are in agreement with these investigations but narrow the size range of the scrapie agent to between 30 and 15 nm. As for PCV, scrapie infectivity was withheld by 15-nm filters, whereas the virus-like particles could be detected in concentrated filtrates.

In general, a minimal particle concentration of 10^6 per mL is necessary to detect a virus by electron microscopy. About this concentration must have been present in the 200-fold concentrated 15-nm filtrates. Therefore, in the unconcentrated filtrates approximately 10^4 particles per mL must have been present of which 1/200 or between 10^2 and 10^3 particles per dose were injected intracerebrally into hamsters. This dose did not result in disease.

There are several possible explanations for this result. First, the particles may not represent a TSE-inducing agent. They could represent a pathological product of infection or a defective form of the agent.

Alternatively, the particle indeed represents the agent. In this case several assumptions could explain the negative outcome of the experiment. For example, the particle to infectivity ratio could be higher than 10^3.

The particle may have been separated from amyloid by the weak urea treatment (2 M). Such a step is involved in obtaining these structures [10]. Indeed, figure 1a indicates that these particles might be enclosed within a coiled amyloid structure which thus may be required for infectivity. In contrast to filtrates passed through 50- or 30-nm filters, SAF have never been found in 15-nm filtrates by electron microscopy, and so far one determination of SAF protein by Western blot analysis with material representing four hamster brains was negative. Heating to 70 °C even for a short period of time may also influence the infectivity of the agent unprotected by amyloid.

Riesner et al [19] consider that the virus-like particles may represent structures obtained by sonication of SAF-containing material. In their study, sonication of SAF fractions resulted in non infectious particles ranging from 8–20 nm consisting of protease-sensitive 'prion' protein. It is obvious that this experiment does not allow any conclusion about the particles described in our publications, where a specific search by electron microscopy resulted in the detection of very characteristic morphological structures with a size range of 10–12 nm. Such a variation in size is well explained by different immersions of the individual particles in the uranyl acetate used for negative staining.

Acknowledgments

The technical assistance of S Lichy and K Dunckelmann is greatly acknowledged. The study was supported by Hertie-Stiftung and EC concerted action Bio 2-CT93-1248.

References

1 Diringer H, Beekes M, Oberdieck U. The nature of the scrapie agent: the virus theory. *Ann NY Acad Sci* 1994; 724:246-58

2 Diringer H, Hilmert H, Simon D, Werner E, Ehlers B. Towards purification of the scrapie agent. *Eur J Biochem* 1983;134:555-60

3 Diringer H, Gelderblom H, Hilmert H, Özel M, Edelbluth C, Kimberlin RH. Scrapie infectivity, fibrils and low molecular weight protein. *Nature* 1983;306:476-8

4 Hilmert H, Diringer H. A rapid and efficient method to enrich SAF-protein from scrapie brains of hamsters. *Biosci Rep* 1984;4:165-70

5 Diringer H, Kimberlin RH. Infectious scrapie agent apparently is not as small as recent claims suggest. *Biosci Rep* 1983;3:563-8

6 Diringer H. Proposed link between transmissible spongiform encephalopathies of man and animals. *Lancet* 1995;346:1208-10

7 Will RG, Ironside JW, Zeidler M et al. A new variant of Creutzfeldt-Jakob disease in the UK. *Lancet* 1996;347:921-5

8 Diringer H. Creutzfeldt-Jakob disease (CJD)–prion disease or virus-induced amyloidosis? *Lancet* 1996 (in press)

9 Özel M, Diringer H. An extraordinarily small, virus-like structure in fractions from scrapie hamster brain. *Lancet* 1994;343:894-5

10 Özel M, Xi YG, Baldauf E, Diringer H, Pocchiari M. Small virus-like structure in fractions from brains of sporadic and familial Creutzfeldt-Jakob disease cases. *Lancet* 1994;344:923-4

11 Tischer I, Gelderblom H, Vettermann W, Koch MA. A very small porcine virus with circular single-stranded DNA. *Nature* 1982;295:64-6

12 Tischer I, Peters D, Pociuli S. Occurrence and role of an early antigen and evidence for transforming ability of porcine circovirus. *Arch Virol* 1995;140:1799-816

13 He LF, Alling D, Popkin T, Shapiro M, Alter HJ, Purcell RH. Determining the size of non-A, non-B hepatitis virus by filtration. *J Infect Dis* 1987;156:636-40

14 Gibbs CJ. Search for infectious etiology in chronic and subacute degenerative disease of the central nervous system. *Curr Top Microbiol Immunol* 1967;40:44-58

15 Kimberlin RH, Millson GC, Hunter GD. An experimental examination of the scrapie agent in cell membrane mixtures. *J Comp Pathol* 1971;81:383-91

16 Eklund CM, Hadlow WJ, Kennedy RC. Some properties of the scrapie agent and its behavior in mice. *Proc Soc Exp Med* 1963;112:974-9

17 Marsh RF, Hanson RP. Physical and chemical properties of the transmissible mink encephalopathy agent. *J Virol* 1969;3:176-80

18 Tateishi J, Koga M, Sato Y, Mori R. Properties of the transmissible agent derived from chronic spongiform encephalopathy. *Ann Neurol* 1980;7:390-1

19 Riesner D, Kellings K, Post K et al. Disruption of prion rods generates 10-nm spherical particles having high α-helical content and lacking scrapie infectivity. *J Virol* 1996;70:1714-22

Transmissible Subacute Spongiform Encephalopathies:
Prion Diseases
L Court, B Dodet, eds
© 1996, Elsevier, Paris

In the community of dinosaurs: the viral view

Laura Manuelidis

Yale University, Section of Neuropathology, School of Medicine, 333 Cedar Street, New Haven, CT 06520-8062, USA

Summary – The vast majority of data on transmissible encephalopathies is inconsistent with the prion hypothesis. The prion protein (PrP or PrP-res) does not correlate with the infectious titer. Moreover, recombinant, purified, refolded and transgenic forms of PrP have no significant infectivity. With increasing purification of these agents, it has become clear that the infectious particle has the physical and molecular characteristics of a virus. Subtractive strategies have begun to delineate cDNA sequences that are present only in infectious fractions.

Introduction

There is often a large divide between speculation and data in the field of transmissible encephalopathies. It is therefore useful to highlight the major biological and molecular data that must be addressed by any hypothesis on the nature of the infectious agent. There are two fundamentally different concepts. The first is embodied by the term prion, a term used to designate a host protein that is infectious and has no nucleic acid [1]. This protein is known as PrP and has a presumed infectious form called PrP-res (protease-resistant prion protein) or PrP^{Sc} (disease-specific prion protein). PrP-res is operationally defined by its slower proteolytic digestion in a test tube. The alternate hypothesis is that the infectious agent contains an independent genome. This genome is either protected by host PrP (the virino hypothesis) or by a nucleocapsid (the viral hypothesis).

Ours was one of the first laboratories to recognize the importance of PrP. We initially considered it could either be part of the infectious agent, or alternatively, was involved in pathogenesis [2, 3]. In 1986, we deduced the correct sugar sequence of PrP [4] to find whether deglycosylation (the major post-translational change) altered the titer of PrP-enriched brain fractions. It did not [5]. During these experiments, we also found that PrP-res was a poor predictor of titer in brain fractions. We therefore turned our attention to defining major attributes of the infectious agent and its further separation from host components. In progressive experiments to test the viral hypothesis, titer has been our ultimate reality. All of our sequential experiments have consistently indicated a viral structure rather than a prion. Nonetheless, the molecular nature of these agents remains open because there is no evidence demonstrating significant infectivity for PrP in any purified, recombinant or transgenic form. Nor has a viral genome been defined by sequence. In some sense I agree that "prions ... have now gained wide

Key words: Creutzfeldt-Jakob disease / bovine spongiform encephalopathy / virus / cDNA / subtraction

recognition (and) are extraordinary." I thank the organizers of this meeting for inviting me, perhaps the most archaic dinosaur, to present the viral view. At the same time, I would first like to acknowledge the many elegant experiments of my colleagues. Regardless of our differences, I think we all concur that PrP is an important molecule in disease expression and suscepti- bility to infection.

Slow viral biology

Creutzfeldt-Jakob disease (CJD) is a fascinating paradigm because it represents a world of slow viruses that remains to be discovered. I suspect that new infectious agents will increasingly be implicated in chronic and neurodegenerative diseases [6]. CJD and scrapie project the biological shape of these infections. These properties are summarized in table I, with original references previously reviewed [7]. The major features of these infections are most simply explained by established viral mechanisms and the postulate of a specific viral genome. Although hypothetical changes in PrP conformation have been proposed to explain some of these properties, there are no current data supporting even one conformation that is diagnostic of titer, and it remains to be seen if 'refolding' of PrP-res can restore infectivity [8–10]. Accepted features that must be addressed include the endemic and epidemic infections (eg, in sheep and cows), the multiplicity of agent strains, and the mechanisms of persistence and spread.

Several additional comments are now most pertinent. Since 1989 I have expressed the concern that an altered virus with an enhanced virulence for humans could be selected in cows (eg, 7). The recent epidemic of BSE underscores the infectious rather than the inherited nature of these diseases. PrP polymorphisms are often thought to cause human CJD. One would presume then that the same holds true for scrapie-infected animals. It is hard to conceive of thousands of cows suddenly mutating their germline (*Prnp*) or somatic PrP, two of the major scenarios in the current prion theory [1]. Inoculation of transgenic mice carrying *PRNP* sequences associated with the familial disease also did not predict the human BSE-linked cases [11, 12]. Interestingly, human infections linked to the bovine outbreak have been found in people without these *PRNP* polymorphisms [13]. Moreover, the vast majority of CJD cases are sporadic, and not linked to inherited *PRNP* polymorphisms. Most importantly, almost all CJD and scrapie can be transmitted from foreign material that is not endogenous to the recipient host. This is a hallmark of an infectious agent, even when its expression is modulat- ed by host genetic factors. Thus, from a virological perspective, these infectious agents are not unique with respect to their combined infectious and host-specific attributes.

Table I. Biology of a slow virus.

Property	Comment
1. Infectious	~90% CJD sporadic with no PrP mutations
2. Unique viral strains	Virulence, incubation time and pathology vary in inbred rodents
3. Exponential replication	Different strain-specific replication rates with same PrP-res
4. Cell specificity for replication	Not predicted by PrP content (eg, spleen versus lung)
5. Latency and persistence	> 20 years dormancy in humans (eg, growth hormone recipients)
6. Evades immune surveillance	Agent hidden within cell
7. Conventional routes of spread	Reticuloendothelial system and white blood cells

The pioneering work of Dickinson and Bruce on the many distinct strains of the scrapie agent [14, 15] poses serious problems for the prion theory [16]. On the other hand, they are readily explained with a viral genome. Although multiple conformations of PrP-res are postulated to explain these strains in the prion theory, PrP-res by Western blotting typically shows species rather than agent specific characteristics. Most scrapie variants in mice cannot be distinguished by this criterion. Similarly, the replication characteristics and titer of CJD and scrapie agents passaged in Syrian hamsters are clearly different, whereas PrP-res is indistinguishable in both instances [4, 17]. Different CJD isolates also display different biological properties. Figure 1A shows a few representative isolates from our many rodent transmissions starting in the 1970s. There are reproducible differences in incubation times for three American isolates passaged in different inbred mice with similar PrP (compare 1–3). More dramatic are the very large differences of all the American isolates as compared to a passaged Asiatic CJD case (the latter being a gift from J Tateishi and P Brown). Additionally, the American CJD isolates can show almost no spongiform changes in clinically ill mice. In contrast, the Asiatic strain produces profound vacuolar pathology (fig 1 bottom, compare the USA and Asia strains). The minimal vacuolar change in some infected animals has always suggested to us that these infections may be underestimated in the human population (reviewed in [7]). Indeed, the phenotype of these infections may be more variable than commonly assumed, and I suspect that there are also infections producing plaques but minimal spongiform change that may be diagnosed as Alzheimer's disease. The chronic course and plaque pathology of the recent human cases are of interest in this context. We have experimentally selected a unique strain of CJD in rodents that bears a remarkable resemblance to the human cases linked to BSE (L Manuelidis and W Fritch, unpublished data). This convenient model should be useful in further defining specific biological and molecular characteristics of a viral strain that is most relevant for emerging human disease.

Most of the other biological features noted in table I are self-explanatory. We know of no straightforward explanation for how a host protein that is rapidly lost from tissue culture cells can survive in a dormant state in peripheral tissues for many years. Most of the data in BSE also points to an oral route of infection, and the survival of a PrP in the gastrointestinal tract is undocumented. In contrast, conventional viruses (enteroviruses for example) can survive the acidic conditions of the stomach and the enzymatic juices of the duodenum to specifically target and invade the Peyer's patches of the distal ileum (see R Bradley, this volume). From this entry, viruses classically spread by way of white blood cells to the reticuloendothelial system. The finding of infectivity only in the distal ileum in BSE parallels this conventional viral route. Moreover, the demonstration of viremia as well as splenic infection in both CJD and scrapie further strengthens this parallel. In contrast, the specific uptake of PrP-res into these tissues has not been shown.

Finally, inflammatory but not immune reactions can be detected in many slow viral infections. Covert, smoldering infections, as in human immunodeficiency virus infections of the central nervous system, educe a cascade of astroglial and microglial (macrophage) responses. Nonetheless, these responses form a more primitive (or less specific) host defense than lymphocyte-mediated mechanisms. The involvement of microglia and astrocytes in CJD and scrapie indicates similar active host responses to infection. Thus, the host retains some ability to recognize these exogenous infections. PrP may also participate in this inflammatory cascade [17]. Some of our recent experimental transmissions also lead me to propose that microglia and astrocytes may be involved in the selection of resistant viral strains, ie, those they cannot destroy.

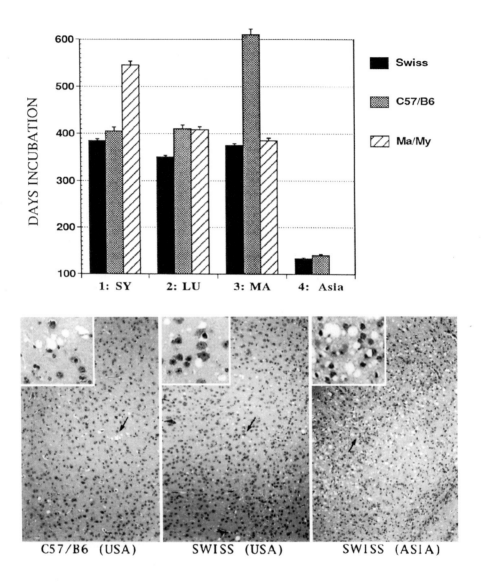

Fig 1. Human CJD passaged in mice. Top graph shows incubation time to clinical disease in three types of inbred mice. Isolates 1–3 are from the western hemisphere (USA, Italy, England), while Asiatic isolate 4 was from Japan. Strain SY was first extensively passaged in hamsters (our standard strain for fractionation studies) before being propagated in mice. Isolates 2 and 3 were directly passaged in mice. MA is derived from a patient with Gerstmann-Sträussler-Scheinker disease (English family). Note a > 200-day difference in incubation time of MA in C57/B6 mice compared to SY. The reverse is true in Ma/My mice, where SY had a > 100-day longer incubation. Isolates 1–3 had unique profiles in inbred mice, indicating that different strains had been selected. Even more dramatic is the uniquely short incubation of the Asiatic strain. Bottom shows the minimal spongiform changes in some terminally ill mice. The USA isolate shown is WEI. Only a few focal vacuoles were seen in the cortex or cerebrum at 566 days (arrow with enlargement inset). In Swiss mice, vacuoles were more obvious at 519 days, and the inset shows a vacuole characteristically impinging on a neuron. In contrast, multiple coalescent vacuoles were seen in many regions of the cerebrum of mice infected with the Asiatic strain by 149 days.

Dimensions of a virus

Because the CJD and scrapie agents are not released extracellularly, their purification will always be compromised by some host components, especially in complex tissues such as brain. The problem is to purify infectivity with minimal losses while reducing host components. We have consistently found that the disaggregation of PrP facilitates the purification of infectivity. Indeed, disaggregation can make all PrP sensitive to protease digestion, but does not reduce titer [18]. At this meeting 10 years ago, we showed that PrP formed insoluble aggregates with limited protease digestion. We therefore avoided protease digestion because these aggregates trap infectious particles and hinder their purification. We also found that commonly used PrP-res-enriched fractions compromise nucleic acid extractions. Other methodological inadequacies can also lead to poor recovery of nucleic acids of viral size.

Figure 2 shows the basic procedure used to purify CJD infectivity. It allows reasonable and reproducible yields of starting titer. Homogenization in Sarkosyl releases only ~15% of the initial brain titer, but thereafter the recovery of infectivity is essentially quantitative [19]. Nuclease treatment is used to reduce host nucleic acids by > 10^6. Underneath this outline is a typical gradient profile titered for both infectivity (top) and the distribution of PrP (bottom). Infectivity separates from most PrP. This shows that PrP is not an indicator of titer. Similar sucrose separations have now been confirmed by scrapie investigators [20], and remarkably one group has shown a sucrose gradient in which 60% of the sedimenting infectivity has no detectable PrP-res [10]. As in our experiments, the majority of PrP (the PrP-res form in scrapie studies) remains with other proteins at the top of the gradient. The separation of non infectious PrP-res suggests that only a few non specific aggregates can sediment. Others have noted the logical and experimental inconsistencies this type of data forces on newer prion models [16].

The homogeneous viral size (120S) of infectious particles was further verified by sedimentation field flow analysis and high-performance liquid chromatography (HPLC), and an operational size of ~27 nm and 10^7 Daltons was determined [21]. An infectious particle of this size can readily accommodate a viral genome of significant length (≥ 1 kb) and perhaps we are the only investigators here to consider a genome size that is not 'small'. Since the 120S particle could consist of protein only, density determinations were used to test for the presence of nucleic acid–protein complexes. Viral cores (particles of nucleic acid bound to capsid proteins) have a higher density than pure protein. CJD infectious particles had a density of 1.27 g/mL, whereas a large proportion of PrP was less dense, as previously summarized [7]. The experimental density determined for CJD infectivity corresponds to that found for several conventional viral cores. Infectious particles maintained the same density after nuclease digestion with no loss of titer. This suggests that the infectious particle consists of nucleic acid protected by capsid protein(s). Although the structure of PrP and its membrane origin made PrP an unlikely component of the infectious particle, at this point we could not rule out a genome protected by a few PrP molecules (the virino hypothesis).

Molecular components of a virus

All more purified infectious preparations of scrapie and CJD contain significant amounts of nucleic acid. Treatments used to destroy nucleic acids in infectious scrapie preparations (eg, nucleases, Zn ions and radiation) have never been shown to destroy protected nucleic acids in these preparations. The abundance as well as the length of these nucleic acids is greater than sometimes acknowledged in scrapie, as determined by independent analyses [reviewed in 7].

1. **Low speed supernatant**

2. **High speed pellet (p215)**

3. **Disaggregage & micrococcal nuclease treatment**

4. **Size sediment for 120S**

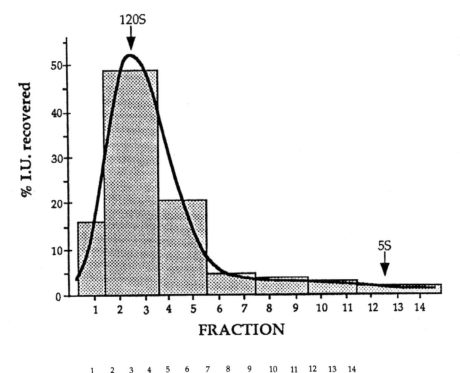

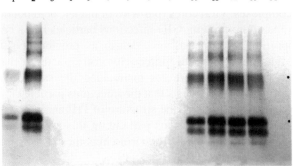

Fig 2. Basic procedure for isolating 120S infectious particles. Titer of gradient fractions is shown in the top graph, with recovery of total titer loaded. Beneath this are the same fractions in a blot stained for PrP. Most PrP remains at the top of the gradient, and some very infectious fractions (eg, 4, 5) contain no detectable PrP. Note multimers of PrP in the 5S fractions even after boiling in sodium dodecyl sulfate (SDS), a treatment that reduces titer by ~3 logs in CJD. Data shown is from reference 19.

Our studies have shown that endogenous viruses with protected nucleic acids of 5-kb cosediment with infectivity [22]. Thus sedimenting CJD particles might harbor nucleic acids of up to 5 kb in length. Although not specific for CJD infection, these productive retroviruses have many interesting parallels for transmissible encephalopathy agents, as they are resistant to many denaturants and do not form unambiguous ultrastructural particles in the brain. Like all retroviruses (and several other RNA viruses), they are also resistant to radiation. For these reasons, we find they provide a useful internal control for monitoring RNA extractions, and for evaluating treatments that destroy infectivity [9]. From the data, we found no compelling reason to exclude a viral genome.

The second component that would be necessary for a viron is a protective capsid protein. Since capsid proteins bind nucleic acids, we evaluated nucleic acid-binding proteins in 120S preparations. Reducing PrP to $\leq 1\%$ of that in the starting homogenate allowed us to visualize many minor proteins in infectious fractions, including some nucleic acid-binding proteins that appeared to be specific for infectious preparations [23]. Some of these proteins were of relatively low abundance, as would be expected if they were to correspond to determined titers. Additionally, some of the nucleic acid-binding proteins had isoelectric points that were comparable to other viral capsid proteins. One of these capsid-like proteins was identified as the endogenous retroviral core protein [22]. Because the remaining uncharacterized capsid-like proteins were of low abundance, they could not be purified for protein sequencing. Their specificity or origin in a viral particle will have to be resolved by nucleic acid strategies (see below). It was however helpful that these components were detected, as they clarified subsequent analyses.

The presence of some PrP in the 120S CJD preparation also allowed us to explore the interaction of PrP with labeled nucleic acids. Even in heavily loaded gel lanes, there was no interaction with PrP on blots. Similarly, we have not detected interaction of recombinant PrP with long nucleic acids in solution, although non specific trapping of nucleic acids in PrP insoluble aggregates have been observed (M Shlyankevich and L Manuelidis, unpublishsed data). Thus, we currently consider the virino hypothesis less likely. In summary, the above analyses demonstrate that PrP is not the sole protein component in more purified infectious preparations. Minor capsid-like proteins may be residues of a viral core, whereas PrP from infected fractions does not have a strong affinity for nucleic acids.

Disruption of viral particles leads to a loss of titer

The preceding molecular characterizations set the stage for testing treatments that selectively destroy viral particles. Viral particles were operationally defined as nucleic acid-protein structures that sediment at 100,000 g in 1 h. We asked if further non denaturing release of PrP would show infectivity in the supernatant, or if infectivity would instead be recovered with the pelleted viral particles. Conditions were found that released $\geq 90\%$ of residual 120S PrP into the supernatant. These conditions preserved the majority of intact viral particles as judged by the recovery of both nucleic acids and capsid-like proteins in the pellet. We also found conditions that destroyed the integrity of viral particles, as evidenced by solubilization of nucleic acids and capsid-like proteins. Figure 3 shows some of the pertinent published data [9]. On the left is a Western blot for PrP with the conditions for the preservation of viral particles (1% SDS at 22 °C). The SDS pellet (P) contains almost no PrP, whereas PrP is recovered in the supernatant (Su). A plus sign indicates that $\geq 75\%$ of the starting RNA was recovered in the pellet. The chosen disruptive condition (2.5 M guanidine hydrochloride, Gdn HCl) is shown in the

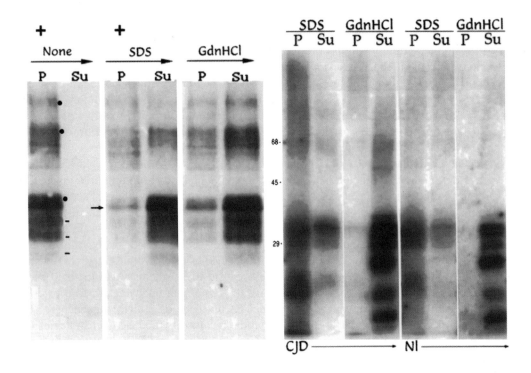

Fig 3. 120S preparations with little PrP were treated to release more PrP (SDS or Gdn HCl). The blot on the left shows PrP in the control 100,000 *g* pellet and supernatant (none) and in the treated fractions. The plus sign indicates a high recovery of RNA in the pellets. After Gdn HCl, most RNA was solubilized. Dots indicate PrP and multimers of the 34-kDa glycoprotein with some lower PrP-derived bands (lines). The arrow indicates a band that was non specifically detected by the PrP antibody, and this band was more prominent than PrP in the SDS pellet. On the right is a blot of the same fractions showing ^{32}P nucleic acids bound to a discrete set of proteins. These binding proteins were retained or solubilized in the same proportion as RNA. Similar release of binding proteins was observed in parallel control fractions (Nl). The majority of infectivity was recovered only in the SDS pellet with intact RNA-binding protein complexes [9]. Work with 263K scrapie has confirmed the lack of infectivity in Gdn HCl-treated preparations even after renaturation [10].

Gdn HCl lanes. For comparison, lanes marked 'none' show a parallel control with no additives (used to assess the recovery of proteins, nucleic acids and infectivity). The second blot (on the right) shows the nucleic acid-binding proteins in the same treated fractions. In both CJD and normal lanes, the retention (in SDS) or release (in Gdn HCl) of capsid-like proteins is shown. Titration data showed close to quantitative recovery only in the control and 1% SDS pellets where the integrity of viral particles was preserved. In contrast, the disruption of viral particles by Gdn HCl, yielding most RNA and capsid-like proteins in the non particulate supernatant, reduced titer by almost three orders of magnitude. The partitioning of infectivity with intact nucleic acid containing particles is a strong indication of the viral structure.

Subtractive approaches may define a viral genome

We concentrated on 120S preparations rather than total RNA because, even in 263K scrapie, only a minuscule population of target molecules should be present in a sea of total nucleic acids. There are two basic obstacles to defining agent-specific sequences in CJD. First, the nanogram amounts of nucleic acid in more purified preparations preclude direct cloning. Second, the intracellular location of these agents prevents their purification to homogeneity. It is therefore unlikely that one will visualize a single or prominent nucleic acid band even in more purified preparations, since a smear of host sequences will always be present. Methods to reduce these background sequences, as in subtractive strategies, are necessary. Subtractive techniques have been successfully used to uncover other slow or persistent viral genomes. For example, with Borna's disease virus (an RNA virus that also shows no diagnostic ultrastructural particles in infected brain tissue) subtractive strategies yielded essential first segments of the viral genome [24].

To address the low levels of nucleic acid in 120S preparations, we began to explore polymerase chain reaction (PCR) strategies to amplify target (CJD) and driver (control) RNAs. RNA extracted from 120S preparations was first converted into cDNA, and designed primers were ligated to amplify these sequences by PCR [19]. Although we had sufficient success to make cDNA libraries, we were unable to representatively amplify longer cDNAs. The smaller fragments were always highly overrepresented, and most clones had small redundant inserts. I therefore was impressed by the new method of representational difference analysis (RDA) [25]. This method was developed for selecting unique fragments of genomic DNA, and has three important elements. First, the use of several designed primers allows one to selectively reamplify subtracted target sequences in a sequential process. Second, the cleavage of DNA with a restriction enzyme makes it possible to amplify fragments of homogeneously small size. Thus, PCR preferences for short pieces are exploited for representative amplification. Third, equivalent cleavage of the driver with the same restriction enzyme yields a balanced subtractor population. Identical sequences present in the driver and target should make a perfect match. In RDA this will prevent amplification of sequences that are homologous to the driver, ie, non specific.

Figure 4 shows a simplified diagram of this process, with the modifications we needed to select unique RNAs in infectious fractions. RNA was chosen because the radiation data suggested that CJD might have an RNA genome. To use RDA, we had to make cDNA. RNA was converted into stable cDNA using random primers because we had no clue if the CJD target genome was polyadenylated. We also chose a more frequent cutter (*Mbo*I or isoschizomers) in an attempt to ensure availability of two cleaved ends on many cDNA molecules. This is necessary for the primer-based amplification.

Most of the RDA results presented at this meeting will be published shortly [26]. We have shown that it is possible to select sequences present only in the CJD preparations by cloning after moderate RDA selection (two rounds of subtraction). Two clones of more than five that did not hybridize to the normal driver were arbitrarily chosen for testing. These two test clones failed to hybridize to blot with either a large amount of control driver or genomic DNA. This could signify their viral origin. The sequences of the cloned inserts were also not present in the data base. Since there was no homology to bacterial or plasmid sequences (the expected contaminants from the enzymes used), the subtractive strategy appeared to work. Furthermore, RDA controls with tiny amounts of added markers have shown it is possible to visualize and clone the control markers (L Manuelidis, unpublished data). Because two CJD-specific clones were retrieved, an additional RDA experiment was undertaken with more stringent subtraction. In this second RDA, obvious discrete bands of 100–400 bp were seen in gels of the CJD preparations after subtraction. Although exhaustive cloning was not done, additional clones

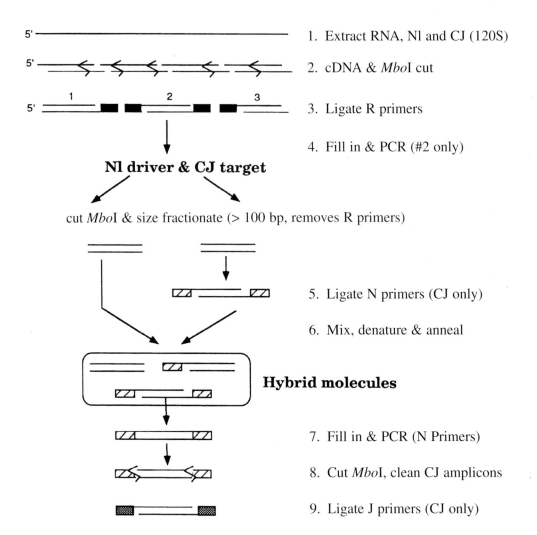

Fig 4. Subtractive strategy to visualize candidate viral sequences. RNA was isolated from 120S preparations and converted into cDNA. Cleavage of cDNA with *Mbo*I (a frequent cutter) facilitated generation of two defined ends for ligation. Some molecules without two cleaved ends will not be exponentially amplified by PCR with the R primers (molecules 1 & 3). Normal driver was generated from 3 g of starting brain to yield more than 120 μg for cycles of subtraction. Excess driver was hybridized to increasingly small amounts of the CJD amplicons in each cycle of subtraction (representational difference analysis or RDA). Steps 5–9 show this RDA strategy [25], with a few of the hybrid forms noted. Only the lower molecule with two ligated ends will be amplified exponentially. After each restriction cleavage, cut fragments were size selected to remove small cDNAs and primer. The cycle is repeated from step 9 by again mixing with excess driver (step 6). In each cycle, a new primer sequence (N or J) is ligated to avoid amplification of residual uncleaved target molecules.

were selected for testing this subtraction, and inserts were sequenced with LA Manuelidis. Nine of the 12 clones had independent sequences. In searching the database, only one of these clones had a clear homology with any submitted sequence. The lack of endogenous retroviral sequences (present in starting control and CJD preparations) again indicated reasonable subtraction of non-specific background sequences.

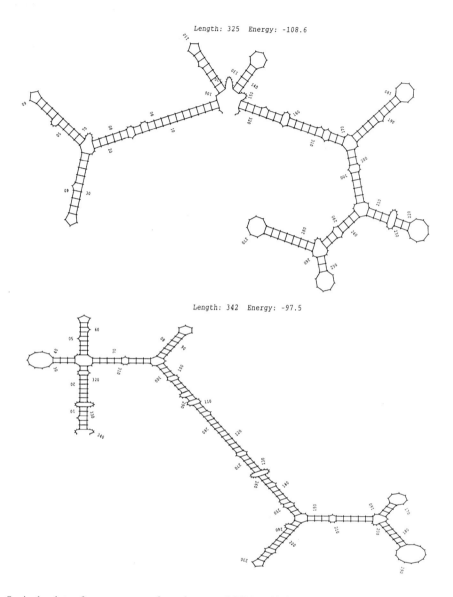

Fig 5. Squiggle plots of two sequences from the second RDA, with lengths of 325 and 342 bp (top and bottom, respectively). Significant folding of RNA was predicted and a representive conformation is shown for each sequence. Neither of these sequences was in the nucleic acid or protein databases.

We cautiously call the selected new sequences viral candidates. Since cleaved cDNA was generated for amplification, it is possible that several of these fragments derive from a longer viral genome. However, at the present time we have no definitive evidence for the origin of these sequences. At the same time, several of the selected clones do have relevant features. Complete open reading frames were found in most of the candidate sequences. Some of the peptides encoded by these open frames were consistent with the predicted isoelectric point of viral capsid proteins (pI ~6), while others had strong positive charges that could facilitate their interaction with nucleic acids. Some of these sequences also had arginine-rich domains similar to those found for other RNA-binding proteins. Additionally, some of the sequences had predicted secondary structure. Figure 5 shows squiggle plots of two clones with the strongest potential secondary structure (energy of –109 and –98, respectively). Finally, the highest homologies matched by the database programs were often viral (both RNA and DNA viruses of eukaryotes), but I consider these homologies too low to be significant (10^{-7}–10^{-8}). Indeed, the lack of strong homology with any known viral sequence makes our task more formidable. If the viral genomes for which we search are completely novel, it will be difficult to select the most meaningful fragments for further analysis. In this context, it should be noted that only two of 60 sequences selected by RDA subtraction were meaningful in another viral search. Strong homology of these two inserts with a known viral sequence in the database was the critical clue to their relative importance [27].

One might presume that all our clones should be immediately tested in reverse transcriptase-PCR (RT-PCR) experiments. However, some of the sequences, especially those with secondary structure, yield no useful primer pairs. Additionally, because RDA gives no indication of the cDNA orientation, and open reading frames were often present in both directions, a large amount of infectious and control material would have to be tested. It would seem more prudent to first evaluate additional RDA experiments on new preparations as well as other independent amplification-subtraction strategies for verification. Completely different directional RNA amplification strategies are being developed here for this purpose.

Our visualization of viral candidates represents the first demonstration of specific sequences in infectious preparations. Even without this data, we consider the accumulating evidence to be most consistent with the concept of an agent genome. Subtractive strategies for defining this genome appear promising. Is the beast a dinosaur or a relic without a brain? It is still an open question. In tracking that beast, we will likely find it more clever than we thought. The footprints, as well as the source, should give us some solutions to the emerging health issues at hand.

Acknowledgments

I thank all my partners in the laboratory for their important contributions to the work cited here. This work was supported by NS 12674 and NS 34569.

References

1 Prusiner S, Baldwin M, Collinge J et al. Subviral agents of spongiform encephalopathies (prions) In: Murphy FA et al, ed. *Virus Taxonomy*. Vienna: Springer Verlag, 1995:498-503

2 Merz PA, Somerville RA, Wisniewski HM, Manuelidis L, Manuelidis EE. Scrapie-associated fibrils in Creutzfeldt-Jakob disease. *Nature* 1983;306:474-6

3 Manuelidis L, Valley S, Manuelidis EE. Specific proteins in Creutzfeldt-Jakob disease and scrapie share antigenic and carbohydrate determinants. *Proc Natl Acad Sci USA* 1985;82:4263-7

4 Sklaviadis T, Manuelidis L, Manuelidis EE. Characterization of major peptides in Creutzfeldt-Jakob disease and scrapie. *Proc Natl Acad Sci USA* 1986;83:6146-50

5 Manuelidis L, Sklaviadis T, Manuelidis EE. Evidence suggesting that PrP is not the infectious agent in Creutzfeldt-Jakob disease. *EMBO J* 1987;6:341-7

6 Manuelidis L. Dementias, neurodegeneration, and viral mechanisms of disease from the perspective of human transmissible encephalopathies. *Ann NY Acad Sci* 1994;724:259-81

7 Manuelidis L. The dimensions of Creutzfeldt-Jakob disease. *Transfusion* 1994;34:915-28

8 Bessen RA, Kocisko DA, Raymond GJ, Nandan S, Lansbury PT, Caughey B. Non-genetic propagation of strain-specific properties of scrapie prion protein. *Nature* 1995;375:698-700

9 Manuelidis L, Sklaviadis T, Akowitz A, Fritch W. Viral particles are required for infection in neurodegenerative Creutzfeldt-Jakob disease. *Proc Natl Acad Sci* 1995;92:5124-8

10 Riesner D, Kellings K, Post K et al. Disruption of prion rods generates 10-nm spherical particles having high helical content and lacking scrapie infectivity. *J Virol* 1996;70:1714-22

11 Collinge J, Palmer MS, Sidle KC et al. Unaltered susceptibility to BSE in transgenic mice expression human prion protein. *Nature* 1995;378:779-83

12 Hope J. Mice and beef and brain diseases. *Nature* 1995;378:761-2

13 Will R, Ironside J, Zeidler M et al. A new variant of Creutzfeldt-Jakob disease in the UK. *Lancet* 1996;347: 921-5

14 Dickinson AG, Fraser H, McConnell I, Outram GW, Sales DI, Taylor DM. Extraneural competition between different scrapie agents leading to loss of infectivity. *Nature* 1975;253:556

15 Bruce ME, Dickinson AG. Biological evidence that scrapie has an independent genome. *J Gen Virol* 1987; 68:79-89

16 Carp R, Kascsak R, Rubenstein R, Merz P. The puzzle of PrPSc and infectivity – Do the pieces fit? *Trends Neurosci* 1994;17:148-9

17 Manuelidis L, Fritch W. Infectivity and host responses in Creutzfeldt-Jakob disease. *Virology* 1996;215:46-59

18 Sklaviadis TL, Manuelidis EE, Manuelidis L. Physical properties of the Creutzfeldt-Jakob disease agent. *J Virol* 1989;63:1212-22

19 Akowitz A, Sklaviadis T, Manuelidis EE, Manuelidis L. Nuclease resistant polyadenylated RNAs of significant size are detected by PCR in highly purified Creutzfeldt-Jakob disease preparations. *Microbiol Pathog* 1990;9:33-45

20 Somerville R, Dunn A. The association between PrP and infectivity in scrapie and BSE infected mouse brain. *Arch Virol* 1996;141:275-89

21 Sklaviadis T, Dreyer R, Manuelidis L. Analysis of Creutzfeldt-Jakob disease infectious fractions by gel permeation chromatography and sedimentation field flow fractionation. *Virus Res* 1992;26:241-54

22 Akowitz A, Sklaviadis T, Manuelidis L. Endogenous viral complexes with long RNA cosediment with the agent of Creutzfeldt-Jakob Disease. *Nucleic Acids Res* 1994;22:1101-7

23 Sklaviadis T, Akowitz A, Manuelidis EE, Manuelidis L. Nucleic acid binding proteins in highly purified Creutzfeldt-Jakob disease preparations. *Proc Natl Acad Sci USA* 1993;90:5713-17

24 Briese T, De la Torre JC, Lewis A, Ludwig H, Lipkin WI. Borna disease virus, a negative-strand RNA virus, transcribes in the nucleus of infected cells. *Proc Natl Acad Sci USA* 1992;89:11486-9

25 Lisitsyn N, Lisitsyn N, Wigler M. Cloning the differences between two complex genomes. *Science* 1993; 259:946-51

26 Dron M, Manuelidis L. Visualization of viral candidate cDNAs in infectious brain fractions from Creutzfeldt-Jakob disease by representational difference analysis. *J Neurovirol* 1996;2:(in press)

27 Simons JN, Pilot-Matias T, Leary TP et al. Identification of two flavivirus-like genomes in the GB hepatitis agent. *Proc Natl Acad Sci USA* 1995;92:3401-5

EPIDEMIOLOGY
AND HUMAN CLINICAL ASPECTS

Transmissible Subacute Spongiform Encephalopathies:
Prion Diseases
L Court, B Dodet, eds
© 1996, Elsevier, Paris

Transmissibility of bovine spongiform encephalopathy from animals to humans

Robert G Will

National Creutzfeldt-Jakob Disease Surveillance Unit, Western General Hospital, Edinburgh EH4 2XU, Scotland

Summary – Epidemiological surveillance of Creutzfeld-Jakob disease (CJD) was instituted in the United Kingdom in 1990 in order to identify any change in this condition that might be linked to the epidemic of bovine spongiform encephalopathy. Twelve cases of a new variant of CJD have been identified in the United Kingdom and one in France, with a novel clinicopathological phenotype. In the absence of any known risk factor for CJD in these cases, it is possible that the new variant of CJD is causally linked to bovine spongiform encephalopathy.

The question, 'Could bovine spongiform encephalopathy (BSE) pose a risk to human health?', has been considered repeatedly since the late 1980s. In the United Kingdom, the Southwood Committee considered such a possibility remote [1] and this view has been supported by a number of other expert bodies including the World Health Organisation [2]. This opinion is based on scientific evidence, particularly that related to the epidemiology of Creutzfeldt-Jakob disease (CJD). Scrapie has been endemic in sheep flocks in many countries for many years, but there is no hard evidence that it has caused disease in the human population [3]. CJD occurs in countries like Australia and New Zealand that are scrapie-free and there is no evidence of an increased risk of CJD through occupational exposure to scrapie. It could be argued that scrapie in cows is therefore unlikely to pose a threat to human health, but this hypothesis may be incorrect. The agents of the spongiform encephalopathies can be transmitted to susceptible laboratory host species and in some of these experiments the characteristics of the infectious agent can change after transmission from one species to another [4]. It is therefore possible that BSE might pose a greater risk to the human population than scrapie and indeed the transmission characteristics of the BSE agent to laboratory mice are distinct from previously studied strains of scrapie [5]. It has therefore been imperative to minimize any potential human exposure to the BSE agent and this has led to a range of legislative measures in the United Kingdom. Of particular importance is the specified bovine offals ban which was introduced in 1989 in order to prevent tissue containing high levels of infectivity, such as brain and spinal cord, from entering the human food chain.

Because of the theoretical possibility that BSE might pose a risk to humans, systematic surveillance of CJD was initiated in the United Kingdom in 1990 in order to identify any

---

Key words: epidemiology / classical Creutzfeldt-Jakob disease / new variant Creutzfeldt-Jakob disease / bovine spongiform encephalopathy

change in this condition following the occurrence of BSE. Data on CJD is available in the United Kingdom going back to 1970 and there is therefore baseline information on which to judge a change. There are, however, major drawbacks to using historical data because of the possibility that external events and scientific advances may influence the results of surveillance, and indeed there is evidence that this has happened [6]. For example, the enormous increase in public awareness of CJD may have improved case identification, and advances in molecular biology have increased the number of genetic cases identified. Through the Biomed 1 programme, it has been possible to harmonize the surveillance of CJD in a number of European countries including France, Germany, Italy, the Netherlands, Slovakia and the United Kingdom, thus allowing the comparison of contemporary data on CJD [7]. One aim of this collaboration has been to identify any change in the human disease that correlates with the incidence of BSE.

Since 1990 in the United Kingdom, a number of changes in the epidemiology of CJD have been identified. The numbers of cases have increased significantly. This is due to several factors including the occurrence of iatrogenic cases, an apparent increase in the number of genetic cases, and in particular, an increase in the number of cases of CJD in the elderly population [6]. However, the incidence of CJD in participating European countries remains very comparable to that of the United Kingdom [8] and indeed there have been similar changes in the incidence of CJD in elderly patients in other countries. A number of apparent dietary risk factors for CJD have been identified in the United Kingdom, including consumption of veal and venison, but detailed analysis suggests that these results are likely to be related to biases inherent in the case–control methodology. A further cause for concern is the identification of four cases among cattle farmers who had BSE in their herds [9]. Comparative information from the European study suggests that the incidence of CJD in farmers is similar in all participating countries and that there is no differential increase in the risk to farmers in the United Kingdom [10, 11]. Overall, until early this year, there was no hard evidence of a change in CJD in the United Kingdom distinct from other countries and therefore likely to be related to BSE.

Two cases of CJD in teenagers were identified in the United Kingdom last year [12, 13] and by March this year, a further eight similar cases had been identified [14]. The clinical features in these cases were unusual for CJD: for example, the duration of illness was prolonged and, most importantly, the neuropathological changes were very unusual with extensive accumulation of disease-associated protein. Of particular concern was that the cases were all remarkably similar both clinically and pathologically, raising the possibility of a novel form of CJD. Crucially, at that time, no similar case had been identified in other European countries, suggesting that this type of CJD was occurring only in the United Kingdom.

Detailed investigation of the ten cases revealed no known risk factor for CJD; for example, none of the patients had a history of potential transmission through medical procedures and none had a mutation of the prion protein gene. There was no common occupational link and no dietary exposure distinct from age-matched control cases. An unexplained apparently novel form of CJD had been identified in a country in which CJD surveillance had been established specifically to search for such a change. It was concluded that a causative link between CJD and BSE was the most plausible explanation for these cases, although this link was not proven.

Subsequently, two further cases of a new variant of CJD (nv-CJD) have been identified in the United Kingdom and one in France, but no other similar cases have yet been identified in other countries despite extensive review of clinicopathological information on CJD. On current evidence, this type of CJD does appear to be novel.

If these cases are linked to BSE, it is probable that the incubation period is between 5 and 15 years by analogy with iatrogenic cases of CJD [15] and kuru [16], indicating an exposure that took place in the 1980s, prior to the specified bovine offals ban. At this time, it is possible that bovine tissues containing high infectivity such as brain and spinal cord may have entered the human food chain. In my opinion, if these novel cases of CJD are causally linked to BSE, the likeliest explanation is exposure to high infectivity bovine tissue in the 1980s. However, a causal link is not proven and it is essential to obtain further evidence.

Laboratory transmission studies have been set up in order to determine whether the characteristics of the transmissible agent in the new variant cases are similar to BSE. These studies will take at least 2 years to complete and prior evidence of a causal link may depend on continued epidemiological surveillance. Although this surveillance system in the United Kingdom is being intensified, comparative information from other countries may be crucial to the interpretation of epidemiological information from the United Kingdom.

References

1   Southwood Committee. *Report of the Working Party on Bovine Spongiform Encephalopathy.* Department of Health and Ministry of Agriculture, Fisheries and Food. ISBN 1989;185197 405 9
2   World Health Organization. *Report of a WHO Consultation on Public Health Issues Related to Animal and Human Spongiform Encephalopathies.* 12–14 November 1991, Geneva, Switzerland
3   Will RG. Editorial: The spongiform encephalopathies. *J Neurol Neurosurg Psychiatry* 1991;54:761-3
4   Gibbs CJ, Gajdusek DC, Amyx H. Strain variation in the viruses of Creutzfeldt-Jakob disease and kuru. In: Prusiner SB, Hadlow WJ, eds. *Slow Transmissible Diseases of the Nervous System.* New York: Academic Press, 1979;87-110
5   Bruce M, Chree A, McConnell I, Foster J, Pearson G, Fraser H. Transmission of bovine spongiform encephalopathy and scrapie to mice: strain variation and the species barrier. *Phil Trans R Soc Lond [Biol]* 1994;343: 405-11
6   The National CJD Surveillance Unit & Department of Epidemiology & Population Sciences, London School of Hygiene & Tropical Medicine. *Creutzfeldt-Jakob Disease in the United Kingdom. Fourth Annual Report.* August 1995
7   Wientjens DPWM, Will RG, Hofman A. Creutzfeldt-Jakob disease: a collaborative study in Europe. *J Neurol Neurosurg Psychiatry* 1994;57:1285-99
8   Alperovitch A, Brown P, Weber T, Pocchiari M, Hofman A, Will RG. Incidence of Creutzfeldt-Jakob disease in Europe 1993. *Lancet* 1994;343:918
9   Sawcer SJ, Yuill GM, Esmonde TFG et al. Creutzfeldt-Jakob disease in an individual occupationally exposed to BSE. *Lancet* 1993;341:642
10 Delasnerie-Lauprêtre N, Poser S, Pocchiari M, Wientjens DPWM, Will RG. Creutzfeldt-Jakob disease in Europe. *Lancet* 1995;346:898
11 Davies PTG, Jahfar S, Ferguson IT. Creutzfeldt-Jakob disease in individual occupationally exposed to BSE. *Lancet* 1993;342:680
12 Britton TC, Al-Sarraj S, Shaw C, Campbell T, Collinge J. Sporadic Creutzfeldt-Jakob disease in a 16-year old in the UK. *Lancet* 1995;346:1155
13 Bateman D, Hilton D, Love S, Zeidler M, Beck J, Collinge J. Sporadic Creutzfeldt-Jakob disease in an 18-year old in the UK. *Lancet* 1995;346:1155-6
14 Will RG, Ironside JW, Zeidler M et al. A new variant of Creutzfeldt-Jakob disease in the UK. *Lancet* 1996;347:921-5
15 Brown P, Preece MA, Will RG. 'Friendly fire' in medicine: hormones, homografts, and Creutzfeldt-Jakob disease. *Lancet* 1992;340:24-7
16 Alpers MP. Epidemiology and ecology of kuru. In: Prusiner SB, Hadlow WJ, eds. *Slow Transmissible Diseases of the Nervous System.* New York: Academic Press, 1979;67-90

Transmissible Subacute Spongiform Encephalopathies:
Prion Diseases
L Court, B Dodet, eds
© 1996, Elsevier, Paris

Epidemiology of Creutzfeldt-Jakob disease in France

Nicole Delasnerie-Lauprêtre*

Inserm U360, Hôpital de la Salpêtrière, 75651 Paris cedex 13, France

Summary – A total of 252 cases of Creutzfeldt-Jakob disease (CJD) that does not include cases related to human growth hormone treatment have been reported to the French Research Group on Human Spongiform Encephalopathies during the 4-year period 1992–1995. According to modified Master's criteria, 210 cases were indexed as definite and probable, with a mean annual incidence of 0.91 per million inhabitants. The *PRNP* gene was studied in 152 patients. We observed 28 patients with a *PRNP* mutation: 19 had a mutation at position 200, 2 at position 102, 4 at position 178 (including two with a methionine at position 129) and 3 had insertions which had not been previously described. Eight of the 19 patients with a codon-200 mutation had a known familial history of CJD and three of these eight patients had a common ancestor. Five patients were classified as possible iatrogenic cases due to a dura mater graft; two of them had received dura mater by the arterial route. In the sporadic CJD group, 23% of the patients lived in rural areas. Among patients for whom the occupational status was known, 12 patients (8%) were farmers or farm workers. Preliminary case–control analysis did not find any significant risk factor for CJD related to occupation or residence.

In 1992, a French research group was created by the French National Institute of Health and Medical Research (Inserm) to carry out several studies on human encephalopathies: epidemiology supervised by A Alpérovitch, neuropathology managed by JJ Hauw, biochemistry by D Dormont and molecular biology conducted by JL Laplanche and P Amouyel (table I).

The epidemiological study monitors the incidence of Creutzfeldt-Jakob disease (CJD) and includes a case-control study on risk factors conducted in association with the European Concerted Action Biomed 1 during the 4-year period 1992–1995. This study does not include the patients who were treated with human growth hormone.

The coordination center, located in the Salpêtrière Hospital in Paris, is informed of all CJD cases by a national network comprising approximately 300 neurologists, neuropathologists and psychogeriatric specialists.

Two hundred and fifty-two cases were recorded between 1 January 1992 and 15 May 1996 (table II). According to modified Master's criteria, 210 cases were indexed as definite and probable, with a mean annual incidence of 0.91 per million inhabitants, slightly more than that calculated by Cathala and Brown in the 1980s [1]. Many cases were recorded a long time after the onset of the disease and some possible cases could become definite on the basis of a neuropathologic examination made long after being reported. As incidence may be regarded

---

*On behalf of the French Research Group on the Epidemiology of Human Spongiform Encephalopathies.

Key words: epidemiology / Creutzfeldt-Jakob disease / prions

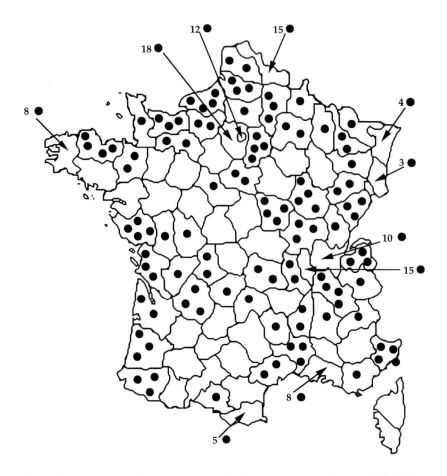

Fig 1. Geographical distribution according to the '*département* of residence', 1992–1995.

Table I. French Research Group on the Epidemiology of Human Spongiform Encephalopathies coordinated by N Delasnerie-Lauprêtre and JP Brandel (Inserm U360).

Field of research	Participants	Location
Epidemiology	A Alpérovitch	Inserm U360, Paris
	JF Dartignes	Inserm U360, Paris
	JM Vallat	CHU, Limoges
	A Ben-Hamida	Institute of Neurology, Tunis
Neuropathology	JJ Hauw	Laboratoire Escourolle, La Salpêtrière, Paris
	V Sazdovitch	Laboratoire Escourolle, La Salpêtrière, Paris
Biochemistry	D Dormont	Laboratoire de Neuro-Virologie, Fontenay-aux-Roses
	JP Deslys	Laboratoire de Neuro-Virologie, Fontenay-aux-Roses
Molecular biology	JM Launay ⎫	Laboratoire de Biologie Cellulaire, Faculté de Pharmacie, Paris
	JL Laplanche ⎬	Laboratoire de Biochimie, Saint-Lazare, Paris
	J Chatelain ⎭	
	D Amouyel	Institut Pasteur, Lille, Inserm CJF 9109
	C Locht	Institut Pasteur, Lille, Inserm CJF 9109

Table II. Number of cases by year of onset according to the year of notification.

Year of notification	Onset 1992		Onset 1993		Onset 1994		Onset 1995	
	Definite and probable	*Possible*	*Definite and probable*	*Possible*	*Definite and probable*	*Possible*	*Definite and probable*	*Possible*
1992	30	4						
1993	17	2	20	2				
1994	1	0	21	4	38	3		
1995	0	1	1	4	18	4	40	8
1996	1	0	2	0	3	5	18	5
Total	49	7	44	10	59	12	58	13
Incidence	0.85		0.76		1.02		1.00	

as practically final 2 years after the year of onset, these incidence values previously published are lower than those shown in table II [2, 3].

One hundred and seventy-seven definite or probable cases (85%) were classified as sporadic. A possible iatrogenic origin was discussed for five patients (2%); three had received a dura mater graft (of which two had been lyophilized) after a neurosurgical procedure, and two others lyophilized dura mater by the percutaneous arterial route. Prion protein gene *(PRNP)* mutations were found in 28 cases, that is, 13% of all cases or 18% of the 152 patients for whom genetic data were available.

Eleven cases show a codon-200 mutation without familial background of CJD or dementia (table III). Three of the eight codon-200 mutations in patients with a familial history of CJD belong to the same family: J Chatelain found that these three patients from the same administrative region *(département)* of Ain have common ancestors who married in 1725. Thus, the codon-200 mutation is present in France since it has been detected in nine native French patients. A woman with a codon-102 mutation presented with Gerstmann-Sträussler-Scheinker (GSS) disease at the age of 40 years, while her father's illness started when he was 65 years old. Two patients have a codon-178 mutation associated with valine on the mutant allele; one is a member of the Wui family described by Cathala et al [4]. A patient with fatal familial insomnia has a codon-178 mutation with methionine/methionine at codon-129, while his mother with methionine/methionine at codon-129 may have CJD. JL Laplanche has detected three new insertions. Two cases without familial history of demential disease have insertions of one and four extra repeats and a late onset (70 and 82 years). Another with a familial history of GSS disease has an insertion of eight extra repeats and early onset at the age of 24 years.

The geographic distribution of definite or probable cases according to their place of residence at the time of onset is shown in figure 1. Two areas seem to deserve special interest: 1) the *département* of Finistère, in Brittany, with eight sporadic cases geographically isolated. The incidence standardized for age is being evaluated to see if a potential focus exists; 2) the *département* of Ain where six patients with a codon-200 mutation are presently living; three belong to the same family and three others, apparently unrelated, have no familial history of CJD, but a search for common ancestors is currently under way.

In the sporadic group, 23% of the patients lived in rural areas. Twelve patients (8%) worked on a farm. What do these rates mean? Do they indicate exposure to animal spongiform encephalopathies? To answer this question on the basis of a case-control study on risk factors,

Table III. Cases with *PRNP* gene mutation.

Mutation	No of cases	Associated with codon-129 genotype on mutant allelle	Occurrence of disease
Glu^{200} Lys	10	Methionine	'Sporadic'
–	1	Valine	–
–	8	Methionine	Familial
Pro^{102} Leu	2	Methionine	Familial
Asp^{178} Asn	2	Valine	Familial
Asp^{178} Asn	2	Methionine	Familial
24	1	Methionine	'Sporadic'
96	1	Valine	'Sporadic'
192	1	Methionine	Familial

we looked at some data on each patient's exposure to animals. When we compared 50 cases with 50 controls matched for sex and age, no difference was found between the two groups for those working in animal husbandry (two cases versus six controls). Similarly, there was no difference whether or not they lived in a rural area (nine cases living in rural area versus 14 controls). Contact with cows and sheep did not differentiate cases from controls (20 cases having contact versus 19 controls). This preliminary analysis showed no difference between cases and controls in exposure to animals.

References

1 Brown P, Cathala F, Rauberyas RF, Gajdusek DC, Castaigne P. The epidemiology of Creutzfeldt-Jakob disease: conclusion of a 15-year investigation in France and review of the world literature. *Neurology* 1987;37:895-904
2 Alpérovitch A, Brown P, Weber T, Pocchiari M, Hofman M, Will RG. Incidence of Creutzfeldt-Jakob disease in Europe. *Lancet* 1994;343:918
3 Delasnerie-Lauprêtre N, Poser S, Pocchiari M, Wientjens DPWM, Will RG. Creutzfeldt-Jakob disease in Europe. *Lancet* 1995;346:898
4 Cathala F, Chatelain J, Brown P, Dumas M, Gajdusek DC. Familial Creutzfeldt-Jakob disease. *J Neurol Sci* 1980;47:313-51

Transmissible Subacute Spongiform Encephalopathies:
Prion Diseases
L Court, B Dodet, eds
© 1996, Elsevier, Paris

Recent findings in the epidemiology of Creutzfeldt-Jakob disease in Hungary

Catherine Majtényi

Department of Neuropathology, National Institute of Psychiatry and Neurology, Hüvösvölgyi u 116, 1281 Budapest, Hungary

Summary – Since 1960, 109 cases of Creutzfeldt-Jakob disease have been diagnosed neurohistologically among approximately 10 million people living in Hungary. In addition, three cases of Gerstmann-Sträussler-Scheinker disease were neurohistologically confirmed in three Hungarian women from the same family.

Creutzfeldt-Jakob disease (CJD) is a relatively rare form of dementia, beginning generally around 50 or 60 years of age. As there are no definitive clinical markers in this disease, the most informative diagnostic confirmation is neurohistological investigation.

At the National Institute of Psychiatry and Neurology (Hungary), all patients who die before 70 years of age after suffering a form of dementia are autopsied and investigated neuro-histologically. The most common cause of dementia in this Institute and, moreover, in the country, is cerebrovascular alteration. Alzheimer's disease is only the second most common cause of dementia in the presenile age and CJD the third.

In order to obtain information about all the Hungarian cases, letters were sent in 1993 to the neurologic and psychiatric wards throughout the country. It seemed important to send letters to the psychiatrists as well as the neurologists, because most of the patients who had been diagnosed previously were treated at one of the psychiatric wards. In total, 143 letters were sent. The letters listed the diagnostic criteria of Will [1] and included a questionnaire to be completed and returned if a case of CJD was suspected. The questionnaire was compiled by the leaders of the project: Surveillance of CJD. Most of the questionnaires were returned, and all of the brains of the suspected cases were sent to the Institute for examination.

Over a period of 3 years, we received information about 27 cases based on these diagnostic criteria. The brains were examined in our laboratory. In 18 of the 27 cases, the CJD diagnosis was neurohistologically confirmed; however, nine of the reported cases were misdiagnosed. The reason for this will be analysed briefly later. This year (1996) seems to be an outstanding one, because there have been ten reported cases of CJD.

Of the 18 confirmed cases of CJD, nine were in men and nine in women. The mean age of the patients was 59 years with a range of 45 to 73 years (three patients).

According to the information obtained from the questionnaires, only one patient had worked at a ranch, another was an agricultural worker. Ten were clerks, one a car driver, one a

---

Key words: Creutzfeldt-Jakob disease / epidemiology / neuropathology

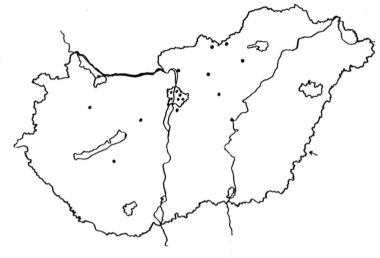

Fig 1. Map of Hungary with the larger cities outlined. The dots represent the residences of the patients. Arrow: residence of a patient living in Transylvania, Roumania.

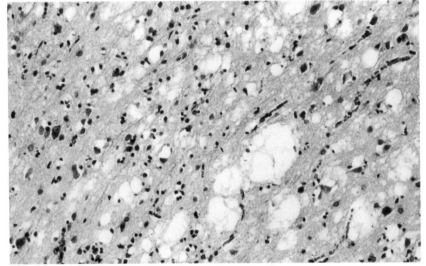

Fig 2. Spongiosis in the frontal cortex (haematoxylin-eosin staining, × 200).

physical education instructor, one a manager of a restaurant, and in three cases the profession remained unknown.

None of the patients had undergone previous organ transplant, corneal transplant, or growth hormone therapy; these known risk factors could not be confirmed.

Two female patients belonged to familial cases of CJD. One of these two patients, who died at 59 years of age, underwent a genetic examination performed by J Collinge, who found a codon-200 lysine mutation. The sister of this patient, a neurologist working at the Intensive Care Unit of the National Neurosurgical Institute, had died previously as a result of CJD. In another case, it was reported that the father had had an illness with dementia and myoclonic jerks at the end of his life.

Regarding the residence of the 16 patients, there was no remarkable clustering during this period. The higher number of patients living in the capital is due to the greater density of the population (fig 1).

Most of the neurohistological alterations were of the classic types described by many authors; Masters and Richardson among others. Widespread spongiform transformation of the cortex (fig 2) could be observed in some of the cases. At the beginning of this transformation, the cavities often formed around the astrocytes or between the fibres of the proliferated astrocytic elements (fig 3). This phenomenon was observed in different cases. In figure 4, by using a

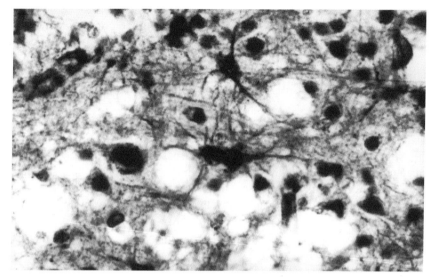

Fig 3. Parietal deep cortical area (Cajàl-Globus method, × 600).

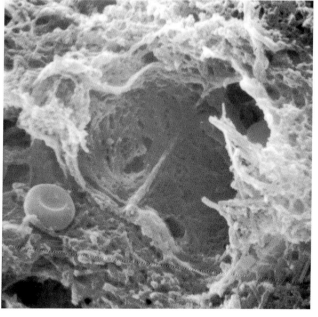

Fig 4. Scanning electron micrograph of holes with astrocytic fibers.

scanning electron microscope, it can be seen that the astrocytic fibres are bridging over the cavities. In other cases, the cavities form focal groups in the deep cortical layers. In two cases, a so-called pseudolaminar spongiosis was observed, but only in the calcarine cortex (fig 5).

In general, the cortical nerve cells remained relatively intact, keeping their original shape and form. When altered, they were ballooned, or stained pale between the cavities. In half of the cases, the alterations extended to the caudate nucleus and putamen. This was a spongiform transformation with nerve cell loss accompanied by astrocytosis. Only one of the 18 cases had a diffuse pallor in the myelin structure, resembling the panencephalopathic type of CJD. The myelin damage in this case was mostly seen in the short association pathways: the long pathways were relatively well preserved.

Thanks to HA Kretschmar and his students, we were able to perform the prion protein (PrP) antibody test, when necessary. However, systematic localization of PrP was performed in only some cases (fig 6).

It seemed beneficial for future investigation to analyse the misdiagnosed cases in order to determine the probable causes of these misdiagnoses. This problem certainly has a subjective part, which depends on the experience of the physicians and especially on that of the electroencephalographer. In fact, most of these misdiagnosed cases had an electroencephalographic (EEG) pattern which resembled that of CJD.

The neuropathological investigation found: Alzheimer's disease of short duration in four cases; alcoholic encephalopathy with pellagroid alteration in the cortical nerve cells in two cases; chronic lymphocytic encephalitis in one case; one patient with meningeal carcinomatosis; and in one patient, who had vascular encephalopathy, the diagnosis was based on the EEG, which showed paroxystic lateral epileptic discharges which were very similar to the spike and wave complexes seen in CJD. However, the original EEG recordings must be analysed in order to confirm the occasional similarities and differences. This is our future plan, which will be achieved in collaboration with the EEG laboratory of our institute.

Fig 5. 'Pseudolaminar' spongiosis in the occipital cortex (haematoxylin-eosin staining, × 100).

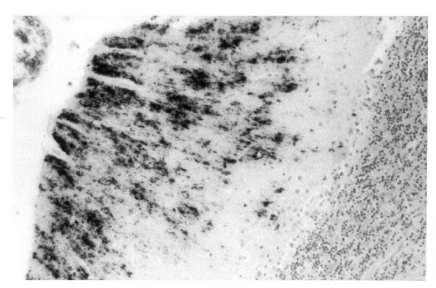

Fig 6. Immunohistochemical staining for PrP in the cerebellar cortex (× 60).

In all nine misdiagnosed cases, a rapid clinical course was observed with fast progression. Intellectual impairment was very severe from the beginning and led to dementia in seven of the nine cases. Hyperkinesia (mostly myoclonic jerks) was observed in six of these cases. The type of EEG alterations resembled that of CJD in seven cases; however, we only had the descriptions, not the original tracings.

In conclusion, after attention had been drawn to the disease, 18 new cases of CJD were diagnosed and neurohistologically confirmed in Hungary over the last 3 years. Including these cases, the number of the neurohistologically investigated cases increased to 109 in our country (of 10 million inhabitants). To date, no iatrogenically-infected patient has been observed. Among the new cases were two patients with positive family histories. In one, a codon-200 lysin mutation was found. The sister of this patient had died previously as a result of CJD. It must be emphasized that only one of the 18 new cases proved to be of the panencephalopathic type of the disease.

Acknowledgements

This work was supported by the EU Biomed I Program "Surveillance of CJD". No: ERBCIPDCT940271.

References

1 Will RG. Epidemiological surveillance of Creutzfeldt-Jakob disease in the United Kingdom. *Eur J Epidemiol* 1991;7:460-5

Transmissible Subacute Spongiform Encephalopathies:
Prion Diseases
L Court, B Dodet, eds
© 1996, Elsevier, Paris

Creutzfeldt-Jakob disease in Australia

Steven Collins, Ashley Fletcher, Trudy De Luise, Alison Boyd,
Colin L Masters

The Australian Creutzfeldt-Jakob Disease Registry, Department of Pathology, University of Melbourne, Parkville, Victoria 3052, Australia

Summary – The Australian Creutzfeldt-Jakob disease (CJD) Registry was established in 1993 in response to the recognition of four CJD deaths related to the contamination of human pituitary hormones (gonadotrophins). The period of surveillance now extends retrospectively to 1970 and will continue prospectively for at least the next 5 years, following recommendations of an official inquiry into the past use of human pituitary hormones within Australia. A wide variety of methods of case ascertainment endeavour to detect all pathologically proven and clinically probable examples of transmissible spongiform encephalopathies occurring within Australia during this surveillance period. Of the 282 persons currently on the Register, seven are iatrogenically related, 11 occurred within a heredofamilial context and the remainder are apparent sporadic CJD. Australia is free of the animal forms of these diseases (bovine spongiform encephalopathy and scrapie), which will prove to be of interest to the worldwide surveillance effort which is examining the possible zoonotic origin of new variant forms of CJD.

Introduction

National surveillance programmes dedicated to monitoring the incidence, risk factors and the genotypic and phenotypic profile of human transmissible spongiform encephalopathies (TSE) now exist in a number of countries [1, 2], although the impetus for their inception varies. The Australian Creutzfeldt-Jakob disease (CJD) Registry was established in October 1993 in response to the recognition of four cases of CJD related to the use of human pituitary hormones (gonadotrophins). One of its objectives is to monitor the Australian population for any further iatrogenic occurrences of this type. To achieve this, a period of detailed scrutiny was planned from 1 January 1988 until 31 December 1997, with this 10-year time frame chronologically overlapping the four index cases. However, following the recommendations of a judicial enquiry [3], the period of surveillance has been extended retrospectively to 1 January 1970 and no completion date has yet been set. Given the long incubation periods now recognized in these diseases, the open-ended nature of this surveillance is appropriate.

Current concerns relating to the possible association of bovine spongiform encephalopathy (BSE) and variant forms of CJD in the United Kingdom [1] are likely to heighten international interest in the findings of the Australian CJD Registry. Australia is free of the domesticated animal forms of BSE and scrapie. The Australian demographic profile and surveillance may therefore assist in the evaluation of possible zoonotic origins of some forms of CJD.

Key words: Creutzfeldt-Jakob disease / epidemiology / case registry / transmissible spongiform encephalopathy / iatrogenic / scrapie / bovine spongiform encephalopathy

Methodology

The various methods of case ascertainment and their relative yields are summarized in table I. Either separately or in combination, these attempt to detect all pathologically proven (brain biopsy and postmortem) and clinically probable or possible cases of CJD occurring after 1 January 1970. Prospective recruitment is to continue for at least the next 5 years. This project and its methods have been approved by the University of Melbourne Human Research Ethics Committee.

Case ascertainment

Personal communications

Of greatest numerical importance has been personal communication of cases by medical practitioners (especially neurologists and pathologists), usually by telephone or correspondence. At the inception of the Registry, all neurologists and pathologists within Australia were requested to inform the Registry of any cases of CJD encountered during their lifetime practice. Appropriate case reviews were then undertaken. Other sources of personal notification of potential cases have included relatives of patients, the Pituitary Hormones (formerly CJD) Taskforce and the CJD Counselling Service. Similar to the Registry, these latter, federally funded bodies were formalized after recognition of the four probable pituitary hormone-related CJD deaths. The Pituitary Hormones Taskforce received enquiries from many interested or concerned relatives regarding deceased persons believed to have died from CJD. Follow-up to confirm the exact cause of death was offered and performed by the Registry if the family desired. Finally, relatives were often referred to the Registry by the CJD Counselling Service, where case review and clarification were offered as appropriate.

Death certificates

A death certificate search through the Australian Health Index commenced at the start of the project, encompassing all states and territories for the years 1980 to 1992. Between 1980 to

Table I. Sources of case ascertainment.

Method	Percentage
Personal communications	51
Neurologists	24
Neuropathologists	15
Pituitary Hormones Task Force[1]	7
Family	3
CJD counselling service	2
Death certificates	29
Hospital and health department searches	15
Hospital medical records	12
Health departments	3
Mailout replies	5
Total	100

[1]Formerly known as the CJD Task Force.

1987, 45 cases were found and between 1988 to 1992, 41 cases were identified. However, to date we are aware of ten Registry patients from this epoch (identified by other methods) who were not discovered by the search, yet have CJD listed on their death certificate. The explanation for this discrepancy has not yet been determined. A similar search is imminent for the years 1970 to 1979 and 1993 and 1994. The usefulness of death certificates is illustrated for the years 1980 to 1987, during which, of the 93 cases on the Register, 44 were ascertained by death certificates, representing 47% of this total. Although of immense help retrospectively, inherent delays in the computerization of certificates usually translate into the ascertainment of prospective cases by means other than death certificate notification.

Hospital medical records

As part of our retrospective analysis, we are undertaking a nation-wide search of the in-patient medical records separation codings from all university-affiliated teaching and tertiary referral hospitals. All state and territory departments of health were contacted at the start of the project to coordinate this task. Official permission and ethical clearances for the Registry and its information procurement methods were obtained from the appropriate authorities prior to commencement. Searches are performed as retrospectively as possible by the individual institutions using the ICD 9 CM separation codes of 046.1 (specific for CJD) and 290.1. The latter is only accepted when it is the principal diagnosis, and is used to identify patients suffering presenile dementia. This inclusion was prompted by our prior awareness that some CJD patients were codified under this less specific classification, and manpower limitations precluded attempts to survey all patients coded as dementia (ie, 290 or 290.0).

Hospital-based searches are now complete in Victoria, Tasmania, South Australia, and Western Australia, and are nearing completion in the other states and territories. In the states where searches are complete, a total of 1,390 hospital files coded as presenile dementia (principal diagnosis) were reviewed, yielding seven probable or definite cases of CJD, all of which were new to the Registry. Eighty-nine cases coded as CJD were identified, but nine were rejected after thorough evaluation. Of the 80 retained patients, 38 were unknown to the Registry prior to the searches. In addition to the nine culled cases initially coded as CJD by hospitals, a further 48 cases of potential CJD notified from other sources have been excluded after detailed follow-up.

Similar searches have been performed within each state's or territory's centralized hospital in-patient morbidity data, and have provided largely comparable, but not identical, information to that obtained from the individual teaching hospitals. This allows follow-up of cases identified at non-teaching hospitals within the appropriate state or territory. The reasons for the slightly complementary nature of these two databases within a state or territory vary, but are not always identifiable.

After cases are identified by their ICD 9 CM separation code, an on-site review of these patients' hospital files is performed by a field research nurse from the Registry. Selected medical records are further reviewed by the clinical coordinator until a decision can be made about the likelihood of the diagnosis, and whether the initiation of a thorough follow-up of the case is justified.

Mailout replies

Following our initial request for retrospective notification of all known CJD cases by neurologists and pathologists within Australia, a reply-paid mailout is posted to these two groups of medical specialists semi-annually, prompting notification of recent or prospective cases. The national average reply rate is 75% for the 80 anatomical and forensic pathologists, but only 56% for the 320 neurologists.

Once a suspected case is identified from any source, the Registry undertakes the completion of a comprehensive medicodemographic questionnaire. This document constitutes the cornerstone of our epidemiological database. The proforma questionnaire usually requires the input of both family members and medical caregivers for adequate completion.

Classification

Classification status is determined after careful review of all available information relating to the patient and centres around the comprehensive questionnaire and relevant investigation results, including central nervous system histological examination. There is no coordinated neuropathological component to the Registry, but in view of the importance of histological confirmation, we encourage and facilitate postmortem brain examination whenever possible and appropriate. Our classification criteria are based on those previously reported [4], but with some modifications. The changes are biased towards maximizing sensitivity of detection of the nonpathologically confirmed patients, and aim to avoid potential 'false-negatives' which may occur if diagnostic criteria (eg, necessitating a 'typical' electroencephalogram, EEG) are too narrow or rigidly enforced [1]. Definite cases are neuropathologically confirmed, whereas clinically very likely (probable) cases must have manifested an invariably progressive, fatal illness of less than 18 months duration, where dementia and/or cerebellar ataxia were the salient or predominant features, and a thorough evaluation did not disclose an alternative explanation. Accompanying features such as a typical electroencephalographic change or myoclonus further strengthen the likelihood and can partly offset other less typical features (eg, slightly longer duration of illness). Possible cases are the same as probable, but concomitant medical illnesses often obscure the more definite 'probable' status. The 'incomplete' status is applied when preliminary investigations have revealed a likely diagnosis of CJD, but further information is known to exist which may impact upon classification status. A final decision is made once all information has been assembled.

Results

Demographic profile

There are currently 282 cases on the Australian CJD Registry: 126 are definite, 86 probable, one possible and 69 incomplete (see table II). Of the 69 incomplete files, 41 relate to patients pre-1988. The data presented are only on the definite, probable and possible cases. The classification of these patients is approximately 92% sporadic, 5% familial and 3% iatrogenic. Further stratification of the Registry cases has been performed according to age at onset and gender (fig 1), while durations of illness are summarized in figure 2. The average age at onset in sporadic cases is 63 years, with a mean illness duration of 8 months. Familial cases are younger (average age at onset, 49 years) with a longer average duration of illness (26 months). Iatrogenic patients have the youngest age at onset (average, 40 years) and an average duration of illness similar to sporadic CJD (7 months). Of the 16 cases who died under the age of 46 years, nine were either iatrogenic ($n = 5$) or familial ($n = 4$). Only 17% of all patients survived longer than 12 months, with the majority (66%) dying within 6 months of onset of symptoms.

Table II. Summary of the classification status of all cases on the register.

Classification	No of cases	Percentage
Definite	126	45
Probable	86	30
Possible	1	< 1
Incomplete	69	25
Total	282	100

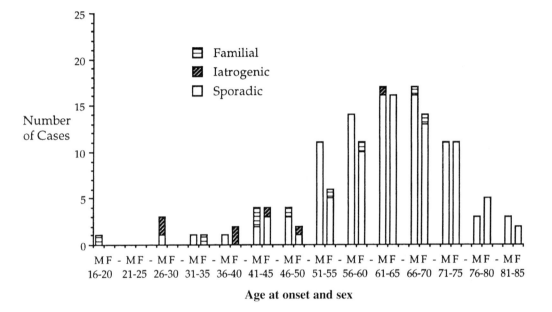

Fig 1. Stratification of cases according to age at onset, gender (M, F) and classification.

Of some relevance to our inclusion criteria, there are 15 definite cases on the Registry with a non characteristic EEG (average recording time 4.3 months prior to death) and 12 probable cases with a negative EEG and a comparable average recording time premortem (4.6 months).

The average annual incidence of CJD in Australia between 1970 and 1987 was 0.379 cases per million, compared to 1.147 cases per million for the population aged 45 years and over. However, for the period 1988 to 1995, for which ascertainment is more complete, these average annual incidences were 0.761 and 2.306 cases per million, respectively. The incidence data are shown graphically in figure 3.

Occupational and travel histories

There are no unusual occupational predilections of Registry cases compared to the Australian population in general [5] (table III). Of the 89 cases known to be born in Australia (table IV),

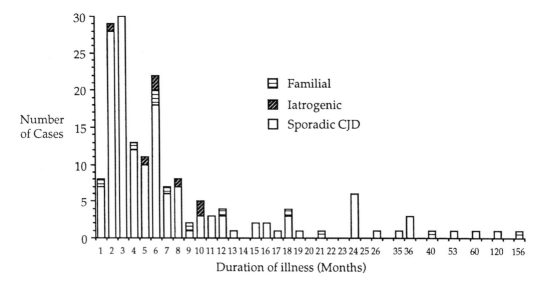

Fig 2. Duration of symptomatic illness in patients.

29 had never travelled overseas and four had travelled only to New Zealand or the Pacific Islands (free of animal forms of spongiform encephalopathies). Only 21 patients were known to have travelled more broadly, and in 35 cases a detailed travel history is unknown.

Iatrogenic cases

Seven of the 282 cases may be related to iatrogenic contamination events, with four occurring after exposure to human cadaveric pituitary gonadotrophin therapy, one possible case subsequent to growth hormone treatment and two cases following dura mater homograft implantations. Figure 4 schematically displays the chronological aspects of the treatment, incubation and symptomatic phases of these patients.

We are currently studying the past surgical/medical histories, blood transfusion (donor and recipient) status and dental histories of the Registry cases in the hope of identifying other iatrogenic risk factors.

Familial cases

Eleven Registry patients (5% of the total) occurred within a familial context. The phenotypic spectrum of the genetically determined cases encompasses seven individuals (two families) with fatal familial insomnia (FFI), three persons (two families) with Gerstmann-Sträussler-Scheinker (GSS) disease, and one case in a pedigree of typical CJD. A single case of a clinically and neuropathologically typical, but apparent non hereditary GSS disease was identified.

Unresolved cases

During the course of surveillance, two patients were identified with a clinical picture very suggestive of CJD, but neuropathological examination revealed non specific changes. Such cases are not included in the Register, but we maintain an interest in them, pending future investigational techniques becoming available.

(a)   Absolute number of cases

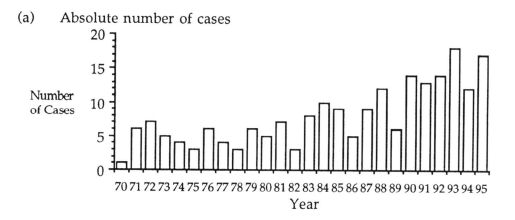

(b)   Crude incidence

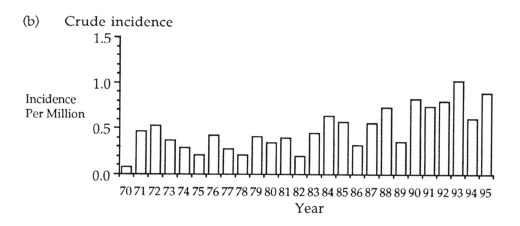

(c)   Incidence, age adjusted (45 and over)

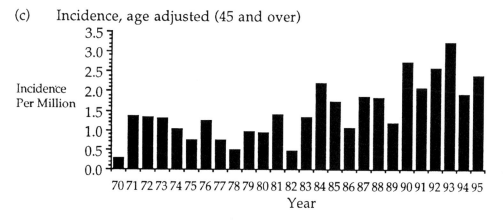

Fig 3. The absolute number of cases (**a**), the incidence in the general Australian population (**b**) and in the Australian population aged 45 years and over (**c**) for the years 1970 to 1995 inclusive.

Table III. Comparison of occupations of Registry patients and the general Australian population according to the Australian Standard Classification of Occupation.

Occupation group	No of cases on register	% cases on register	% working Australian population[1]
Managers and administrators	17	13.4	9.4
Health professionals	17	13.4	11.3
Health diagnosis and treatment practitioners	2[2]		
Para-professionals	6	4.7	4.9
Registered nurses	1[3]		
Tradespersons	16	12.6	12.2
Clerks	15	11.8	13.4
Salespersons and personal service workers	5	4.0	12.6
Plant and machine operators, and drivers	12	9.4	5.6
Labourers and related workers	17	13.4	12.0
Home duties/childcare[4]	22	17.3	18.7
Total	127	100	100

[1]1993 Labour Statistics Australia [6]; [2]two general practitioners; [3]one of the four human-derived pituitary gonado-trophin recipients; [4]home duties/childcare is not an official grouping according to the Australian Standard Classification of Occupation.

Table IV. Country of birth of Registry patients.

Country	Cases
Australia	89
United Kingdom	20
Italy	9
Yugoslavia	5
Greece	4
Hungary	3
Germany	3
Russia	1
USA	1
Papua New Guinea[1]	1
India	1
Chile	1
Poland	1
Portugal	1
Estonia	1
Unknown	71

[1]Born in Papua New Guinea to Australian parents. Lived there for for first 2 years, then resided in Australia and died at 73 years of age.

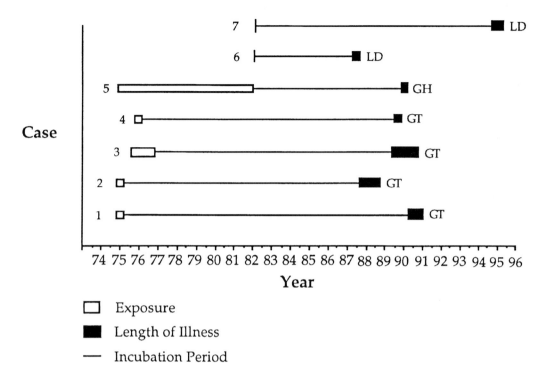

Fig 4. Chronological profile of the Australian iatrogenic Creutzfeldt-Jakob disease cases. LD: Lyodura® recipients; GH: growth hormone recipient; GT: gonadotrophin recipients.

Discussion

Completeness of case ascertainment is a primary concern for surveillance registries, but the difficulties of the successful investigation of retrospective cases, especially those dating back more than 20 years are significant. This is partly illustrated by our current sizeable 'incomplete' category, of which 60% relate to cases pre-1988. Nevertheless, the remaining death certificate search by the Australian Health Index (for the years 1970 to 1979, and 1993 thereafter) in combination with the completion of the national, hospital-based searches should further aid retrospective case detection. However, it is unlikely that this will increase the pre-1988 annual incidences to the present levels. We doubt there has been a true doubling of the annual occurrence of spongiform encephalopathies since 1988, and analogous to the experience of other CJD surveillance programmes [1], believe that any increased incidence most likely reflects case ascertainment bias stemming from improved case recognition, confirmation and reporting. Any true increase in the incidence should become clearer over the next few years, once the Registry's monitoring methods and notification profile stabilize. The post-1988 average annual incidence of 0.761 cases per million, and the composition of these patients (92% sporadic, 5% familial and 3% iatrogenic) is comparable to previously published estimates [6–8]. However, the development of CJD after pituitary gonadotrophin therapy appears unique to Australia [9].

The application of molecular biological techniques to the study of TSE has profoundly altered our understanding of this group of disorders, and stimulated new attempts at clinico-pathologic subtyping based on protease-resistant protein genotyping and immunoblotting results [10]. Given our evolving insight into these diseases, any clinical and pathological classification of the TSE is contentious, including the relative importance of electroencephalographic results. The present diagnostic criteria adopted by the Registry are biased towards the detection of iatrogenic cases, which often differ clinically from sporadic CJD by manifesting progressive cerebellar ataxia as the salient clinical abnormality [11]. Previous classification requirements (including those in various European surveillance programmes) appear more aligned to the ascertainment of sporadic cases, whereas cases of iatrogenic CJD may not show definite dementia until relatively late in the illness, nor do these patients usually show typical EEG changes. Until a sensitive, specific, premortem diagnostic test becomes available, neuro-pathological examination remains the most reliable method of case confirmation, but this is not always achievable even in prospective patients.

Recent developments in the United Kingdom [1] are likely to heighten international interest in the findings of the Australian CJD Registry. Australia is fortunately free of the animal forms of TSE. Detailed demographic information sought for all Australian CJD patients may allow the detection of imported zoonotic or other forms of disease. Comparative studies of this type may prove useful in research strategies aimed at resolving this challenging epidemiological problem. At the time of writing, the Registry is not aware of any Australian cases matching the new variant CJD profile reported in Europe [1, 12], which has evoked the crisis concerning possible zoonotic transmission of BSE. Our experience with younger patients manifesting CJD (ie, < 46 years) indicates an increased probability of association with familial and iatrogenic forms of the disease.

Acknowledgments

The authors wish to thank Dr J Kaldor, Deputy Director and Head, National Centre in HIV Epidemiology and Clinical Research, for constructive discussions relating to the epidemiological aspects of the Registry's functions. Funding for these epidemiological studies is provided by a grant from the Commonwealth Department Human Services and Health. We also thank all physicians and other health professionals who have assisted us with case ascertainment.

References

1 Will RG, Ironside JW, Zeidler M et al. A new variant of Creutzfeldt-Jakob disease in the UK. *Lancet* 1996; 347:921-5

2 Delasnerie-Lauprêtre N, Poser S, Pocchiari M, Wientjens DPWM, Will R. Incidence of Creutzfeldt-Jakob disease in Europe. *Lancet* 1995;346:898

3 Allars M. *Report of the Inquiry into the Use of Pituitary-derived Hormones in Australia and Creutzfeldt-Jakob Disease.* Australian Government Publishing Service Canberra, June 1994

4 Masters CL, Harris JO, Gajdusek DC, Gibbs CJ, Bernoulli C, Asher DM. Creutzfeldt-Jakob disease: patterns of worldwide occurrence and the significance of familial and sporadic clustering. *Ann Neurol* 1979;5:177-88

5 Australian Bureau of Statistics. 1993 *Labour Statistics Australia.* The Australian Government Publishing Service, Canberra 1994

6 Brown P, Cathala F, Raubertas RF, Gajdusek DC, Castaigne P. The epidemiology of Creutzfeldt-Jakob disease: conclusion of a 15 year investigation in France and review of the world literature. *Neurology* 1987;37:895-904

7 Irving WL, Cummins DS, Masters CL, Cunningham AL. Creutzfeldt-Jakob disease and slow infections: a review. *Aust NZ J Med* 1990;20:283-90

8 Palmer MS, Collinge J. Human prion diseases. *Curr Opin Neurol Neurosurg* 1992;5:895-901

9 Collins S, Masters C. Iatrogenic and zoonotic Creutzfeldt-Jakob disease: the Australian perspective. *Med J Aust* 1996;164:598-602

TRANSMISSIBLE SUBACUTE SPONGIFORM ENCEPHALOPATHIES:
PRION DISEASES

10 Parchi P, Castellani R, Capellari S et al. Molecular basis of phenotypic variability in sporadic Creutzfeldt-Jakob disease. *Ann Neurol* 1996;39:767-78
11 Brown P, Preece MA, Will RG. "Friendly fire" in medicine: hormones, homografts and Creutzfeldt-Jakob disease. *Lancet* 1992;340:24-7
12 Chazot G, Broussolle E, Lapras CI, Blättler T, Aguzzi A, Kopp N. New variant of Creutzfeldt-Jakob disease in a 26-year-old French man. *Lancet* 1996;347:1181

Transmissible Subacute Spongiform Encephalopathies:
Prion Diseases
L Court, B Dodet, eds
© 1996, Elsevier, Paris

The varied manifestations of E200K Creutzfeldt-Jakob disease

Amos D Korczyn, Joab Chapman

Department of Neurology, Sackler School of Medicine, Tel Aviv University, Ramat Aviv 69978, Israel

Summary – The largest genetically determined cohort of Creutzfeldt-Jakob disease is found among Jews of Libyan origin. We have found the disease in this cohort to be linked to a point mutation of the *PRNP* gene. The penetrance of this mutation was estimated to approach 100% at the age of 80. A striking feature of the E200K mutation is its phenotypic variability, including features of all types of genetic prion disease such as Gerstmann-Sträussler-Scheinker disease and fatal familial insomnia. Unusual clinical manifestations included excruciating itching in three patients and demyelinating polyneuropathy in two. The focal nature of the clinical disease, especially at onset, may reflect a focal pathogenic mechanism involving the prion protein.

The Libyan-Jewish cohort of Creutzfeldt-Jakob (CJD) patients is an important source of information for understanding of this disease. Being the largest pool of genetically determined cohort, it has been extensively described. The original belief that Libyan CJD was transmitted from scrapie persisted for several years, but had to be discarded when an Israeli-born man was diagnosed with CJD [1]. Although he was of Libyan origin, the idea of a link with scrapie was untenable since scrapie does not exist in Israel. Our search for an alternate explanation for the high prevalence of CJD in a small ethnic group of Jews who had migrated from Libya and Tunisia to Israel was rewarded when we identified a mutation at codon 200 of the *PRNP* gene (E200K) in all Libyan patients who presented with the disease [2, 3]. On the other hand, the same mutation was never identified in a large number of Jewish patients in Israel who did not come from the Mediterranean basin (ie, Ashkenazi Jews). Moreover, the Libyan CJD variant was transmissible to experimental animals [4]. Libyan CJD is thus identical to the sporadic form of CJD in its clinical manifestations, pathology and transmissibility.

In an attempt to exploit this large cohort for further understanding of this unusual disease, we examined the penetrance of the genetic mutation. Using data from all available patients, we were able not only to confirm the age-dependent penetrance of the disease, but also to demonstrate that, by age 80, practically all carriers expressed neurological manifestations [5]. The most senior subjects did not fulfill the clinical criteria for CJD, but rather had slowly progressive dementia with parkinsonian signs. In the absence of histological confirmation, the exact diagnosis can only be speculated upon, but it must be recalled that several patients with pathologically proven CJD had, in fact, dementia without distinguishing features. It is thus possible that, in some subjects, CJD develops at a late age, lacking specific features [6].

Key words: Creutzfeldt-Jakob disease / genetics / epidemiology / Libya

Even outside this issue of progressive prion dementia, one of the most surprising aspects of CJD associated with the E200K mutation is its phenotypic variability. Some patients present with a cerebellar syndrome [7], which is quite reminiscent of Gerstmann-Sträussler-Scheinker disease (although its progression may be faster), while others present with cortical disorders, but never develop cerebellar manifestations. Cortical blindness (Heidenhain's syndrome) is just one such possible initial manifestation [8].

One of our patients presented with severe insomnia [9] and, at autopsy, had changes similar to those observed in fatal familial insomnia [10]. Although later the focus of much attention because of its association with codon 178 mutation, sleep impairment – either hyper- or insomnia – has been recognized for a long time as a manifestation of CJD [11], perhaps ever since the first descriptions by Heidenhain [12]. Interestingly, selective thalamic damage at autopsy was not always associated with sleep disturbances [13].

Not previously reported in CJD is excruciating itching, observed by us in 3 E200K patients, in two of them preceding other symptoms by several weeks. The mechanism involved is unknown, but we have provided support for the notion that the pruritus results from brain-stem involvement [14]. While pruritus was not reported in non-E200K CJD patients, it is reminiscent of scrapie, in which sheep and goats severely scratch themselves against fences, finally losing much of their wool.

Two of our E200K patients had demyelinating polyneuropathy [15]. Although there is no direct proof for the association of neuropathy with the underlying prion disease, other patients were subsequently reported [16].

This remarkable heterogeneity of the clinical manifestations of the E200K mutation [17], as opposed to the more uniform clinical course of iatrogenic CJD [18], is intriguing. While no complete explanation is yet available, the most plausible one is that the E200K mutation predisposes the prion protein (PrP) to assume a different conformation, the so-called disease-specific PrPSc. This conformational change [19] may occur spontaneously and spread to other parts of the central nervous system (CNS). This change can also occur in any part of the CNS, but perhaps some areas are more vulnerable. PrP is not distributed equally throughout the brain and, totally by chance, the cortical and cerebellar cortices are most likely to bear the burden. In most cases it is in these areas that the change will first start (sheep and goats have a less developed telencephalon), and the initial symptoms therefore occur in the cerebellum or the brain stem. In many E200K CJD patients with cortical onset, the initial symptoms are focal (eg, aphasia or paroxysmal lateralized epileptiform discharges (PLED) [13, 20], and generalized discharges typical of CJD have also been demonstrated to have a focal onset [21]. The progression of PrPSc from this focal onset focus probably takes place through axons and trans-synaptically.

References

1 Nisipeanu P, El-Ad B, Korczyn AD. Spongiform encephalopathy in an Israeli born to immigrants from Libya. *Lancet* 1990;336:686

2 Goldfarb LG, Korczyn AD, Brown P, Chapman J, Gajdusek DC. Mutation in codon 200 of scrapie amyloid precursor gene linked to Creutzfeldt-Jakob disease in Sephardic Jews of Libyan and non-Libyan origin. *Lancet* 1990;336:637

3 Korczyn AD, Chapman J. A mutation in the prion protein gene in Creutzfeldt-Jakob disease in Jewish patients of Libyan, Greek and Tunisian origin. *Ann NY Acad Sci* 1991;640:171-6

4 Chapman J, Brown P, Rabey JM et al. Transmission of spongiform encephalopathy from a familial CJD patient of Jewish Libyan origin carrying the *PRNP* codon 200 mutation. *Neurology* 1992;42:1249-50

5 Chapman J, Ben-Israel J, Goldhammer Y, Korczyn AD. The risk of developing Creutzfeldt-Jakob disease in subjects with the Prnp gene codon 200 point mutation. *Neurology* 1994;44:1683-6
6 Collinge J, Owen F, Poulter M et al. Prion dementia without characteristic pathology. *Lancet* 1990; 336:7-9
7 Kott E, Bornstein B, Sandbank U. Ataxic form of Creutzfeldt-Jakob disease: its relation to subacute spongiform encephalopathy. *J Neurol Sci* 1967;5:107-13
8 Meyer A, Leigh D, Bagg CE. A rare presenile dementia associated with cortical blindness (Heidenhain's syndrome). *J Neurol Neurosurg Psychiatry* 1954;17:129-33
9 Chapman J, Arlazoroff A, Goldfarb LG et al. Fatal insomnia in a case of familial Creutzfeldt-Jakob disease with the codon 200Lys mutation. *Neurology* 1996;46:758-61
10 Medori R, Tritschler H-J, LeBlanc A et al. Fatal familial insomnia, a prion disease with a mutation at codon 178 of the prion protein gene. *N Engl J Med* 1992;326:444-9
11 Roos R, Gajdusek DC, Gibbs CJ Jr. The clinical characteristics of transmissible Creutzfeldt-Jakob disease. *Brain* 1973;96:1-20
12 Meyer A, Leigh D, Bagg CE. A rare presenile dementia associated with cortical blindness (Heidenhain's syndrome). *J Neurol Neurosurg Psychiatry* 1954;17:129-33
13 Heye N, Cervos-Navarro J. Focal involvement and lateralization in Creutzfeldt-Jakob disease: Correlation of clinical, electroencephalographic and neuropathological findings. *Eur Neurol* 1992; 32:289-92
14 Shabtai H, Nisipeanu P, Chapman J, Korczyn AD. Pruritus in Creutzfeldt-Jakob disease. *Neurology* 1996; 46:940-1
15 Neufeld MY, Josiphov J, Korczyn AD. Demyelinating peripheral neuropathy in Creutzfeldt Jakob disease. *Muscle Nerve* 1992;15:1234-9
16 Antoine JC, Laplanche JL, Mosnier JF, Beaudry P, Chatelain J, Michel D. Demyelinating peripheral neuropathy with Creutzfeldt-Jakob disease and mutation at codon 200 of the prion protein gene. *Neurology* 1996;46:1123-7
17 Chapman J, Brown P, Goldfarb LG, Arlazoroff A, Gajdusek DC, Korczyn AD. Clinical heterogeneity and unusual presentations of Creutzfeldt-Jakob disease in Jewish patients with the PRNP codon 200 mutation. *J Neurol Neurosurg Psychiatry* 1993;56:1109-12
18 Brown P, Gajdusek C, Gibbs CJ, Asher DM. Potential epidemic of Creutzfeldt-Jakob disease from human growth hormone therapy. *N Engl J Med* 1985;313:728-31
19 Korczyn AD. Creutzfeldt-Jakob disease among Libyan Jews. *Eur J Epidemiol* 1991;7:490-3
20 Heye N, Henkes H, Hansen ML, Gosztonyi G. Focal-unilateral accentuation of changes observed in the early stage of Creutzfeldt-Jakob disease. *J Neurol Sci* 1990;95:105-10
21 Neufeld MY, Korczyn AD. Topographic mapping of the periodic discharges in Creutzfeldt-Jakob disease. *Brain Topogr* 1992;4:201-6

Transmissible Subacute Spongiform Encephalopathies:
Prion Diseases
L Court, B Dodet, eds
© 1996, Elsevier, Paris

Fatal familial insomnia: clinical and neuropathological features

Danielle Seilhean, Françoise Lazarini, Charles Duyckaerts, Jean-Jacques Hauw*

Laboratoire de Neuropathologie Raymond Escourolle, Groupe Hospitalier Pitié-Salpêtrière 47, boulevard de l'Hôpital, 75013 Paris, France

Summary – Fatal familial insomnia (FFI) has recently enlarged the group of prion diseases. It is distinct from Creutzfeldt-Jakob disease on clinical, histopathological and molecular grounds. The disease starts between 35 and 60 years of age, is inherited as an autosomal dominant trait and leads to death within 7 to 32 months. Clinical symptoms and signs include insomnia, dysautonomia, cognitive and motor alteration. Atrophy, neuronal loss and gliosis are prominent in the anterior and dorsomedial nuclei of the thalamus. Spongiosis is usually absent in FFI, in contrast to most other prion diseases. The accumulation of an abnormal prion protein (PrP) (the disease-specific PrP, PrPSc) varies among different brain regions. It is found in the dorsomedial thalamic nucleus and in the brain stem regardless of the disease duration, whereas in the neocortex, limbic system and caudate nucleus, the amount increases with the disease duration. Codon-178 of the gene encoding this protein is mutated, and the phenotypic expression of the disease is associated with a polymorphism at codon 129 of the same gene. The disease has been successfully transmitted to experimental animals. FFI provides new information about the genetics of prion diseases which share the characteristics of being altogether inherited and, in most cases, transmissible disorders. In addition, coexistence in FFI of a discrete topography of the lesions with dysregulation of the autonomic system and of the sleep-wake cycle opens a field of research in neurophysiology.

Introduction

Since the description of the first family of fatal insomnia [1, 2], eight other kindreds have been found in Italy, France, Great Britain and the United States [3–7]. The disease which is inherited as an autosomal dominant trait, starts between 35 and 60 years of age, and leads to death within 7 to 32 months. Clinical symptoms and signs include insomnia, dysautonomia, cognitive and motor alteration. Atrophy, neuronal loss and gliosis are prominent in the anterior and dorsomedial nuclei of the thalamus. Abnormal prion protein (PrP) is heterogeneously distributed in the brain. Fatal familial insomnia (FFI) is associated with a mutation of codon-178 of the *PRNP* gene, as some families of Creutzfeldt-Jakob diseases (CJD). The phenotype and the duration of the disease are linked to a polymorphism of codon 129. Case IV-37 from the first family described has been extensively studied [2, 8–14]. We have had the opportunity to perform neuropathological examination of this typical case [10, 13], which illustrates the steps

---

*Correspondence and reprints

Key words: prion / polymorphism / thalamus / sleep

which have led to a better understanding of FFI, from clinical diagnosis and neuropathology to biochemistry and genetics.

Clinical study

The patient, who has been followed up in La Salpêtrière Hospital, was a 60-year-old woman who developed a quite typical clinical course including insomnia, dysautonomia and motor dysfunction, which led to death within 18 months [10, 12]. The first symptoms were sleep disturbances associated with delusion, hallucinations and mood changes with depression. The second stage of the disease was characterized by daytime sleepiness associated with enacted dreams and dysautonomia. After 8 months, insomnia was total and neurological disorders developed, characterized by myoclonus and ataxia. Dementia was a late event during the last stage when the patient was stuporous. The 24-h recording of the patient showed the disappearance of slow waves and rapid eye movements (REM) sleep [8]. The sleep disturbance is in most cases detected by polysomnography at an early stage of the disease [15]. Impairment of autonomic function is also an early sign and is associated with the involvement of the endocrine system, including reduction of circadian oscillations of sleep-coupled as well as sleep-uncoupled hormones [16].

Neuropathology

The main macroscopic finding in case IV-37 was prominent atrophy of the dorsomedial and anterior nuclei of the thalamus, whereas the lateral thalamus was spared [13]. There was also a milder involvement of the body of the caudate nucleus. On microscopic examination, these regions were characterized by neuronal loss and astrogliosis. In contrast with classical CJD, there was spongiosis neither in the neocortex nor in the head of the caudate nucleus. In addition, there was neither amyloid deposit nor spongiosis in the cerebellum. Similar lesions were found in patients belonging to the same family who have been autopsied [10]. Metabolic brain studies in three other members of the same family have shown that thalamic hypometabolism was an early and prominent feature [17]. The discrete topography of the lesions contrasts with the spectrum of clinical signs and symptoms, including dementia, dysregulation of the autonomic system and of the sleep-wake cycle, and opens a field of research in neurophysiology [15, 18, 19]. A recent paper has shown that the distribution of protease-resistant PrP (PrP 27-30) in brains of patients with FFI varies according to the duration of the disease [14]. Whatever the duration, abnormal PrP was detected in the thalamus and the brain stem, whereas it was detected in the neocortex and limbic system only in long duration cases. This detection corresponds to the presence of spongiosis in the same area.

FFI is associated with a mutation at codon-178 of the *PRNP* gene [2]. This mutation is also found in some families with CJD. The various phenotypes are associated with a polymorphism at codon 129 [9, 11]. Methionine determines a FFI phenotype, whereas valine determines a CJD phenotype. In both cases heterozygosity leads to longer disease duration and the appearance of spongiosis in FFI cases.

The last event in the history of FFI is that it is not only familial, but also transmissible to experimental animals [20, 21]. These transmissions lead to clinical and neuropathological features very similar to those obtained after inoculation of CJD brains, including spongiosis and amyloid deposits.

FFI reveals unresolved questions in neurophysiology about the regulation of the autonomic system and of the sleep-wake cycle and is a paradigm for studying the influence of genetic polymorphism upon the expression of another gene.

Acknowledgments

We thank G Rancurel, L Garma and I Arnulf for providing clinical information and documents concerning case IV-37.

References

1 Lugaresi E, Medori R, Montagna P et al. Fatal familial insomnia and dysautonomia with selective degeneration of thalamic nuclei. *N Engl J Med* 1986;315:997-1003

2 Medori R, Tritschler HJ, LeBlanc A et al. Fatal familial insomnia, a prion disease with a mutation at codon 178 of the prion gene. *N Engl J Med* 1992;326:444-9

3 Gambetti L, Parchi P, Petersen RB, Chen SG, Lugaresi E. fatal familial insomnia and familial Creutzfeldt-Jakob disease: clinical, pathological and molecular features. *Brain Pathol* 1995;5:43-51

4 Medori R, Montagna P, Tritschler HJ et al. Fatal familial insomnia: a second kindred with mutation of prion protein gene at codon 178. *Neurology* 1992;42:669-70

5 Julien J, Vital C, Deleplanque B, Lagueny A, Ferrer X. Atrophie thalamique subaiguë familiale. Troubles mnésiques et insomnie totale. *Rev Neurol* 1990;146:173-8

6 Petersen RB, Tabaton M, Berg L et al. Analysis of the prion protein gene in thalamic dementia. *Neurology* 1992;42:1859-63

7 Mednick AS, Reder AT, Spire JP et al. Fatal familial insomnia (FFI). *Neurology* 1994;44:285

8 Garma L, Rancurel G, Hauw JJ, Gambetti P, Medori R, Lugaresi E. Familial thalamic degeneration with fatal insomnia: similar clinico-pathologic and polygraphic data in a newly affected family member. In: Horne J, ed. *Sleep' 90*. Bochum: Pontenagel Press, 1990;242-4

9 Goldfarb LG, Petersen RB, Tabaton M et al. Fatal familial insomnia and familial Creutzfeldt-Jakob disease: disease phenotype determined by a DNA polymorphism. *Science* 1992;258:806-8

10 Manetto V, Medori R, Cortelli P et al. Fatal familial insomnia: clinical and pathological study of five new cases. *Neurology* 1992;42:312-9

11 Monari L, Chen SG, Brown P et al. Fatal familial insomnia and familial Creutzfeldt-Jakob disease: different prion proteins determined by a DNA polymorphism. *Proc Natl Acad Sci USA* 1994;91:2839-42

12 Rancurel G, Garma L, Hauw JJ et al. Familial thalamic degeneration with fatal insomnia: clinicopathological and polygraphic data on a French member of Lugaresi's Italian family. In: Guilleminault C, Lugaresi E, Montagna P, Gambetti P, eds. *Fatal Familial Insomnia: Inherited Prion Diseases, Sleep and the Thalamus.* New York: Raven Press, 1994;15-25

13 Seilhean D, Rancurel G, Garma L, Lugaresi E, Gambetti P, Hauw JJ. Fatal familial insomnia: neuropathological study. In: Guilleminault C, Lugaresi E, Montagna P, Gambetti P, eds. *Fatal Familial Insomnia: Inherited Prion Diseases, Sleep and the Thalamus.* New York: Raven Press, 1994;33-6

14 Parchi P, Castellani R, Cortelli P et al. Regional distribution of protease-resistant prion protein in fatal familial insomnia. *Ann Neurol* 1995;38:21-9

15 Lugaresi E. The thalamus and insomnia. *Neurology* 1992;42 (Suppl 6):628-33

16 Lugaresi A, Baruzzi A, Cacciari E et al. Lack of vegetative and endocrine circadian rhythms in fatal familial thalamic degeneration. *Clin Endocrinol* 1987;26:573-80

17 Perani D, Cortelli P, Lucignani G et al. [18F]FDG PET in fatal familial insomnia: the functional effects of thalamic lesions. *Neurology* 1993;43:2565-9

18 Seilhean D, Duyckaerts C, Hauw JJ. Insomnie fatale familiale et maladies à prions. *Rev Neurol* 1995;151:225-30

19 Galassi R, Morreale A, Montagna P et al. Fatal familial insomnia: behavioral and cognitive features. *Neurology* 1996;46:935-9

20 Tateishi J, Brown P, Kitamoto T et al. First experimental transmission of fatal familial insomnia. *Nature* 1995; 376:434-5

21 Collinge J, Palmer MS, Sidle KCL et al. Transmission of fatal familial insomnia to laboratory animals. *Lancet* 1995;346:569-70

Transmissible Subacute Spongiform Encephalopathies:
Prion Diseases
L Court, B Dodet, eds
© 1996, Elsevier, Paris

Genotype–phenotype correlations in familial spongiform encephalopathies associated with insert mutations

Lev G Goldfarb[1], Larisa Cervenáková[2], Paul Brown[2], D Carleton Gajdusek[2]

[1]Clinical Neurogenetics Unit, and [2]Laboratory of Central Nervous System Studies, National Institute of Neurological Disorders and Stroke, National Institutes of Health, Bethesda, MD 20892, USA

Summary – A number of point and insert mutations responsible for familial forms of spongiform encephalopathy (SE) have been identified in the human *PRNP* gene. Twelve point mutations appear on 16 alleles, and 19 distinct alleles carry insert mutations. In addition, there are 10 polymorphic variants due to point substitutions or deletion of a single 24-nucleotide repeat. Analysis of the phenotypic expression of these alterations may significantly improve diagnostic capabilities and provide insight into the disease mechanisms. A 24-nucleotide repeat expansion in the *PRNP* gene is attracting much attention due to the uniqueness of the associated syndromes and a growing number of reported families. To date, 19 families of various ethnic origins with clinical, neuropathologic and transmission features of SE have been shown to carry one to nine additional repeats of 24-base pair (bp) variant sequences inserted between codons 51 and 91 of the *PRNP* open reading frame. The disease in this group of patients starts on average 20 to 30 years earlier and lasts 4 to 8 years longer than in the most common forms of familial SE, Gerstmann-Sträussler-Scheinker (GSS) disease associated with a P102L mutation or in familial Creutzfeldt-Jakob disease (CJD) patients with an E200K mutation. Clinical features include abnormal behavior, depression and aggressiveness, apraxia and visual impairment early in the course of illness. Cortical atrophy, localization of spongiform change predominantly in frontal and occipital cortical areas, and the presence of congophilic and non-congophilic PrP accumulations in uncharacteristic regions are distinct neuropathologic features. The inserted repeat sequences are stable in male and female meioses, but the age at disease onset significantly correlates with the number of 24-bp repeats. The phenomenon of anticipation is not a feature in the *PRNP* insert families. Transmission of familial disease to experimental animals has been achieved from patients with 5, 7 and 8 extra repeats.

Pathogenic alleles identified in the *PRNP* gene: an update

Point mutations

Familial spongiform encephalopathies (SE) are autosomal dominant disorders associated with mutations in the *PRNP* gene. A number of pathogenic point mutations and common polymorphisms have been identified. A polymorphism at codon 129 coding for methionine/valine has been shown to determine the disease phenotype and predispose to sporadic and

---

Correspondence: LG Goldfarb, Building 36, Room 4D03, National Institutes of Health, Bethesda, MD 20892, USA.

Key words: *spongiform encephalopathy / prion diseases / PRNP gene / mutation / 24-nucleotide repeat expansion / genotype–phenotype correlation*

iatrogenic Creutzfeldt-Jakob disease (CJD). Table I lists all reported pathogenic alleles due to point mutations. The mutation at codon D178N appears on three alleles; two with methionine at position 129 are responsible for fatal familial insomnia, and the one with valine is associated with the CJD phenotype. The E200K mutation appears on two alleles, one coupled with methionine at position 129 is causing CJD in more than 100 known families including all families originating from CJD clusters in Slovakia, Libyan Jews, and a local population in central Chile. The E200K mutation that appears on the valine allele was found very recently in our laboratory in a patient who was phenotypically very similar to the E200K/129M phenotype. Two different alleles with the P102L mutation have been reported, of which one encoding glutamic acid at codon 219 has been shown to be associated with Gerstmann-Sträussler-Scheinker (GSS) disease; the other, coding for lysine has been described in a Japanese family with comparatively minor cerebellar signs as compared with typical GSS disease [1].

24-nucleotide repeat expansion

The *PRNP* gene has five repeating sequences between codons 51 and 91, of which the first 27-nucleotide sequence is followed by four 24-base pair (bp) variant repeats, each coding for a P(H/Q)GGG(-/G)WGQ octapeptide. Owen et al [2–4] reported six extra 24-bp repeats in a British family with atypical dementia. Among the multiple families with SE studied at the NIH, members of 8 families with clinical, neuropathological and transmission features of SE had an expanded number of repeat units: from the normal 5 to 7, 9, 10, 12, or 13 repeats [5–10]. Several additional families with a 24-bp repeat expansion were later reported in France (6 and 9 repeats) [11], UK (4, 6 and 9 repeats) [12-16], Germany (14 repeats) [17], the

Table I. Pathogenic alleles associated with familial spongiform encephalopathies.

Codon	Amino acid change	Polymorphisms			Spongiform encephalopathy
		Codon 129	Deletion	Codon 219	
102	Pro → Leu	Met		Glu	GSS
102	Pro → Leu	Met		Lys	GSS
105	Pro → Leu	Met		Glu	GSS
117	Ala → Val	Val		Glu	GSS
145	Tyr → Stop	Met		Glu	GSS
178	Asp → Asn	Met		Glu	FFI
178	Asp → Asn	Met	R3/R4	Glu	FFI
178	Asp → Asn	Val		Glu	CJD
180	Val → Ile	Met		Glu	CJD
198	Phe → Ser	Val		Glu	GSS
200	Glu → Lys	Met		Glu	CJD
200	Glu → Lys	Val		Glu	CJD
208	His → Arg	Met		Glu	CJD
210	Val → Ile	Met		Glu	CJD
217	Gln → Arg	Val		Glu	GSS
232	Met → Arg	Met		Glu	CJD

Netherlands (9 repeats) [18] and Japan (4, 6 and 7 repeats) [19–21]. Altogether, 19 families with 92 affected members have been characterized clinically, and 21 neuropathologically.

Table II lists all known families with a 24-nucleotide repeat expansion. Each family is characterized by its own allele differing by the repeat number, as well as by the order of inserted repeats. In addition, some of the repeats had point substitutions that allowed recognition of a unique inserted element. In 4 families, the inserts were found on the allele encoding valine at position 129, while in 15 other families the inserts were present on the allele coding for methionine.

In this analysis, we combined the updated information on our 8 families with a 24-nucleotide repeat expansion with data on 10 other families reported in the literature. The results were compared to published data on the original large British family [12, 13].

Analysis of genotype–phenotype correlations

Many attempts have been undertaken to ascribe a discrete phenotype to each of the known *PRNP* alleles, but the phenotypes are variable, even within a single family. Research progress is bringing new insight into the mechanisms determining the phenotype and there is a chance that in the future it will be possible to make better predictions, based on the knowledge of genotype–phenotype interactions.

Table II. Alleles with a 24-nucleotide repeat expansion.

No of additional repeats	Insert	Ethnic origin	Amino acid at codon 129	No of patients	Reference
1	R2	French	Met	1	11
2	R2a,R2a	American	Met	1	5
4	R2,R3,R2,R3	American	Met	0	6
4	R2,R2,R2,R2	British	Met	1	14
4	R3,R2,R2,R2	French	Val	1	11
4		Japanese	Met	1	19
5	R3,R2,R3g,R2,R2	American	Met	2	6
5	R2a,R2,R2a,R2,R2	American	Val	2	10
5	R3,R2,R2,R2,R2	American	Met	4	9
6	R2,R3,R2,R3g,R2,R2	British	Met	47	2
6	R3g,R2,R2,R3g,R2,R2	Japanese	Met	6	20
6	R3,R2,R3g,R2,R3g,R2	British	Met	2	15
6	R3g,R2,R2,R3g,R2,R2	American	Met	7	10
7	R2c,R3,R2,R3,R2,R3,R2,R3g	American	Met	3	8
7		Japanese	Met	1	21
8	R2,R2,R2,R2,R2,R2,R2,R2a	French	Val	4	7
8	R3g,R3,R2,R2,R2,R2,R2,R2	Dutch	Val	6	18
9	R2,R3g,R2a,R2,R2,R2,R3g,R2,R3	British	Met	1	4
9	R3,R2,R3,R3g,R2,R2a,R2,R3,R2	German	Met	2	17

Total: families 19, patients 92.

The age at onset and duration of illness

The median and average age at disease onset in patients with extra 24-bp repeats were significantly lower than in patients with other types of familial SE. Most of the patients with inserts were diagnosed in their early thirties, while patients with point mutations were diagnosed in their fifties and sixties, indicating that the disease associated with repeat expansion is the most aggressive of all familial forms. In contrast, the disease duration in the repeat-expansion patients was much longer: 9 years on average as compared to 1 to 2 years in familial CJD and 5 years in GSS disease [22]. Our data clearly show that, in addition, the age at disease onset (and age at death) correlate inversely and the disease duration correlates directly with the number of repeats (fig 1). This finding is reminiscent of the trinucleotide repeat expansion recently characterized as the disease mutation in Huntington's disease and a number of other neurological disorders [23]. Huntington's alleles are highly unstable when transmitted from parents to children, especially through male meioses [23]. This is not the case with the *PRNP* repeat expansion. The inserted sequences were exactly the same in the descendants of different lines in the six-generation British family [12] as well as in four two-generation families [5, 7, 8, 20]. Intergenerational instability of the repeat number or anticipation was not observed in the repeat-expansion families.

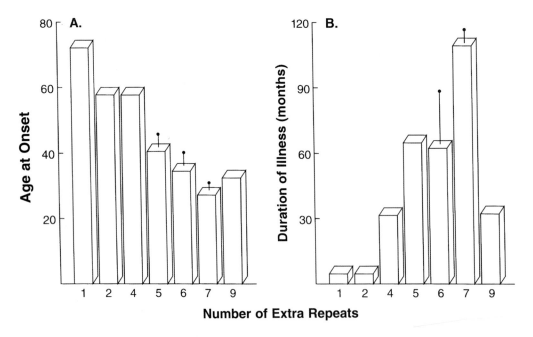

Fig 1. Analysis of the age at disease onset and disease duration in patients with a varying number of 24-nucleotide repeats appearing on an allele with methionine at position 129. **A.** Relationship between the age at onset of disease and the number of repeats. The age at disease onset is lower ($P < 0.01$) in patients with 5 additional repeats as compared to those with 1 to 4 repeats. It is also lower in patients with 6 extra repeats as compared to those with 5 ($P < 0.05$), and in patients with 7 repeats as compared to those with 6 repeats ($P < 0.05$). **B.** Relationship between the disease duration and the number of extra repeats. The disease duration is significantly longer in patients with 5 and 6 repeats as compared to those with 1 to 4 repeats ($P < 0.05$), but is shorter than in patients with 7 repeats ($P < 0.001$). The number of studied patients with 9 additional repeats was too small for statistical analysis.

Polymorphism at codon 129 has attracted much attention in the last several years and became the subject of intensive research [24–26]. Mounting evidence has now proven that neither amino acid (methionine or valine) at position 129 causes any disease, but in many cases the presence of one or the other amino acid changes the effects of other mutations [24] and may determine predisposition to SE in patients with no mutations [25, 26]. Patients with a repeat expansion on the valine allele had a later age at disease onset and prolonged disease duration as compared to patients with the expansion on the methionine allele.

Clinical characteristics

Onset of the disease and its evolution in patients with a 24-nucleotide repeat expansion was typically gradual and slow. In many cases, it was difficult to determine when the disease had actually started, which is very different from the subacute onset in many of the patients with point mutations. In 11 of 19 families (20 of 43 studied cases), a clear premorbid personality disorder was observed (table III). Premorbid personality changes were best characterized in the original British family [13]. Symptoms commonly described were: difficulty in concentrating, excessive mood swings, irritability, life-long depression, learning disability, social insensitivity, inability to keep a steady job, a history of long-time psychiatric care, clumsiness, and poor coordination. Long-lasting premorbid personality change was not observed in patients with point mutations.

Early symptoms marking the disease onset consisted of deficit of short-term memory, progressive intellectual slowing, inappropriate behavior, incoordination and slurred speech.

Table III. Clinical and neuropathological characteristics of the repeat-expansion patients as compared to patients with P102L and E200K mutations.

Characteristic features	24-bp repeat expansion	Mutations	
		P102L	E200K
Premorbid phase			
Premorbid personality change or behavioral abnormalities	20/43	0/20	0/43
Early disease			
Cognitive impairment	24/43	4/20	35/43
Incoordination	15/43	16/20	23/43
Developed disease			
Dementia	42/43	20/20	43/43
Cerebellar ataxia	30/42	20/20	34/43
Spasticity	9/43	13/20	25/43
Myoclonus	16/43	5/20	38/43
Periodic synchronous discharge on EEG	9/30	0/20	32/43
Neuropathology			
Brain atrophy (neuroimaging)	17/17	Not done	Not done
Brain atrophy (at autopsy)	11/15	3/20	0/20
Widespread spongiform change/focal spongiosis/no spongiosis	12/7/2	4/5/4	10/10/0
Plaques: Congophilic/non Congophilic/no plaques	6/2/11	20/0/0	0/0/20

Typically, these symptoms lasted for years and then a sudden and fast decline brought patients to death within a year or less. Signs and symptoms of the fully developed encephalopathy, ie, progressive dementia, ataxia, seizures, spasticity, myoclonus and electroencephalographic periodic complexes, were basically very similar to those in patients with point mutations and sporadic disease.

Neuropathological features

Neuropathological features were studied in 21 well characterized cases (12 of our own). Brain atrophy, predominantly in the frontal and occipital cortices, was reported in 73% of the cases (table III). Diffuse spongiform change was observed in 57% of the cases, and regional and 'patchy' spongiosis in another 33%; two patients did not show spongiform degeneration. This is similar to the GSS phenotype in patients with the P102L mutation, but different from the E200K phenotype in which spongiform change is a consistent and strong feature. Diffuse astrogliosis was observed in 29% of the cases. The number of patients with diffusely spread Congophilic and non Congophilic PrP plaques was unexpectedly high in patients with inserts: PrP plaques were observed in eight patients, and in at least three cases PrP plaques were non Congophilic. In contrast, patients with the P102L mutation always show Congophilic plaques and patients with the E200K mutation never have plaques. Brain suspension from three patients with an increased number of repeats (5, 7 and 8 extra repeats) transmitted disease to experimental primates after intracerebral inoculation [27].

Implications for the pathogenesis

Recent data have confirmed that the area of repeats is located away from the predicted α-helix domains that are thought to be the site of transformation from PrP^C to PrP^{Sc} [28]. Furthermore, the repeat area is cleaved by proteases and absent in the PrP27–30 that accumulates in the brain of patients with SE. This probably means that the repeat expansion does not promote the transformation process from PrP^C to PrP^{Sc} or amyloid formation as other mutations probably do. We propose that the 24-nucleotide repeat expansion is an inactivating mutation and that functionally disabled mutated PrP is the cause of the premorbid personality disorder and prolonged early disease through the loss of function mechanism. One study demonstrated early pathological changes in synapses [29]. Analysis of knock-out mice have shown that normal cellular PrP supports the integrity of GABA$_A$ receptor–mediated fast inhibition [30], and the neocortex, particularly the frontal cortex, is rich in synapses. This hypothesis needs further development on the basis of genetic analysis and pathogenetic studies of the insert-associated diseases, including modeling in transgenic mice and tissue cultures.

References

1 Kitamoto T, Tateishi J. Human prion disease with variant prion protein. *Philos Trans Biol Sci* 1994;343:391-8
2 Owen F, Poulter M, Lofthouse R et al. Insertion in prion protein gene in familial Creutzfeldt-Jakob disease. *Lancet* 1989;1:51-2
3 Owen F. Poulter M, Shan T et al. An in-frame insertion in the prion protein gene in familial Creutzfeldt-Jakob disease. *Mol Brain Res* 1990;7:273-6
4 Owen F, Poulter M, Collinge J et al. A dementing illness associated with a novel insertion in the prion protein gene. *Mol Brain Res* 1992;13:155
5 Goldfarb LG, Brown P, Little BW et al. A new (two-repeat) octapeptide coding insert mutation in Creutzfeldt-Jakob disease. *Neurology* 1993;43:2392-4
6 Goldfarb LG, Brown P, McCombic WR et al. Transmissible familial Creutzfeldt-Jakob disease associated with five, seven, and eight extra octapeptide coding repeats in the *PRNP* gene. *Proc Natl Acad Sci USA* 1991;88:

10926-30

7 Goldfarb LG, Brown P, Vrbovska A et al. An insert mutation in the chromosome 20 amyloid precursor gene in a Gerstmann-Sträussler-Scheinker family. *J Neurol Sci* 1992;111:189-94

8 Brown P, Goldfarb LG, McCombie WR et al. A typical Creutzfeldt-Jakob disease in an American family with an insert mutation in the *PRNP* amyloid precursor gene. *Neurology* 1992;42:422-7

9 Cochran EJ, Bennett DA, Cervenáková L et al. Familial Creutzfeldt-Jakob disease with a five-repeat octapeptide insert mutation. *Neurology*, 1996;47:727-33

10 Cervenakova L, Goldfarb LG, Brown P et al. Three new *PRNP* genotypes associated with familial Creutzfeldt-Jakob disease. Abstract. *Am J Hum Genet* 1995;57:A209

11 Laplanche JL, Delasnerie-Laupretre N, Brandel JP, Dussaucy M, Chatelain J, Launay JM. Two novel insertions in the prion protein gene in patients with late-onset dementia. *Hum Mol Genet* 1995;4:1109-11

12 Poulter M, Baker HF, Frith CD et al. Inherited prion disease with 144 base pair gene insertion. 1. Genealogical and molecular studies. *Brain* 1992;115:675-85

13 Collinge J, Brown J, Hardy J et al. Inherited prion disease with 144 base pair gene insertion. 2. Clinical and pathological features. *Brain* 1992;115:687-710

14 Campbell TA, Palmer MS, Will RG, Gibb WR, Luthert PJ, Collinge J. A prion disease with a novel 96-base pair insertional mutation in the prion protein gene. *Neurology* 1996;46:761-6

15 Nicholl D, Windl O, de Silva R et al. Inherited Creutzfeldt-Jakob disease in a British family associated with a novel 144 base pair insertion of the prion protein gene. *J Neurol Neurosurg Psychiatry* 1995;58:65-9

16 Duchen LW, Poulter M, Harding AE. Dementia associated with a 216-base pair insertion in the prion protein gene. Clinical and neuropathological features. *Brain* 1993;116:555-67

17 Krasemann S, Zerr I, Weber T et al. Prion disease associated with a novel nine octapeptide repeat insertion in the *PRNP* gene. *Mol Brain Res* 1995;34:173-6

18 van Gool WA, Hensels GW, Hoogerwaard EM, Wiezer JHA, Wesseling P, Bolhuis PA. Hypokinesia and presenile dementia in a Dutch family with a novel insertion in the prion protein gene. *Brain* 1995;118:1565-71

19 Isozaki E, Miyamoto K, Kagamihara Y et al. CJD presenting as frontal lobe dementia associated with a 96-base pair insertion in the prion protein gene *(in Japanese)*. *Dementia* 1994;8:363-71

20 Oda T, Kitamoto T, Tateishi J et al. Prion disease with 144 base pair insertion in a Japanese family line. *Acta Neuropathol (Berl)* 1995;90:80-6

21 Mizushima S, Ishii K, Nishimaru T. A case of presenile dementia with a 168 base pair insertion in prion protein gene *(in Japanese)*. *Dementia* 1994;8:380-90

22 Brown P. The phenotypic expression of different mutations in transmissible human spongiform encephalopathy. *Rev Neurol (Paris)* 1992;148:317-27

23 La Spada AR, Paulson HL, Fischbeck KH. Trinucleotide repeat expansion in neurological disease. *Ann Neurol* 1994;36:814-22

24 Goldfarb LG, Petersen RB, Tabaton M et al. Fatal familial insomnia and familial Creutzfeldt-Jakob disease: disease phenotype determined by a DNA polymorphism. *Science* 1992;258:806-8

25 Collinge J, Palmer MS, Dryden AJ. Genetic predisposition to iatrogenic Creutzfeldt-Jakob disease. *Lancet* 1991;337:1441-2

26 Palmer MS, Dryden AJ, Hughes JT, Collinge J. Homozygous prion protein genotype predisposes to sporadic Creutzfeldt-Jakob disease. *Nature* 1991;352:340-2

27 Brown P, Gibbs CJ, Rodgers-Johnson P et al. Human spongiform encephalopathy: The National Institutes of Health series of 300 cases of experimentally transmitted disease. *Ann Neurol* 1994;35:513-29

28 Kaneko K, Peretz D, Pan KM et al. Prion protein (PrP) synthetic peptides induce cellular PrP to acquire properties of the scrapie isoform. *Proc Natl Acad Sci USA* 1995;92:11160-4

29 Kitamoto T, Shin RW, Doh-ura K et al. Abnormal isoform of prion proteins accumulates in the synaptic structures of the central nervous system in patients with Creutzfeldt-Jakob disease. *Am J Pathol* 1992;140:1285-94

30 Collinge J, Whittington MA, Sidle KC, Palmer MS, Clarke AR, Jeffreys JG. Prion protein is necessary for normal synaptic function. *Nature* 1994;370:295-7

Transmissible Subacute Spongiform Encephalopathies:
Prion Diseases
L Court, B Dodet, eds
© 1996, Elsevier, Paris

24-nucleotide deletion in the *PRNP* gene: analysis of associated phenotypes

Larisa Cervenáková[1], Paul Brown[1], Pedro Piccardo[3], Jeffrey L Cummings[4], James Nagle[2], Harry V Vinters[4], Priti Kaur[1], Bernardino Ghetti[3], Joab Chapman[2], D Carleton Gajdusek[1], Lev G Goldfarb[2]

[1]*Laboratory of Central Nervous System Studies and* [2]*Clinical Neurogenetics Unit, National Institute of Neurological Disorders and Stroke, National Institutes of Health, Bethesda, MD 20892;* [3]*Department of Pathology and Laboratory Medicine, Indiana University School of Medicine, Indianapolis, IN 46202;* [4]*Department of Neurology, UCLA School of Medicine, Los Angeles, CA 90095, USA*

Summary – Multiple point and insert mutations in the prion protein gene (*PRNP*) on chromosome 20p are responsible for the development of familial spongiform encephalopathies. A common polymorphism at codon 129 of this gene encoding methionine (Met) or valine (Val) was shown to determine phenotypic features and influence susceptibility in sporadic and infectious forms. The 5' region of the *PRNP* gene contains five variant repeats that expand up to 14 repeats in some families with spongiform encephalopathy. A deletion of a single repeat is another rare polymorphism currently thought to be neutral. We have analyzed a series of 23 individuals (from 15 families) carrying a one-repeat deletion and found that of four different alleles resulting from a one-repeat deletion, the R3–R4 type on an allele encoding Met at the polymorphic codon 129 was present in five cases of familial or environmentally acquired Creutzfeldt-Jakob disease (CJD). The same R3–R4/Met allele was also identified in a patient with an atypical phenotype of Machado-Joseph's disease (MJD), and in a patient with early onset familial presenilin 1 (PS 1) Alzheimer's disease (AD) with both CJD and AD neuropathology. A structurally different R3/Met allele was present in two patients with atypical familial dementia and dementia with ataxia. Deletion of the R2cR3–R4 type on the allele encoding Val at position 129 was found in two families with phenotypic characteristics of MJD and Unverricht-Lundborg-Lafora disease. An unaffected member of the MJD family was found to be a compound heterozygote R3-R4/Met:R2cR3–R4/Val genotype. Finally, an R2/Met allele was identified in five healthy control individuals. These data suggest that some types of one-repeat deletion in the *PRNP* gene may predispose to spongiform encephalopathy or affect the phenotypic expression of other neurological conditions.

Introduction

The *PRNP* gene codes for the prion protein (PrP) that is transformed in patients with transmissible spongiform encephalopathy (TSE) into an abnormal isoform producing amyloid deposition and neurodegeneration; therefore, any structural alteration in this gene needs to be carefully evaluated.

Most of the alterations so far found in the *PRNP* gene on chromosome 20p are associated with phenotypically diverse forms of familial spongiform encephalopathy [1]. One group of

---

Key words: spongiform encephalopathy / prion diseases / PRNP gene / mutation / genotype–phenotype correlation / deletion

alterations is the expansion of the number of 24-base pair (bp) repeats in the 5' portion of the gene [2–7]. Conversely, the presence of a deletion in the repeat area in three healthy members of a Moroccan family has been demonstrated [8]. Incidental findings of a 24-bp deletion located between codons 51 and 91 of the *PRNP* gene in a cosmid library construct obtained from the human HeLa cell line S3 and a human brain cDNA library were reported [9]. This deletion was not found in 30 unrelated (apparently healthy) individuals.

There has been no agreement between investigators on the possible role of the one-repeat deletion in the development or phenotypic modification of spongiform encephalopathy. Some authors have suggested that the deletion may predispose to neurological disease [10, 11], have a pathogenic role in the development of atypical dementias [12], or influence the disease phenotype [13]. Others have denied any association between the deletion and spongiform encephalopathy [14–17].

We describe here four different variants of 24-bp deletions in the *PRNP* gene represented on five different alleles (one type in combination with a point mutation at codon 178 on the same allele) found in a series of 24 individuals, and discuss phenotypic features of neurodegenerative disease in each particular case at which the mutation was detected.

Materials and methods

Patients and controls

Multiple cases of spongiform encephalopathy and other neurological disorders from the US and other countries referred to the National Institutes of Health were studied for the presence of deletions in the *PRNP* gene. Control samples were randomly selected from a population bank and anonymous blood bank donors. All patients were evaluated clinically, and electroencephalograms (EEG) and magnetic resonance imaging (MRI) examinations were performed in most of the analyzed cases, using standard procedures. DNA was extracted from frozen or formalin-fixed, paraffin-embedded brain tissue of deceased patients and fresh or frozen anticoagulated whole blood of living patients and control subjects.

Amplification, subcloning and sequencing of the PRNP coding region

Two partially overlapping fragments of the *PRNP* coding region were amplified in polymerase chain reaction (PCR) using two sets of oligonucleotide primers: 5'-TACTGAGCGGCCGCGCAGTCATTATGG-CGAACCTTG-3' and 5'-TACTGAGTCGACAATGTATGATGGGCCTGCTCA-3'; 5'-TACTGAGCG-GCCGCCAACATGAAGCACATGGCTGGT-3' and 5'-TACTGAGTCGACCCTTCCTCATCCCACT-ATCAGG-3' [4] and cloned using the pCR-Script™ SK(+) Cloning Kit (Stratagene, USA) according to the manufacturer's instructions. Plasmid DNA from six to nine clones was prepared using the alkaline lysis method and sequenced using the Prism Cycle Sequencing Kit (Applied Biosystems-Perkin Elmer, USA) on automated 373A sequence analyzer (Applied Biosystems-Perkin Elmer). Analysis of sequenced fragments was performed using the software from Genetics Computer Group, Inc (Wisconsin, USA).

5' fragment visualization and restriction analysis

The PCR produced DNA fragments of the 5' part of the *PRNP* coding region from all 718 tested individuals were screened in a 2% agarose gel containing ethidium bromide. The 24-bp deletion was clearly identifiable by the presence of a double band. The type of deletion could be roughly determined by the use of the *Nco*I restriction enzyme [8] that recognizes the CCATGG sequence (fig 1). Restriction analysis for the determination of the codon 129 polymorphism was performed using the *Mae*II restriction enzyme. Mutations in the *PS1* and *MJD1* genes were identified using previously described methods [18, 19].

Neuropathology and immunohistopathology

Brain tissue was fixed in 4% formaldehyde solution, embedded in paraffin and stained for neuropathologic examination. Samples of neocortex, basal ganglia, hippocampus, thalamus, midbrain, pons, medulla and cerebellum were studied. Sections selected for immunohistochemistry were incubated with a 1:500 dilution of monoclonal antibody (MAb) 3F4 which recognizes an epitope corresponding to residues 109 to 112 of the *PRNP* gene-encoded PrP [20], or a 1:50 dilution of an antiserum to glial fibrillary acidic protein (GFAP) from BioGenex. Immunodetection was performed using unlabeled peroxidase-antiperoxidase from Sternberger Monoclonals with subsequent development by diaminobenzidine (DAB) substrate. Before immunostaining, sections to be labeled with Mab 3F4 were treated with 88% formic acid and hydrolytic autoclaving as previously described [21, 22].

Results

Types of deletions

A total of 718 individuals, 228 patients with familial TSE and their first-degree relatives from 83 families, 40 sporadic and 22 iatrogenic Creutzfeldt-Jacob disease (CJD) cases, 36 kuru cases, 167 patients with non spongiform encephalopathy (SE) neurological disorders from 102 families and 225 healthy control individuals were tested for the number of repeats in the *PRNP* gene. In the majority of subjects, the repeat region of the *PRNP* gene consisted of an initial 27-bp element starting at codon 51 followed by four 24-bp sequences (fig 2). Using the terminology of Owen et al (1990), the 27-bp element is designated as R1, and the following 24-bp repeats as R2, R2, R3, and R4 [3]. The second and third R2 elements are identical, and R3 and R4 code for the same octapeptide as R2, but two of the eight codons have consistent changes in the third nucleotide. These differences between the individual repeating elements help to establish the location of a deleted 24-bp repeat.

Allele with deletion of R2 type and methionine at position 129 (R2/Met)

In this type of mutation, a complete R2 sequence is deleted. Because of their identity, it is impossible to determine whether the first or the second R2 repeat is deleted. No irregular nucleotide substitutions occur in any of the remaining repeats.

Allele with deletion of R3 type and methionine at position 129 (R3/Met)

A complete deletion of the R3 element occurs, the remaining four repeats do not have irregular alterations.

Allele with deletion of R3–4 type and methionine at position 129 (R3–R4/Met)

The 3' part of R3 and the 5' part of R4 repeats are deleted. Crossing-over inside the normal R3 and R4 repeats may result in a structure in which invariant triplets GGA in R3 and CCT in R4 are not present. The nature of the repeats does not allow a distinction between a disruption just before GGA or several codons upstream. No irregular nucleotide substitutions were observed in this type of deletion.

Allele with deletion of R2cR3–R4 type and valine at position 129 (R2cR3–R4/Val)

The same area as in the R3–R4 type is deleted, although because the breakpoint cannot be determined, there is a possibility that a different sequence is deleted. A clear difference between R3–R4 and R2cR3–R4 alleles stems from the presence of a CCT to CCC transition at codon 68 on R2cR3–R4 and, probably more importantly, from the fact that the R3–R4

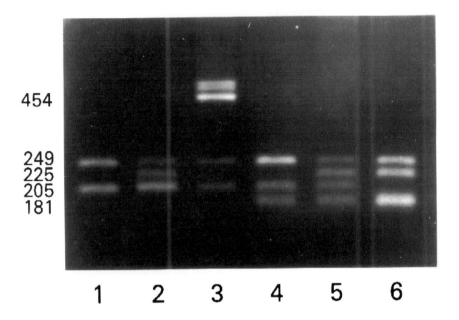

Fig 1. Agarose gel electrophoresis of PCR-amplified 5' region of the *PRNP* open reading frame after digestion with *Nco*I. *Nco*I normally cleaves the 454 bp wild-type allele into 249 bp and 205 bp fragrnents (lane 1). An R2 deletion produces an additional fragment of 225 bp (lane 2). An R3 deletion eliminates the restriction site and produces a 430-bp fragment (lane 3), which is 24 bp shorter than the wild-type fragment. An R3–R4 deletion with a deleted 3' part of R3 (following the restriction site) and the 5' part of R4 repeat produce an additional 181-bp fragment (lane 4). An R2cR3–R4 deletion has two restriction sites for *Nco*I (one is the result of silent nucleotide substitution of CCT by CCC at codon 68) and produces fragments of 225 bp and 181 bp (lane 5). Combination of type R2cR3–R4 and R3–R4 deletions produces a pattern characteristic of a compound heterozygote (lane 6).

deletion occurs on the allele coding for Met, whereas the R2cR3–R4 deletion is always found on the allele coding for Val at codon 129. The T to C transition in codon 68 does not change the predicted amino acid.

A total of twenty-three individuals from 15 families had a one-repeat deletion. Of these, five presented with TSE, nine had other neurological disorders, four were unaffected members of the affected families and five were healthy controls from four unrelated families.

Analysis of phenotypes (table I)

Spongiform encephalopathy patients with R3–R4/Met deletion

Case 1

A 54-year-old female in a family of Libyan origin suffered from mental decline followed by progressive gait and limb ataxia, spasticity with brisk reflexes, abnormal plantar responses and myoclonus. The EEG repeatedly showed generalized slowing and periodic sharp wave discharges. Brain computerized tomography (CT) was normal. A brain biopsy performed 5 months after the onset of illness showed spongiform degeneration and astrocytosis characteristic of CJD; she died 1 month later. Molecular genetic analysis revealed her to be a compound heterozygote carrying an E200K mutation on one allele encoding Met at codon 129

Wild-type allele

R1   R2   R2
CCT CAG GGC GGT GGT GGC TGG GGG CAG CCT CAT GGT GGT GGC TGG GGG CAG CCT CAT GGT GGT GGC TGG GGG CAG
pro gln gly gly gly gly trp gly gln pro his gly gly gly trp gly gln pro his gly gly gly trp gly gln

R3   R4
CCC CAT G GT GGT GGC TGG GGA CAG CCT CAT GGT GGT GGC TGG GGT CAA
pro his gly gly gly trp gly gln pro his gly gly gly trp gly gln

Deletion R2

R1   R2
CCT CAG GGC GGT GGT GGC TGG GGG CAG CCT CAT GGT GGT GGC TGG GGG CAG
pro gln gly gly gly gly trp gly gln pro his gly gly gly trp gly gln

R3   R4
CCC CAT G GT GGT GGC TGG GGA CAG CCT CAT GGT GGT GGC TGG GGT CAA
pro his gly gly gly trp gly gln pro his gly gly gly trp gly gln

Deletion R3

R1   R2   R2
CCT CAG GGC GGT GGT GGC TGG GGG CAG CCT CAT GGT GGT GGC TGG GGG CAG CCT CAT GGT GGT GGC TGG GGG CAG
pro gln gly gly gly gly trp gly gln pro his gly gly gly trp gly gln pro his gly gly gly trp gly gln

R4
CCT CAT GGT GGT GGC TGG GGT CAA
pro his gly gly gly trp gly gln

Deletion R3–R4

R1   R2   R2
CCT CAG GGC GGT GGT GGC TGG GGG CAG CCT CAT GGT GGT GGC TGG GGG CAG CCT CAT GGT GGT GGC TGG GGG CAG
pro gln gly gly gly gly trp gly gln pro his gly gly gly trp gly gln pro his gly gly gly trp gly gln

R3–R4
CCC CAT G GT GGT GGC TGG GGTCAA
pro his gly gly gly trp gly gln

Deletion R2cR3–R4

R1   R2   R2c
CCT CAG GGC GGT GGT GGC TGG GGG CAG CCT CAT GGT GGT GGC TGG GGG CAG CCC CAT G GT GGT GGC TGG GGG CAG
pro gln gly gly gly gly trp gly gln pro his gly gly gly trp gly gln pro his gly gly gly trp gly gln

R3–R4
CCC CAT G GT GGT GGC TGG GGTCAA
pro his gly gly gly trp gly gln

Fig 2. Wild-type allele nucleotide and amino acid sequence of the *PRNP* gene repeat area and localization of four types of 24-bp deletions. The initial 27-bp sequence is designated as R1 and the following 24-bp repeats as R2, R2, R3, and R4 [3]. R2 deletion: a complete R2 sequence is deleted; R3 deletion: a complete R3 sequence is deleted. R3–R4 deletion: a 3' part of R3 and 5' part of R4 repeat are deleted (distinctive elements of R3 and R4 repeats are underlined). R2cR3–R4 deletion: the same area as in the R3–R4 is deleted, in addition a transition of CCT to CCC (underlined) at codon 68 creates an irregular R2c repeat. Restriction site CCATG for endonuclease *Nco*I is shown in bold.

and an R3–R4/Met deletion on the other allele. Of the patient's six unaffected children, five carried the E200K mutation but not the deletion, and the sixth had the deletion but not the E200K mutation.

Table I. Patients with transmissible spongiform encephalopathy or other neurological disorders carrying deletions in the *PRNP* gene.

| Family | Disease status | Diagnosis | PRNP gene alterations | | | Mutations in other genes |
			Deletion	Other mutations	Codon 129 polymorphism	
Transmissible spongiform encephalopathy cases						
Case 1	affected	Familial CJD[1]	R3–4	E200K	Met/Met	
Case 2	affected	FFI	R3–4	D178N	Met/Met	
Case 3	affected	Familial CJD	R3–4	None	Met/Met	
Case 4	affected	Iatrogenic	R3–4	None	Met/Met	
Case 5	affected	Conjugal CJD	R3–4	None	Met/Met	
Group of patients with other neurological disorders						
Case 6	affected	Familial AD	R3–R4	None	Met/Met	*PS1*
Case 7	affected	Familial AD	R3	None	Met/Met	
Case 8	affected	Dementia with ataxia	R3	None	Met/Met	
Case 9	affected	Machado-Joseph disease	R3–R4	None	Met/Met	*MJD1*
Case 10	affected	Machado-Joseph disease[2]	R2cR3–R4	None	Val/Val	*MJD1*
Case 11	affected	Machado-Joseph disease[2]	R2cR3–R4	None	Val/Val	*MJD1*
Case 12	affected	Lafora disease	R2cR3–R4	None[3]	Val/Met	
Case 13	affected	Lafora disease	R2cR3–R4	Nt	Val/Met	
Case 14	affected	Lafora disease	R2cR3–R4	Nt	Val/Met	

[1]One unaffected member of this family had the same deletion; [2]two unaffected members of this family had the same deletion and one was a compound heterozygote (see text); [3]only a 5' portion (420 bp) of the *PRNP* gene was sequenced; Nt: not tested; FFI: fatal familial insomnia; AD: Alzheimer's disease.

Case 2

A 41-year-old male developed insomnia, impotence and visual problems, then dementia, dysmetria, spastic gait and myoclonus. Polysomnography performed in the mid-stage of his illness confirmed severe insomnia. The patient died after a 14-month course of progressive encephalopathy. At autopsy, severe neuronal loss and prominent astrogliosis were found in the thalamus [23]. A molecular study showed the presence of a D178N mutation in the *PRNP* gene and a deletion on the same allele. Clinical, pathological and molecular data were consistent with the diagnosis of FFI. The family history was remarkable for the presence of five distant relatives having the same syndrome and the same allele with a D178N mutation and an R3–R4/Met deletion in the *PRNP* gene [24].

Case 3

A 55-year-old female developed a memory deficit, mild personality change and a subtle language disturbance. Two months later, she developed a transcortical aphasia and prominent echolalia which progressed to severe aphasia, apathy and rigidity. Brain MRI was normal. The EEG showed generalized slowing. The patient died after an 8-month course; at autopsy, extensive spongiform degeneration with vacuoles of irregular shape and size, severe neuron loss and gliosis were present. The neocortex, caudate nucleus, putamen, globus pallidus, thalamus and hypothalamus were all involved in varying degrees. In sections of the thalamus and basal ganglia, spongiform changes were more prominent in the ventral anterior nuclei, putamen and amygdala (fig 3D). PrP-immunopositivity was frequently seen in these areas and characterized by a fine punctate pattern and seldom by coarse positive deposits (fig 3E). In the cerebellum, spongiform changes were present in the molecular layer (fig 3F), with PrP-immunopositivity in the molecular and granule cell layers (fig 3G). The patient's paternal grandmother had dementia with a rapid progressive course. The disease in this family is consistent with the diagnosis of CJD with an unusual speech disorder. Except for the R3–R4/Met deletion, no changes were identified in the *PRNP* gene.

Case 4

A 26-year-old man with congenital panhypopituitarism was repeatedly treated with human growth hormone (hGH) between ages 6 and 21. He grew an additional foot (from 3'9" to 4'9"), and did not have significant medical problems until he slowly developed short-term memory impairment, clumsiness, slurred speech, severe truncal and gait ataxia, and myoclonus. A brain biopsy confirmed the presence of spongiform changes in the frontal cortex and established the diagnosis of iatrogenic CJD that developed as a result of inoculation of contaminated hGH. No family history of neurological disease is known. This patient had an R3–R4/Met deletion and no other structural alterations in the *PRNP* gene.

Case 5

A 55-year-old female with a history of hypertension, diabetes mellitus and congestive heart failure had a subacute onset of agitation, confusion and mental slowness. The EEG showed diffuse slowing and periodic discharges characteristic of CJD. Very soon, the patient developed ataxic gait and myoclonic jerks. A brain biopsy taken 1 month after the onset of her cognitive disorder showed widespread spongiform degeneration (fig 3A), PrP immunoreactivity appearing as fine punctate deposits (fig 3B), and severe gliosis (fig 3C) consistent with the diagnosis of CJD. The R3–R4/Met deletion was the only change detected in the completely sequenced *PRNP* coding region. This patient's 54-year-old husband had died of CJD 9 months after a progressive course of mental and physical decline 5 years before the onset of clinical symptoms in his wife. Extensive spongiform changes, neuronal loss and astrocytosis were found in his cerebral cortex, consistent with the diagnosis of CJD. Molecular analysis failed to detect any alterations in the husband's *PRNP* gene and there was no history of familial disease. Conjugation of CJD cases in a husband–wife pair with a time difference of 5 years may represent a case of human-to-human transmission, apparently at the time when the husband was symptomatic.

All the above described cases were homozygous for methionine at codon 129 of the *PRNP* gene.

Patients with other neurological disorders

Case 6

A 43-year-old man had a four-year illness characterized by progressive dementia, ataxia and myoclonus. The EEG showed periodic discharges. In addition to numerous neurofibrillary tangles and senile amyloid plaques characteristic of Alzheimer's disease (AD), mild gliosis and mild spongiform degeneration were found in the frontal cortex. His father and younger brother had similar clinical syndromes. Moreover, his brother had a very similar pathology,

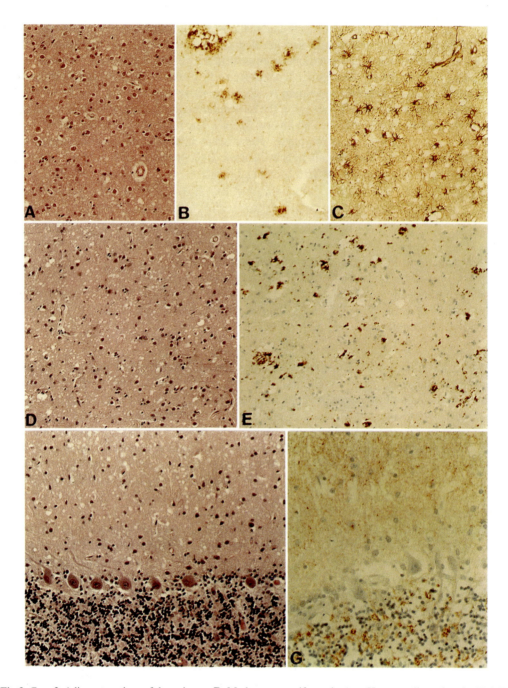

Fig 3. Case 3. Adjacent sections of the striatum. **D.** Moderate spongiform changes. Hematoxylin and eosin (H & E) staining (152×). **E.** PrP deposits. Immunolabeling with monoclonal antibody (mAb) 3F4 (3F4 mAB) (156×). Adjacent sections of the cerebellum. **F.** Spongiform changes in the molecular layer (H & E, 192×). **G.** PrP deposits in the granular and molecular layers (3F4 mAb, 290×). Case 5. Brain biopsy. Adjacent sections of the cerebral cortex. **A.** Moderate spongiform change (H & E, 160×). **B.** PrP deposits in the neuropil (3F4 mAb, 152×). **C.** Severe astrocytosis. Immunolabeling for GFAP (152×).

except for somewhat milder spongiform changes. Molecular analysis established the presence of an H163R mutation in the *PS1* gene for familial AD on chromosome 14q and an R3–R4/Met deletion in the *PRNP* gene. His brother had the *PS1* gene mutation, but not the deletion [25].

Case 7

A 59-year-old woman developed memory deficit, intermittent agitation, word-finding difficulty and non-fluent aphasia that later progressed to loss of ambulation, increasing confusion and global aphasia. She also developed myoclonus. The EEG showed generalized slowing, and CT scan of the head revealed marked cortical atrophy. The patient was still alive and in her 8th year of illness. The patient's mother died of a similar 5-year-long illness with progressive dementia. Molecular study did not identify any known mutation in the *PS1* gene, and no mutations were found in the *PRNP* gene, except for the R3/Met deletion. The patient was homozygous for methionine at codon 129.

Case 8

A 41-year-old woman developed a slowly progressive scanning dysarthria, dysmetria, and limb and gait ataxia followed by cognitive dysfunction and dementia. At the time this study was completed, she was in the 6th year of her illness. Her mother and half-sister had similar problems. No mutations were found in the patient's *SCA1* or *MJD1* genes, but an R3/Met deletion was detected in the *PRNP* gene.

Case 9

A 26-year-old patient of Italian origin developed a slowly progressive ataxia and moderate dementia. The EEG indicated significant slowing and the MRI was negative. The patient was alive in her 8th year of illness at completion of this study. Her grandmother, mother, two uncles and an older brother had ataxia and died at young age. Molecular analysis included a search for mutations in the *PRNP* gene because the illness was suggestive of the Gerstmann-Sträussler-Scheinker disease, as well as mutations in the recently identified genes associated with autosomal dominant cerebellar ataxias, *SCA1* and *MJD1*. A trinucleotide repeat expansion in the *MJD1* gene on chromosome 14q was identified, establishing the diagnosis of SCA3/Machado-Joseph's disease (MJD). In addition, an R3–R4 deletion in the *PRNP* gene was identified on an allele with methionine at codon 129.

Cases 10 and 11

Two patients in this family of African–American ancestry, a 49-year-old male and his 34-year-old nephew developed limb ataxia and poor coordination that slowly progressed until they became wheel-chair bound by the 6th–8th year of illness. No mental impairment was observed. The patients were alive at completion of this study. Five other members of this family had ataxia and none was demented. Molecular diagnosis of MJD was established in these patients by finding an expansion of the CAG repeat number in the *MJD1* gene. A deletion of the 2cR3–R4/Val type was detected in both patients and three of their unaffected relatives. All were homozygous for valine at the polymorphic codon 129, except for one of the nieces who was a compound heterozygote for the deletion: R2cR3–R4/Val of paternal origin and R3–R4/Met apparently coming from her mother.

Cases 12, 13 and 14

Three members of a family afflicted with Unverricht-Lundborg-Lafora disease, pathologically characterized by the presence of inclusion bodies containing polyglucosan in neuronal cytoplasm, developed a progressive myoclonic epilepsy. All three had a 2cR3–R4/Val deletion in the *PRNP* gene. There were no phenotypic differences between these three patients and other affected family members who did not carry the deletion.

Discussion

Among 23 individuals with a one-repeat deletion in the *PRNP* gene, we encountered four types of 24-bp deletion: R2/Met, R3/Met, R3–R4/Met, and R2cR3–R4/Val. We did not observe in our group of patients the fifth type, ie, the R2–R3 deletion that has been previously described [15]. Analysis of medical conditions associated with each allele indicates that the R3–R4/Met variant was associated with all 5 cases of spongiform encephalopathy in our series. Two patients had other mutations in the *PRNP* gene (E200K and D178N) that are pathogenic and could themselves have been the cause of the disease – but the possibility that the deletion predisposed to CJD in these mutation carrying individuals cannot be excluded. E200K mutation carriers have a 56% to about 80% risk of developing CJD in their lifetime [26]. Case 2, with a pathogenic D178N mutation and a R3–R4 deletion on the allele with methionine at position 129 is genotypically identical to affected members of the Tennessee family [24]; this patient has recently been identified as a member of that family [23]. Our patient had all classical features of FFI [23]. Of six studied affected members of the Tennessee family, all had dementia, four had insomnia and dysautonomia; of two examined neuropathologically, both had major pathology in the thalamus and a few individuals showed spongiform changes in the neocortex and deep nuclei. Thus, the presence or absence of the deletion appears to be an inconsequential coincidence.

Our observation of a familial disease (case 3) with the R3–R4 deletion on a methionine allele as the only structural change in the *PRNP* gene may indicate that in some instances this allele alone is responsible for CJD development. The presence of spongiform degeneration and PrP immunoreactivity confirmed beyond doubt that this patient had CJD and the unusual speech disorder makes this case distinct from sporadic CJD. A similar speech disorder was observed in two other patients in this series.

The existence of genetically determined susceptibility to environmentally acquired CJD was previously established in patients with valine homozygosity and later in patients with homozygosity for either amino acid at codon 129 [29, 31–34]; the R3–R4/Met allele may have had a similar effect in cases 4 and 5. In case 5, the disease may have been transmitted from the patient's CJD-affected husband. On the basis of clinical, pathological and immunocytochemical data, our patient and her husband both had CJD. Although there is no specific evidence that horizontal transmission has occurred, it is interesting that our patient was the only caregiver during the course of her husband's illness. Moreover, the presence of the deletion in the *PRNP* gene may confer an increased susceptibility to CJD.

In a series of patients with non TSE neurological disorders, the presence of the R3–R4/Met allele is likely to influence the phenotypes by adding features characteristic of TSE. In cases 6 and 9, the condition was clearly identified as early onset familial AD or MJD. Nevertheless, the phenotypes included spongiform changes in case 6 and dementia in case 8. A similar influence from the R3/Mct allele may have contributed to the phenotypes in cases 7 and 8.

Descriptions of atypical dementia with features of TSE in a patient with the R3–R4/Met allele [11] or a deletion 'upstream of codon 76' (that could be R2 or R2–R3) in another similar case [12] confirm that this is a widespread phenomenon. Significantly, the R2cR3–R4/Val allele did not introduce any phenotypic modification in cases 10 and 11 presenting with MJD and Lafora disease cases 12, 13 and 14. An unaffected member of one MJD family was a compound heterozygote for R3–R4/Met and R2cR3–R4/Val alleles. A family analyzed by Perry et al also segregated two distinct deletions [17]. A patient with dementia and found to be homozygous for a *PRNP* deletion has been reported [12].

We did not find the R3–R4/Met or R3/Met alleles in 225 healthy controls. Other investigators have reported R3–R4 deletions subjects in control [8, 9, 14–16], but it was not always clear whether the R3–R4 deletion was present on the methionine allele, or if these individuals were at risk for some familial neurological diseases. In general, deletions are common polymorphisms with an estimated allele frequency in the control population of 0.5 to 2.5% [14, 16, 17]. Our data suggest that the presence of some types of these deletions in the *PRNP* gene modifies the phenotype of TSE and other neurological disorders. Examples of such influence have been demonstrated for the *PRNP* gene codon 129 polymorphism [30] and polymorphisms in other genes [35–38].

In conclusion, we are currently at an early stage of data accumulation, and even this large series does not provide reliable answers to the question of the possible role of *PRNP* deletions in spongiform encephalopathies and non-spongiform encephalopathy disorders. Our account provides a reasonable doubt that the deletions are not as neutral as sometimes regarded. Allelic association studies must be continued to detect relatively small gene effects that contribute to complex phenotypes.

References

1 Goldfarb LG, Brown P. The transmissible spongiform encephalopathies. *Ann Rev Med* 1995;46:57-65
2 Owen F, Poulter M, Lofthouse R et al. Insertion in prion protein gene in familial Creutzfeldt-Jakob disease. *Lancet* 1989;1:51-2
3 Owen F. Poulter M, Shan T et al. An in-frame insertion in the prion protein gene in familial Creutzfeldt-Jakob disease. *Mol Brain Res* 1990;7:273-6
4 Goldfarb LG, Brown P, McCombie WR et al. Transmissible familial Creutzfeldt-Jakob disease associated with five, seven, and eight extra octapeptide coding repeats in the *PRNP* gene. *Proc Natl Acad Sci USA* 1991;88: 10926-30
5 Goldfarb LG, Brown P, Vrbovska A et al. An insert mutation in the chromosome 20 amyloid precursor gene in a Gerstmann-Sträussler-Scheinker family. *J Neurol Sci* 1992;111:189-94
6 Owen F, Poulter M, Collinge J et al. A dementing illness associated with a novel insertion in the prion protein gene. *Mol Brain Res* 1992;13:155
7 Goldfarb LG, Brown P, Little BW et al. A new (two-repeat) octapeptide coding insert mutation in Creutzfeldt-Jakob disease. *Neurology* 1993;43:2392-4
8 Laplanche JL, Chatelain J, Launay JM, Gazengel C, Vidaud M. Deletion in prion protein gene in a Moroccan family. *Nucleic Acids Res* 1990;18:6745
9 Puckett C, Concannon P, Casey L, Hood L. Genomic structure of the human prion protein gene. *Am J Hum Genet* 1991;49:320-9
10 Goldfarb LG, Brown P, Gajdusek C. The molecular genetics of human transmissible spongiform encephalopathy. In: Prusiner SB, Collinge J, Powell J, Anderton B, eds. *Prion Diseases of Humans and Animals.* New York: Ellis Horwood Limited, 1992;139-53
11 Diedrich JF, Knopman DS, List JF et al. Deletion in the prion protein gene in a demented patient. *Hum Mol Genet* 1992;6:443-4
12 Masullo C, Salvatore M, Macchi G, Genuardi M, Pocchiari M. Progressive dementia in a young patient with a homozygous deletion of the *PRP* gene. *Ann NY Acad Sci* 1994;724:358-60

13 Yamada M, Itoh Y, Fujigasaki H et al. A deletion in the prion protein gene in a Japanese family. *Biomed Res* 1994;15:131-3

14 Vnencak-Jones CL, Phillips JA. Identification of heterogeneous *PRP* gene deletions in controls by detection of allele-specific heteroduplex (DASH). *Am J Hum Genet* 1992;50:871-2

15 Palmer MS, Mahal SP, Campbell TA et al. Deletions in the prion protein gene are not associated with CJD. *Hum Mol Genet* 1993;2:541-4

16 Pocchiari M, Salvatore M, Cutruzzola F et al. A new point mutation of the prion protein gene in Creutzfeldt-Jakob disease. *Ann Neurol* 1993;34:802-7

17 Perry RT, Go RCP, Harrell LE, Acton RT. SSCP analysis and sequencing of the human prion protein gene (PRNP) detects two different 24 bp deletions in an atypical Alzheimer's disease family. *Am J Med Genet* 1995;60:12-8

18 Sherrington R, Rogaev EI, Liang Y et al. Cloning of a gene bearing missense mutations in early-onset familial Alzheimer's disease. *Nature* 1995;375:754-60

19 Kawaguchi Y, Okamoto T, Taniwaki M et al. CAG expansion in a novel gene for Machado-Joseph disease at chromosome 14q32.1. *Nat Genet* 1994;8:221-7

20 Kascsak RJ, Rubenstein R, Merz PA et al. Mouse polyclonal and monoclonal antibody to scrapie-associated fibril proteins. *J Virol* 1987;61:3688-93

21 Kitamoto T, Osomori K, Tateishi J, Prusiner SB. Methods in laboratory investigation. Formic acid pretreatment enhances immunostaining of cerebral and systemic amyloids. *Lab Invest* 1987;57:230-6

22 Shin RW, Iwaki T, Kitamoto T, Tateishi J. Hydrated autoclave pretreatment enhances TAU immunoreactivity in formalin-fixed normal and Alzheimer's disease brain tissue. *Lab Invest* 1991;5:693-702

23 Reder AT, Mednick AS, Brown P et al. Clinical and genetic studies of fatal familial insomnia. *Neurology* 1995;45:1068-75

24 Bosque PJ, Vnencak-Jones CL, Johnson MD, Whitlock JA, McLean MJ. A PrP gene codon 178 base substitution and a 24-bp interstitial deletion in familial Creutzfeldt-Jakob disease. *Neurology* 1992;42:1864-70

25 Hachimi KHE, Cervenakova L, Brown P et al. Mixed features of Alzheimer's disease and Creutzfeldt-Jakob disease in a family with a presenilin 1 mutation in chromosome 14. *Amyloid* 1996; in press

26 Chapman J, Ben-Israel J, Goldhammer Y, Korczyn AD. The risk of developing Creutzfeldt-Jakob disease in subjects with the *PRNP* gene codon 200 point mutation. *Neurology* 1994;44:1683-6

27 Medori R, Tritschler HJ, LeBlanc A et al. Fatal familial insomnia, a prion disease with a mutation at codon 178 of the prion protein gene. *N Engl J Med* 1992;326:444-9

28 Manetto V, Medori R, Cortelli P et al. Fatal familial insomnia: clinical and pathological study of five new cases. *Neurology* 1992;42:312-9

29 Goldfarb LG, Brown P, Goldgaber D et al. Patients with Creutzfeldt-Jakob disease and kuru lack the mutation in the *PRNP* gene found in Gerstmann-Sträussler syndrome, but they show a different double-allele mutation in the same gene [abstract]. *Am J Hum Genet* 1989;45(suppl): A189

30 Goldfarb LG, Petersen RB, Tabaton M et al. Fatal familial insomnia and familial Creutzfeldt-Jakob disease: disease phenotype determined by a DNA polymorphism. *Science* 1992;258:806-8

31 Collinge J, Palmer MS, Dryden AJ. Genetic predisposition to iatrogenic Creutzfeldt-Jakob disease. *Lancet* 1991;337:1441-2

32 Palmer MS, Dryden AJ, Hughes JT, Collinge J. Homozygous prion protein genotype predisposes to sporadic Creutzfeldt-Jakob disease. *Nature* 1991;352:340-2

33 Deslys JP, Marce D, Dormont D. Similar genetic susceptibility in iatrogenic and sporadic Creutzfeldt-Jakob disease. *Gen Virol* 1994; 75:23-7

34 Brown P, Cervenakova L, Goldfarb LG et al. Iatrogenic Creutzfeldt-Jakob disease: an example of the interplay between ancient genes and modern medicine. *Neurology* 1994;44:291-3

35 Corder EH, Saunders AM, Strittmatter WJ et al. Gene dose of apolipoprotein E type 4 allele and the risk of Alzheimer's disease in late onset families. *Science* 1993;261:921-3

36 Ruiz J, Blanche H, Cohen N et al. Insertion/deletion polymorphism of the angiotensin-converting enzyme gene is strongly associated with coronary heart disease in non-insulin-dependent diabetes mellitus. *Proc Natl Acad Sci USA* 1994;91:3662-5

37 Sommer SS, Lind TJ, Heston LL, Sobell JL. Dopamine D4 receptor variants in unrelated schizophrenic cases and controls. *Am J Med Genet* 1993;48:90-3

38 Ebstein PE, Novick O, Umansky R et al. Dopamine D4 receptor (D4DR) polymorphism associated with the human personality trait of novelty seeking. *Nature Genet* 1996;12:78-80

IATROGENIC CREUTZFELDT-JAKOB DISEASE: RISKS AND CONTROLS

Transmissible Subacute Spongiform Encephalopathies:
Prion Diseases
L Court, B Dodet, eds
© 1996, Elsevier, Paris

The risk of blood-borne Creutzfeldt-Jakob disease

Paul Brown

Laboratory of Central Nervous System Studies, Building 36, Room 5B21, National Institute of Neurological Disorders and Stroke, National Institutes of Health, Bethesda, MD 20892, USA

Summary – Because 10 to 15% of patients dying of Creutzfeldt-Jakob disease (CJD) have donated blood at some point in their lives, and because the transmissible agent of CJD has been detected in the blood of some human patients and experimentally infected animals, the question of risk of iatrogenic disease from the administration of blood or blood products has come under increasing scrutiny. Accumulated information gained from experimental isolation attempts and epidemiological studies suggests that although the potential for iatrogenic transmission exists, the occurrence of such an event has not been documented. Further studies aimed at establishing the presence or absence of infectivity and, if present, its distribution in experimentally infected blood, together with continuing epidemiological surveillance of human blood donor and recipient populations should provide a definitive resolution of the issue within the near future.

Physicians and medical scientists are sometimes caught unawares by problems that leave them awash in a sea of uncontrollable consequences. In this unenviable posture we have witnessed the rising tide of iatrogenic Creutzfeldt-Jakob disease (CJD), confronting us first in ripples and then waves of recognition that corneal grafts, neurosurgical instruments, cadaveric human pituitary hormones, and dura mater grafts could each be subject to inadvertant or unsuspected contamination, all of which were recognized after the fact during the 1970s and 1980s [1]. These therapeutic misfortunes have prompted us to take a much closer look at other possible sources of contamination in order to avert further episodes of iatrogenic disease. Blood and blood products have received particular attention because their widespread use would amplify any potential contamination to the first rank of importance.

Information gained from attempted isolations of infectivity from naturally and experimentally infected animal models (including both CJD and scrapie, an analogous disease of sheep and goats), and from clinically ill humans, provide some idea of the frequency, amount, and distribution of infectivity in blood [2–18; for review, see 19]. These data show that infectivity is rather unpredictably present in comparatively low titer in the blood of experimentally inoculated animals; that the titer may remain stable, decrease or increase during the course of the incubation period; and that it is most probably confined to the white cell component. It is worth noting that all the tests for infectivity have been conducted using the highly efficient intracerebral inoculation route rather than the less efficient, but more appropriate, intravenous route, and that infectivity has not been detected in the blood of naturally infected sheep or goats.

Key words: Creutzfeldt-Jakob disease / spongiform encephalopathy / prion / iatrogenic disease / blood product

Isolation attempts from the blood or blood components of humans with CJD suggest that circulating white cells of clinically ill patients may also be irregularly infectious, in that four different laboratories have independently reported success in recovering the infective agent from the blood or buffy coat of patients with either sporadic or iatrogenic forms of CJD [14–17]. However, as in the animal experiments, these successes were all accomplished using intra-cerebrally inoculated assay animals, and it is worth emphasizing that the only attempts to transmit the disease by intravenous inoculation, including transfusion into chimpanzees of whole units of blood from three patients with sporadic CJD, have failed [18, 19]. Notwithstanding this record of imperfect success, we are obliged to conclude that blood, and particularly white cells, may at times contain the infectious agent of CJD, however low the infective titer, and represents a legitimate concern with respect to the use of blood components and derivatives in human medicine.

The question then becomes whether there is any evidence that blood donated by a patient who later died from CJD has actually been responsible for transmission of the disease to recipient individuals, a question that will not be answered by further laboratory experimentation, but by the time-honored if not especially fashionable discipline of epidemiology. Two types of approach to the question have been used: one takes advantage of ongoing CJD surveillance studies in an effort to determine risk factors associated with disease, and makes statistical comparisons between diseased and control populations; the other approach focuses on individual episodes in which CJD patients have donated blood in an effort to track down and obtain medical information for all recipients of the donations. The population approach has the Achilles heel of uncertainty that a few blood-borne cases might occur but be insufficiently numerous to create a statistically significant result; the individual donor approach suffers from the twin liability of never being able to track down every single recipient individual, and of some recipients dying of non-neurological causes before a long enough period of time has elapsed for them to develop symptoms of CJD.

Although we do not yet have enough information from either type of study to provide definitive answers, such information as we do have strongly suggests that blood-borne transmission is not an important cause of CJD. Four separate case control studies conducted during the 1980s in the United States [20, 21], the United Kingdom [22] and Japan [23] all found the percentage of patients with CJD who had received blood transfusions to be no higher than in healthy control groups. The UK study also found no increase in the incidence of CJD in areas where multiple blood donations from CJD patients had been distributed. Newer case control studies have been incorporated into the European CJD surveillance program, and preliminary analysis from the first 3-year period again finds no significant differences between CJD and control populations in the proportion of individuals who had either donated or received blood (personal communication, RG Will). One report from Australia described four patients who had received transfusions 5 years before the onset of iatrogenic-type illnesses (cerebellar signs with little or no dementia), but no information was provided about the comparable frequency of transfusions in a non-CJD control population, or about the existence of CJD patients among the blood donors [24].

In Germany, an effort was made to trace all recipients of blood from a regular donor who later died of CJD [25]. Among 35 traceable individuals from the 55 identified recipients, the majority had died from non-CJD illnesses (up to 22 years after having received transfusions), and the remainder were still alive without evidence of neurologic disease after an average survival of 12 years. Similar efforts are presently being pursued for several more such incidents by the American Red Cross and the European CJD surveillance study, including a recently identified US patient who had donated a total of 93 units of blood during the 35-year period before the onset of CJD, and several members of a French family with CJD who also had donated blood on multiple occasions.

Finally, it is worth recalling that no case of CJD has yet been identified among patients whose diseases require repeated administration of whole blood, blood components or blood derivatives. Such potentially high risk groups include patients with marrow transplants, congenital anemias and clotting deficiencies (sickle cell disease, thalassemia major, hemophilia), immune deficiency or suppression syndromes, multiple sclerosis, α_1-anti-trypsin deficiency, and multiple surgical procedures.

Summarizing the experimental and epidemiological evidence bearing on blood-borne CJD, we can say that although the potential for iatrogenic transmission exists, the occurrence of such an event has not been documented. However, we will need to continue and extend existing studies both in the laboratory and in the field in order to be able to make any final statement about the subject. In particular, we need to confirm in animal models that most if not all infectivity is confined to white cells, and carefully examine the possibility that a small amount of infectivity might be present in plasma when separated from other blood components by the same methods used for human blood processing. If so, we will need to investigate its distribution in plasma fractions and therapeutically useful derivatives, about which there is at present no information.

In humans, we need to continue efforts to isolate the infectious agent from the blood of patients with CJD, again with an emphasis on the plasma derivatives of commercially processed donations, and we need to emphasize the importance of necropsy examinations in dying members of potentially high risk groups (particularly hemophiliacs) who are subjected to repeated administration of blood products.

Information from such studies will provide a rational basis for recall policy decisions about products found to have been 'tainted' by a CJD donor, and knowledge of the presence or absence of infectivity in various blood products will allow us to predict the comparative risks for patients receiving these products. It will then be a matter for epidemiologists to determine whether these predictions have been translated into reality. 'Look-back' studies of individual donor-recipient incidents in the United States and Europe, and case control analyses from the smoothly operating European CJD Surveillance Program should provide a timely resolution of this urgent issue.

References

1 Brown P, Preece MA, Will RG. 'Friendly fire' in medicine: hormones, homografts, and Creutzfeldt-Jakob disease. *Lancet* 1992;340:24-7
2 Hadlow WJ, Kennedy RC, Race RE, Eklund CM. Virologic and neurohistologic findings in dairy goats affected with natural scrapie. *Vet Pathol* 1980;17:187-99
3 Hadlow WJ, Kennedy RC, Race RE. Natural infection of Suffolk sheep with scrapie virus. *J Infect Dis* 1982;146: 657-64
4 Pattison IH, Millson GC. Distribution of the scrapie agent in the tissues of experimentally inoculated goats. *J Comp Pathol* 1962;72:233-44
5 Gibbs CJ Jr, Gajdusek DC, Morris JA. Viral characteristics of the scrapie agent in mice. In: Gajdusek DC, Gibbs CJ Jr, Alpers M, eds. *Slow, Latent, and Temperate Virus Infections*. NINDB Monograph No. 2, PHS Publication No. 1378. Washington, DC: US Government Printing Office, 1965:195-202
6 Clarke MC, Haig DA. Presence of the transmissible agent of scrapie in the serum of affected mice and rats. *Vet Rec* 1967;80:504

7 Eklund CM, Kennedy RC, Hadlow WJ. Pathogenesis of scrapie virus infection in the mouse. *J Infect Dis* 1967; 117:15-22

8 Dickinson AG, Meikle VMH. Genetical control of the concentration of ME7 scrapie agent in the brain of mice. *J Comp Pathol* 1969;79:15-22

9 Hadlow WJ, Eklund CM, Kennedy RC, Jackson TA, Whitford HW, Boyle CC. Course of experimental scrapie virus infection in the goat. *J Infect Dis* 1974;129:559-67

10 Diringer H. Sustained viremia in experimental hamster scrapie. *Arch Virol* 1984;82:105-9

11 Casaccia P, Ladogana A, Xi YG, Pocchiari M. Levels of infectivity in the blood throughout the incubation period of hamsters peripherally injected with scrapie. *Arch Virol* 1989;108:145-9

12 Manuelidis EE, Gorgacz EJ, Manuelidis L. Viremia in experimental Creutzfeldt-Jakob disease. *Science* 1978; 200:1069-71

13 Kuroda Y, Gibbs CJ Jr, Amyx HL, Gajdusek DC. Creutzfeldt-Jakob disease in mice: persistent viremia and preferential replication of virus in low-density lymphocytes. *Infect Immun* 1983;41:154-61

14 Manuelidis EE, Kim JH, Mericangas JR, Manuelidis L. Transmission to animals of Creutzfeldt-Jakob disease from human blood. *Lancet* 1985;ii:896-7

15 Tateishi J. Transmission of Creutzfeldt-Jakob disease from human blood and urine into mice. *Lancet* 1985; ii:1074

16 Tamai Y, Kojuma H, Kitajima R, et al. Demonstration of the transmissible agent in tissue from a pregnant woman with Creutzfeldt-Jakob disease. *N Engl J Med* 1992;327:649

17 Deslys JP, Lasmézas C, Dormont D. Selection of specific strains in iatrogenic Creutzfeldt-Jakob disease. *Lancet* 1994;343:848-9

18 Brown P, Gibbs CJ Jr, Rodgers-Johnson P et al. Human spongiform encephalopathy: the National Institutes of Health series of 300 cases of experimentally transmitted disease. *Ann Neurol* 1994;35:513-29

19 Brown P. Can Creutzfeldt-Jakob disease be transmitted by transfusion? *Curr Opin Hematol* 1995; 2:472-7

20 Davanipour Z, Alter M, Sobel E, Asher DM, Gajdusek DC. A case-control study of Creutzfeldt-Jakob disease. Dietary risk factors. *Am J Epidemiol* 1985;122:443-51

21 Little BW, Mastrianni J, DeHaven AL, Brown P, Goldfarb L, Gajdusek DC. The epidemiology of Creutzfeldt-Jakob disease in eastern Pennsylvania (Abstract). *Neurology* 1993;43:A316

22 Esmonde TFG, Will RG, Slattery JM et al. Creutzfeldt-Jakob disease and blood transfusion. *Lancet* 1993;341: 205-7

23 Kondo K, Kuroiwa Y. A case control study of Creutzfeldt-Jakob disease: association with physical injuries. *Ann Neurol* 1982;11:77-81

24 Klein R, Dumble LJ. Transmission of Creutzfeldt-Jakob disease by blood transfusion. *Lancet* 1993;341:768

25 Heye N, Hensen S, Müller N. Creutzfeldt-Jakob disease and blood transfusion. *Lancet* 1994;343:298-9

Transmissible Subacute Spongiform Encephalopathies:
Prion Diseases
L Court, B Dodet, eds
© 1996, Elsevier, Paris

Clinical and pathological aspects of iatrogenic Creutzfeldt-Jakob disease

Thierry Billette de Villemeur[1,2], Jean-Philippe Deslys[3], Antoinette Gelot[4], Jean-Guy Fournier[5], Olivier Robain[4], Dominique Dormont[3], Yves Agid[1,6]

[1]*Centre National de Référence de la Maladie de Creutzfeldt-Jakob Iatrogène, Hôpital de la Salpêtrière, 47, boulevard de l'Hôpital, 75651 Paris;* [2]*Service de Neuropédiatrie, Hôpital Trousseau, Paris;* [3]*Laboratoire de Neuropathologie Expérimentale et Neurovirologie, CEA, Fontenay-aux-Roses;* [4]*Laboratoire de Neuropathologie, Hôpital Saint-Vincent-de-Paul, Paris;* [5]*Inserm U 153, Hôpital Saint-Vincent-de-Paul, Paris;* [6]*Inserm U 289, Hôpital de la Salpêtrière, Paris, France*

Summary – Thirty-four patients were referred for iatrogenic Creutzfeldt-Jakob disease after growth hormone therapy. The clinical and pathological symptoms were homogeneous, with, at disease onset, unsteady gait and diplopia, progressive neurological deterioration, and after a few months dementia, myoclonic jerks and bedridden condition. Pathological examination performed in three cases was typical of spongiform encephalopathy; particularly numerous amyloid plaques were present in the cortex and subcortical white matter of the cerebrum and cerebellum. Plaques were strongly labeled with anti-PrP in immunohistochemistry. In France, all the patients were treated with extractive growth hormone between January 1985 and July 1986.

Introduction

It is known that iatrogenic Creutzfeldt-Jakob disease (CJD) occurs after surgical procedures [1] or dural grafts [2–12]. Since 1985, iatrogenic CJD has also been reported after injection of pituitary hormones from human cadavers. A few cases were recorded after human pituitary-derived gonadotrophin treatment [13–15]. The most frequent cause of iatrogenic CJD appears to be growth hormone derived from human cadaveric pituitary glands. In France, the number of CJD cases related to this treatment has been unusually high [16–22]. The preparation of cadaveric growth hormone used in France, as elsewere, included pooling of human pituitary glands. Transmission is presumed to result from intramuscular or subcutaneous injections of human growth hormone (hGH) during childhood [13]. The disease appears after an incubation time of more than 5 years and clinical onset occurs earlier than in sporadic cases. Between 1991 and 1995, CJD was identified in 34 of the 2305 patients treated with hGH between 1970 and 1988 [23], when recombinant hGH became available. We present the clinical and pathological features of these patients.

Materials and methods

A nation-wide network (seven regional centers and one national reference center) was set up in 1992 to collect data from pediatricians, endocrinologists, neurologists and neuropathologists about patients

---

Key words: *Creutzfeldt-Jakob disease / children / iatrogenic / growth hormone / plaques*

previously treated with hGH who developed CJD. The type of hGH extracted at the Pasteur Institute (Paris, France) and the dosage and duration of treatment reported for each patient were cross-checked with the records of the Association France Hypophyse and those of the Pharmacie Centrale des Hôpitaux. Additional data were obtained via a survey of the numerous physicians treating those patients. The diagnosis was recorded as definite, probable, possible or ruled out.

The diagnosis of definite CJD required one or more of the following: neuropathological evidence (gray matter spongiosis, neuronal loss, amyloid plaques, astrocyte proliferation, prion protein [PrP] immunoreactivity after proteinase K treatment), abnormally high concentrations of proteinase K resistant PrP on Western blots [24], and successful transmission to experimental animals.

The diagnosis was probable if neurological deterioration over a 3-month period included dementia, according to the DSM-IV [25] and if at least three of the following were present: cerebellar, pyramidal and extrapyramidal signs, myoclonus, oculomotor disorders, periodic or pseudoperiodic electro-encephalographic pattern, but normal magnetic resonance imaging and computerized tomography (CT) scans, and normal cerebrospinal fluid (CSF) parameters.

The diagnosis was considered possible if the patients presented at least one of the probable criteria and if exclusion criteria could not be rejected.

The diagnosis was ruled out if a relapse or complication of the underlying disorder (tumor) or radio-necrosis was observed.

In three cases, pathological examination was performed. Brains were removed 6, 2 and 36 h after death, respectively, and fixed in formalin. Paraffin-embedded sections were stained with Bodian, periodic acid Schiff (PAS) Congo red, hematoxylin-eosin, cresyl violet and luxol. Deparaffined slides were dipped into 100% formic acid for 7 min as described by Kitamoto et al [26]. Immunolabeling was performed with or without a 10-min proteinase K pretreatment (10 mg/mL) at laboratory temperature. Immunostaining was done using a monoclonal antibody directed against 263 K hamster PrP (kindly provided by P Brown, NIH, Bethesda, MD). The antibody was used at a 1/50 dilution overnight in a 4 °C chamber. The second antibody was applied. Samples were then reacted with diaminobenzidine (DAB). Frontal, cerebellar and spinal samples were post-fixed in 1% paraformaldehyde–1% glutaraldehyde 0,1 M phosphate buffer for further electron microscopical examination. Frontal and cerebellar samples were frozen at – 20 °C for Western blot analysis.

Results

Clinical findings

Forty-four patients were included in the analysis. Seven patients were excluded because of recurrent tumors (2 craniopharyngiomas, 1 germinoma), radionecrosis (1 case) or suspected psychogenic disease (3 cases). Three patients with possible CJD were also excluded: two of them with a disease duration of less than 3 months and one in whom relapse or complications of a craniopharyngioma were likely.

Diagnosis was definite in 16 cases and probable in 18. All 34 cases were retained for the assessment. The hGH deficiency for which the patients had been treated was cryptogenic in 23 patients and acquired in 11 (6 craniopharyngiomas, 1 medulloblastoma, 1 sarcoma of the cavum, 1 leukemia, 1 lymphosarcoma and 1 head trauma). Eight patients had undergone neurosurgery (6 craniopharyngiomas, 1 medulloblastoma, 1 spina bifida), but none had received dura mater grafts. The patients had been treated with hGH for various periods of time (mean ± SD: 6.4 ± 2.9 years; range 1.9–13.3 years), but all of them had received at least one batch of extract from France Hypophyse during the possible 18-month 'at-risk period', ie, between January 1984 and June 1985 inclusive. The mean hGH dose received by the 1136 patients during the 'at-risk period' was 326 ± 104 mg (range 144–540 mg). Male patients predominated with a sex ratio of 4 male/1 female patients. The age at onset of the disease was 20.7 ± 4.9

years (range: 9.9–26.8 years) and the duration of the disease was 17 ± 9 months (range: 3–30 months). None of the patients had a family history of CJD. The 33 patients who were genotyped were homozygous for a polymorphism at codon 129 of the *PRNP* gene (Met/Met or Val/Val), a genotype reported to predipose to sporadic CJD [27] that may also constitute a risk factor in the 34 hGH-treated patients. In definite cases, the diagnosis of CJD was confirmed by neuropathological examination [18, 20, 22] and/or Western blot [24]. These examinations were not performed in the other cases which were therefore diagnosed as probable on the basis of clinical criteria. In addition, inoculation of hamsters with white blood cells taken from patient number 2 two months after the appearance of symptoms caused spongiform encephalopathy in one animal 21 months later [19, 28].

The first clinical manifestations of definite and probable CJD were not different between the two groups. Ataxic gait and visual disorders – basically, diplopia (53%) and nystagmus (47%) – were the most common initial complaints. Several other symptoms, such as hyperphagia with weight gain of more than 10%, asthenia and sleep disorders, were frequently detected at initial clinical examination. Tremor, memory disorders and headaches were sometimes present.

Neurological deterioration progressed rapidly during the first 6 months. At the time of clinical examination, major symptoms were dementia, cerebellar, pyramidal and oculomotor disorders, myoclonic jerks, and sensory or visual loss. Brain images (33 by nuclear magnetic resonance imaging, 25 by CT scans) however remained normal or unchanged compared to examinations at onset of symptoms. Fifteen patients underwent further neuro-imaging examinations and scans remained unchanged in eight of them after 3 to 8 months. Cerebellar atrophy (mainly at the level of the vermis) was seen in seven patients after 3 to 11 months and diffuse cerebral atrophy was observed in two others after 18 months. Electroretinograms, although not systematically recorded, were frequently altered (9 out of 11) during the first 3 months of the disease. Electroencephalograms were practically normal, except for slow background activity in 11 of the 31 patients initially recorded. They were also normal in 18 out of 22 patients 4 months later. Pseudoperiodic bursts were present in nine patients after 5 to 13 months, but characteristic triphasic slow waves were found in only two cases after 6 and 20 months.

CSF analyses, normal in 20 out of 26 patients 1 and 7 months after onset of the disease, were also uninformative. Moderate pleiocytosis was recorded in one patient, and high protein levels (0.55-1.18 g/L) in five, none of whom had oligoclonal bands.

Forty-one per cent of the patients died within less than 1 year; 19% survived more than 2 years and only four of them were still alive in May 1995. The incubation period between the termination of hGH treatment and the first symptoms increased chronologically, but no correlation that might explain the increase (dose of hGH, duration of treatment, age) could be evidenced.

Neurohistology

Spongiosis was seen in the cerebral cortex. Vacuoles (6 to 15 μm in diameter) were mainly present in layers III and IV. In some areas, necrotic destruction of the medial portion of the cortex was complete, with few macrophagic cells. At high magnification, vacuoles could be seen in the soma and dendritic emergence of the remaining cortical neurons. In the cortex, neuronal loss was moderate to severe. Gliosis was always present. Numerous reactive astrocytes with a large swollen eosinophilic cytoplasm and stellar disposition of processes were mainly present in the molecular layer, although they could be observed throughout the cortex and white matter in the three cases.

Spongiosis, neuronal loss and glial proliferation were marked in the striatum and thalamus in all three cases. In two cases, they were visible in the locus niger, brain stem nuclei, medulla, olivary nuclei and spinal cord gray matter. Spongiosis was present in the hippocampus in two cases.

In all three cases, plaques were observed. They were strongly PAS-positive, but only weakly Congo red-stained. They were also visible after Bodian silver impregnation, but more weakly stained with hematoxylin-eosin and cresyl violet, however. Each plaque consisted of a homogeneous core surrounded by a halo of delicate, arranged fibrils. The plaques (10 to 60 µm in diameter) were either isolated (kuru type) or multicentric (Gerstmann-Sträussler-Scheinker (GSS) disease type). No argentophilic neurites and no tangles were observed in the vicinity of the plaques. These were scattered in the cerebral cortex and subcortical white matter. Their abundance was 2 to 20 plaques per mm². In all three cases, they were widely distributed, particularly in the frontal and parietal lobes, and were never seen in the hippocampus and basal ganglia. In two cases, numerous plaques were also visible in the cerebellum. They were mainly distributed in the granule cell and molecular layers, but were also present in the Purkinje cell layer and subcortical white matter. In one case, plaques were not seen in the cerebellum.

Electron microscopical examination showed that spongiosis consisted of numerous intra-neuronal vacuoles bordered by a membrane. Some of them were in contact with a presynaptic *bouton*, indicating that vacuoles are mainly located within dendrites. Small vacuoles were also visible in neuronal soma. Some of them contained curled membrane fragments. In neurons, mitochondria were frequently swollen and embodied dense inclusions.

Discussion

Because of the young age of the patients, the time course of the disease, the clinical picture and its exceptionally high frequency, CJD in hGH-treated patients without a family history of CJD is distinguishable from other forms of the disease.

The clinical picture in the hGH-treated patients was homogeneous and began, respectively, in 94 and 88% of the patients with ataxia and visual disorders unexplained by brain images, whereas in patients with sporadic or familial CJD, onset of the disease at the level of the cerebellum occurred only in one-third of them and visual impairment in only one-fifth [29]. Conversely, dementia and myoclonic jerks – considered to be characteristic of CJD – only occurred 6 months after onset of the disease in hGH-treated patients. The polymorphism at codon 129 of the *PRNP* gene did not affect the clinical picture in these patients (nor does it in sporadic CJD). Plaques were present in our three histologically evaluated cases and were also recorded in other cases of CJD occurring in pituitary growth hormone recipients [20, 30, 31]. The abundance of plaques is not correlated with disease duration. This result differs from the findings relating to sporadic CJD in which the presence of plaques is rare [32] and has been correlated with the long duration of the disease [33]. By contrast, plaques are frequent in genetically determined prion diseases such as GSS disease [9, 34] and kuru [35, 36] disease.

In the general population, the incidence of CJD has been estimated at 1/106 per year [37]. On the basis of the present study, the risk of CJD in the 2305 patients treated with hGH between 1959 and 1988 is 1.5%, reaching 3% if calculated for the 1136 patients treated with hGH during the putative 'at-high risk period', ie, between January 1984 and June 1985. The risk is higher than that observed under comparable conditions in the United States [38] and Great Britain [39] where, to our knowledge, only 15 cases in each country, plus two related patients in New Zealand [40] and Brazil [41], have been reported. It is unlikely that under-diagnosis would explain this discrepancy.

Can strain-related differences in incubation time be causative? The route of administration should not be incriminated, as it was the same in France and other countries. A possible host susceptibility factor [42, 43] may contribute to the apparent virulence of the infection in the 34 patients contaminated in France. It has been demonstrated that the incubation period of some strains of scrapie is under genetic control [44] and that the human *PRNP* gene corresponds to the scrapie incubation-time gene [45]. One might speculate that homozygosity at codon 129 of the *PRNP* gene may also have predisposed the 34 hGH-treated patients presented in this study to develop the disease, as in sporadic CJD [27]. French hGH extract may have been contaminated by a particularly virulent strain of the CJD agent [28], but this cannot be ascertained since retrospective analysis failed to detect a particular batch of extract that was common to the 34 patients.

Contamination of hGH recipients with the CJD agent poses a difficult public health problem. The 2305 young subjects treated with hGH extracts up to January 1988 are a potential reservoir of infection. In France, starting in 1993, physicians were directed to decide, on an individual basis, when to inform the patients who had been treated, particularly during the putative 'at-high risk period', of the consequences for themselves and others of possible contamination. To contain the spread of iatrogenic contamination, measures have been taken [46, 47] and are as follows: a) to exclude hGH recipients as blood, tissue and organ donors; b) to specify disinfection protocols for surgical material; c) to suspend cadaveric dura mater collection and albumin extraction from placentas.

Acknowledgments

We are indebted to the neurologists and neuropathologists of the regional reference centers of Lille (Prof Leys, Dr Ruchoux), Lyon (Prof Confavreux, Prof Kopp), Marseille (Prof Khalil, Prof Pellissier), Paris (Dr Aubourg), Rennes (Prof Chauvel, Dr Darcel), Nancy (Prof Weber, Prof Floquet) and Toulouse (Dr Fabre, Prof Delisle); the neuropathologists, Dr Mikol, Dr Robain, Dr Sazdovitch, Prof Henin and Dr Fallet Bianco; the French National Pharmacovigilance network; and to all the physicians who referred their patients and medical records; to Drs Coquin and Capek from the Direction Générale de la Santé and to Mrs Hallain from the Grandir Association, for assistance in data collection.

Supported by a grant from the French Health Ministry, Programme Hospitalier de Recherche Clinique: AOA 94005.

References

1 Duffy P, Wolf J, Collins G, De Voe AG, Streeten B, Cowen D. Possible person-to-person transmission of Creutzfeldt-Jakob disease. *N Engl J Med* 1974;290:692-3
2 Esmonde T, Lueck CJ, Symon L, Ducken LW, Will RG. Creutzfeldt-Jakob disease and lyophilised dura mater grafts: report of two cases. *J Neurol Neurosurg Psychiatry* 1993;56:999-1000
3 Lane KL, Brown P, Howell DN et al. Creutzfeldt-Jakob disease in a pregnant woman with an implanted dura mater graft. *Neurosurgery* 1994;34:737-40
4 Martinez-Lage JF, Poza M, Sola J et al. Accidental transmission of Creutzfeldt-Jakob disease by dural cadaveric graft. *J Neurol Neurosurg Psychiatry* 1994;57:1091-4
5 Masullo C, Pocchiari M, Macchi G, Alema G, Piazza G, Panzera MA. Transmission of Creutzfeldt-Jakob disease by dural cadaveric graft. *J Neurosurg* 1989;71:954-5
6 Miyashita K, Inuzuka T, Kondo H et al. Creutzfeldt-Jakob disease in a patient with a cadaveric dural graft. *Neurology* 1991;41:940-1
7 Nisbet TJ, McDonaldson I, Bishara SN. Creutzfeldt-Jakob disease in a second patient who received a cadaveric dura mater graft. *JAMA* 1989;261:11-8

8   Pocchiari M, Masullo C, Salvatore M, Genuardi M, Galgani S. Creutzfeldt-Jakob disease after non-commercial dura mater graft. *Lancet* 1992;340:614-5

9   Takayama S, Hatsuda N, Matsumura K, Nakasu S, Handa J. Creutzfeldt-Jakob disease transmitted by cadaveric dural graft: a case report. *No Shinkei Geka* 1993;21:167-70

10 Thadani V, Penar PL, Partington J et al. Creutzfeldt-Jakob disease probably acquired from a cadaveric dura mater graft. Case report. *J Neurosurg* 1988;69:766-9

11 Willison HJ, Gale HN, McLaughlin JE. Creutzfeldt-Jakob disease following cadaveric dura mater graft. *J Neurol Neurosurg Psychiatry* 1991;54:940

12 Yamada S, Aiba T, Endo Y, Hara M, Kitamoto T, Tateishi J. Creutzfeldt-Jakob disease transmitted by a cadaveric dura mater graft. *Neurosurg* 1994;34:740-4

13 Brown P, Preece MA, Will RG. 'Friendly fire' in medecine: hormones, homografts, and Creutzfeldt-Jakob disease. *Lancet* 1992;340:24-7

14 Cochius JI, Burns RL, Blumbergs PC, Mack K, Alderman CP. Creutzfeldt-Jakob disease in a recipient of human pituitary-derived gonadotrophin? *Aust N Z J Med* 1990;20:592-3

15 Klien R, Dumble LJ. Transmission of Creutzfeldt-Jakob disease by blood transfusion. *Lancet* 1993;341:768

16 Billette de Villemeur T, Beauvais P, Gourmelen M, Richardet JM. Creutzfeldt-Jakob disease in children treated with growth hormone. *Lancet* 1991;337 864-5

17 Billette de Villemeur T, Deslys JP, Pradel A et al. Creutzfeldt-Jakob disease from contaminated growth hormone extracts in France. *Neurology* 1996;47:690-5

18 Billette de Villemeur T, Gelot A, Deslys JP et al. Iatrogenic Creutzfeldt-Jakob disease in three growth hormone recipients. Pathological findings. *Neuropathol Appl Neurobiol* 1994;20:111-7

19 Billette de Villemeur T, Gourmelen M, Beauvais P et al. Maladie de Creutzfeldt-Jakob chez quatre enfants traités par hormone de croissance. *Rev Neurol (Paris)* 1992;148:328-34

20 Delisle MB, Fabre N, Rochiccioli P, Doerr-Schott J, Rumeau JL, Bes A. Maladie de Creutzfeldt-Jakob après traitement par l'hormone de croissance extractive humaine: étude clinicopathologique. *Rev Neurol (Paris)* 1993; 149:524-7

21 Labauge P, Pages M, Blard JM, Chatelain J, Laplanche JL. Valine homozygous 129 *PrP* genotype in a French growth-hormone-related Creutzfeldt-Jakob disease patient. *Neurology* 1993;43:447

22 Masson C, Delalande I, Deslys JP et al. Creutzfeldt-Jakob disease after pituitary-derived human growth hormone therapy: two cases with valine 129 homozygous genotype. *Neurology* 1994;44:179

23 Job JC, Maillard F, Goujard J. Epidemiologic survey of patients treated with growth hormone in France in the period 1959–1990: Preliminary results. *Horm Res* 1992;38 Suppl 1:35-43

24 Brown P, Coker-Vann M, Franko M, Asher DM, Gibbs CJ, Gajdusek DC. Diagnosis of Creutzfeldt-Jakob disease by Western blot identification of marker protein in human brain tissue. *N Engl J Med* 1986;314:547-51

25 American Psychiatric Association. *Diagnostic and Statistical Manual of Mental Disorders.* 4th ed: DSM-IV. Washington, DC: American Psychiatric Association, 1994

26 Kitamoto T, Ogomori K, Tateishi J, Prusiner SB. Formic acid pretreatment enhances immunostaining of cerebral and systemic amyloids. *Lab Invest* 1987;57:230-6

27 Prusiner SB, Hsiao KK. Human prion disease. *Ann Neurol* 1994;35:385-95

28 Deslys JP, Lasmezas C, Dormont D. Selection of specific strains in iatrogenic Creutzfeldt-Jakob disease. *Lancet* 1994;43:3848-9

29 Brown P, Gibbs CJ Jr, Rodgers-Johnson P et al. Human spongiform encephalopathy: the National Institute of Health series of 300 cases of experimentally transmitted disease. *Ann Neurol* 1994;35:513-29

30 Ellis CJ, Katifi H, Weller RO. A further British case of growth hormone induced Creutzfeldt-Jakob disease. *J Neurol Neurosurg Psychiatry* 1992;55:1200-2

31 Tintner R, Brown P, Hedley-Whyte, Rappaport EB, Piccardo CP, Gajdusek DC. Neuropathologic verification of Creutzfeldt-Jakob disease in the exhumed American recipient of human pituitary growth hormone; epidemiologic and pathogenic implications? *Neurology* 1986;36:932-6

32 Miyazono M, Kitamoto T, Doh-Ura K, Iwaki T, Tateishi J. Creutzfeldt-Jakob disease with codon 129 polymorphism (valine): a comparative study of patients with codon 102 point mutation or without mutations. *Acta Neuropathol* 1992;84:349-54

33 Kitamoto T, Tateishi J, Sato Y. Immunohistochemical verification of senile and kuru plaques in Creutzfeldt-Jakob disease and allied disorders. *Ann Neurol* 1988;24:537-42

34 Carlson GA, Hsiao K, Oesch B, Westaway D, Prusiner SB. Genetics of prion infections. *Trends Gent* 1991;7: 61-65

35 Masters CL, Gajdusek DC, Gibbs CR. Creutzfeldt-Jakob disease virus isolations from the Gerstmann-Sträussler syndrome. With an analysis of the various forms of amyloid plaque deposition in the virus induced spongiform encephalopathies. *Brain* 1981;104:559-88

36 Neuman MA, Gajdusek DC, Zigas V. Neuropathologic findings in exotic neurologic disorder among natives of highlands of New Guinea. *J Neuropathol Exp Neurol* 1964;23:486-507

37 Brown P, Cathala F, Raubertas RF, Gajdusek DC, Castaigne P. The epidemiology of Creutzfeldt-Jakob disease: conclusion of a 15-year investigation in France and review of world literature. *Neurology* 1987;37:895-904

38 Frasier SD, Foley TP. Creutzfeldt-Jakob disease in recipients of pituitary hormones. *J Clin Endocrinol Metab* 1994;78:1277-9

39 Adlard P, Preece MA. Safety of pituitary growth hormone. *Lancet* 1994;344:612-3

40 Croxson M, Brown P, Synek B et al. A new case of Creutzfeldt-Jakob disease associated with human growth hormone therapy in New Zealand. *Neurology* 1988;38:1128-30

41 Macario ME, Vaisman M, Buescu A, Neto VM, Araujo HMM, Chagas C. Pituitary growth hormone and Creutzfeldt-Jakob disease. *Br Med J* 1990;302:924

42 Collinge J, Palmer MS, Dryken AJ. Genetic predisposition to iatrogenic Creutzfeldt-Jakob disease. *Lancet* 1991;337:1441-2

43 Deslys JP, Marce D, Dormont D. Similar genetic susceptibility in iatrogenic and sporadic Creutzfeldt-Jakob disease. *J Gen Virol* 1994;75: 23-7

44 Dickinson AG, Meikle VMH, Fraser H. Identification of a gene which controls the incubation period of same strains of scrapie agent in mice. *J Comp Pathol* 1968;78:293-9

45 Telling GC, Scott M, Mastrianni J et al. Prion propagation in mice expressing human and chimeric *PrP* transgenes implicates the interaction of cellular PrP with another protein. *Cell* 1995;83:79-90

46 Farrington M. Use of surgical instruments in Creutzfeldt-Jakob disease. *Lancet* 1995;345:194

47 McGreevy Steelman V. Creutzfeldt-Jakob disease: recommendations for infection control. *Am J Infect Control* 1994;22:312-8

Transmissible Subacute Spongiform Encephalopathies:
Prion Diseases
L Court, B Dodet, eds
© 1996, Elsevier, Paris

Iatrogenic Creutzfeldt-Jakob disease acquired after neurosurgery and dura mater grafts

Juan F Martínez-Lage, Maximo Poza

Regional Service of Neurosurgery, 'Virgen de la Arrixaca' University Hospital, E-30120 Murcia, Spain

Summary – We report four cases of acquired Creutzfeldt-Jakob disease (CJD) following the implant of cadaveric dural grafts, thus substantiating this mechanism of propagation of prion infection in humans. We review other reported and unreported cases of dura mater-related CJD that occurred worldwide. In the same way, we analyze the means and ways of CJD propagation via contaminated instruments, contaminated EEG electrodes, corneal and tympanic grafts in patients who underwent surgery or neurosurgery. We think that better understanding of the mechanisms involved in its transmission will contribute to stopping CJD spread in the surgical and neurosurgical environment.

Introduction

Major advances have been recently made in the knowledge of prion diseases. The transmissibility of prion disease was proven by a series of experiments conducted by Gajdusek and coworkers [1]. Although the pathogenesis of the spontaneous form of Creutzfedlt-Jakob disease (CJD) is still unknown, experimental transmission and iatrogenic propagation of CJD and other prion diseases are now better understood. Available data suggest that prion diseases must be combated on several fronts. Clinical and epidemiological facts regarding iatrogenic CJD associated with the administration of native hormones will be discussed in another section of this book. Here, we summarize four cases of acquired CJD following the implant of dura mater grafts in the course of neurosurgical procedures. Our previous report definitively established the involvement of this mechanism of propagation [2]. We also review other published and unpublished examples of this complication related to neurosurgery, dural implants, and ocular and tympanic grafts of cadaveric origin. We think that this information should contribute to the disappearance of the iatrogenic dissemination of CJD among patients undergoing neurosurgical procedures and homografts.

CJD after homografts

In 1974, Duffy et al reported the first evidence of person-to-person transmission of CJD in a 55-year-old woman who had received a corneal transplant [3]. The donor was diagnosed with CJD shortly after his death. The corneal graft recipient developed the disease about 18 months later. The duration of the illness was 8 months. In 1990, Tange et al described a man aged 54 who developed fatal CJD 4 years after tympanoplasty with homograft pericardium [4]. The

Key words: Creutzfeldt-Jakob disease / dura mater grafts / prion infection / slow-virus infection / spongiform encephalopathies

early death of the donor, 2 days after cardiac surgery, precluded postmortem examination when CJD manifested in the graft recipient.

CJD after contact with contaminated instruments

In 1977, Bernoulli et al described transmission of CJD to two young patients by means of stereotactic EEG electrodes [5]. The contaminated electrodes used in these patients were shown to transmit the disease to a chimpanzee [6]. Other reports refer to the possibility of prion infection by contaminated instrument [7–11]. In this group of cases in which contaminated instruments were considered responsible for prion infection, the mean patients' age was 44.8 years, the average incubation period was 29 months, and the mean duration of the disease was 7 months. For all these patients, the diagnosis of CJD was made by biopsy or necropsy.

CJD acquired from cadaveric dura mater grafts

In 1988, Thadani et al reported a case of CJD in a 28-year-old woman who had received a cadaveric dural graft 19 months earlier after resection of a cholesteatoma [12]. This was a Lyodura® graft manufactured by B Braun Melsungen AG, Germany. By that time, we already had suffered the chilling experience of two iatrogenic cases of CJD following Lyodura® implants. While we were performing an epidemiological investigation of these incidents, we saw our third and fourth cases [2, 13, 14].

Case reports

Case 1

A 17-year-old boy underwent removal of a grade 1 cerebellar astrocytoma which included a Lyodura® patch. This patient also received a ventriculoperitoneal shunt. The duration of the incubation period was 16 months. Initial complaints were unsteadiness and blurred vision, followed shortly thereafter by dementia and myoclonus. Initially, the EEG showed generalized slow activity, later followed by periodic activity. Computerized tomography (CT) scan was unremarkable. The diagnosis of CJD was confirmed by brain biopsy. The patient died 21 months after clinical onset of the disease.

Case 2

A 53-year-old woman underwent a posterior fossa decompression that ended with a Lyodura® implant for treatment of a Chiari malformation and syringomyelia. Forty-three months later, the patient showed ataxia and dysarthria, rapidly followed by dementia and myoclonus. CT scan showed incipient cerebral atrophy. The EEG demonstrated generalized slowing that evolved to periodic activity. Brain biopsy confirmed the diagnosis of CJD. She died 25 months after clinical onset of the disease.

Case 3

A 10-year-old boy received a Lyodura® graft after completion of removal of a grade 2 cerebellar astrocytoma. He also underwent ventriculoperitoneal shunting, blood transfusion and cobalt therapy. Seventy-nine months later, he was admitted with gait unsteadiness, dysarthria and a psychotic-like behavior. After 2 weeks, he was mute, showed a decorticate attitude and had myoclonic jerks. CT scan and magnetic resonance imaging (MRI) ruled out tumor recurrence and demonstrated subtle cerebral atrophy. The EEG showed the classical triphasic waves. Diagnosis of spongiform encephalopathy was confirmed by brain biopsy. The boy died 3 months after clinical onset of the disease.

Case 4

A 25-year-old man was treated with posterior fossa decompression and a Lyodura® patch for a Chiari malformation associated with syringomyelia. One hundred and five months later, he was admitted with ataxia and dysarthria, with rapidly ensuing dementia and myoclonus. CT scan and MRI were unremarkable at the initial stages of the disease. These investigations showed generalized cerebral atrophy and subdural collections of cerebrospinal fluid (CSF) in successive studies. The EEG showed marked slow activity that evolved to periodic complexes. A cerebral biopsy confirmed the diagnosis of CJD. In March 1996, the patient was still alive.

Clinical features common to all four individuals were: 1) their young age; 2) a rapid clinical course that started with cerebellar symptoms and was followed by dementia and myoclonus within 2 weeks of onset; 3) the presence of optic pallor during the initial neurological examination; 4) all four patients had benign diseases; 5) the four individuals were operated on during the years 1983 and 1984; 6) in all cases, a Lyodura® graft (lot number not known) was used to accomplish dural closure; 7) no patient had a history of previous surgery, of administration of native growth hormone, or of familial dementia; and 8) all four individuals were homozygous for methionine at codon 129, as in all the cases previously studied by Brown and coworkers [15]. Differential diagnosis in our patients was established with hydrocephalus, CSF shunt complications, viral encephalitis, posterior fossa arachnoiditis, tumor recurrence, psychoses and radiation necrosis.

Epidemiological survey of our cases

Operation protocols corresponding to all surgeries performed at our hospital between January 1983 and December 1984 were investigated. During this period, 1,052 surgical procedures were performed at our hospital. Thirty-seven patients had received a Lyodura® graft. Four of them developed CJD, as opposed to none of the remaining 1,015 patients not subjected to dural grafting, who have not shown any evidence of prion infection up to now. Accordingly, although we lack experimental proof that Lyodura® is a vector for CJD, we think that we have demonstrated a causal relationship between the Lyodura® implant and the subsequent occurrence of CJD, at least from an epidemiological point of view.

CJD following dura mater grafts elsewhere in the world

Altogether, there have been 24 cases of CJD attributed to the use of cadaveric dural grafts: 18 published cases [2, 16–26] and six unpublished cases (CJD Surveillance Unit at Edinburgh, AM Geue, P Brown, unpublished data, 1995–1996). In two cases, CJD developed after the implantation of non commercial dura mater grafts [19, 21]. In both cases, the dural grafts took place between 1969 and 1981. In the remaining 22 iatrogenic cases, CJD developed after implantation of Lyodura® and, in all the cases, the years of implants were 1981 through 1989. A graft implanted in 1989 was probably from a lot prepared before 1987 (Clavel and Clavel, personal communication, 1995), the year in which the Lyodura® manufacturers changed their processes of graft sterilization [18]. Common clinical features of the dura mater-related CJD were: 1) the mean age of the patients (35.3 years, range: 17–62 years), 2) the mean incubation time which was 5.7 years, 3) the mean duration of the disease (19.8 months, range: 2–158 months), and 4) the fact that all the patients received their dural implant during surgery for benign neurosurgical conditions (congenital malformations, benign cerebral and spinal tumors, and head injury). Four patients, including one in our series, are still alive. The fact that all the patients with dura mater related CJD had benign diseases suggests that other patients might have died from their malignant disease before having time to develop overt signs of CJD. Brown et al have demonstrated in two recent reports [15, 27] that most patients with dura mater-related CJD were homozygous for methionine at polymorphic codon 129, and that homozygosity per se at this codon enhances the susceptibility of the patients to this form of exogenous prion infection.

We are currently conducting a follow-up study of the 37 patients who received a Lyodura® graft during the years at risk at our hospital (1983–1984). Nineteen of these individuals have died of their original diseases or of unrelated causes. One was lost to follow-up. The survivors are being submitted to yearly investigations. A study of codon 129 in these 17 survivors has shown that 7 of them were homozygous for methionine, 4 were homozygous for valine, and 6 were heterozygous. Altogether 11 (65%) out of 17 patients were homozygous. A study of the normal Spanish population showed that, among 37 control subjects, 54% were homozygous and 46% heterozygous at codon 129.

Risks of exposure to CJD in the workplace

Although there is no convincing proof of an increased risk of contracting CJD among health-care workers, several disquieting reports have suggested that these persons are overrepresented in relation to lay people. There are altogether 14 physicians who have developed CJD, including 3 dentists, 2 neurosurgeons, 1 general surgeon, 1 orthopedic surgeon, 1 pathologist, and 1 general surgeon [11, 28–35]. Weber et al reported the case of an orthopedic surgeon who had previously worked for 20 years with sheep and human dura mater for research [36]. Samples were sent to the company that had sold the dura mater grafts through which CJD was transmitted in several instances. This patient died of CJD and was homozygous for methionine at codon 129. Several publications have also reported CJD in 8 nurses, 4 auxiliary nurses, and 2 histopathology technicians [30, 32, 35, 37–39]. However, medical personnel have the ethical and legal duty to perform the necessary diagnostic procedures (including biopsy or autopsy), using safety measures to avoid contracting the disease. Health-care workers should be appropriately trained and aware of the dangers and means of prevention. Precautions in handling patients and pathological specimens of proven or suspected CJD cases have been amply publicized [28, 40–43].

Prevention of CJD spread in the neurosurgical milieu

Precautions should be taken not only when managing proven cases of CJD, such as in performing cerebral biopsies, but also in patients whose antecedents include being receivers of native hormones or dura mater grafts and in patients who have a familial history of spongiform encephalopathies. We think that brain biopsy should be strictly restricted to cases of uncertain diagnosis, and always if a subsequent therapeutic decision depends on the biopsy. Confirmation of CJD with clinical and EEG evidence should await postmortem verification. Guidelines for proper management of these patients with suspected or proven CJD have been discussed elsewhere [28, 44, 45]. Neurosurgeons should strictly avoid the use of dura mater grafts from cadavers and instead use autologous tissues (pericranium, fascia temporalis, fascia occipitalis and fascia lata) or even synthetic implants, even though these are suboptimal for dura mater repair. To avoid the use of foreign tissues, we are currently developing a technical procedure for the use of occipitalis fascia for posterior fossa procedures which account for the majority of dural graft indications. Neurosurgeons and surgeons specializing in other area fields, who are still using dural grafts from cadavers, should record in their operation protocols the lot number of the implanted tissue to allow further investigation of potential future cases of CJD or other diseases. Finally, we would like to recall the words of Janssen and Schonberger about the risks and safety of dural grafts: "What remains unclear is the number of patients who received allogeneic dural grafts from these banks and lived long enough to be at risk of both developing related symptoms of CJD and being seen by medical care personnel who would appropriately diagnose and report them" [46].

Acknowledgments

We would like to acknowledge Paul Brown and Larissa Cervenáková for genetic molecular analysis of our CJD patients, control subjects and persons at risk of contracting CJD. Information about unpublished cases of dural-related CJD was provided by the CJD Surveillance Unit at Edinburgh and by Alexandra M Geue from the NHMRC, Department of Health and Family Services, Canberra (Australia).

References

1 Gajdusek DC, Gibbs CJ Jr, Alpers M. Experimental transmission of a kuru-like syndrome in chimpanzees. *Nature* 1966;209:794-6
2 Martínez-Lage JF, Poza M, Sola J et al. Accidental transmission of Creutzfeldt-Jakob disease by dural cadaveric grafts. *J Neurol Neurosurg Psychiatry* 1994;45:1091-4
3 Duffy P, Wolf J, Collins G, De Voe AG, Streeten B, Cowen D. Possible person-to-person transmission of Creutzfeldt-Jakob disease. *Lancet* 1974; 290:692-3
4 Tange RA, Troost D, Limburg M. Progressive fatal dementia (Creutzfeldt-Jakob disease) in a patient who received homograft tissue for tympanic membrane closure. *Eur Arch Otorhinolaryngol* 1990;247:199-201
5 Bernouilli C, Siegfried J, Baumgartner G et al. Danger of accidental person-to-person transmission of Creutzfeldt-Jakob disease by surgery. *Lancet* 1977;i:478-9
6 Nevin S, McMenemey WH, Behrman S, Jones DP. Subacute spongiform encephalopathy – a subacute form of encephalopathy attributable to vascular dysfunction (spongiform cerebral atrophy). *Brain* 1960;83:519-64
7 Brown P, Gibbs CJ Jr, Rodgers-Johnson P et al. Human spongiform encephalopathy: the National Institutes of Health series of 300 cases of experimentally transmitted disease. *Ann Neurol* 1994;35:513-29
8 Gaches J, Supino-Viterbo V, Oughourlian JM. Unusual electroencephalographic and clinical evolution of a case of meningioma of the left temporo-occipital convexity. *Eur Neurol* 1971;5:155-64
9 Matthews WB. Epidemiology of Creutzfeldt-Jakob disease in England and Wales. *J Neurol Neurosurg Psychiatry* 1975;38:210-3
10 Foncin JF, Gaches J, Cathala F, El Sherif E, Le Beau J. Transmission iatrogène interhumaine possible de la maladie de Creutzfeldt-Jakob avec atteinte des grains du cervelet. *Rev Neurol (Paris)* 1980;136:280
11 Will RG, Matthews WB. Evidence for case-to-case transmission of Creutzfeldt-Jakob disease. *J Neurol Neurosurg Psychiatry* 1982;45:235-8
12 Thadani V, Penar PL, Partington J et al. Creutzfeldt-Jakob disease probably acquired from a cadaveric dura mater graft. *J Neurosurg* 1988;69:766-9
13 Martínez-Lage JF, Sola J, Poza M, Esteban JA. Pediatric Creutzfedlt-Jakob disease: probable transmission by a dural graft. *Child Nerv Syst* 1993;9:239-42
14 Martínez-Lage JF, Poza M, Tortosa JG. Creutzfeldt-Jakob disease in patients who received a cadaveric dura mater graft – Spain, 1985-1992. *MMWR* 1993;42:560-3
15 Brown P, Cervenaková L, Goldfarb LG et al. Iatrogenic Creutzfeldt-Jakob disease: an example of the interplay between ancient genes and modern medicine. *Neurology* 1994;44:291-3
16 Nisbet TS, McDonaldson Y, Bishara SN. Update: Creutzfeldt-Jakob disease in a second patient who received a cadaveric dura mater graft. *JAMA* 1989;261:1118
17 Massullo C, Pocchiari M, Maschi G, Alema G, Piaza G, Panzera MA. Transmission of Creutzfeldt-Jakob disease by dural cadaveric graft. *J Neurosurg* 1989;71:954-5
18 Miyashita K, Inuzuka T, Kondo H et al. Creutzfeldt-Jakob disease in a patient with a cadaveric dural graft. *Neurology* 1991;41:940-1

19 Pocchiari M, Massullo C, Salvatore M, Genuardi M, Galgani S. Creutzfeldt-Jakob disease after non-commercial dura mater graft. *Lancet* 1992;340:614-5

20 Willison HJ, Gale A, McLaughlin JE. Creutzfeldt-Jakob disease following cadaveric dura mater graft. *J Neurol Neurosurg Psychiatry* 1991;54:940

21 Esmonde T, Lueck CJ, Symon L, Duchen LW, Will RG. Creutzfeldt-Jakob disease and lyophilised dura mater grafts: report of two cases. *J Neurol Neurosurg Psychiatry* 1993;56:999-1000

22 Takayama S, Hatsuda N, Matsumara K, Nakasu S, Handa J. Creutzfeldt-Jakob disease transmitted by cadaveric dural graft: a case report (Jpn). *No Shinkei Geka* 1993;23:167-70

23 Lane KL, Brown P, Howell DN et al. Creutzfeldt-Jakob disease in a pregnant woman with an implanted dura mater graft. *Neurosurgery* 1994;34:737-40

24 Yamada S, Aiba T, Endo Y, Hara M, Kitamoto T, Tateishi J. Creutzfeldt-Jakob disease transmitted by cadaveric dura mater graft. *Neurosurgery* 1994;34:740-4

25 Lang CJG, Schüller P, Engelhardt A, Spring A, Brown P. Probable Creutzfeldt-Jakob disease after a cadaveric dural graft. *Eur J Epidemiol* 1995;11:79-81

26 Clavel M, Clavel P. Creutzfeldt-Jakob disease transmitted by dura mater graft. *Eur Neurol* (in press)

27 Brown P, Preece MA, Will RG. 'Friendly fire' in medicine: hormones, homografts, and Creutzfedlt-Jakob disease. *Lancet* 1992;340:24-7

28 Gajdusek DC, Gibbs CJ Jr, Asher DM et al. Precaution in medical care of, and in handling materials from, patients with transmissible virus dementia (Creutzfeldt-Jakob disease). *N Engl J Med* 1977;297:1253-8

29 Masters CL, Harris JO, Gajdusek DC, Gibbs CJ Jr, Bernoulli C, Asher DM. Creutzfeldt-Jakob disease: patterns of world-wide occurrence and the significance of familial and sporadic clustering. *Ann Neurol* 1979;5:177-88

30 Harries-Jones R, Knight R, Will RG, Cousens S, Smith PG, Matthews WB. Creutzfeldt-Jakob disease in England and Wales, 1980–1984: a case-control study of potential risks factors. *J Neurol Neurosurg Psychiatry* 1988;51:1113-9

31 Schoene WC, Masters CL, Gibbs CJ Jr et al. Transmissible spongiform encephalopathy (Creutzfeldt-Jakob disease). Atypical clinical and pathological features. *Arch Neurol* 1981;38:473-7

32 Brown P, Cathala F, Raubertas RF, Gajdusek DC, Castaigne P. The epidemiology of Creutzfeldt-Jakob disease: conclusion of a 15-year investigation in France and review of the world literature. *Neurology* 1987;37:895-904

33 Gorman DG, Benson DF, Vogel DG, Vinters HV. Creutzfeldt-Jakob disease in a pathologist. *Neurology* 1992; 42:463

34 Berger JR, David NJ. Creutzfeldt-Jakob disease in a physician: a review of the disorder in health care workers. *Neurology* 1993;43:205-6

35 Berger JR, David NJ. Creutzfeldt-Jakob disease in health care workers. *Neurology* 1993;43:2421

36 Weber T, Tumani H, Holdorff B et al. Transmission of Creutzfeldt-Jakob disease by handling of dura mater. *Lancet* 1993;341:123-4

37 Ruiz Bremón A, Plitt Gómez C, DePedro Cuesta J. Vigilancia epidemiólogica de la enfermedad de Creutzfeldt-Jakob. *Bol Epidemiol Semanal (Spain)* 1995;3:129-32

38 Miller DC. Creutzfeldt-Jakob disease in histopathology technicians. *N Engl J Med* 1988;318:853-4

39 Sitwell L, Lach B, Atack E, Atack D, Izukawa D. Creutzfeldt-Jakob disease in histopathology technicians. *N Engl J Med* 1988;318:854

40 Traub RD, Gajdusek DC, Gibbs CJ Jr. Precautions in conducting biopsies and autopsies on patients with presenile dementia. Technical note. *J Neurosurg* 1974;41:394-5

41 Bell JE, Ironside JW. How to tackle a possible Creutzfeldt-Jakob disease necropsy. *J Clin Pathol* 1993;46: 193-7

42 Budka H. Aguzzi A, Brown P et al. Tissue handling in suspected Creutzfeldt-Jakob disease (CJD) and other human spongiform encephalopathies (prion diseases). *Brain Pathol* 1995;5:319-22

43 MacMurdo SD, Jakymec AJ, Bleyaert AC. Precautions in the anesthetic management of a patient with Creutzfeldt-Jakob disease. *Anesthesiology* 1984;60:590-2

44 Brown P, Gibbs CJ Jr, Amyx HL et al. Chemical disinfection of Creutzfeldt-Jakob disease virus. *N Engl J Med* 1982;306:1279-82

45 Wight A, Metters JS. Neuro and ophthalmic surgery procedures on patients with or suspected to have, or at risk of developing, Creutzfeldt-Jakob disease (CJD), or Gerstmann-Sträussler-Sheinker syndrome (GSS). *J Public Health Med* 1993;15:209-10

46 Janssen RS, Schonberger LB. Discussion to Marx RE, Carlsson ER: Creutzfeldt-Jakob disease from allogeneic dura: a review of risks and safety. *J Oral Maxillofac Surg* 1991;49:275

TRANSMISSIBLE SUBACUTE SPONGIFORM ENCEPHALOPATHIES:
PRION DISEASES

Transmissible Subacute Spongiform Encephalopathies:
Prion Diseases
L Court, B Dodet, eds
© 1996, Elsevier, Paris

Growth hormone-related Creutzfeldt-Jakob disease: genetic susceptibility and molecular pathogenesis

Jean-Philippe Deslys, Corinne Ida Lasmézas, Alexandre Jaegly, Dominique Marce, François Lamoury, Dominique Dormont

Service de Neurovirologie, Commissariat à l'Energie Atomique, DSV/DRM/CRSSA, 60–68, avenue Division Leclerc, BP 6, 92265 Fontenay-aux-Roses cedex, France

Summary – The French cohort of extractive growth hormone-linked iatrogenic Creutzfeldt-Jakob disease (CJD) cases is unique: 1) by its size: about 1,000 patients have been exposed to the CJD agent; and 2) by the defined period of exposure (1984–1985), which makes it possible to determine the incubation times. There is now a 7-year period of observation since the occurrence of the first cases in 1989. We have studied 40 patients among the 42 confirmed or highly probable cases of iatrogenic CJD. The distribution of the *PRNP* genotypes coding for the prion protein clearly demonstrates that heterozygosity at codon 129 lengthens incubation time and probably lowers susceptibility. Moreover, our data suggest the involvement of a factor of resistance to the disease which is not linked to the *PRNP* gene, and perhaps not to the host genotype.

Inroduction

Creutzfeldt-Jakob disease (CJD) belongs to the group of transmissible spongiform encephalopathies (TSE). These fatal neurodegenerative diseases are linked to the multiplication in the central nervous system of TSE agents also called 'prions', the nature of which remains elusive. The susceptibility to these diseases is governed mainly by a host gene present as a single copy on chromosome 20 in man *(PRNP)* [1, 2]. The whole coding region is included in a single exon, excluding the possibility of splicing. It codes for prion protein (PrP), a protein mainly expressed in neurons, which is the precursor of an abnormal isoform called PrP-res (protease-resistant PrP) [3]. The latter is resistant to proteolytic degradation and accumulates in the brain of infected individuals [4, 5]. Conversion of normal PrP into PrP-res is observed only in TSE and the unusual resistance to degradation of TSE agents seems to be linked to the properties of PrP-res. Moreover, different strains of TSE agents, harboring specific characteristics, can multiply in the same host (more than 20 strains have been characterized in syngeneic mice) [6]. As in other TSE, the development of CJD in man depends on two main factors: the host and the agent.

We have studied the involvement of the host *PRNP* genotype in the development of CJD in the French cohort of patients contaminated with extractive human growth hormone (hGH) [7].

Key words: Creutzfeldt-Jakob disease / growth hormone / prion protein (PrP) / prion protein gene (PRNP)

The whole coding region of the *PRNP* gene has been sequenced for 17 of the first hGH-linked iatrogenic cases: we found no mutation and the sequences were identical to those of the normal population [8]. The sole abnormality was observed at codon 129 which is naturally polymorphic in man (fig 1). About 50% of the Caucasian population is heterozygous at this codon: one *PRNP* allele encodes a methionine (Met) and the other one a valine (Val) [9]. We verified that this proportion was similar in the population of children treated with hGH for hypopituitarism. In contrast, until 1995, none of the iatrogenic cases was heterozygous at codon 129. Iatrogenic CJD cases linked to hGH have been reported in different countries [10, 11] and the French cohort is the largest and the most homogeneous. In fact, epidemiological studies have determined that all the contaminations occurred within a limited period of time between 1984 and 1985 and that the whole treated population (1,000 patients) was exposed to the risk [12]. Therefore, the evolution of the genotype distribution in this cohort could provide new information on the role of codon 129 in humans.

Methods

PrP-res detection

PrP-res was purified according to previously published methods for scrapie-associated fibrils [13–15]. Samples were treated with 10 μg/mL proteinase K for 1 h at 37 °C and thereafter they were denatured with sodium dodecyl sulfate–polyacrylamide gel electrophoresis (SDS–PAGE) sample buffer. The equivalent of 0.5 mg of brain was loaded onto the wells of a 12% SDS–PAGE gel. Electrophoresis, electrotransfer and Western blot detection were performed according to standard procedures using enhanced chemiluminescence. For immunodetection, we used an anti-hamster PrP monoclonal antibody

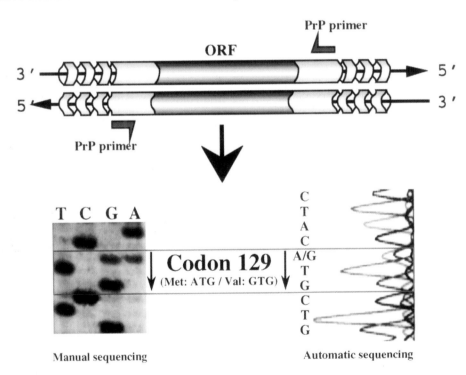

Fig 1. PrP sequencing: example of heterozygosity at codon 129. ORF: open reading frame.

kindly provided by RJ Kascsak (NYS Institute for Basic Research in Developmental Disabilities, Staten Island, NY, USA).

PRNP *amplification*

Genomic DNA was extracted using standard techniques. The *PRNP* gene was amplified by polymerase-chain reaction between bases 142 and 819 (including the coding region between codon 48 and the end of the open reading frame). The primers used were 5'-CGCTACCCACCTCAGGGCGG-3' (sense) and 5'-CCGCCTCCCTCAAGCTGGAA-3' (antisense). Samples of 50 μL, each containing a standard buffer (1.5 mM $MgCl_2$), 200 mM of dNTP, 250 nM of each primer, 100 ng of DNA and 1 unit of Taq polymerase (Boehringer) were subjected to 35 cycles of amplification (94 °C for 30 s, 59 °C for 1 min and 72 °C for 1 min).

Allele-specific oligonucleotide hybridization

Samples (10 μL) were run on a 1% Tris-acetate ethylenediaminetetracetic acid (TAE) Agarose gel and visualized by ethidium bromide staining. After Southern blotting under vacuum with alkaline denaturing conditions, nylon filters (Hybond-N Amersham) were fixed by exposure to ultraviolet light for 5 min. Filters were prehybridized for 1 h at 37 °C in 5X SSPE (1X SSPE is 0.15 M NaCl, 10 mM sodium phosphate, pH 7.4, and 1 mM EDTA), 5X Denhardt's solution, 0.5% SDS and 100 μg/mL denatured salmon sperm. Allele-specific oligonucleotides (Met[129]: 5'-CGGCTACATGCTGGG-3', Val[129] 5'-CGGCTAC-GTGCTGGG-3') were radiolabeled with T4 kinase (solution comprising 25 pmol oligonucleotide, 8 pmol [γ-32P]ATP 185 TBq/mmol and 5 units of T4 kinase for 20 μL). Hybridization was performed for 2 h under the same conditions with 10^6 cpm/mL of 32P-labeled oligonucleotide. Filters were washed twice for 5 min at room temperature in 2X SSPE/0.1% SDS and once for 15 min at high stringency temperature (52 °C for Met[129] and 54 °C for Val[129]) with a final 5-min wash at room temperature. Autoradiography was performed with 2-h exposure.

Results

Forty of the 42 confirmed or highly suspected iatrogenic CJD cases have now been studied. They were characterized biochemically by the predominent accumulation of PrP-res in the cerebellum as compared to the forebrain [16] (fig 2). The case distribution based on the genotype at codon 129 was as follows (table I, fig 3): in the Met/Val group, the first cases did not appear until mid-1995, ie, 6 years after the two other groups; in the Val/Val group, the number

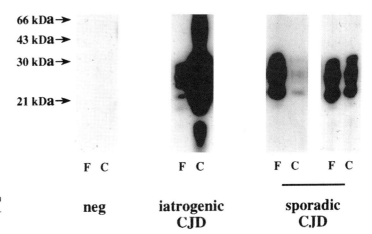

Fig 2. Distribution of PrP-res in the cerebellum (C) and the forebrain (F). neg: negative control.

Table I. Polymorphism of PrP codon 129.

Subjects	Met/Val^{129}	Val/Val^{129}	Met/Met^{129}	Total
Healthy controls	54% (37)	10% (7)	36% (25)	100% (69)
Controls given hGH	44% (61)	12% (16)	44% (61)	100% (138)
CJD patients given hGH	8% (3)	30% (12)	62% (25)	100% (40)

Numbers are given between parentheses.

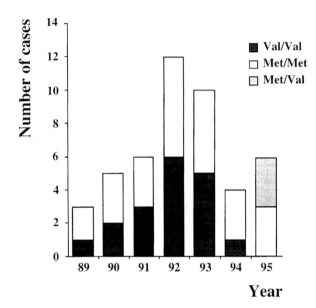

Fig 3. Evolution of genotype at codon 129 of iatrogenic CJD cases in France.

of cases peaked in 1992, and decreased dramatically in 1994 (8 years after contamination) when a single case was observed and no more cases thereafter; in the Met/Met group there was also a peak in 1992, but new cases are still appearing steadily.

This cohort constitutes a well-differentiated group when compared to patients with sporadic CJD: 1) The age groups are strictly different [17] (fig 4) and 2) the risk of developing sporadic CJD has been evaluated to be about 1 per million and the risk of developing sporadic CJD before the age of 40 years is about 1 per 20 million. On the other hand, the risk of developing CJD in patients treated with hGH during the 1984–1985 period is about 1 per 25 in France.

An important issue for the interpretation of the data is assessment of the contamination level. Although much evidence exists which suggests that it has certainly been high, it cannot be evaluated accurately. It is well known that, for peripheral inoculation to be efficient, a higher number of infectious particles is required (eg, with the 139A strain in mice, the infective dose is 25,000 times higher by the subcutaneous route than by the intracerebral route). Together with the short incubation times of the first iatrogenic cases and the large number of cases, this fact suggests the presence of a high number of infectious particles in the incriminated batches of hormone.

Epidemiological studies have shown the contamination of at least five batches of hGH which were distributed to almost 1,000 patients [12]. Hence, a large population has been

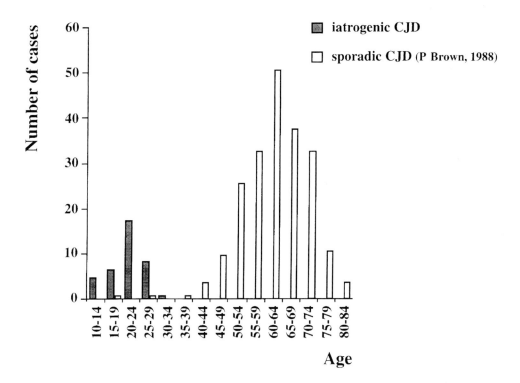

Fig 4. Age (years) at onset of iatrogenic versus sporadic CJD.

potentially exposed with an overlapping distribution of contaminated batches. The number of infectious hypophyses, the clinical presentation and, of course, the PrP sequence of the contaminating donors are unknown. It is very unlikely that the children would have been infected with a particular strain of CJD agent, as they received different, and, some of them, several contaminated batches. Therefore, the differences in the incubation periods we observed are clearly due to the host genotype. The data on the heterozygous patients provide the following evidence. As heterozygous patients represent 50% of the contaminated population, statistically 37 heterozygous patients received the same infectious doses as the 37 homozygous patients who developed the disease (table I). However, the incubation time was 6 years longer and the number of cases was 10 times lower in this group. Thus, the presence of two types of PrP in these patients prolonged the incubation period and probably lowered susceptibility. Thus, the two types of PrP found in humans are not biologically equivalent. It is important to note that, in mice, two types of PrP are also found which have different effects and induce the selection of different strains of TSE agents.

Many patients of the Val/Val group have been potentially contaminated (ie, 10–12% of the 1,000 patients of the cohort according to the distribution of this genotype in the normal population) and a high level of contamination is suggested by the short incubation period of the first cases (4–5 years). In contrast to this high exposure, only a limited number of cases appeared in Val/Val carriers and the evolution curve suddenly broke (only one case has appeared since the beginning of 1993). If this evolution were to be confirmed over the next few years, one interpretation could be that few Val/Val patients have been exposed to infectious hGH, but it does not correspond to previous epidemiological data evidencing the absence of any common batch distributed to a particular population. The alternative explanation would be that up to 90% of Val/Val patients have a protective factor that is not linked to the *PRNP* gene (as we found no abnormality in the PrP sequences of these patients) and which remains to be determined.

In conclusion, polymorphism at codon 129 is responsible for the expression of two PrP molecules with different biological properties in humans. Theoretically, they may favor the replication of different strains of TSE agents, as is observed in experimental mouse models. Moreover, it is now clear that homozygosity at codon 129 increases the transmission rate and shortens the incubation period in iatrogenic CJD, and probably also in familial and sporadic CJD. Finally, our data suggest the existence of a protective factor present in the majority of the population and different from the *PRNP* gene.

Acknowledgments

We thank the professors and medical doctors from Angers, Bordeaux, Lille, Marseille, Montpellier, Nevers, Périgueux, Paris, Saint-Méloir, Thionville and Toulouse for providing tissue and blood from patients. This work was supported by a grant from DRET (France).

References

1 Pruiner SB. Novel properties and biology of scrapie prions. In: Chesebro BW, ed. *Transmissible Spongiform Encephalopathies: Scrapie, BSE and Related Diseases.* 1991;172:233-57
2 Chesebro B. PrP and the scrapie agent. *Nature* 1992;356:560
3 Chesebro B, Race R, Wehrly K et al. Identification of scrapie prion protein-specific mRNA in scrapie infected and uninfected brain. *Nature* 1985;315:331-3
4 Pruiner SB, Bolton DC, Groth DF, Bowman KA, Cochran SP, McKinley MP. Further purification and characterization of scrapie prions. *Biochemistry* 1982;21:6942-50
5 Bolton DC, McKinley MP, Pruiner SB. Identification of a protein that purifies with the scrapie prion. *Science* 1982;218:1309-11
6 Bruce ME. Scrapie strain variation and mutation. *Br Med Bull* 1993;49:822-38
7 Deslys JP, Marcé D, Dormont D. Similar genetic susceptibility in iatrogenic and sporadic Creutzfeldt-Jakob disease. *J Gen Virol* 1994;75:23-7
8 Jaegly A, Boussin F, Deslys J, Dormont D. Human growth hormone related iatrogenic Creutzfeldt-Jakob disease: limits of a genetic susceptibility detection based on the analysis of the *PRNP* coding region. *Genomics* 1995;27:382-3
9 Owen F, Poulter M, Collinge J, Crow TJ. Codon 129 changes in the prion protein gene in Caucasians. *Am J Hum Genet* 1990;46:1215-6
10 Collinge J, Palmer MS, Dryden AJ. Genetic predisposition to iatrogenic Creutzfeldt-Jakob disease. *Lancet* 1991;339:1441-2
11 Brown P, Preece MA, Will RG. Friendly fire in medicine – hormones, homografts, and Creutzfeldt-Jakob disease. *Lancet* 1992;340:24-7
12 Alpérovitch A. Traitement par hormone de croissance extractive et maladie de Creutzfeldt-Jakob: analyse préliminaire de la relation entre les lots reçus et la survenue de la maladie. Paris: INSERM, Report 1993
13 Diringer H, Hilmert H, Simon D, Werner E, Ehlers B. Towards purification of the scrapie agent. *Eur J Biochem* 1983;134:555-60.
14 Kascsak RJ, Rubenstein R, Merz PA, Cap RL, Wisniewski HM, Diringer H. Biochemical differences among scrapie-associated fibrils support the biological diversity of scrapie agents. *J Gen Virol* 1985;66:1715-22
15 Hope J, Morton LJD, Farquhar CF, Multhaup G, Beyreuther K, Kimberlin RH. The major polypeptide of scrapie-associated fibrils (SAF) has the same size, charge distribution and N-terminal protein sequence as predicted for the normal brain protein. *EMBO J* 1986;5:2591-7
16 Deslys JP, Lasmézas C, Dormont D. Selection of specific strains in iatrogenic Creutzfeldt-Jakob disease. *Lancet* 1994;343:848-9
17 Brown P. The clinical neurology and epidemiology of Creutzfeldt-Jakob disease, with special reference to iatrogenic cases. In: *Ciba Foundation Symposium. Novel Infectious Agents and the Central Nervous System.* Chichester: John Wiley & Sons, 1988;135:3-23

Transmissible Subacute Spongiform Encephalopathies:
Prion Diseases
L Court, B Dodet, eds
© 1996, Elsevier, Paris

The management of risk from exposure of biomedical products to the spongiform encephalopathy agent(s)

Robert G Rohwer

Medical Research Service, Veterans Affairs Medical Center, Baltimore, and Department of Neurology, University of Maryland at Baltimore, Baltimore, MD 21201, USA

Summary – The transmissible spongiform encephalopathy agents causing Creutzfeldt-Jakob disease, scrapie and bovine spongiform encephalopathy are among the most difficult pathogens to detect and remove from products containing human- or animal-derived components. As such, they constitute an especially stringent challenge for viral clearance. Factors that must be considered in the management of risk from exposure to this class of agents include 1) the end use of the product (routes of administration and exposure); 2) sourcing: the medical and familial history of human donors and the geographical origins, secondary exposures, health, husbandry and maintenance history of animals; 3) the tissue(s) utilized in the product and their likelihood of harboring or being contaminated by agent(s); 4) the methods by which the tissues or fluids are collected (in the case of animals, slaughter and dissection); 5) the production process including steps that remove agent(s) either by inactivation or physical separations, the disposal of agent(s) that is removed but not inactivated, flow control, barriers and between-batch cleaning; 6) batch size; and 7) validations and quality assurance monitoring.

Introduction

The ultimate goal of viral clearance should be the removal of all viruses, known or unknown, that may contaminate the raw materials used in production. Since the spectrum of viral properties is vast, generic strategies must be based on the ability to clear those agents with the most extreme characteristics.

The spongiform encephalopathy (SE) agents are among the most cumbersome to detect, difficult to remove and laborious to validate and as such constitute an especially stringent challenge for viral clearance [1, 2]. Nevertheless, significant clearance can be demonstrated for many processes. Here we will discuss several of the most important factors that influence risk from these agents.

End use

End use dictates the level of concern for animal-derived products. In general, parenterals are of greater concern than are foods, which in turn pose greater risks than do topical exposures or incidental contacts from non biological uses. Higher risk is also more likely to be tolerated for essential/critical applications affecting small high-need populations than for luxury or

Key words: transmissible spongiform encephalopathies / scrapie / bovine spongiform encephalopathy / risk analysis / virus removal / virus inactivation / virus validation

casual uses or for those exposing large segments of the population. We also tolerate much higher levels of risk for traditional uses of animal products than are sanctioned for new, untried, or experimental applications.

Route of exposure

Risk of infection is strongly affected by the route of exposure. In the case of the scrapie agent, an intracerebral inoculation is 10 times more potent than an intravenous inoculation, 10 to 100 times more potent than an intraperitoneal inoculation and 1,000 to 10,000 times more potent than a subcutaneous exposure [3–5]. Gastrointestinal (GI) exposure may be even less efficient [6, 7]. Nevertheless, consumption of rendered ruminant protein is almost certainly the cause of epidemic bovine spongiform encephalopathy (BSE) in cattle and is the likely source of feline spongiform encephalopathy and the spongiform disease in exotic zoo ruminants and felidae that has occurred concurrently with the BSE epidemic [8]. Natural scrapie in sheep is also presumed to be transmitted by the GI route despite the inefficiency of this route and what is presumed to be a very low level of exposure [9]. Apparently, chronic exposure compensates for the other deficiencies of this route. Of course, the efficiency of various routes of exposure may also be agent strain-specific, and this may account for the apparent virulence of the oral route for the BSE agent. As yet, this possibility has not been tested experimentally.

Sourcing

When a manufacturing process lacks intrinsic clearance potential, sourcing may be the only means of managing viral contamination. However, it is important to recognize the limitations of the sourcing strategy. Sourcing can be used only to avoid specifically identified contaminants and a source that is free of one contaminant may be at risk for others.

Closed herds offer the potential for greatly limiting exposure to pathogens that circulate at large; however, the term 'closed' is unregulated and is often used in contexts that are irrelevant to pathogen control.

If the objective is to avoid BSE, the most important consideration is the feeding history of the herd. The desired trait is a herd that has never been exposed to feed supplements containing ruminant-derived protein. Other considerations are lack of contact with sheep or other animals at risk for spongiform disease such as cats, deer and mink, as well as a breeding program that avoids physical contact with outside cattle. While such herds exist, they are an unrealistic source for large quantities of tissue.

Since the United States (US) still permits the feeding of ruminant-derived protein to ruminants, and since many foreign markets penalize the sourcing of bovine-derived biological products from countries that permit this practice, some US producers of exported biologicals are now sourcing bovine and ovine materials from Australia and New Zealand, which are the only countries that are widely accepted to be free of sheep scrapie. The rational for this is based on the unproven assumption that BSE is caused by exposure of cattle to sheep scrapie. If instead it turns out that BSE originates from some other source, for example a wild animal reservoir, exposure to sheep scrapie may prove irrelevant to the etiology of BSE, thereby invalidating this sourcing rational. To date, attempts to transmit scrapie to cattle have resulted in a neurological disease that is clinically and pathologically distinct from BSE [10].

There is no country currently claiming to be free of ruminant spongiform disease that has not been exposed to imported animals at risk. That such animals can carry BSE has now been demonstrated by outbreaks in Canada, Oman, the Falkland Islands, Portugal, France, Germany,

Italy, Denmark and Switzerland in imported stock [11]. If such an animal were rendered during preclinical disease, the stage would be set for a recurrence of the BSE epidemic.

If the prion hypothesis is correct there could be no cattle population that is totally free of SE. It is suggested that prion disease occurs spontaneously at low frequency in any population susceptible to prion protein (PrP) amyloidosis [12, 13]. Such rare spontaneous conversions to the amyloidotic state might require years to progress to a clinically recognizable disease. Thus, affected animals might never be recognized. Slaughter and rendering into high protein feed supplements would again set the stage for an eventual epidemic of BSE.

Whatever the origins of BSE, whether interspecies transfer of an exogenous pathogen, or intraspecies expansion of a spontaneously generated intrinsic pathogen (as proposed by the prion hypothesis), the principal defense against a recurrence in other cattle populations is validated sterilization of foodstuffs and biologicals derived and administered intraspecifically or bans on their use. It should be evident that any species that is similarly exposed to intraspecific recycling of foodstuffs and biologicals may also be at risk. In the case of biologicals, this includes man.

Tissues

Viruses often exhibit stringent tissue tropisms. For a given virus, tissues that are not infected should pose far less risk of contamination than the targeted tissues. However, this isolation is seldom complete. Blood, nerves and lymph provide global access to all tissues for viruses that use those avenues of infection. Viremias are frequent adjuncts of virus infection and result in systemic contamination. Finally, careless collection methods can result in the contamination of otherwise virus-free or virus-reduced tissue.

In the case of BSE, the British, WHO and European regulatory authorities have published lists of tissues ranked according to their presumed level of contamination by BSE. As expected, brain and other central nervous system (CNS) tissues occupy category I, the highest infectivity, highest risk category. The ileum, lymph nodes, proximal colon, spleen, tonsil, dura mater, pineal gland, placenta, cerebrospinal fluid, pituitary and adrenal glands occupy category II, medium infectivity. The distal colon, nasal mucosa, sciatic nerve, bone marrow, liver, lung, pancreas and thymus occupy category III, low infectivity. Finally, blood clots, feces, heart, kidney, mammary gland, milk, ovary, saliva, salivary gland, seminal vesicle, serum, skeletal muscle, testis, thyroid, uterus, fetal tissue, bile, bone, cartilage, connective tissue, hair, skin and urine are in category IV, no detectable infectivity [14].

It is important to keep in mind that most of these assessments are based on titers from sheep and goats measured across a species barrier in mice with a great loss in sensitivity. Most involve inoculation of 50 μL of a 10% tissue homogenate into four to eight mice for a total representation of the parent tissue of approximately 20 mg. No detectable infectivity in category IV means ≤ 100 LD_{50}/mL. Moreover, many of these estimates are based on only one or ·at most two determinations. In some cases rodent-to-rodent inoculations are at odds with some of these recommendations. Titration of bovine tissues in mice have revealed no infectivity outside of the CNS (R Bradley, this volume). However, it is likely that this is due to the insensitivity of the cross-species titrations rather than to the complete absence of infectivity in peripheral organs.

Slaughter

Slaughter, while highly regulated, is directed toward the efficient collection of clean, safe food – not biological products. Animal-derived foods are presumed to be contaminated by microbes and it is the proper storage and preparation of foods that ensures their safety. However, most of this contamination occurs during the slaughter. The viscera of a healthy animal are essentially sterile and yet this attribute is wasted in the typical slaughter environment. In contrast, slaughter provides multiple opportunities for contamination. Animals from many locations are crowded together, ensuring a high level of contact exposure. The slaughter itself exposes the desired tissue both to other organs from the same body and to the tissues of other animals.

In the US the typical slaughter sequence is kill, bleed, delimb, decapitate, skin, eviscerate, split, wash and chill [15]. Accidental slaughter of a BSE-infected animal would likely result in widespread contamination of the facility. The vast majority of spongiform infectivity is found in CNS tissue with lesser amounts in the lymphoreticular organs. Imperfect use of captive bolt killing frequently results in skull penetration, the release of CNS tissue and contamination of the slaughterhouse from the kill floor forward with CNS tissue. The head is also removed early in the process with resultant contamination by the spinal cord. If it is further processed additional contamination would be inevitable. The evisceration step would expose the facility to the spleen and lymphatics. The carcass is split by bisecting the spinal column with a saw, resulting in the aerosolization of spinal cord tissue and widespread contamination

Collection

The challenge of tissue collection is to overcome the obstacles to sterility in the slaughter environment. If the tissue requirements are modest and the product is valuable it may be feasible to collect the tissue under surgical conditions in a more controlled environment and dispose of the carcass as waste. Alternatively, if the manufacturer supplies the labor, it may be possible to retrieve a particular organ under more stringent conditions without unduly interrupting the slaughter. Unfortunately, these are not practical options for manufacturers with large mass requirements where lot sizes representing hundreds of thousands of animals are not uncommon.

Whatever the collection circumstances, much of the contamination of non-CNS tissues, if present, may be from exogenous sources, and superficial in nature. In the case of category IV, low-risk tissues from cattle, simple surface washing of the tissue could be used to remove superficial CNS contamination. If one is dealing with a solid organ it may be possible to improve this wash by use of NaOH as a disinfectant.

Inactivation

Virus removal can either be by inactivation or separation and the manufacturing process may contain incidental or specific virus removal steps. The most efficacious of these are inactivations which result in the destruction of virus infectivity. Because viral sensitivities to heat, pH, desiccation, solvents, surfactants and chemical disinfectants vary widely, there are probably few processes that will not inactivate at least some viruses. The corollary is that there are also very few agents that are effective against all viruses and those that are, tend to be destructive of biological products as well. Nevertheless, if the goal is to eliminate a particular virus, conditions can sometimes be found that result in its selective inactivation.

All inactivations proceed kinetically. Thus, if it is discovered that a step in the manufacturing process incidentally inactivates a virus of interest, it may be possible to achieve a much higher level of clearance with little consequence to the product by simply extending the exposure at that step.

The SE agents are notoriously resistant to disinfection. However, where this has been studied, it appears that this is largely due to small subpopulations, representing 0.1% or less of the infectivity, that are refractory to killing. The vast majority of the infectivity is inactivated with kinetics similar to conventional viruses [1, 16, 17]. Sterility requires steam sterilization for at least 1 h at 132 °C, exposure to 1 N NaOH for 1 h or incineration [18, 19]. Nevertheless, significant inactivation is achieved by phenols, lipophilic solvents, some detergents, strong chaotrophs and, in general, reagents that denature proteins and thereby viruses [20].

Separation

Viruses can also be removed from a product by filtration, chromatography, absorption, extraction, precipitation or other mechanical means. Such separations can often achieve high levels of viral clearance. However, they create downstream problems because they concentrate the infectivity and the viruses are still alive. In the case of filters and chromatography columns, the infectivity may still be in the process stream. To avoid cross-contaminations of the facility by sequestered virus, potentially contaminated filters and columns must, where possible, be sterilized in place before disassembly. When virus has been removed from the process stream by precipitation or extraction, it may be necessary to dispose of the virus-bearing phase as infectious waste.

Virus aggregation results in a loss of titer which can be confused with inactivation or mechanical removal of infectivity when in fact virus numbers have remained constant. If unrecognized, this could lead to a very dangerous presumption of clearance where in fact none was achieved. The SE agents are prone to aggregation. Moreover, according to the prion hypothesis, the amyloid fibrils that accumulate in these diseases are huge linear aggregations of the unit pathogen. The stability of aggregates often varies greatly with agent concentration, buffer conditions, the presence of surfactants and many other factors. A later change in buffer conditions might disperse the aggregates and reintroduce the infectivity into the production stream.

Cross-contamination

The risk of contaminating the facility is especially great when clearance is by removal rather than inactivation. In the case of SE such contaminations would persist indefinitely unless specifically removed by disinfection. Well planned barriers and personnel and materials flow can reduce the likelihood of carrying infectivity forward past the removal steps. This is a practical daily consideration in the scrapie laboratory where high titer agent is used intentionally. Inadequate attention to this issue in the testing laboratory could compromise the results of validation studies.

Cleaning

Rigorous between-batch cleaning and disinfection of process equipment and the facility is essential to limit the consequences of any viral contamination that may have escaped notice. This could easily happen in the case of the spongiform agents and apparently did happen during the production of cadaveric human growth hormone (hGH) [21, 22]. In most instances the disinfection options for equipment can be much harsher than those for the biological materials themselves. High pressure steam is optimum for cleaning and sterilization of vessels and lines. Work surfaces and floors can be disinfected with 1 N NaOH. Maximal security is achieved when chemical and physical methods of decontamination are combined.

Batch size

In general, the more animals that are represented in a batch the greater the likelihood that one or more of them will harbor adventitious agents. If such contaminations have high titers and are highly dispersed, a large fraction of product will be contaminated. Approximately 10,000 cadaveric human pituitaries were used to produce a single lot of hGH with catastrophic results [21, 22].

Small batch sizes are rarely compatible with current manufacturing approaches. However, when means can be found to implement such strategies, they go a long way toward limiting consumer exposure to rare sporadic contamination of a product. Nevertheless, the effectiveness of this approach depends on the adequacy of the decontamination cycle between lots.

Validation

As noted earlier, the ultimate goal of viral clearance is the removal of all viruses known and unknown. However, since viruses vary greatly in their physical properties and sensitivities to inactivating agents, no single virus can represent the entire spectrum. Rather it is necessary to test numerous viruses representing extremes of behavior that subsume the rest of virology. When specific viruses are of concern they should, if possible, be tested directly.

In the case of BSE this is not practical. A validation utilizing the BSE agent itself and titration in cattle would require 8–12 years to complete. Alternatively, the process could be challenged by the BSE agent and titrated in mice, but in this case, the host barrier effect would greatly reduce the sensitivity of detection, thereby reducing the clearance that could be demonstrated [23]. Moreover, the cross-species infection would itself have to be validated. A mouse-adapted BSE strain might be used but has not been well characterized [24].

To date, spongiform agent removal has been validated exclusively using rodent-adapted laboratory strains isolated from sheep scrapie. This use presumes that the inactivation and separation properties of the rodent-adapted strains adequately represent the agents of concern. Use of rodent-adapted scrapie strains is also perceived as posing far less hazard to both agriculture and personnel than would the BSE agent or the human agents.

The mouse-adapted scrapie strains develop agent titers of $\leq 10^9$ ID_{50}/g of brain. Spiking at the level of 1% brain homogenate and inoculating 30 μL, 5 logs of clearance can be demonstrated in 12 months. The hamster-adapted 263K strain has been favored for research, due to its high titer and shorter incubation time. The hamster 263K strain develops agent titers of $\leq 10^{11}$ ID_{50}/g of brain, and a 1% spike can demonstrate 7 logs of clearance in 6 months.

A contamination can be either intrinsic or extrinsic. In an intrinsic contamination the agent is naturally present in the tissue of interest either because it replicates or is sequestered there. Extrinsic contaminations result from exposure during collection or processing to extraneous sources of infectivity. Extrinsic contamination is easily modeled by spiking. Since CNS titers are orders of magnitude greater than those for any other tissue, CNS tissue is the most likely source of extrinsic contamination. Thus, high-titer brain homogenate may constitute a perfect challenge with which to represent an extrinsic contamination.

In contrast, except in the rare instances in which brain or other CNS tissues are being processed, spiked, brain-derived infectivity may constitute a very imperfect representation of intrinsic contamination of infected tissues. First, it may be difficult to incorporate the spike into the tissue in a way that realistically recreates the in vivo association. Second, it is conceivable, though not established, that there may be differences in the properties of infectivity associated with non CNS tissues that would affect its fractionation and inactivation properties. For example, the low infectivity titer associated with blood may represent infectivity that has been modified by exposure to the antigen elimination pathways. When brain homogenate is used as the challenge material, the validation is of brain contamination of the process. High-titer inoculum prepared from purified PrP amyloid or detergent-lipid-protein-complexes [2] are subject to the same caveats. If the tissue of interest contains only 1 or 2 logs of infectivity per gram, one cannot hope to demonstrate significant clearances by employing the infected tissue itself as the challenge. As a consequence, there is usually no other choice than to spike with brain-derived infectivity. One hopes in that case that brain-associated infectivity represents a worst case challenge.

In an ideal manufacturing process, clearance is realized by more than one mechanism and at several different stages of production. It is generally, but not always the case that each removal step is additive. While the objective is to demonstrate the safety of the overall procedure, one cannot by this means demonstrate more clearance than is present in the challenge inoculum. To adequately assess the maximal inactivation potential of the process, each step with removal potential must be challenged independently with high-titer infectivity. Adding the stepwise clearances provides an estimate of the clearance potential of the entire process which, in many cases, is far in excess of the challenge titer, thereby providing a large margin of safety and attendant confidence in the product. The additivity of the stepwise clearances assumes that the surviving infectivity at each step is randomly selected and does not represent a refractory subpopulation that will resist all steps. It is to eliminate this possibility that the entire process is tested for its ability to clear the agent.

It has been argued that dilution of infrequent, low-titer contamination into huge tissue pools might disperse a virus contamination beyond the point of infectiousness [25, 26]. In the case of the spongiform agents, this argument is based on the observation that peripheral routes of inoculation are far less efficient than intracerebral inoculation. The presumption is that this reduction in efficiency is due to a requirement for multiple infections to establish disease. High dilution should greatly diminish the possibility of multiple infections of individuals. However, if the reduction in infection efficiency represents a reduction in the probability of a single productive infection in a background of abortive exposures, there is no a priori reason to expect that this probability would differ for a population and for an individual. In this case the consequence of dilution would be to disperse and thereby increase the number of infected individuals. The iatrogenic transmissions of Creutzfeldt-Jakob disease by contaminated cadaveric hGH warn against dilution as a means of removal [21, 22].

A distinction is sometimes attempted between the terms 'clearance' and 'sterility' with greater value attached to the latter. However, both terms must be operationally defined and to the same end-point. Whether 'clearance' or 'sterility' is used, its significance is constrained by the titer of the challenge. It is always possible that a larger challenge will produce survivors.

Conclusion

Effective means for reducing the risk of TSE contamination to acceptable minimums are available for even the most delicate biological products. Unfortunately, they can seldom be imposed without considerable expense and inconvenience. Once such measures are in place, however, the resulting product is often significantly safer with respect to all adventitious agents. Nevertheless, there are many potential pitfalls and even when a validation study indicates high levels of clearance for a process, the realization of that potential depends critically upon its implementation.

References

1 Rohwer RG. The scrapie agent: "a virus by any other name". *Curr Top Microbiol Immunol* 1991;172: 195-232

2 Prusiner SB. Natural and experimental prion diseases of humans and animals. *Curr Opin Neurobiol* 1991;2: 638-47

3 Kimberlin RH, Walker CA. Pathogenesis of mouse scrapie: effect of route of inoculation on infectivity titers and dose-response curves. *J Comp Pathol* 1978;88:39-47

4 Gibbs CJ Jr, Gajdusek DC, Amyx HL. Strain variation in the viruses of Creutzfeldt-Jakob disease and kuru. In: Prusiner SB, Hadlow WJ, eds. *Slow Transmissible Diseases of the Nervous System*. New York: Academic Press, 1979;2:87-110

5 Outram G. The pathogenesis of scrapie in mice. In: Kimberlin RH, ed. *Slow Virus Diseases of Animals and Man*. Amsterdam: North-Holland Publishing Co, 1976;325-57

6 Kimberlin RH, Walker CA. Pathogenesis of scrapie in mice after intragastric infection. *Virus Res* 1989;12: 213-20

7 Gibbs CJ Jr, Amyx HL, Bacote A, Masters CL, Gajdusek DC. Oral transmission of kuru, Creutzfeldt-Jakob disease, and scrapie to non human primates. *J Infect Dis* 1980;142:205-8

8 Bradley R, Wilesmith JW. Epidemiology and control of bovine spongiform encephalopathy (BSE). *Br Med Bull* 1993;49:932-59

9 Hunter N. Natural transmission and genetic control of susceptibility of sheep to scrapie. *Curr Top Microbiol Immunol* 1991;172:165-80

10 Cutlip RC, Miller JM, Race RE et al. Intracerebral transmission of scrapie to cattle. *J Infect Dis* 1994;169:814-20

11 Bovine Spongiform Encephalopathy (BSE). *Anim Health Rep* 1996;(spring)12-3

12 Cohen FE, Pan KM, Huang Z, Baldwin M, Fletterick RJ, Prusiner SB. Structural clues to prion replication. *Science* 1994;264:530-1

13 Gajdusek DC. The transmissible amyloidoses: genetical control of spontaneous generation of infectious amyloid proteins by nucleation of configurational change in host precursors: kuru-CJD-GSS-scrapie-BSE. *Eur J Epidemiol* 1991;7:567-77

14 World Health Organization Consultation on Public Health. Issues related to animal and human spongiform encephalopathies. In: WHO Report: *WHO/CDS/VPH/92. 104*. Geneva, Switzerland, 12-14 November 1991

15 United States Department of Agriculture, Food Safety and Inspection Service. *Sanitation Handbook for Meat and Poultry Inspectors*. Washington, DC: US Government Printing Office, 261-442/20210, 1990

16 Rohwer RG. Scrapie infectious agent is virus-like in size and susceptibility to inactivation. *Nature* 1984;308: 658-62

17 Rohwer RG. Virus like sensitivity of the scrapie agent to heat inactivation. *Science* 1984;223:600-2

18 Rutala WA. APIC guideline for infection control practice. APIC guideline for selection and use of disinfectants. *Am J Infect Control* 1990;18:99-117

19 Brown P, Rohwer RG, Gajdusek DC. Sodium hydroxide decontamination of Creutzfeldt-Jakob disease virus [letter]. *N Engl J Med* 1984;310:727

20 Millson GC, Hunter GD, Kimberlin RH. The physico-chemical nature of the scrapie agent. In: Kimberlin RH, ed. *Slow Virus Diseases of Animals and Man*. Amsterdam: North-Holland Publishing Co, 1976;243-66

21 Brown PW, Gajdusek C, Gibbs CJ Jr, Asher DH. Potential epidemic of Creutzfeldt-Jakob disease from human growth hormone therapy. *N Engl J Med* 1985;313:728-31

22 Brown P. Human growth hormone therapy and Creutzfeldt-Jakob disease: a drama in three acts. *Pediatrics* 1988;81:85-92

23 Kimberlin RH, Cole S, Walker CA. Temporary and permanent modifications to a single strain of mouse scrapie on transmission to rats and hamsters. *J Gen Virol* 1987;68:1875-81

24 Fraser H, McConnell I, Wells GA, Dawson M. Transmission of bovine spongiform encephalopathy to mice. *Vet Rec* 1988;123:472

25 Schrieber R, Seybold U. Gelatin production, the six steps to maximum safety. *Dev Biol Stand* 1993;80:195-8

26 Transcript of the 46th meeting of the Blood Products Advisory Committee, US Department of Health and Human Services, Food & Drug Administration, Rockville, MD, 15 December 1994

Transmissible Subacute Spongiform Encephalopathies:
Prion Diseases
L Court, B Dodet, eds
© 1996, Elsevier, Paris

Transmissible subacute spongiform encephalopathies: practical aspects of agent inactivation

David M Taylor

BBSRC and MRC Neuropathogenesis Unit, Institute for Animal Health, West Mains Road, Edinburgh EH9 3JF, UK

Summary – The unconventional agents that cause transmissible subacute spongiform encephalopathies are relatively resistant to decontamination by procedures which inactivate conventional microorganisms. In the studies described in this paper, it was shown that the agent of bovine spongiform encephalopathy (BSE) was inactivated by exposure to sodium hypochlorite, but not to sodium dichloroisocyanurate. Using BSE and scrapie agents, exposure to 1 M or 2 M sodium hydroxide for up to 2 h was shown to be an unreliable decontamination procedure, as was porous-load autoclaving at 134–138 °C for up to 1 h. The resistance of scrapie agent to porous-load autoclaving was enhanced by prior exposure to ethanol. Gravity-displacement autoclaving at 132 °C for 1 h was also ineffective. Exposure to gravity-displacement at 121 °C for 30 min was effective with scrapie agent if the test material was exposed to 2 M sodium hydroxide during autoclaving. The amount of scrapie infectivity in paraformaldehyde-lysine-periodate fixed tissue treated with concentrated formic acid for 1 h was about 2 logs higher than that reported for similarly treated scrapie-infected tissue which had been fixed in formol saline. After BSE-spiked raw materials were processed through facsimiles of European tendering processes, infectivity was recoverable in the meat and bone meal produced by four of these procedures.

Introduction

The unconventional agents that cause transmissible subacute spongiform encephalopathies (TSSE) are known to be relatively resistant to decontamination by procedures which inactivate conventional microorganisms [1, 2]. The unwitting use of inadequate decontamination procedures has resulted in iatrogenic transmission of Creutzfeldt-Jakob disease between humans via surgical equipment [3, 4]. The failure of formalin to inactivate scrapie agent during the preparation of a louping-ill vaccine resulted in the transmission of scrapie to many sheep which received the contaminated vaccine [5]. Bovine spongiform encephalopathy (BSE) is considered to have resulted from the transmission of scrapie to cattle through the use of scrapie-contaminated meat and bone meal (MBM) in cattle feed, scrapie agent having survived the heating procedures used during the manufacture of this product [6].

During the 1980s, there were increased efforts to establish satisfactory decontamination standards for TSSE agents, particularly with regard to the human agents. The resulting recommendations in the United Kingdom (UK) were to use sodium hypochlorite solutions containing 20,000 ppm available chlorine for 1 h [7] or porous-load autoclaving at 134–138 °C for 18 min [8]. In the USA the preferred procedures were exposure to 1 M sodium hydroxide for 1 h or gravity-displacement autoclaving at 132 °C for 1 h [9].

Key words: bovine spongiform encephalopathy agent / scrapie / agent inactivation

It was considered prudent to determine whether or not the BSE agent shared the property of inactivation resistance with other TSSE agents. Additional experiments were conducted with scrapie agent. The results available to date from these studies are presented in this publication.

Chemical procedures

Sodium hypochlorite and sodium dichloroisocyanurate

Two homogenates of BSE-infected cow brain containing $10^{2.6}$ or $10^{3.1}$ of BSE infectivity per mL were mixed with equal volumes of sodium hypochlorite or sodium dichloroisocyanurate solutions to give from 8,250–16,500 ppm available chlorine. Exposure times ranged from 30 up to 120 min. None of the mice injected with the hypochlorite-treated samples showed spongiform encephalopathy (SE). However, SE was detected in mice injected with BSE-infected brain treated with dichloroisocyanurate solutions containing up to 16,500 ppm available chlorine ; it was found that the rate of chlorine release from this compound was slower than for hypochlorite [2].

Sodium hydroxide

Homogenates of mouse, hamster or cow brain (10%) infected with scrapie or BSE agents were exposed to 1 M or 2 M sodium hydroxide for periods of up to 120 min. None of the exposures were effective with all of the agents tested. Although 5 logs of infectivity were lost, the surviving titre of the 263K strain of hamster-passaged scrapie agent was $10^{4.2}$ ID_{50}/g after a 120-min exposure to 2 M sodium hydroxide [2]. These findings are in contrast to an earlier report showing that 1 M sodium hydroxide was completely effective [10]. However, in that study it was found necessary to considerably dilute the samples to avoid toxicity for the recipient mice. In the presently described studies, the sensitivity of the bioassay was enhanced because it was possible to inject mice with undiluted samples following careful neutralization [2]. These present data add to previous concerns regarding the incomplete effectiveness of sodium hydroxide [11–13].

Autoclaving

Two sources of BSE-infected cow brain with infectivity titres of $10^{3.6}$ and $\geq 10^{5.2}$ ID_{50}/g were exposed to gravity-displacement autoclaving at 132 °C for 30 or 60 min, either as 10% saline homogenates or as 350-mg samples of undiluted, macerated brain. These experiments are still in progress, but have shown so far that neither the homogenate nor the macerate have been completely inactivated by the 30-min process. With the 60-min cycle, the homogenate appears to have been inactivated, but the macerate has not (Taylor et al, unpublished data).

Macerates of cow, hamster or mouse brain (341 mg average weight) infected with BSE or scrapie agents were exposed to porous-load (PL) autoclaving at temperatures between 134 and 138 °C for periods of up to 60 min [2]. None of the cycles inactivated all agents; the titre of the hamster-passaged scrapie agent was reduced from $10^{9.3}$ to 10^2 ID_{50}/g following PL autoclaving at 134 °C for 30 min. These data contrast with those from earlier studies in which 50-mg macerates of scrapie-infected mouse brain were inactivated by PL autoclaving at 136 °C for ≥ 4 min [7]. These differences are considered to be explicable by the amount of smearing and drying onto glass surfaces which would have occurred with the 341-mg samples. Enhanced resistance to autoclaving has been reported for scrapie agent as a consequence of infected brain tisssue being dried onto glass or metal surfaces [14]. It is impossible to assess whether the 341-mg tissue challenge was excessive because no advice has been available as to what the maximum weight of infected tissue is that might have to be dealt with by autoclaving in the course of human or veterinary healthcare.

It has been shown previously that scrapie-infected mouse brain which has been fixed in formol saline is resistant to inactivation by an autoclaving process which is effective with unfixed brain tissue [15]. The same phenomenon has now been found to occur when similar tissues are immersed in ethanol for 48 h prior to autoclaving (Taylor et al, unpublished data).

A previous report records the successful inactivation of CJD agent when gravity-displacement autoclaving at 121 °C for 30 min was preceded by a 1-h treatment with 1 M sodium hydroxide [16]. Using 2 M sodium hydroxide, it has been found that scrapie agent is inactivated by this autoclaving cycle without prior holding-period in the hydroxide (Taylor et al, unpublished data).

Precaution in the pathology laboratory

Because TSSE agents are not inactivated by the procedures used to produce sections of central nervous system tissue for investigation in the laboratory [17], there is the possibility of occupational exposure to infectivity during the processing of such tissues in pathology departments. A considerable improvement in this situation was achieved by the finding that treatment of formol-fixed tissues with concentrated formic acid reduced substantially the infectivity titres of CJD and scrapie agent [18]. However, in pilot studies on the effect of formic-acid treatment on mouse brain infected with the 301V strain of BSE agent fixed in paraformaldehyde-lysine-periodate, the observed degree of inactivation has been about 2 logs less than previously reported [19]. Further studies are in progress to determine whether this is related to the strain of agent or to the fixation process.

BSE-spiked rendering studies

In studies designed to mimic the rendering procedures used throughout the European Union (EU) to manufacture MBM (the source of BSE in the UK), tissues discarded by abattoirs were spiked with BSE-infected cattle brain. Although the infectivity titre of the spiked materials was only $10^{1.7}$ ID_{50}/g, infectivity was recoverable by mouse bioassay in the MBM produced by four of the 15 rendering procedures tested (20]). As a consequence, the minimum combinations of time and temperature for rendering within the EU have been revised [21].

Discussion

These studies have shown that BSE agent has a general resistance to inactivation, similar to that for scrapie agent. The previous recommendation to use sodium hypochlorite solution (containing 20,000 ppm available chlorine for 1 h) to inactivate TSSE agents was also found to be appropriate for BSE agent. Based on studies with scrapie agent, the only other apparently completely effective procedure was autoclaving in 2 M sodium hydroxide at 121 °C for 30 min. Treatment with sodium hydroxide alone produced a considerable loss in titre, but permitted survival of a significant amount of infectivity (about 4 logs).

In view of the remarkable resistance of TSSE agents to inactivation, every effort should be made to minimize contamination of both the equipment and the workplace. Use of disposables which can be incinerated is to be encouraged.

References

1 Taylor DM. Inactivation of unconventional agents of the transmissible degenerative encephalopathies. In: Russell AD, Hugo WB, Ayliffe GAJ, eds. *Principles and Practice of Disinfection, Preservation and Sterilization.* Oxford: Blackwell Scientific Publications, 1992;171-9

2 Taylor DM, Fraser H, McConnell I, Brown DA, Brown KA, Lamza KA, Smith GRA. Decontamination studies on the agents of bovine spongiform encephalopathy and scrapie. *Arch Virol* 1994;139:313-26

3 Bernoulli C, Siegfried J, Baumgartner G, Regli F, Rabinowicz T, Gajdusek DC, Gibbs CJ. Danger of person-to-person transmission of Creutzfeldt-Jakob disease by surgery. *Lancet* 1977;i:478-9

4 Foncin JF, Gaches J, Cathala F, El Sherif E, Le Beau J. Transmission iatrogene interhumaine possible de la maladie de Creutzfeldt-Jakob avec atteinte des grains du cervelet. *Rev Neurol* 1980;136:280

5 Gordon WS, Brownlee A, Wilson DR. Studies in louping-ill, tick-borne fever and scrapie. *Report of the Proceedings of the Third World Congress for Microbiology.* Baltimore: Waverley, 1940;362-3

6 Wilesmith JW, Wells GAH, Cranwell MP, Ryan JBM. Bovine spongiform encephalopathy: epidemiological studies. *Vet Rec* 1988;123:638-44

7 Kimberlin RH, Walker CA, Millson, GC, Taylor DM, Robertson PA, Tomlinson AH, Dickinson AG. Disinfection studies with two strains of mouse-passaged scrapie agent. *J Neurol Sci* 1983;59:355-69

8 Department of Health and Social Services. Management of patients with spongiform encephalopathy (Creutzfeldt-Jakob disease [CJD]). DHSS Circular DA 1984;84:16

9 Rosenberg RN, White CL, Brown P, Gajdusek DC, Volpe JJ, Dyck PJ. Precautions in handling tissues and other contaminated materials from patients with documented or suspected Creutzfeldt-Jakob disease. *Ann Neurol* 1986;19:75-7

10 Brown P, Rohwer RG, Gajdusek DC. Newer data on the inactivation of scrapie virus or Creutzfeldt-Jakob disease virus in brain tissue. *J Infect Dis* 1986;153:1145-8

11 Ernst DR, Race RE. Comparative analysis of scrapie agent inactivation. *J Virol Meth* 1993;41:193-202

12 Tamai Y, Taguchi F, Miura S. Inactivation of the Creutzfeldt-Jakob disease agent. *Ann Neurol* 1988;24:466-7

13 Tateishi J, Tashima T, Kitamoto T. Inactivation of the Creutzfeldt-Jakob disease agent. *Ann Neurol* 1988;24:466

14 Asher DC, Pomeroy KL, Murphy L, Rohwer RG, Gibbs, Gajdusek DC. Practical inactivation of scrapie agent on surfaces. *Abstracts of the IXth International Congress of Infectious and Parasitic Diseases* 1986, Munich

15 Taylor DM, McConnell I. Autoclaving does not decontaminate formol-fixed scrapie tissues. *Lancet* 1988;i:1463-4

16 Taguchi F, Tamai Y, Uchida K et al. Proposal for a procedure for complete inactivation of the Creutzfeldt-Jakob disease agent. *Arch Virol* 1991;119:297-301

17 Brown P, Rohwer RG, Green EM, Gajdusek DC. Effect of chemicals, heat and histopathological processing on high-infectivity hamster-adapted scrapie virus. *J Infect Dis* 1982;145:683-7

18 Brown P, Wolff A, Gajdusek DC. A simple and effective method for inactivating virus infectivity in formalin-fixed tissue samples from patients with Creutzfeldt-Jakob disease. *Neurology* 1990;40:887-90

19 Taylor DM. Survival of mouse-passaged bovine spongiform encephalopathy agent after exposure to paraformaldehyde-lysine-periodate and formic acid. *Vet Microbiol* 1995;44:111-2

20 Taylor DM, Woodgate SL, Atkinson MJ. Inactivation of the bovine spongiform encephalopathy agent by rendering procedures. *Vet Rec* 1995;137:605-10

21 Commission Decision. On the approval of alternative heat treatment systems for processing animal waste of ruminant origin, with a view to the inactivation of spongiform encephalopathy agents. Commission Decision 94/382/EC. *Off J Europ Commun* 1994;L172:25-7

Transmissible Subacute Spongiform Encephalopathies:
Prion Diseases
L Court, B Dodet, eds
© 1996, Elsevier, Paris

Inactivation of prions in daily medical practice

Jacqueline Paul†

Laboratoire Extérieur Inserm, Prévention des Risques Professionnels, ADR5 Rhône-Alpes, 162, avenue Lacassagne, 69394 Lyon cedex 03, France

When the French Health Ministry decided to establish recommendations to prevent patient-to-patient transmission of Creutzfeldt-Jakob disease (CJD) by medical treatment, little was known about the disease. It therefore undertook to summarise the available information on the disease, to group organs and tissues according to their infectivity during the clinical phase, and to evaluate the three procedures considered by the World Health Organization as most effective for inactivating CJD-contaminated tissues and other materials:

– autoclaving at 134–138 °C for 18 min for porous load and at 132 °C for 1 h for gravity displacement;

– treatment with sodium hydroxide (preferably at 1 M for 1 h at 20 °C);

– treatment with sodium hypochlorite (preferably as a solution containing at least 2% available chlorine, for 1 h at 20 °C).

There were some rules for general care of patients with CJD, but a lack of national research on inactivation procedures led us to adopt reliable methods described in the international literature, taking into account a number of controversies. The task was complicated by differences in the methods used: the autoclaves described in studies from the United Kingdom and the USA differ from those now in use; concentrations of sodium hypochlorite, or bleach or Chlorox, were expressed as dilutions, percentages or available chlorine. As some publications questioned the efficacy of porous load autoclaving for 18 min at 136 °C for inactivating CJD-contaminated material, we decided to increase the time to 30 min for security in clinical situations.

The first French recommendations, issued in July 1994, came as a bombshell to hospital staff and perplexed sterilisation teams for several reasons: (i) the recommended combination of time (30 min) and temperature (136 °C) for autoclaving differed from all other established standards for sterilising. (ii) Although sodium hypochlorite is currently used in French hospi-

†*In memoriam:* Jacqueline Paul died suddenly on 30 April 1996. She was a research director at the Institut National de la Santé et de la Recherche Médicale (Inserm). In 1991, she began work at the office for the prevention of occupational risks on an ambitious new mission: to identify the biological risks present in the Inserm laboratories, then to inform the personnel, and to recommend measures for prevention and protection. She was an authority in her field, and shortly before her death she was nominated as a member of the expert committee on spongiform encephalopathies under the aegis of the ministries of Agriculture, Research and Health.

Key words: Creutzfeld-Jakob disease / inactivation / precautions

tals (as *eau de Javel*), the minimal concentration recommended for prion destruction (dilution of 1:2, resulting in 6% chloride, equivalent to about 2% NaOCl and 1.8% available chlorine) is not compatible with most metals. As nurses had never used 1 N sodium hydroxide and clear information on its use was not available, several dramatic incidents occurred. One hospital department had to be evacuated by firemen because a large amount of smoke was generated after immersion of several boxes of instrument made of various metals, including aluminium, in a 40-L tub of sodium hydroxide. All of the boxes were damaged beyond repair by electrolysis. (iii) Some sophisticated, expensive instruments, such as ophthalmic instruments, fibroscopes and electrodes, are used to examine patients who have not yet received a diagnosis. As these instruments would be damaged by the recommended methods, they are often re-used. (iv) Medical and surgical specialties in which there is contact with the tissues that have the highest infectious titres (central nervous system, including optic tissues and cerebrospinal fluid) are relatively easy to define. Many others involve contact with tissues with lower infectivity but where the risk of transmission cannot be discounted; these include liver, lung, lymph nodes, kidney, placenta and leucocytes. In most maternity and gastroenterology wards, for example, systematic adherence to the recommendations appeared to be unfeasible. (v) Controversy still exists concerning the infectivity of urine, for example. This would have implications for the care of incontinent patients with CJD, either in hospital or at home.

In order to allay these concerns, revised recommendations (December 95) were drawn up for two groups of patients. The 'at risk' group comprises patients with documented or suspected CJD and those with personal risk factors such as treatment with pituitary growth hormone from cadavers, dura mater grafts and familial CJD or unexplained neurodegenerative disease. The second group comprises all other patients, who are considered 'at virtual risk'.

The new recommendations also define two categories of invasive treatment. 'At risk' treatments are those in which there is contact with the eye or central nervous system, and treatments 'at virtual risk' are those that involve contact with other organs, including those described above that present a medium level of infectivity at the clinical phase, even if no transmission has yet been demonstrated. The statistically estimated risk in such cases is too low to justify modification of internationally agreed safety measures.

The three basic chemical and physical methods (including 18-min autoclaving) are reiterated, but are used either alone or in combination, depending on the risks for the patient and the treatment. If the patient has CJD and the treatment is considered at risk, all contaminated material must be incinerated. For other combinations of patients and treatments, the material used must be resistant to both chemical immersion and steam autoclaving; otherwise, it must be destroyed by incineration. If there is only one risk factor, one of the methods, preferably autoclaving, should be used. For situations in which there is no obvious risk, the recommendations take into account doubts about the infectivity of some tissues and the impossibility of diagnosing CJD during the latent period. The usual validated methods for washing, decontamination (with products that do not contain aldehydes) and sterilisation must still be strictly adhered to. The recommendations specify that disposable materials should be used for people at risk when available, and that examinations and treatment involving expensive thermosensitive instruments be used only if they are the sole means for providing therapeutic benefit for the patient.

If a patient is sent to another department of a hospital, information must be provided in advance. All infectious specimens sent to a laboratory must be clearly labelled, and adequate information must be given about the disposal of contaminated wastes. Many hospitals in

France have adopted methods that provide an alternative to incineration, but none is effective for the decontamination of prions. Special measures must be taken to ensure incineration of prion-contaminated material and wastes. When a patient dies, cremation rather than burial is advised. Although cases of transmission of CJD from blood transfusions were reported by Klein and Dumble in Australia, experimental data show that the agent of CJD cannot be transmitted from humans to monkeys by intravenous transfusion (only leucocyte transfusion was effective in one case). For absolute security, however, patients with CJD are prohibited from donating not only organs but also blood. Furthermore, blood products from CJD patients taken before the clinical phase have been traced and recalled, and the interview preceding blood donation excludes people considered 'at risk'.

These recommendations may be overcautious. Newer procedures that are less aggressive for instruments have been described to avoid contamination with the agent of scrapie; however, the CJD prion is difficult to test because it is found so rarely, and its susceptibility may differ from that of scrapie.

It is to be hoped that despite the scarcity of CJD, more knowledge, the results of a large epidemiological study and acquired experience will lead to standardisation of inactivation methods and evaluation of their efficacy on the agent of the disease in humans, optimising the preventive measures best adapted to the real risk.

Transmissible Subacute Spongiform Encephalopathies:
Prion Diseases
L Court, B Dodet, eds
© 1996, Elsevier, Paris

Bovine spongiform encephalopathy and public health: some problems and solutions in assessing the risks

Richard H Kimberlin

Scrapie and Related Diseases Advisory Service (SARDAS), 27 Laverockdale Park, Edinburgh EH13 OQE, Scotland, UK

Summary – The advent of bovine spongiform encephalopathy (BSE) made it essential to develop an objective method to compare the relative risks from a large number of medicinal products that were being manufactured using bovine tissues. In its simplest form, this method depends on only five factors. One is a matter of fact: the weight of each bovine tissue that is represented in a maximum treatment regime with a given product. Another is the capacity of the manufacturing process to remove any contaminating BSE infectivity. Potential clearance factors can be deduced, conservatively, from the known physicochemical properties of the scrapie agent or, when necessary, measured by validation studies. However, the other three factors could only be derived by making assumptions based mainly on scrapie. Recent information on BSE is rapidly improving this situation. First, it was particularly difficult to estimate the prevalence of BSE infection in affected source countries until good evidence was found that it is directly proportional to the incidence of disease. Apart from the United Kingdom, in only three major source countries has the incidence of BSE been greater than one per million dairy cattle per annum. It is unlikely that this situation will change significantly in the future. Second, a much simpler classification of tissues (according to infectivity titre) is now possible compared to that originally based on scrapie. Bioassays in mice have not detected any infectivity in any tissue from confirmed cases of BSE except the central nervous system (CNS) and retina. Therefore, the risks from all non-CNS tissues can be based on the limits of detectability. Concerns about the limited sensitivity of bioassays in mice have been reduced by studies of transgenic mice that express the human *PRNP* gene which indicate that the cattle/human species barrier for BSE is probably greater than the cattle/mouse barrier. Third, there are now data to support the assumption that the relative efficiency of BSE infection of mice by the oral route is about 100,000-fold less than by the intracerebral route. However, as little difference was found between the intracerebral and intraperitoneal routes of injection of BSE, an efficiency of infection significantly lower than that by the intracerebral route can only be assumed for some of the other parenteral routes, for example, intramuscular and subcutaneous. Quantitative risk assessment is invaluable in identifying many medicinal products as safe, in guiding decisions to increase the safety of other products and in the design of process validation studies. However, there are situations where the estimated risks may be greatly exaggerated because of the need to make several 'worst case' assumptions, including the extent of the species barrier that limits the transmission of BSE from cattle to man.

Introduction

Bovine spongiform encephalopathy (BSE) was first recognized in November 1986 in the United Kingdom as a new, scrapie-like disease in cattle [1]. The implications of this discovery

---

Key words: bioassays of infectivity / bovine spongiform encephalopathy / classification of tissues / medicinal products / prevalence of infection / public health / risk assessment / routes of infection / species barriers

to medicinal products were not immediately obvious because the scale of the forthcoming BSE epidemic in the United Kingdom (UK) was not known, and most people were unaware that bovine tissues are used in the manufacture of many hundreds of different products, either as a source of devices and biologically active ingredients, or as excipients.

One approach to minimizing the risk of transmitting BSE via medicinal products was to avoid using bovine material from the UK, but, this did not address the potential risks from other countries which later reported cases of BSE, or from those in which BSE might be present, either undeclared or unrecognized [2]. Similarly, focusing mainly on potentially high risk tissues such as the brain ignored the fact that some brain-derived products would be of very low risk to man because of the ability of the manufacturing process to remove any contaminating BSE infectivity [3, 4], and because brain obtained from neonates or calves would be among the lowest risk of all tissues in terms of infectivity titre.

In practice, the risk from any given product depends on several interacting factors, all of which have to be considered. A simple, quantitative method was needed to compare the relative risks from a large number of medicinal products so that the decisions about these risks, and the measures to reduce them, could be made objectively and consistently, by the many different regulatory authorities [5].

A method was developed based on the assumption that BSE is potentially transmissible to humans if the effective exposure is sufficiently high [6]. Therefore, the relative risks from different products can be compared, directly, in terms of the effective exposure. For each product, the total exposure will be the sum of the exposures from each bovine tissue used in its manufacture. However, for the purposes of illustration, it is convenient to take the case where a single bovine tissue is the source of a given biological product.

In its simplest form, the method for calculating the effective exposure requires only five factors (table I). Factor A is the weight of each bovine tissue that is represented in a maximum treatment regime with a given product. Factor B depends on information about the epidemiology of BSE, specifically, the relationship between the prevalence of infection and the incidence of disease. Initially, factors C and D were based on appropriate extrapolations from the literature, particularly on scrapie, until experimental data were obtained for BSE. Factor E can be deduced, conservatively, from the physicochemical properties of the agents causing transmissible spongiform encephalopathy (TSE). For example, these agents can be removed, to varying degrees, by such commonly used procedures as centrifugation [7–9], filtration [10–14] and column chromatography [15–18]. When necessary, factor E can be measured directly by validation studies. A discussion of how to perform validation studies is outside the scope of

Table I. Factors affecting the exposure to BSE from a biological product made from one type of bovine tissue.

A. Weight (g) of source tissue represented in a maximum treatment regime.

B. Prevalence of BSE infection in the source country.

C. BSE infectivity (mouse intracerebral/intraperitoneal LD_{50}) per gramme of tissue.

D. Relative efficiency of BSE infection by the route of administration of the product compared to the combined intracerebral and intraperitoneal routes used in the mouse bioassays.

E. Scrapie/BSE removal factor during manufacture.

this paper, except to say that they are usually performed with well characterized models of mouse or hamster scrapie by adding known amounts of infectivity at different stages of an accurate facsimile of the manufacturing process, and then measuring the removal of infectivity [19, 20].

The product of factors A to D divided by factor E gives the exposure to BSE from treatment with a given product, from which can be derived an estimate of the risk of fatal disease. The purpose of this paper is two-fold: to describe in detail the derivation of each of the factors A to D, particularly in the light of recent information about BSE, and then to discuss the applications and limitations of the subsequent calculations of effective exposure and risk.

Factor A: the tissue equivalent of a maximum treatment

The weight of bovine tissue that is represented in a maximum treatment regime is calculated as the product of the total amount of source tissue used in one production batch and the proportion of a batch used in a maximum treatment. The latter might be a single dose of vaccine or a chronic treatment with a biological product requiring, for example, two doses per week for 1 year.

Studies of experimental scrapie in mice and hamsters suggest that infection occurs as an 'all-or-none' event, which is established in less than 30 min after intraperitoneal (ip) injection [21]. When two infecting events were induced 4 h apart, pathogenesis was determined by whichever produced the highest effective dose [21]. When the same ip dose, sufficient to cause infection, was repeated 13 times at 3 week intervals, pathogenesis was initiated by the first infecting dose [22]. It follows that the number of doses in a maximum treatment regime becomes irrelevant if a single dose is sufficient to establish infection. However, the calculation of exposure must consider the total number of doses because, if the risk of infection from one dose is only 1 in 100, then after 100 doses, the probability of infection approaches 1. The concept of 'multiple potential exposures' to low doses of BSE, when applied to a large cattle population, accurately describes the epidemic in the UK: over 160,000 cases of BSE had occurred by the end of May 1996 [23] but, characteristically, the average incidence within the affected herds was never greater than 3% of adults per year [24, 25].

A quite different scenario is theoretically possible, however, which also requires that the total number of doses should be used to determine exposure. Depending on the particle-to-infectivity ratio of the BSE agent (see Calculations of effective exposure and risk), a single exposure to a given low dose may never cause infection, but repeated 'non infectious' doses might conceivably accumulate, in vivo, over weeks or months, until an infectious dose is achieved.

Direct evidence against the 'accumulated dose' hypothesis comes from titration studies of a standard scrapie brain homogenate in mice inoculated by the intragastric route. Three parallel titrations were carried out [26]. In the first and second titrations, mice were given 0.2 and 1.0 mL of inoculum, respectively. A total of 1.0 mL of inoculum was also given in the third titration, but as five consecutive, daily doses of 0.2 mL. The scrapie titres per gram of brain were the same whether mice were given one 0.2-mL dose or five consecutive daily doses: there was no accumulation effect. However, the titre was about ten times higher when mice were given the equivalent of 5×0.2 mL of inoculum as a single 1.0-mL dose [26]. Similarly, there was no evidence for an accumulating dose, in terms of a reduced incubation period, when mice were injected intraperitoneally 13 times at 3-week intervals with an isolate of hamster scrapie [22]. Claims to the contrary, based on feeding similar low doses of 263K scrapie to ten hamsters every day for 30 consecutive days, are invalid because the correct control for this experiment, 300 hamsters fed once each, was not carried out [27]. Nevertheless, the calculation of factor A as the total number of doses allows for the theoretical possibility of an 'accumulated dose' in vivo, even though there is no evidence for this.

Factor B: prevalence of BSE infection in source countries

Relationship between infection and disease

In the absence of a sensitive laboratory test with which to identify BSE-infected animals, it was difficult to estimate the prevalence of infection in affected source countries until good evidence was found that it is directly proportional to the incidence of disease. This is almost certainly not true of natural scrapie because variations in the strain of agent and in the sheep *Sip/Prnp* gene [28] may result in many animals being genetically unsusceptible to the disease within their commercial life span, while being infected carriers and sources of infection to other sheep [29]. In contrast, the BSE agent seems to occur as a single strain [30] and there is almost no variation in the coding region of the bovine *Prnp* gene apart from the presence of either five or six copies of the octapeptide repeat sequence [31–33]. However, BSE is not associated with this polymorphism [34, 35], and there is no clear evidence from epidemiological studies of BSE to implicate genetic factors [24, 36].

Other evidence suggests that host genetic variation is not important for the occurrence of BSE [37]. First, there is a remarkably (compared to scrapie) consistent pattern of severity and distribution of vacuolar lesions in BSE which suggests a uniformity of pathogenesis in terms of route of infection, strain of agent and genotype of host [38, 39]. Second, 100% incidence and highly uniform incubation periods were seen in a group of eight Jersey and eight Holstein-Friesian cattle that had been injected with BSE [40]. Comparable experiments with scrapie in sheep gave extremely variable results [41].

These findings provide strong evidence that all breeds of cattle are probably equally susceptible to BSE; thus, all cattle would develop the clinical disease if they lived long enough after infection. Therefore, the incidence of disease among adults (of an age when clinical signs develop) reflects the prevalence of infection [25]. Calculation of the latter depends on the age at which cattle are infected, the incubation period of BSE and the proportion of animals that are culled at different ages. In the UK, a high proportion of BSE-affected cattle were infected in calfhood and the median incubation period is about 5 years [42, 43]. The estimated prevalence of BSE infection in a given birth cohort of cattle going to slaughter at more than 2 years of age would be up to three times the incidence of disease in that cohort [44]. A similar correction factor would apply to other countries with a comparable age structure of the cattle population and calfhood infection with BSE.

Incidence of BSE in affected countries

Ten other countries have reported BSE. Six have experienced cases only in cattle imported from the UK, namely, Canada, Denmark, the Falkland Islands (Islas Malvinas), Germany, Italy and the Sultanate of Oman. Of the other four countries, Ireland and Portugal have also had some cases in cattle imported from the UK, but they as well as France and Switzerland have had cases of BSE in native-born cattle. Although difficult to prove, the most likely cause of the majority of the latter cases is meat and bonemeal exported from the UK, as discussed elsewhere [25, 37, 45].

Table II shows the maximum annual incidence of BSE for each country in relation to the size of the dairy cattle population (because BSE is predominantly a disease of cattle in, or originating from, dairy herds [36, 42]). At the peak of the UK epidemic in 1992, just over 1% of dairy cattle developed BSE. In comparison, the risk of a medicinal product being sourced from a BSE-infected animal was 100 to 1,000 times lower if it came from Ireland, Portugal or Switzerland, and 10,000 times lower from all the other countries listed in table II. Therefore, in only three major source countries (other than the UK) is the incidence of BSE likely to be greater than about one per million dairy cattle per annum. In view of the continuing rapid decline in the UK epidemic [23], and the hypothesis that most cases of BSE in other countries were directly or indirectly associated with the UK, this situation is unlikely to change much in the future.

Risks of BSE in apparently unaffected countries

The situation was much less clear at the beginning of the BSE epidemic, however, because the absence of BSE infection or disease could not be assumed from the fact that a given country had not reported cases. This was a major problem that could only be addressed by risk assessment and surveillance.

Feed-borne infection of cattle with scrapie is the most likely hypothesis for the origin of BSE because sheep are the only known reservoir of TSE infection in animals in the UK [37, 46]. Studies in the UK indicate that BSE was the result of the simultaneous presence of four main factors: 1) a large sheep population, relative to that of cattle; 2) a sufficient prevalence of endemic scrapie; 3) the use of ovine-derived (and later bovine-derived) meat and bonemeal in cattle feeds; and 4) rendering conditions that allowed the survival of some infectivity, which depended on the amount of contamination [25, 47, 48].

Many countries have formally assessed the food-borne risks of infection of cattle with scrapie, based particularly on factors 1 and 3, which are the easiest to quantify [49]. The main conclusion is that no other country has the same combination of risk factors as existed in the UK. The unique occurrence of a major BSE epidemic in the UK strongly supports the scrapie origin of BSE [37, 46].

Table II. Maximum annual incidence of BSE reported to OIE (the World Organisation for Animal Health) (May 1996) by countries (excluding the Falkland Islands and Oman) in relation to the total dairy cattle population.

Country	Peak year	No of BSE cases	No of dairy cattle[a]	Cases/10[6] dairy cattle per annum
United Kingdom	1992	36,681	2,790,000	13,147
Switzerland	1995	68	780,000	87
Portugal	1995	14	404,000	35
Ireland	1994	19	1,387,000	14
Denmark	1992	1	747,000	1.3
France	1996	6	5,271,000	1.1
Italy	1994	2	2,881,000	0.7
Canada	1993	1	1,359,000	0.7
Germany	1994	3	6,061,000	0.5

[a]Data from *FAO/OIE/WHO Yearbook* for 1993.

Two of the most detailed risk assessments have been made outside Europe: in Argentina [50] and the United States [51, 52]. These countries are important sources of bovine material for medicinal products because of their very large cattle populations and efficient state veterinary services. Both risk assessments have been followed by targeted surveillance in which over 1,000 brains from Argentina [53] and from the United States [54, 55] were examined neuro-pathologically and found to be negative for BSE. A conservative estimate of the maximum prevalence of inapparent BSE in the United States was one in 1 million of all cattle [56]. This is no higher than the annual incidence in countries such as Canada, Denmark, Germany and Italy with only imported cases of BSE (table II).

The actual risk of BSE infection in Argentina is likely to be even lower because, in contrast to the United States, Argentina does not have endemic scrapie and does not feed cattle with meat and bonemeal (the vehicle of food-borne infection [36]) [53]. Exceptionally low risks of BSE infection can be assumed for some other South American countries that use extensive production systems based on grass, and for Australia which has a well established scrapie-free status.

There is good evidence that any natural transmission of TSE infection to cattle, other than via feeds, would not be sufficient to initiate or sustain an epidemic to a significant extent [24, 36, 37, 44]. Therefore, the simplest way to reduce the risks of BSE infection in countries with endemic scrapie is to ban the feeding of ruminant-derived meat and bonemeal to ruminants. Such a ban was introduced throughout the European Union in 1994 [57]. The ban also prevents the cattle-to-cattle recycling of any BSE infection that might already be present which, in Britain, was primarily responsible for dramatically increasing the epidemic between 1986 and 1992 [48]. At the same time, the ban guards against the theoretical possibility that the origin of BSE was a naturally transmitted TSE infection of cattle that rarely caused disease until cattle were exposed via feed which then permitted the rapid selection of a pathogenic BSE agent. This hypothesis has been discussed in detail elsewhere [41]. Although it does not explain the worldwide distribution of BSE, it is difficult to exclude totally and can only be preempted by a ruminant-to-ruminant feed ban [37, 46].

Factor C: BSE infectivity in different tissues

Data on scrapie

The characteristically long incubation periods of the TSEs are mainly due to the interactions between the genome of each strain of agent and the genotype of the host, particularly with regard to the *Prnp* gene. These interactions restrict multiplication of the agent to a limited number of nondividing cell populations in the nervous system and in certain tissues of the lymphoreticular system (LRS); they also limit the cell-to-cell spread of infection [29, 58]. The consequence is that infectivity titres vary enormously according to the tissue and the stage of incubation period.

For the purposes of risk assessment, it is convenient to classify tissues according to their maximum infectivity titre at the clinical stage of disease. Corrections can be made, when necessary, for the fact that titres may be substantially lower (particularly in the central nervous system, CNS) in preclinically infected animals [59, 60].

Before data on BSE became available, natural scrapie was the appropriate 'worst case' analogue of BSE for two reasons.

First, it was important to extrapolate from natural scrapie, which reproduces the time scale of BSE, and not exclusively from the short incubation period models of scrapie in rodents where parenteral infection can give exceptionally high infectivity titres, particularly in the brain [61]. The peak age-specific incidence of natural scrapie (3 to 4 years) is reasonably close to that of BSE (4 to 5 years) and, with both diseases, this peak reflects the incubation period [29, 36, 42, 43].

Second, scrapie is the appropriate 'worst case' analogue of BSE because infectivity titres outside the CNS are usually higher in scrapie than with the other diseases, such as transmissible mink encephalopathy (TME). It is likely that the presence of substantial amounts of agent in non-CNS tissues of preclinically infected sheep is one reason why scrapie occurs as a natural endemic infection [26]. This is not the case with human kuru and TME, both of which are 'dead end' diseases [62].

Bioassays of scrapie infectivity in sheep and goat tissues are usually performed in mice. The most appropriate data come from studies of natural scrapie in goats and in Suffolk sheep in which the species barrier is relatively small [59, 63]. Data from some other breeds of sheep with scrapie indicate very large species barrier effects which obscure potential differences in titre among tissues with no detectable infectivity [64].

A system was devised for classifying tissues into four categories based on the average infectivity titres in tissues from up to nine Suffolk sheep (34–57 months old) and up to three goats (38–49 months old), at the clinical stage of natural scrapie [37]. This classification has been widely used as the scientific basis of the specified bovine offals ban in the UK [65], the recommendations of the Office International des Épizooties (the World Organisation for Animal Health) concerning trade in live cattle and bovine products from countries with cases of BSE [66], European guidelines on the use of bovine tissues in the manufacture of medicinal products [5, 67] and advice from the World Health Organization on public health issues related to animal and human TSEs [68, 69].

New classification of bovine tissues

A much simpler classification of tissues is now possible compared to that based on scrapie. In general, bioassays are carried out using the intracerebral (ic) route of injection because studies of experimental scrapie in rodents show it to be more efficient than peripheral routes [58]. In addition, the incubation period is shorter, partly due to the effective dose being higher and also because ic injection bypasses the need for the LRS and the autonomic nervous system in the multiplication and spread of infection to the spinal cord and brain [58]. BSE infectivity titres have been measured by a combination of ip and ic routes of injection. This was because primary transmissions of some sources of sheep scrapie to mice were found to be just as efficient by the ip as by the ic route. A study of one of these transmissions showed that ic injection failed to establish infection of the brain directly, and disease was the result of some inoculum infecting the LRS and neuroinvasion occurring later, exactly as happened after ip injection [70]. Table III summarizes the published data on BSE infectivity titres in the brain from histologically confirmed cases and suggests that the same is true for the transmission of BSE to RIII mice. With sample 3, the titre in the brain was similar when measured intracerebrally or intraperitoneally. Moreover, a combination of ic and ip injections gave a slightly higher titre than either route alone.

The combination of ic and ip injections has been used to test more than 50 different tissues and body fluids from confirmed cases of natural BSE. Infectivity has been detected only in the CNS and in the retina which, anatomically, is an extension of the brain [23].

In order to quantify the risks from CNS tissues, an appropriate 'worst case' is to use the highest titre in the brain stem measured by the combination of routes, that is, about $10^{4.0}$ mouse ic/ip LD_{50} units per g (table III). The tissues in which BSE infectivity has not been detected are listed in table IV. Since all tissues were tested by injecting mice with 0.02 mL intracerebrally and 0.1 mL intraperitoneally of a 10% homogenate [74], the limit of detectability is $10^{1.4}$ per g of tissue. This value can be used as a 'worst case' in calculating the risks from any BSE-negative tissue that is used in the manufacture of medicinal products.

The difference between the BSE titre of the brain ($10^{4.0}$) and the limit of detectability for all the non-CNS tissues is $10^{2.6}$, or 400-fold. However, this difference is a minimum because bioassays across the cattle/mouse species barrier will tend to underestimate the titres in the brain and may overestimate the amount of undetected agent, because of a reduction in sensitivity. If bioassays were carried out in cattle, the BSE titre in the brain would probably be close to $10^{6.0}$ LD_{50} units per g. If the non-CNS tissues remained negative and the calculated limit of detectability was the same as in mice, then the difference in risk between the use of CNS and non-CNS tissues in medicinal products would be about 40,000-fold. Because larger volumes can be injected into cattle than mice, the limit of detectability could be reduced to $10^{0.5}$. If the non-CNS tissues were still negative, then the difference in risk, relative to the CNS, would increase to more than 300,000-fold. These studies are in progress (GAH Wells, personal communication).

The proposed new classification of bovine tissues ignores the finding of BSE infectivity in the ileum of cattle that were experimentally infected by the oral route [69, 75, 76]. This is excluded because it was necessary to ensure that all exposed cattle became infected by giving an experimental dose that was several orders of magnitude greater than the average natural exposure of cattle via contaminated feeds [75]. Such a large dose will have exaggerated the risks from bovine ileum by enabling a far greater uptake of infection (and subsequently, multiplication of agent) than occurs naturally. This is the 'Heineken effect' which has been found in several other studies: for example, the uptake of agent by spleen in mouse scrapie is much more efficient after intravenous (iv) injection [77] than after intragastric administration [26]. It is analogous to the problem, mentioned earlier, of estimating titres in the brain by extrapolating from short incubation models of experimental scrapie in rodents, which greatly exaggerates the risks [61].

Table III. Infectivity titres in the brains of confirmed cases of BSE measured by titration in RIII mice (*Sinc* s7s7) injected intracerebrally (ic), intraperitoneally (ip) or by both routes.

Brain sample [reference]	No of brains in the pool	ic		ip		ic + ip	
		Vol[a]	Titre[b]	Vol	Titre	Vol	Titre
1. Brain stem [71]	1	20	5.2				
2. Brain stem [72]	6					120	4.1
3. Whole brain [72]	1	20	2.9	100	3.3	120	3.6
4. Whole brain [73]	861					520	2.7

[a]Volume in μL; [b]titres in log LD_{50} units/g. Sample 1 was obtained from a clinically advanced case of BSE. Samples 3 and 4 had part of the hind brain removed for histology. Only 85% of the brains in sample 4 were from histologically confirmed cases of BSE.

Table IV. Tissues from confirmed cases of BSE in which no infectivity was detected by bioassay in mice injected both intracerebrally and intraperitoneally.

Nervous tissues	*Lymphoreticular tissues*
Cerebrospinal fluid	Spleen
Cauda equina	Tonsil
Peripheral nerves:	Lymph nodes:
– sciaticus	– prefemoral
– tibialis	– mesenteric
– splanchnic	– retropharyngeal
Alimentary tract	*Reproductive tissues*
Oesophagus	Testis
Reticulum	Prostate
Rumen (pillar)	Epididymis
Rumen (oesophageal groove)	Seminal vesicle
Omasum	Semen
Abomasum	Ovary
Proximal small intestine	Uterine caruncle
Distal small intestine	Placental cotyledon
Proximal colon	Placental fluids:
Distal colon	– amniotic fluid
Rectum	– allantoic fluid
	Udder
	Milk
Other tissues	
Blood:	
– buffy coat	Liver
– clotted	Lung
– foetal calf	Muscle:
– serum	– semitendinosus
Bone marrow	– diaphragma
Fat (midrum)	– longissimus
Heart	– masseter
Kidney	Pancreas
	Skin
	Trachea

Data from [23] and [74].

Factor D: efficiencies of different routes of BSE infection

A substantial body of data on the relative efficiencies of different routes of infection has been obtained from studies of the 139A strain of mouse-adapted scrapie agent in Compton white mice [58]. Order-of-magnitude comparisons can be made in terms of the number of ic LD_{50} units of infectivity required to give 1 LD_{50} unit by different routes of administration of medicinal products. The following are guideline values that have been used in past risk assessments [25]: iv, 10; ip, 100; subcutaneous (sc) and intramuscular (im), 10,000; and oral/ intragastric, 100,000.

However, the assumption had to be made that similar proportional differences between routes would apply when exposure occurred across a species barrier. Table III shows little or no difference between the ic and ip routes of injection of mice with BSE. Therefore, for the purposes of risk assessment, it would be prudent to treat the iv and ip routes as being equivalent to ic injection, and to regard the sc and im routes as being about 100 times less efficient.

With regard to the relative efficiency of oral transmission of BSE to mice, a previous estimate [37] was based on 1) the first published titre (by the ic route) of BSE in the brain stem (not whole brain) from a confirmed clinical case of BSE (sample 1 in table III), and 2) the very conservative assumption that a limiting infectious dose by the oral route was contained in 1 g of a pool of whole brains from four cases of BSE [78]. The ic/oral difference was about 200,000-fold [37].

A different calculation can now be made in light of more recent information. First, there seems to be more BSE infectivity in the brain stem than in whole brain (table III). This is not surprising given that the brain stem is where the occurrence of vacuolar lesions [38, 39] and the deposition of modified prion protein [79] are consistently greatest. Second, the exceptionally high BSE titre in sample 1 may have been because this material was obtained from an animal at a very advanced stage of BSE (table III). A third point is that the original estimate of titre in whole brain by the oral route was a 'worst case' because it ignored the fact that the mice ate an average of 9.1 g of whole brain over a period of 7 days [78]. Since 30% of these mice did not develop disease, the oral titre can be estimated from the total amount of brain consumed. This gives a value of about $10^{-0.8}$ oral LD_{50} units/g. If this titre is compared with the previously suggested 'worst case' titre for whole brain of $10^{4.0}$ ic/ip LD_{50}/g, the oral route is still about 60,000 times less efficient. A conservative value would be a difference of 10,000-fold.

Calculations of effective exposure and risk

To simplify the discussion, it is assumed that a given medicinal product is manufactured on a large scale for the treatment of many thousands of patients. This allows the effects of batch size to be ignored, even though there may be situations where, in theory, risks could be reduced by manufacturing many small batches instead of one large one.

From the factors shown in table I, the maximum exposure of a patient to a product can be calculated as $A \times B \times C \times D$ divided by E, and the result is expressed as mouse LD_{50} units per patient. A calculated exposure of, for example, 10 mouse LD_{50} units would mean that all patients were at risk of fatal disease (but see the assumptions given later). However, an average exposure of 10^{-3} would indicate a risk of one in 1,000 patients (the reciprocal of the exposure) divided by 2 (because infectivity is expressed as LD_{50} units), which is one in 2,000 patients. This is quite a low risk, but if 10,000 patients were exposed, five would be infected. However, it is extremely important to recognize that this calculation of absolute risk necessarily contains three 'worst case' assumptions.

The first assumption is that the particle to infectivity ratio is 1; in other words, 1 LD_{50} unit is physically indivisible and would always be associated with a 50% chance of causing infection. In the example above, a total of 10 mouse LD_{50} units of BSE agent contaminated a batch of product sufficient to treat 10,000 patients with one dose each. The result was that ten patients were exposed to 1 LD_{50}, and five were infected. However, if 1 mouse LD_{50} unit contained 100 particles that were uniformly distributed throughout the batch, then 1,000 patients would receive one particle which would not be infectious and there would be no risk. A particle to infectivity ratio of 1 is most unlikely whether the agent is an 'unconventional virus' [80], a 'prion' [81] or a 'virino' [82]. However, until the nature of the TSE agents is precisely defined, there is no alternative but to assume a ratio of 1.

The second assumption is that all patients infected with BSE will develop fatal disease. This is clearly a 'worst case' because no account is taken of the age of exposure to BSE or what the incubation period might be in patients of different *PRNP* genotypes, particularly with respect to the polymorphisms at codon 129 [83].

The third assumption arises from the fact that bioassays in mice tend to underestimate BSE infectivity titres because of the effects of the species barrier. However, any exposure of humans to BSE will also be across a species barrier and this will reduce the effective exposure. The assumption is made that the cattle/human species barrier is no higher than the barrier between cattle and mice. Almost certainly this assumption is too pessimistic. It has been shown that the expression of multiple copies of a human *PRNP* transgene in mice expressing the endogenous *Prnp* gene did not reduce the incubation period of ic-injected BSE [83]. Since the BSE agent seems to occur as a single strain [30], this finding implies that the cattle/human species barrier is greater than that between cattle and mice, but by how much is not known.

The extent of the species barrier

An illustration of the possible extent of the barrier to the transmission of BSE to man can be obtained by considering the situation with scrapie [25]. A substantial body of epidemiological evidence, collected worldwide, shows that the occurrence of sporadic Creutzfeldt-Jakob disease (CJD) in man is not associated with scrapie, sheep or the consumption of sheep products [84, 85]. This is not because man has not been exposed to scrapie: on the contrary, for decades, a very large number of people in many countries have been exposed to many strains of scrapie through eating infected sheep tissues, including the brain [86]. The absence of a link between scrapie and CJD must be due to the species barrier.

A calculation of how big the barrier might be can be made by considering the exposure received by anyone who, once in their life, consumes a cooked brain from an incipient scrapie case. This calculation is shown in table V. It is based on an estimate of the scrapie titre in the sheep brain if it were measured with no species barrier [25], but includes the other two 'worst case' assumptions described earlier. If the lifetime risk of sporadic CJD for people receiving a dose of 10,000 oral LD_{50} units of scrapie infectivity is the same as for everyone else in the world, namely, 1 in 10,000 [87], then the species barrier must reduce the effective exposure by about 100 million-fold.

This figure is not without precedent. A comparable species barrier effectively prevents the transmission of the 263K strain of scrapie from Syrian golden hamsters to mice [88, 89]. This is calculated as follows: the average titre in clinically affected hamster brain is $10^{9.8}$ ic LD_{50}/g [61]. The failure to transmit disease by injecting 0.03 mL of a 5% homogenate of affected brain [88, 89] is equivalent to a limit of detectability of $10^{2.3}$ ic LD_{50}/g. The difference between these two values indicates a species barrier of at least 30 million-fold.

Unfortunately, it cannot be assumed that the exposure of man to BSE would be across a species barrier of the same magnitude as for sheep scrapie: it might be greater or smaller. The reason is that the species barrier depends on the interaction of two independent factors, both of which are likely to be different for BSE.

The first factor is the 'donor species effect' [90] which is related to the degree of *Pnrp* gene sequence homology of the species infected and the species exposed [91]. The *Prnp* gene sequences of sheep and cattle (the species infected) are similar but not identical [92], which implies a different donor species effect. However, it should be noted that both have *Prnp* gene sequences which are substantially different from that in man [92].

The second factor is the strain of agent; for example, the apparent nontransmissibility of 263K hamster scrapie to mice is in contrast to the comparative ease of transmission of Me7-

Table V. An approximate estimate of the extent of the sheep/human species barrier for scrapie based on the consumption of one brain from a clinical or incipient case of natural scrapie.

Dose: weight of tissue (g) × infectivity titre per g

10^2	×	10^7	=	10^9 ic LD_{50} units
weight		titre		dose

Route of exposure: oral route is 100,000-fold less efficient than ic injection

10^9	×	10^{-5}	=	10^4 oral LD_{50} units
dose		route		exposure

Extent of the species barrier: 100 million for the lifetime risk of CJD to be 1 in 10,000

10^4	×	10^{-8}	=	10^{-4}
exposure		1/species		risk of CJD
		barrier		

ic: intracerebral; CJD: Creutzfeldt-Jakob disease.

H hamster scrapie to mice of the same *Prnp* genotype [89]. The BSE agent seems to exist as a single strain [30] and, although it was probably derived from a common strain of sheep scrapie [41, 46], it is distinct from all known strains of scrapie agent [30].

A comparison was made of the transmission of BSE and of a single isolate of sheep scrapie to marmosets (a species of primate). Two animals were injected intracerebrally with each inoculum and all four developed disease. The fact that the incubation periods were broadly similar was taken as evidence of a similarity of the species barriers for the transmission of BSE and scrapie to primates, including man [93]. However, this study is difficult to interpret because 1) the titres of the agent in the two inocula are not known and a different 'donor species effect' would be expected (both of which affect the incubation period); 2) a different strain of scrapie may have given a different result; and 3) there are several differences between the prion protein gene sequence of marmosets and man [92].

Given that the *Prnp* gene seems to be the major factor in the 'donor species effect', a better approach might be to study the transmission of BSE to transgenic mice in which the endogenous *Prnp* gene has been replaced by the human gene. Ideally, the studies should include different polymorphisms at codon 129 (which affect the incubation period and/or susceptibility to disease). Comparisons of the effective titre of a standard BSE inoculum in these 'human' mice, and in the mice used for bioassays of BSE infectivity, would give a measure of how much greater the cattle/human species barrier is than the barrier between cattle and mice. These studies are in progress [83].

Practical applications and limitations of risk assessment

Quantitative risk assessment provides a simple way of comparing the relative risks from a very wide variety of different types of medicinal products. In this situation, the assumptions mentioned earlier about the particle-to-infectivity ratio, the relationship between infection and disease in man and the extent of the species barrier, can be ignored because they are the same for all products.

However, it is important to be able to assess absolute risks, and, because of the 'worst case' nature of the three assumptions given earlier, the method described is pessimistic. This has the advantage of giving considerable confidence about the safety of a product if the calculated risks are satisfactorily low. But when an estimate of absolute risk is higher, judgements have to be made about the efficacy of the product (ie, the risks to patients of not using it), and the possibilities of using nonbovine alternatives.

If the risk assessment is highly dependent on any one factor, for example, the source country (table I), then the effects of obtaining bovine material from a lower risk country (table II) can be calculated very quickly. The method is of great value in designing cost-effective validation studies for estimating factor E. If the risk from a given product is calculated only from factors A to D, a minimum removal of BSE during the manufacturing process can be determined that will achieve an acceptable level of risk. The validation study can then be designed so that it is capable of demonstrating the predetermined removal factor with a sufficient margin of safety.

The major limitation of absolute risk assessment becomes apparent when it is dominated by one invariant factor. A classic example is the use of a given tissue in food products in quantities far higher than its equivalent use in a chronic treatment with a medicinal product developed from it. In this situation, a combination of 1) the three inherent 'worst case' assumptions described earlier with 2) the requirement to estimate the BSE infectivity in a non-CNS tissue from the limit of detectability of the bioassay, can bias the calculation of risk by several orders of magnitude. There is a danger that decisions based solely on quantitative risk assessment could have medical, social and economic consequences which are unjustifiable. Therefore, the interpretation of some quantitative risk assessments have to be tempered by other information about the TSE, particularly from studies of the pathogenesis of disease in experimental animal models.

References

1 Wells GAH, Scott AC, Johnson CT et al. A novel progressive spongiform encephalopathy in cattle. *Vet Rec* 1987;121:419-20
2 Schreuder BEC, Straub OC. BSE: a European problem. *Vet Rec* 1996;138:575
3 Di Martino A, Safar J, Ceroni M, Gibbs CJ Jr. Purification of non-infectious ganglioside preparations from scrapie-infected brain tissue. *Arch Virol* 1992;124:111-21
4 Di Martino A, Safar J, Gibbs CJ Jr. The consistent use of organic solvents for purification of phospholipids from brain tissue effectively removes scrapie infectivity. *Biologicals* 1994;22:221-5
5 German Federal Ministry of Health. *Guidelines on Safety Measures in Connection with Medicinal Products Containing Body Materials Obtained from Cattle, Sheep or Goats for Minimizing the Risk of Transmission of BSE and scrapie.* Berlin: Federal Bulletin, 1994;40
6 Kimberlin RH. An overview of bovine spongiform encephalopathy. In: Horaud F, Brown F, eds. *Developments in Biological Standardization.* Basel: Karger, 1991;75:75-82
7 Millson GC, Hunter GD, Kimberlin RH. The physicochemical nature of the scrapie agent. In: Kimberlin RH, ed. *Slow Virus Diseases of Animals and Man.* Amsterdam: North-Holland, 1976;243-66
8 Prusiner SB, Hadlow WJ, Eklund CM, Race RE, Cochran SP. Sedimentation characteristics of the scrapie agent from murine spleen and brain. *Biochemistry* 1978;17:4987-92
9 Prusiner SB, Bolton DC, Groth DF, Bowman KA, Cochran SP, McKinley MP. Further purification and characterization of scrapie prions. *Biochemistry* 1982;21:6942-50
10 Gibbs CJ Jr. Search for infectious etiology in chronic and subacute degenerative diseases of the central nervous system. *Curr Topics Microbiol Immunol* 1967;40:44-58
11 Marsh RF, Hanson RP. Physical and chemical properties of the transmissible mink encephalopathy agent. *J Virol* 1969;3:176-80
12 Kimberlin RH, Millson GC, Hunter GD. An experimental examination of the scrapie agent in cell membrane mixtures. III. Studies of the operational size of the scrapie agent. *J Comp Pathol* 1971;81:383-91
13 Pocchiari M, Macchi G, Peano S, Conz A. Can potential hazard of Creutzfeldt-Jakob disease infectivity be reduced in the production of human growth hormone? Inactivation experiments with the 263K strain of scrapie. *Arch Virol* 1988;98:131-5
14 Pocchiari M, Peano S, Conz A et al. Combination ultrafiltration and 6M urea treatment of human growth hormone effectively minimizes risk from potential Creutzfeldt-Jakob disease virus contamination. *Horm Res* 1991;35:161-6

15 Hunter GD, Millson GC. Studies on the heat stability and chromatographic behaviour of the scrapie agent. *J Gen Microbiol* 1964;37:251-8

16 Hunter GD. Progress toward the isolation and characterisation of the scrapie agent. In: Gajdusek DC, Gibbs CJ Jr, Alpers M, eds. *Slow, Latent and Temperate Virus Infections.* Washington, DC: US Government Printing Office, NINDB Monograph no 2, 1965;259-62

17 Diringer H, Kimberlin RH. Infectious scrapie agent is not as small as recent claims suggest. *Biosci Rep* 1983;3:563-8

18 Taylor DM, Dickinson AG, Fraser H, Robertson PA, Salacinski PR, Lowry PJ. Preparation of growth hormone free from contamination with unconventional slow viruses. *Lancet* 1985;ii:260-2

19 European Commission, Committee for Proprietary Medicinal Products. *Notes for Guidance on Virus Validation Studies: The Design, Contribution and Interpretation of Studies Validating the Inactivation and Removal of Viruses.* Brussels: Document no CPMP/BWP/268, 1995

20 Pocchiari M. Problems in the evaluation of theoretical risks for humans to become infected with BSE-contaminated bovine derived pharmaceutical products. In: Gibbs CJ Jr, ed. *Bovine Spongiform Encephalopathy: The BSE Dilemma.* New York: Springer-Verlag, 1996;375-83

21 Kimberlin RH, Walker CA. Intraperitoneal infection with scrapie is established within minutes of injection and is non-specifically enhanced by a variety of different drugs. *Arch Virol* 1990;112:103-14

22 Kimberlin RH. Early events in the pathogenesis of scrapie in mice. In: Prusiner SB, Hadlow WJ, eds. *Slow Transmissible Diseases of the Nervous System.* New York: Academic Press, 1979;2:33-54

23 *Bovine Spongiform Encephalopathy in Great Britain: A Progress Report.* London: United Kingdom Ministry of Agriculture, Fisheries and Food, May 1996

24 Wilesmith JW. Bovine spongiform encephalopathy: epidemiological factors associated with the emergence of an important new animal pathogen in Great Britain. *Semin Virol* 1994;5:179-87

25 Kimberlin RH, Wilesmith JW. Bovine spongiform encephalopathy (BSE): epidemiology, low dose exposure and risks. In: Bjornsson J, Carp RI, Love A, Wisniewski HW, eds. *Slow Infections of the Central Nervous System: the Legacy of Dr Bjorn Sigurdsson. Ann NY Acad Sci* 1994;724:210-20

26 Kimberlin RH, Walker CA. Pathogenesis of scrapie in mice after intragastric infection. *Virus Res* 1989;12: 213-20

27 Diringer H, Beekes M, Oberdieck U. The nature of the scrapie agent: the virus theory. In: Bjornsson J, Carp RI, Love A, Wisniewski HW, eds. *Slow Infections of the Central Nervous System: the Legacy of Dr Bjorn Sigurdsson. Ann NY Acad Sci* 1994;724:246-58

28 Goldmann G, Hunter N, Smith G, Foster J, Hope J. *PrP* genotype and agent effects in scrapie: change in allelic interaction with different isolates of agent in sheep, a natural host of scrapie. *J Gen Virol* 1994;75:989-95

29 Kimberlin RH. Unconventional "slow" viruses. In: Collier LH, Timbury MC, eds. *Topley and Wilson's Principles of Bacteriology, Virology and Immunity.* London: Edward Arnold, 8th Edn, 1990;4:671-93

30 Bruce ME, Chree A, McConnell I, Foster J, Pearson G, Fraser H. Transmission of bovine spongiform encephalopathy and scrapie to mice: strain variation and the species barrier. *Philos Trans R Soc Lond [Biol]* 1994;343: 405-11

31 Goldmann W, Hunter N, Martin T, Dawson M, Hope J. Different forms of the bovine *PrP* gene have five or six copies of a short, G-C-rich element within the protein-coding exon. *J Gen Virol* 1991;72:201-4

32 McKenzie DI, Cowan CM, Marsh RF, Aiken JM. *PrP* gene variability in the US cattle population. *Anim Biotechnol* 1992;3:309-15

33 Ryan AM, Womack JE. Somatic cell mapping of the bovine prion protein gene and restriction fragment length polymorphism studies in cattle and sheep. *Anim Genetics* 1993;24:23-6

34 Hunter N, Goldmann W, Smith G, Hope J. Frequencies of *PrP* gene variants in healthy cattle and cattle with BSE in Scotland. *Vet Rec* 1994;135:400-3

35 Neibergs HL, Ryan AM, Womack JE, Spooner RL, Williams JL. Polymorphism analysis of the prion gene in BSE-affected and unaffected cattle. *Anim Genetics* 1994;25:313-7

36 Wilesmith JW, Wells GAH, Cranwell MP, Ryan JMB. Bovine spongiform encephalopathy: epidemiological studies. *Vet Rec* 1988;123:638-44

37 Kimberlin RH. A scientific evaluation of research into BSE. In: Bradley R, Marchant B, eds. *Transmissible Spongiform Encephalopathies.* Brussels: European Commission Document VI/4131/94-EN, 1994;455-77

38 Wells GAH, Hawkins SAC, Hadlow WJ, Spencer YI. The discovery of bovine spongiform encephalopathy and observations on the vacuolar changes. In: Prusiner SB, Collinge J, Powell J, Anderton B, eds. *Prion Diseases of Humans and Animals.* New York: Ellis Harwood, 1992;256 74

39 Simmons MM, Harris P, Jeffrey M, Meek SC, Blamire IWH, Wells GAH. BSE in Great Britain: consistency of the neurohistopathological findings in two random annual samples of clinically suspect cases. *Vet Rec* 1996;138: 175-7

40 Dawson M, Wells GAH, Parker BNJ et al. Transmission studies of BSE in cattle, pigs and domestic fowl. In: Bradley R, Marchant B, eds. *Transmissible Spongiform Encephalopathies*. Brussels: European Commission Document VI/4131/94-EN, 1994;161-7

41 Kimberlin RH. Bovine spongiform encephalopathy: an appraisal of the current epidemic in the UK. *Intervirology* 1993;35:208-18

42 Wilesmith JW, Ryan JBM, Hueston WD, Hoinville LJ. Bovine spongiform encephalopathy: epidemiological features 1985–1990. *Vet Rec* 1992;130:90-4

43 Wilesmith JW, Ryan JBM, Hueston WD. Bovine spongiform encephalopathy: case-control studies of calf feeding practices and meat and bone meal inclusion in proprietary concentrates. *Res Vet Sci* 1992;52:325-31

44 Hoinville LJ, Wilesmith JW, Richards MS. An investigation of risk factors for cases of bovine spongiform encephalopathy born after the introduction of the "feed ban". *Vet Rec* 1995;136:312-8

45 Hörnlimann B, Guidon D, Griot C. Risk assessment on the importation of BSE. *Dtsch Tierärztli Wochenschrift* 1994;101:295-8

46 Kimberlin RH. Speculations on the origin of BSE and the epidemiology of CJD. In: Gibbs CJ Jr. *Bovine Spongiform Encephalopathy: The BSE Dilemma*. New York: Springer-Verlag, 1996;155-75

47 Wilesmith JW, Ryan JBM, Atkinson MJ. Bovine spongiform encephalopathy: epidemiological studies on the origin. *Vet Rec* 1991;128:199-203

48 Wilesmith JW. The epidemiology of bovine spongiform encephalopathy. *Semin Virol* 1991;2:239-45

49 Bradley R, Marchant B, eds. *Transmissible Spongiform Encephalopathies*. Brussels: European Commission Document VI/4131/94-EN, 1994

50 Cané B, Gimeno E, Manetti JC, van Gelderen C, Ulloa E, Schudel A. *Analysis of BSE Risk Factors in Argentina*. Buenos Aires: Secretaria de Agricultura Ganaderia y Pesca de la Nacion, 1991

51 Walker KD, Hueston WD, Hurd HS, Wilesmith JW. Comparison of bovine spongiform encephalopathy risk factors in the United States and Great Britain. *J Am Vet Med Assoc* 1991;199:1554-61

52 Hueston WD, Bleem AM, Walker KD. Bovine spongiform encephalopathy (BSE): risk assessment and surveillance in the Americas. *Animal Health Insight*, Fall 1992. Fort Collins, CO, USA: USDA:APHIS: Veterinary Services, 1992;1-7

53 Schudel AA, Carrillo BJ, Weber EL et al. Risk assessment and surveillance for bovine spongiform encephalopathy (BSE) in Argentina. *Prev Vet Med* 1996;25:271-84

54 Francy B, Hueston W, Davis A, Jenny A, Miller L. Retrospective surveillance for bovine spongiform encephalopathy (BSE) in the United States. *Animal Health Insight*, Winter 1991. Fort Collins, CO, USA: USDA:APHIS: Veterinary Services, 1991;11-6

55 Bleem AM, Crom RL, Francy DB, Hueston WD, Kopral C, Walker K. Risk factors and surveillance for bovine spongiform encephalopathy in the United States. *J Am Vet Med Assoc* 1994;204:644-51

56 *Bovine Spongiform Encephalopathy (BSE): Implications for the United States*. Fort Collins, CO, USA: USDA: APHIS:Veterinary Services, July 1993;1-25

57 European Commission Decision 94/381 of 27 June 1994

58 Kimberlin RH, Walker CA. Pathogenesis of experimental scrapie. In: Bock G, Marsh J, eds. *Novel Infectious Agents and the Central Nervous System*. Ciba Foundation Symposium no 135. Chichester: John Wiley & Sons, 1988;37-62

59 Hadlow WJ, Kennedy RC, Race RE. Natural infection of Suffolk sheep with scrapie virus. *J Infect Dis* 1982;146: 657-64

60 Kimberlin RH, Walker CA. Incubation periods in six models of intraperitoneally injected scrapie depend mainly on the dynamics of agent replication within the nervous system and not the lymphoreticular system. *J Gen Virol* 1988;69:2953-60

61 Kimberlin RH, Walker CA. Transport, targeting and replication of scrapie in the CNS. In: Court L, Dormont D, Brown P, Kingsbury D, eds. *Unconventional Virus Diseases of the Central Nervous System*. Fontenay-aux-Roses: Commissariat à l'Energie Atomique, 1989;547-62

62 Hadlow WJ, Race RE, Kennedy RC. Temporal distribution of transmissible mink encephalopathy virus in mink inoculated subcutaneously. *J Virol* 1987;61:3235-40

63 Hadlow WJ, Kennedy RC, Race RE, Eklund CM. Virologic and neurohistologic findings in dairy goats affected with natural scrapie. *Vet Pathol* 1980;17:187-99

64 Hadlow WJ, Race RE, Kennedy RC, Eklund CM. Natural infection of sheep with scrapie virus. In: Prusiner SB, Hadlow WJ, eds. *Slow Transmissible Diseases of the Nervous System*. New York: Academic Press, 1979;2:3-12

65 The Bovine Offal (Prohibition) Regulations 1989. Statutory Instrument 1989 no 2061. London: Her Majesty's Stationery Office

66 Office International des Épizooties, Animal Health Code Commission. *Bovine Spongiform Encephalopathy (BSE)*. Chapter 3.2.13. Paris: May 1996

67 European Commission, Committee for Proprietary Medicinal Products. *Notes for Guidance for Minimizing the Risk of Transmitting Agents Causing Spongiform Encephalopathy via Medicinal Products.* Brussels: Document no 111 3298/91, 1991

68 Report of a WHO consultation on public health issues related to animal and human spongiform encephalopathies. *WHO Bull* 1992;70:183-90

69 Report of a WHO Consultation on Public Health Tissues Related to Human and Animal Transmissible Spongiform Encephalopathies. Geneva, 17–19 May 1995. WHO/CDS/VPH/95.145

70 Kimberlin RH. Modified pathogenesis of scrapie at first intracerebral passage from sheep to mice. In: *Abstracts of the IXth International Congress of Virology.* Glasgow, 1993; P59-1:318

71 Fraser H, Bruce ME, Chree A, McConnell I, Wells GAH. Transmission of bovine spongiform encephalopathy and scrapie to mice. *J Gen Virol* 1992;73:1891-7

72 Taylor DM, Fraser H, McConnell I et al. Decontamination studies with the agents of bovine spongiform encephalopathy and scrapie. *Arch Virol* 1994;139:313-26

73 Taylor DM, Woodgate SL, Atkinson MJ. Inactivation of the bovine spongiform encephalopathy agent by rendering procedures. *Vet Rec* 1995;137:605-10

74 Fraser H, Foster JD. Transmission to mice, sheep and goats and bioassay of bovine tissues. In: Bradley R, Marchant B, eds. *Transmissible Spongiform Encephalopathies.* Brussels: European Commission Document VI/4131/94-EN, 1994;145-59

75 Wells GAH, Dawson M, Hawkins SAC et al. Infectivity in the ileum of cattle challenged orally with bovine spongiform encephalopathy. *Vet Rec* 1994;135:40-1

76 Wells GAH, Dawson M, Hawkins SAC et al. Preliminary observations on the pathogenesis of experimental bovine spongiform encephalopathy. In: Gibbs CJ Jr. *Bovine Spongiform Encephalopathy: The BSE Dilemma.* New York: Springer-Verlag, 1996;28-44

77 Kimberlin RH, Walker CA. The role of the spleen in the neuroinvasion of scrapie in mice. *Virus Res* 1989;12:201-12

78 Middleton DJ, Barlow RM. Failure to transmit bovine spongiform encephalopathy to mice by feeding them with extraneural tissues of affected cattle. *Vet Rec* 1993;132:545-7

79 Haritani M, Spencer YI, Wells GAH. Hydrated autoclave pretreatment enhancement of prion protein immunoreactivity in formalin-fixed bovine spongiform encephalopathy affected brain. *Acta Neuropathol (Berl)* 1994;87:86-90

80 Rohwer RG. The scrapie agent: "A virus by any other name". In: Chesebro BW, ed. *Transmissible Spongiform Encephalopathies. Curr Top Microbiol Immunol* 1991;172:195-232

81 Prusiner SB. Novel properties and biology of scrapie prions. In: Chesebro BW, ed. *Transmissible Spongiform Encephalopathies. Curr Topic Microbiol Immunol* 1991;172:233-57

82 Kimberlin RH. Scrapie and possible relationships with viroids. *Semin Virol* 1990;1:153-62

83 Collinge J, Palmer MS, Sidle KCL et al. Unaltered susceptibility to BSE in transgenic mice expressing the human prion protein gene. *Nature* 1995;378:779-83

84 Brown P, Cathala F, Raubertas RF, Gajdusek DC, Castaigne P. The epidemiology of Creutzfeldt-Jakob disease: conclusion of a 15-year investigation in France and review of the world literature. *Neurology* 1987;37:895-904

85 Will RG, Matthews WB, Smith PG, Hudson C. A retrospective study of Creutzfeldt-Jakob disease in England and Wales 1970–1979. II. Epidemiology. *J Neurol Neurosurg Psychiatry* 1986;49:749-55

86 Sigurdarson S. Epidemiology of scrapie in Iceland and experience with control measures. In: Bradley R, Savey M, Marchant BA, eds. *Sub-acute Spongiform Encephalopathies.* Dordrecht, the Netherlands: Kluwer Academic Publishers, 1991;233-42

87 Brown P, Gajdusek DC, Gibbs CJ Jr, Asher DM. Potential epidemic of Creutzfeldt-Jakob disease from human growth hormone therapy. *New Engl J Med* 1985;313:728-31

88 Kimberlin RH, Walker CA. Evidence that the transmission of one source of scrapie agent to hamsters involves separation of agent strains from a mixture. *J Gen Virol* 1978;39:487-96

89 Kimberlin RH, Walker CA, Fraser H. The genomic identity of different strains of mouse scrapie is expressed in hamsters and preserved on reisolation in mice. *J Gen Virol* 1989;70:2017-25

90 Kimberlin RH, Cole S, Walker CA. Temporary and permanent modifications to a single strain of mouse scrapie on transmission to rats and hamsters. *J Gen Virol* 1987;68:1875-81

91 Scott M, Foster D, Mirenda C et al. Transgenic mice expressing hamster prion protein produce species-specific scrapie infectivity and amyloid plaques. *Cell* 1989;59:847-57

92 Schätzl HM, Da Costa M, Taylor L, Cohen FE, Prusiner SB. Prion protein gene variation among primates. *J Mol Biol* 1995;245:362-74

93 Baker HF, Ridley RM, Wells GAH. Experimental transmission of BSE and scrapie to the common marmoset. *Vet Rec* 1993;132:403-6

Transmissible Subacute Spongiform Encephalopathies:
Prion Diseases
L Court, B Dodet, eds
© 1996, Elsevier, Paris

ABBREVIATIONS

ABC: avidin-biotin complex
APP: amyloid precursor protein
BSE: bovine spongiform encephalopathy
CD: circular dichroism spectroscopy
CHO: Chinese hamster ovary
CJD: Creutzfeldt-Jakob disease
CMC: critical micelle concentration
CNS: central nervous system
CSF: cerebrospinal fluid
CT: computerized tomography
DAB: diaminobenzidine
dLGN: dorsal lateral geniculate nucleus
DMSO: dimethyl sulfoxide
DLPC: detergent-lipid-protein complex(es)
dpi: days post-infection
EEG (noun): electroencephalogram
EEG (adj): electroencephalographic
EM: electron microscopy
FDC: follicular dendritic cells
FFI: fatal familial insomnia
FTIR: Fourier transform infrared spectroscopy
GABA: gamma-aminobutyric acid
Gdn HCl: guanidine hydrochloride
GFAP: glial fibrillary acidic protein
GPI: glycosyl-phosphatidylinositol
GSS: Gerstmann-Sträussler-Scheinker disease
HFIP: hexafluoroisopropanol
HIV: human immunodeficiency virus
hGH: human pituitary-derived growth hormone or human growth hormone
HPLC: high performance liquid chromatography
Hsp: heat shock protein
ID_{50} : 50% infectious dose
ISEL: in situ-end-label(l)ing
KLH: key hole limpet hemocyanin
LD_{50}: 50% lethal dose
LLME: L-leucine-methyl ester
LOD: log of the odds
LRS: lymphoreticular system
LTP: long-term potentiation
NCR: nitrogen catabolic repression
mAb: monoclonal antibody
MBM: meat and bone meal
Mo: mouse
Mo*Prnp* gene: mouse prion protein gene
MRI: magnetic resonance imaging
MTT: 3[4,5-dimethylthiazolide-2yl]-2,5 diphenyltetrazolium bromide
NSE: non specific enolase

NMR: nuclear magnetic resonance
ORF: open reading frame
PAP: peroxydase-anti peroxydase
PAS: periodic acid-Schiff
PBS: phosphate-buffered saline
PIPLC: phosphatidylinositol-specific phospholipase C
PK: proteinase K
PCR: polymerase chain reaction
PCV: porcine circovirus
PLP: paraformaldehyde protein
PLP: periodate-lysine paraformaldehyde
PrP: prion protein
PrP 27–30, PrP-res: protease-resistant prion protein
PrP*: unfolded intermediate prion protein
PrP^C, PrP-sen: protease-sensitive prion protein
PrP-CAA: PrP cerebral amyloid angiopathy
Prp^{CJD}: Creutzfeldt-Jakob disease prion protein
PrP^{Sc} : disease-specific prion protein
RDA: representational difference analysis
REM: rapid eye movements (sleep, REM sleep)
RER: rough endoplasmic reticulum
ROS: reactive oxygen species
SAF: scrapie-associated fibrils
SBM: specified bovine material
SBO: specified bovine offal
SCID (noun): severe combined immunodeficiency
SCID (adj): severe combined immunodeficient
SDS: sodium dodecyl sulfate
SDS-PAGE: sodium dodecyl sulfate-polyacrilamide gel electrophoresis
SE: spongiform encephalopathy
SHa : Syrian hamster
SHa*Prnp* gene: Syrian hamster prion protein gene
Sp1: transcription factor Sp1
Tg mice: transgenic mice
TAE: tris acetate EDTA (ethylenediaminetetracetic acid)
TCA: transmissible cerebral amyloidoses
TME: transmissible mink encephalopathy
TNF: tumor necrosis factor
TSSE: transmissible subacute spongiform encephalopathy(ies)
TSE: transmissible spongiform encephalopathy
TUNEL: transferase uridine-triphosphate nick end labeling
TVS: tubulovesicular structures
UMP: uridine-monophosphate
UAS: upstream activating sequence
USA: ureidosuccinate

Transmissible Subacute Spongiform Encephalopathies:
Prion Diseases
L Court, B Dodet, eds
© 1996, Elsevier, Paris

AUTHOR INDEX

Transmissible Subacute Spongiform Encephalopathies:
Prion Diseases
L Court, B Dodet, eds
© 1996, Elsevier, Paris

KEY WORD INDEX